Principles of Plant Disease Management

Principles of Plant Disease Management

Edited by Ian Brock

SYRAWOOD
PUBLISHING HOUSE
New York

Published by Syrawood Publishing House,
750 Third Avenue, 9th Floor,
New York, NY 10017, USA
www.syrawoodpublishinghouse.com

Principles of Plant Disease Management
Edited by Ian Brock

International Standard Book Number: 978-1-68286-763-1 (Hardback)

Cataloging-in-Publication Data

Principles of plant disease management / edited by Ian Brock.
  p. cm.
Includes bibliographical references and index.
ISBN 978-1-68286-763-1
1. Phytopathogenic microorganisms--Control. 2. Plant diseases. I. Brock, Ian.
SB731 .P75 2019
632.3--dc23

TABLE OF CONTENTS

Preface..IX

Chapter 1   **A Comparative Study of the Commercially Available Fungicides to Control**
Sheath Blight of Rice in Lahore..1
Ahmad Ali Shahid, Muhammad Shahbaz and Muhammad Ali

Chapter 2   **Emergent Potato Leaf Spot Diseases in the Highland and Lowland Regions**
of Bolivia..4
Coca Morante M

Chapter 3   **Biocontrol Potentiality of Active Ingredients from Endophytic *Bacillus subtilis***
Isolated from *Alhagi pseudalhagi* Desv on Maize Spot Diseases............................10
Mahmutjan Dawut, Abdukadir Abliz, Wenera Ghopur, Pezilet Behti,
Burabiyem Obulkhasim and Ghopur Mijit

Chapter 4   **Determination of Suitability of Deoiled Cakes of Neem and Jatropha for Mass**
Multiplication of *Pseudomonas luorescens*..17
Ajay Tomer, Ramji Singh and Manoj Maurya

Chapter 5   **Effect of Climate Change Resilience Strategies on Common Bacterial Blight**
of Common Bean (*Phaseolus vulgaris* L.) in Semi-Arid Agro-Ecology of Eastern
Ethiopia..21
Negash Hailu, Chemeda Fininsa, Tamado Tana and Girma Mamo

Chapter 6   **Antimicrobial Behavior of Intracellular Proteins from Two Moderately**
Halophilic Bacteria: Strain J31 of *Terribacillus halophilus* and Strain M3-23
of *Virgibacillus marismortui*..31
Badiaa Essghaier, Cyrine Dhieb, Hanene Rebib, Saida Ayari, Awatef Rezgui
Abdellatif Boudabous and Najla Sadfi-Zouaoui

Chapter 7   **Enzymatic Responses of Ginger Plants to *Pythium* Infection after SAR Induction**...................38
Rajyasri Ghosh

Chapter 8   **Distribution and Importance of Maize Grey Leaf Spot *Cercospora zeaemaydis***
(Tehon and Daniels) in South and Southwest Ethiopia...44
Alemu Nega, Fikre Lemessa and Gezahegn Berecha

Chapter 9   **Evaluation of Rice Germplasm against Bacterial Leaf Streak Disease Reveals**
Sources of Resistance in African Varieties...51
Issa Wonni, Gustave Djedatin, Léonard Ouédraogo and Valérie Verdier

Chapter 10   **Biosurfactant-Mediated Biocontrol of *Macrophomina phaseolina* causing**
Charcoal Rot in *Vigna mungo* by a Plant Growth Promoting *Enterococcus*
sp. BS13..56
Sumit Kumar, Dubey RC and Maheshwari DK

Chapter 11   **Antifungal Properties of Phytoextracts of Certain Medicinal Plants against Leaf Spot Disease of Mulberry, *Morus* spp.**...64
Ul-Haq S, Hasan SS, Dhar A, Mital V and Sahaf KA

Chapter 12   **Genetic Diversity of *Alternaria alternata* Isolates causing Potato Brown Leaf Spot, using ISSR Markers in Iran**...69
Shima Bagherabadi, Doustmorad Zafari and Mohammad Javad Soleimani

Chapter 13   **Coffee Thread Blight (*Corticium koleroga*): A Coming Threat for Ethiopian Coffee Production**...75
Kifle Belachew, Demelash Teferi and Legese Hagos

Chapter 14   **Biochemical Approach for Virulence Factors' Identification in *Xanthomonas Oryzae* Pv. *Oryzae***..81
Sylvestre Gerbert Dossa C, Petr Karlovsky and Kerstin Wydra

Chapter 15   **Efficacy and Economics of Fungicides and their Application Schedule for Early Blight (*Alternaria solani*) Management and Yield of Tomato at South Tigray, Ethiopia**...88
Mehari Desta and Mohammed Yesuf

Chapter 16   **Bio Control Potential of *Pseudomonas fluorescens* against Coleus Root Rot Disease**...94
S Vanitha and R Ramjegathesh

Chapter 17   **Induction of Systemic Resistance in Sugar-Beet against Root-Knot Nematode with Commercial Products**...98
Mostafa Fatma AM, Khalil AE, Nour El Deen AH, Ibrahim and Dina S

Chapter 18   **The Effect of Planting Time on Phytophthora Blight Disease Incidence and Severity on Cucumber (*Cucumis sativus*) in Nsukka, Derived Savanna, Agro Ecology, South Eastern Nigeria**...105
Patience Ukamaka Ishieze, Ugwuoke Kelvin I and Aba Simon C

Chapter 19   **Maize (*Zea Mays* L.) Growth and Metabolic Dynamics with Plant Growth-Promoting Rhizobacteria under Salt Stress**...113
Abd El-Ghany TM, Masrahi YS, Mohamed A, Al Abboud, Alawlaqi MM and Nadeem I Elhussieny

Chapter 20   **Occurrence and Importance of *Xanthomonas axonopodis* pv. *Phaseoli* in Common Bean (*Phaseolus vulgaris* L) Seed Produced under Different Seed Production System in Central Rift Valley of Ethiopia**...122
Leta A, Lamessa F and Ayana G

Chapter 21   **Shikimic and Salicylic Acids Induced Resistance in Faba Bean Plants against Chocolate Spot Disease**...127
Heshmat Aldesuquy, Zakaria Baka and Nahla Alazab

Chapter 22   ***In Vitro* Evaluation of Botanicals, Bio-Agents and Fungicides against Leaf Blight of *Etlingera linguiformis* caused by *Curvularia lunata* Var. Aeria**.............135
Chijamo Kithan and Daiho L

Chapter 23   **The Effect of Tartaric Acid-Modified Enzyme-Resistant Dextrin from Potato Starch on Growth and Metabolism of Intestinal Bacteria**..........................141
Slizewska Katarzyna, Barczynska Renata, Kapusniak Janusz and Kapusniak Kamila

Chapter 24   **Phytochemical Analysis and Antibacterial and Cytotoxic Properties of *Barleria lupulina Lindl.* Extracts**..........................147
Reshma Kumari and Ramesh Chandra Dubey

Chapter 25   **Three-Dimensional Coating of Porous Activated Carbons with Silver Nanoparticles and its Scale-Up Design for Plant Disease Management in Greenhouses**..........................153
Oleksandra Savchenko, Jie Chen, Yuzhi Hao, Xiaoyan Yang, Xiujie Li and Jian Yang

Chapter 26   **New Way to Develop Mixture of Lactic Leavens and Cardoon Flower Powder (*Cynaraca rdunculus*) in Producing Yoghurt: Approach to Immobilization**..........................161
Benahmed Djilali Adiba, Ouelhadj Akli, Derridj Arezki, Bedrani Fatiha, Belkhir Fatiha and Raman Yakout

Chapter 27   ***In Vitro* and *In Vivo* Management of *Alternaria* Leaf Spot of *Brassica campestris* L.**..........................165
Aqeel Ahmad and Yaseen Ashraf

Chapter 28   **Virulence Diversity in *Rhizoctonia Solani* causing Sheath Blight in Rice Pathogenicitya**..........................171
Ramji Singh, Shiw Murti, Mehilal, Ajay Tomer and Durga Prasad

Chapter 29   **Nutrigenic Efficiency Change and Cocoon Crop Loss due to Assimilation of Leaf Spot Diseased Mulberry Leaf in Silkworm, *Bombyx Mori L***..........................179
Sajad-Ul-Haq, S. S Hasan, Anil Dhar and Vishal Mittal

Chapter 30   **Isolation, Identification and Antifungal Activities of *Streptomyces aureoverticillatus* HN6**..........................186
Lanying Wang, Mengyu Xing, Rong Di and Yanping Luo

Chapter 31   **Study of the Potential Application of Lactic Acid Bacteria in the Control of Infection caused by *Agrobacterium tumefaciens***..........................191
Limanska N, Korotaeva N, Biscola V, Ivanytsia T, Merlich A, Franco BDGM, Chobert JM, Ivanytsia V and Haertlé T

Chapter 32   **Isolation, Identification, *In Vitro* Antibiotic Resistance and Plant Extract Sensitivity of Fire Blight causing *Erwinia amylovora.***..........................200
Mohammed Amirul Islam, Md Jahangir Alam, Samsed Ahmed Urmee, Muhammed Hamidur Rahaman, Mamudul Hasan Razu and Reaz Mohammad Mazumdar

Permissions

List of Contributors

Index

PREFACE

The scientific study of diseases in plants falls under the domain of plant pathology. Diseases can be caused by pathogens and due to environmental conditions. Plant disease management is an important aspect of plant pathology. It involves the study of pathogen identification, disease cycles, pathosystem genetics, etc. Control of plant diseases is vital for the production of food and reduction in agricultural use of water, land and fuel. It is achieved by cultivating plants that have been bred with high resistance to diseases. Other approaches of disease management include crop rotation, use of pathogen-free seed, control of field moisture, appropriate pesticide use, etc. The book studies, analyzes and upholds the pillars of plant disease management and its utmost significance in modern times. It strives to provide a fair idea about this discipline and to help develop a better understanding of the latest advances within this field. This book will assist researchers and students in this field.

This book is a result of research of several months to collate the most relevant data in the field.

When I was approached with the idea of this book and the proposal to edit it, I was overwhelmed. It gave me an opportunity to reach out to all those who share a common interest with me in this field. I had 3 main parameters for editing this text:

1. Accuracy – The data and information provided in this book should be up-to-date and valuable to the readers.

2. Structure – The data must be presented in a structured format for easy understanding and better grasping of the readers.

3. Universal Approach – This book not only targets students but also experts and innovators in the field, thus my aim was to present topics which are of use to all.

Thus, it took me a couple of months to finish the editing of this book.

I would like to make a special mention of my publisher who considered me worthy of this opportunity and also supported me throughout the editing process. I would also like to thank the editing team at the back-end who extended their help whenever required.

Editor

A Comparative Study of the Commercially Available Fungicides to Control Sheath Blight of Rice in Lahore

Ahmad Ali Shahid[1], Muhammad Shahbaz[2] and Muhammad Ali[1]*

[1]Institute of Agricultural Sciences, University of the Punjab, Lahore, Pakistan
[2]Department of Pest warning and Quality control of pesticides, Govt. of Punjab, Lahore, Pakistan

Abstract

A comparative study of commercially available fungicides (Cordate, Precurecombi, Curon, Bendict, Nativo, Valedamycin and Tilt) was made to find out the best fungicide against sheath blight fungi (*Rhizoctoniasolani*). Experimental field was divided into nine treatment units (T1, T2, T3, T4, T5, T6, T7, T8 and T9) and each unit was treated with each fungicide and data was collected on the basis of different agronomic traits such as number of tillers per hill, number of grains per spike, 1000 grain weight, disease incidence and yield of crop. As a result fungicides Nativo and Tilt were proved to be best control for sheath blight of rice.

Keywords: Sheat blight of rice; *Rhizoctonia solani;* Rice; Chemical control

Introduction

Rice (*Oryza sativa* L.), a member of the family Graminae is widely grown in tropical and subtropical regions [1]. Approximately 90% of the world's rice is grown in the Asian continent and constitutes a staple food for 2.7 billion people worldwide [2]. In Pakistan it is the second most important staple food after wheat (*Triticum aestivum*) and is also one of the main export items of the country. During 2005-2006 the crop was grown over an area of 2621.4 thousands hectares with 5547.2 thousands tones production [3]. Pakistan is on the fourth number among rice exporting countries. The Punjab province produces 69% of the total rice production of the country. Area under cultivation of rice is getting decreased due to urbanization and now it is requirement of time to improve cultivation practices along with plant protection practices.

Sheath blight, caused by *Rhizoctonia solani* Kühn, is one of the most destructive rice (*Oryza sativa* L.) diseases worldwide [4]. Asexual stage of *R. solani* belonging to Fungi imperfecti, hyphomycetes, Agonomycetales was called *R. solani* while sexual stage belonging to Basidiomycota, hymenomycetes, homobasidiomycetida, Tulasnellale was called *Thanatephorus cucumeris*. Sclerotium stage belongs to *Rhizoctonia, Tulasnellaceae,* and *ThanatepHorus. R. solani* is one type of filamentous fungi. It is pathogenic to economically important crops such as rice, wheat, maize, cotton, potato, bean, vegetables and grasses [5-10]. This fungus is responsible for the root rot, stem-foot rot as well as seedling blight [11,12].

In occurrence fields, rice yield usually reduced by 10-30%, even up to 50%, owing to this disease [13]. Sheath blight often prevails in rice field with high plant density and high application rate of nitrogen fertilizer. Besides, with the extension of semidwarf, high-yielding, and multitiller cultivars, this disease has been aggravated in recent years, and becomes the most important disease in rice regions [14,15].

Following field trials were made to evaluate the commercially available best fungicide with potential to control sheath blight of rice and later on transfer this information to farmer level in order to get maximum yield of the crop.

Materials and Methods

Identification of infected field

A field of rice variety KSK-133 was identified as infected with sheath blight at Tatlly Malian, Lahore. Field was identified on the basis of visual symptoms on the standing crop plants. Photography was done in order to keep the records of visual symptoms. Samples of infected plants were also collected for the microscopic authentication of pathogenic fungal organism (Plate 1).

Pathogen verification

Infected samples were brought to the laboratory of Institute of Agricultural Sciences, University of the Punjab, Lahore. These samples were surface sterilized and inoculated on the Malt Extract Agar (MEA) and Potato Dextrose Agar (PDA) medium containing petriplates. Plates were then incubated at 27°C for four days. After the germination of mycelia growth the slides were prepared and organism was identified by means of several morphological features as well as identification manuals and keys.

Experimental design

Field trial was conducted during the rice crop of year 2012. Experiment was laid out in randomized complete block design with 4 blocks and 9 treatments. Area under experiment was 2 acre.

Pretreatment data

Before treatment of fungicide in the field the disease incidence in the field was scattered and was evaluated by means of following formula (Table 1):

***Corresponding author:** Muhammad Ali, Institute of Agricultural Sciences, University of the Punjab, Lahore, India
E-mail: Muhammadali.mycologist@gmail.com

A&B: Rice field infected with sheath blight. **C&D:** Infected root zone of the diseased plants. **E:** Best recovered unit after treatment

Plate 1: Pictorial presentation of the experiment.

S. No	Observations	T1	T2	T3	T4	T5	T6	T7	T8	T9
1	Av. No of tillers per hill	4	5	7	9	12	8	11	9	4
2	Av. No of grains	85	95	90	110	126	100	117	100	75
3	1000 grain weight (gm)	22	26	28	33	31	25	29	23	22
4	Disease incidence (%)	65	58	45	40	38	65	62	58	55

Table 1: Pretreatment data in each of the selected blocks.

$$\text{Disease incidence} = \frac{\text{Number of infected plants}}{\text{Total Number of plants visited}} \times 100$$

Along with the disease incidence the following crop parameters were also observed and recorded for each plot.

a. Average number of tillers per hill

b. Average number of grains per spike

c. 1000 grains weight

d. Disease incidence

Fungicide treatment

The fungicidal treatment was given to each infected plot in order to evaluate the comparative potential of each fungicide to control the sheath blight (Table 2).

Results and Discussion

Pathogen identification

Pathogenic fungi was isolated at the Institute of Agricultural Sciences, University of the Punjab, Lahore. Pathogen was identified as *Rhizoctoniasolani.*

Disease incidence

Disease incidence before treatment of fungicide in the field was infested highly with disease. Disease incidence after treatment of fungicides, the observation collected about disease in the field in response to fungicide application was as follows:

The best control against sheath blight was found in T5 and T7, T4 (Figure 1) stood at second number while T3, T6 and T8 results were almost same in controlling the disease and T2 and T1 gave less control along with no control of disease in the experimental unit of T9.

Number of tillers per hill

The average number of tillers were calculated after treatment of fungicides in the field were as follows: Maximum number of tillers were found in T5 along with T4 and T7 while other treatment gave average number of tillers, minimum number of tillers were found in T9.

Number of grains per spike

Maximum average number of grains per spike were observed in T5, T4, and T7 respectively. T3, T6 and T2 gave medium number of grains per spike, minimum number of grains were found in T9.

1000 grain weight in grams

1000 grains were collected from the spikes of each experimental units and their weight was recorded in grams and it is given in table (Table 2 and Figure 1).

Yield of crop

Yield of rice crop was recorded and explained in table. The best results were found in T5 (Table 3).

Treatments	Name of fungicide	Dose per Acre
T1	Cordate	75 gm
T2	Precurecombi	62.5 gm
T3	Curon	50 ml
T4	Bandict	62.5 ml
T5	Nativo	15 gm
T6	Valedamycin	25 ml
T7	Tilt	50 ml
T8	Curon (Flooding)	100 ml
T9	Control	

Table 2: Treatment details of each infected plot.

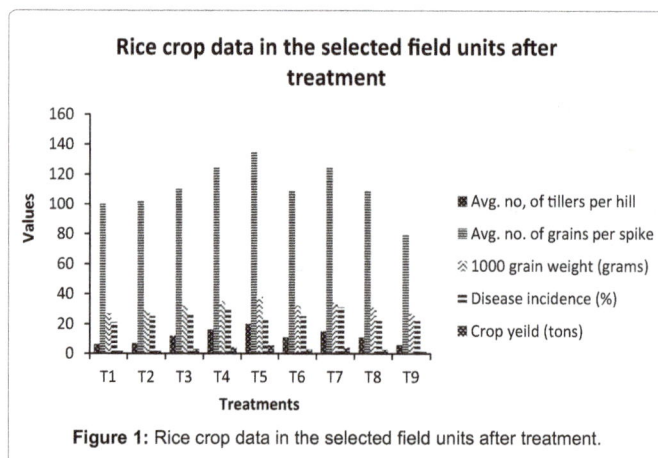

Figure 1: Rice crop data in the selected field units after treatment.

S.no	Observations	T1	T2	T3	T4	T5	T6	T7	T8	T9
1	Av. No of tillers per hill	6	7	12	16	20	11	15	11	6
2	Av. No of grains	100	102	110	125	135	109	125	109	80
3	1000 grain weight (gm)	27	29	32	35	38	32	34	31	27
4	Disease incidence (%)	21	25	27	30	23	25	31	22	22
5	Crop Yield (Kg)	1439	1594	2838	3930	5462	2577	3793	2503	1360

Table 3: Rice crop data in the selected field units after treatment.

Conclusion

Sheath blight of rice is one of the most devastating diseases of rice. In Punjab, Pakistan the losses due to this infection varies from 30 to 50 %. Nativo was found to be the best fungicide used against sheath blight of rice. Benomyl was reported to be mostly used for this disease but Nativo is another new addition to the group of fungicides used to control sheath blight of rice. In the recent research work Nativo, Bandict and Tilt have shown better result in the field against this disease. These pesticides have improved the agronomic features of the rice crop such as Average number of tillers per hill, Average number of grains, weight of 1000 grains in grams, disease incidence and crop yield. These pesticides was also recommended to local farmers to be applied for sheath blight of rice and the results were quite satisfactory. The economics of the recommended pesticides was also evaluated and found that Nativo require only 350 ml/ha, Bandict 520 ml/ha and Tilt 550 ml/ha. Whereas the required concentration of rest of the pesticides lies in the range of 1000ml/ha. The cost of Nativo, Bandict and tilt is 350, 300, 430 respectively which is very economical for the local farmer. Cordate is an old fungicide which is used in rice canopy and causes resistance in pathogen. Hence it is time to replace the old fungicides with new and alternate fungicides especially in a monocropping system like rice.

References

1. Achmadi P, Verma E, Escopalao, Naomi G, Tangonan, et al. (2001) Characterization of a New Subgroup of *Rhizoctonia solani* Anastomosis Group 1 (AG-1-ID), Causal Agent of a Necrotic Leaf Spot on Coffee. Ecol Popul Biol 91: 1054-1061.

2. Anonymous (2006) *Agricultural Statistics of Pakistan.* Government of Pakistan, Ministry of food, agricultural and livestock (Economic Wing), Islamabad. pp. 13-17.

3. Ezuka A, Kaku H (2000) A historical review of bacterial blight of rice. *National Institute of Agro-biological Research Bulletin,* Japan. pp. 207.

4. He Z, Li HX, Tang HR (2010) Effect of *Arbuscular Mycorrhizal* Fungi on Endogenous in Cucumber after *Rhizoctonia solani* Inoculation. Chinese Agr Sci Bull 26: 187-190.

5. Hu C, Wei YW, Huang SL, Shi G, Li YR (2010). Identification and Characterization of Fungal Strains Involved in Rice Sheath Blight Complex in Guangxi Province. Acta Agr Boreali-occidentalis Sinica 19: 45-51.

6. Huang JH, Zeng R, Luo SM (2004) Studies on disease resistance of maize toward sheath blight induced by arbuscular mycorrhizal fungi. Chin J Eco-Agr 14: 167-169.

7. Liao HNL, Xiao S, Wang HS (1997) Analysis of developing annals and evolving causation of rice sheath blight. Guangxi Plant Prot 1997: 35-38.

8. Li F, Cheng LR, Xu MR, Zhou Z, Zhang Y, et al. (2009) QTL mining for sheath blight resistance using the backcross selected introgression lines for grain quality in rice. Acta Agron Sin 35: 1729-1737.

9. Salim M, Akram M, Akhtar ME, Ashraf M (2003) Rice, A production Hand Book. Pakistan Agricultural Research Council, Islamabad. pp. 70.

10. Shi RC, Shang HS, Zhang JZ (2007). Nucleus number of *Rhizoctonia* mycelium cells from turf-grasses in China. Mycosystema, 26: 221-225.

11. Slaton NA, Cartwright RD, Meng J, Gbur EE, Norman RJ (2003) Sheath blight severity and rice yield as affected by nitrogen fertilizer rate, application method, and fungicide. Agron J, 95: 1489-1496.

12. Tan ZJ, Hao SZ (2007) Potato *Rhizoctonia* canker as well as prevention and cure. China Potato 21(2): 108-109.

13. Willocquet LL, Fernandez, Savary S (2000) Effect of various crop establishment methods practised by Asian farmers on epidemics of rice sheath blight caused by Rhizoctoniasolani. Plant Pathol 49: 346-354.

14. Xiao Y, Li SC, Chu MG, Zhou P, Guan P (2008) Cloning, Expression and Characterization of G Protein β-subunit Gene in Rhizoctonia solani from Rice. Chinese J Rice Sci 22: 541-544.

15. Yang JH, Guo QY, Ji L (2005) Study on anastomosis groups Rhizoctonia solani isolated from six Leguminaceous crops in Xinjiang. Xinjiang Agr Sci 42: 382-385.

Emergent Potato Leaf Spot Diseases in the Highland and Lowland Regions of Bolivia

Coca Morante M*

Plant Science and Plant Production Department, University of San Simón, Cochabamba, Bolivia

Abstract

In Bolivia, potato (*Solanum tuberosum* L.) leaf spot diseases have traditionally been regarded as of little importance especially in the Andean highlands. In recent years, however, new types of leaf spot disease have appeared, their distribution widened, and their incidence and severity increased. The present work identifies the main types of leaf spot present in the traditional highland growing regions of the Departments of La Paz (around 4350 m) and Cochabamba (2900-4100 m), and in the north of the Department of Santa Cruz, the new lowland area of potato production (around 235 m). Five causal agents were identified in the highland region: *Alternaria solani, Septoria lycopersici, Cercospora solanicola, Passalora concors* and *Botrytis cinerea*. These affected several types of native potato. In the lowland region, *A. solani* and *Stagonospora* spp. were found to cause leaf spot disease on cv. Desireé. In both agroecosystems, the diseases sometimes appeared alongside late blight (caused by *Phytophthora infestans*). The leaf spot disease caused by *A. solani*, was very destructive, while that caused by *S. lycopersici* was only destructive in the highlands.

Keywords: Native potatoes; Disease intensity; Destructive disease

Introduction

Potato (*Solanum tuberosum* L.) leaf spot diseases are caused by different necrotrophic phytopathogens. They are sometimes confused with late blight, which is caused by *Phytophthora infestans*. Leaf spot diseases are most frequently seen in Bolivia's cold, damp Andean highlands (they are far less common in the warmer regions of the country), usually affecting native crops such as *imillas* (*Solanum tuberosum* subsp. *andigena*), bitter (*Solanum×juzepczukii*) and *phurejas* (*Solanum phureja*) potatoes [1]. Leaf spot diseases have also been recorded in the highland regions of Peru, Ecuador and Colombia [2-4]. According to Hooker [2], the causal agents in high altitude areas include *Septoria lycopersici, Cercospora* spp., *Phoma andina, Ulocladium atrum* and *Botrytis cinerea*, while *Chaenophora cucurbitarum* affects plants in the lowlands. Leaf spot diseases in the highland regions have traditionally been regarded as of little importance. However, in recent years their distribution and intensity (incidence and severity) have increased, as has the impact of late blight [5]. Other types of leaf spot disease are caused by *Phoma andina* and *Phoma huancayense* [4]. The leaf spot caused by *S. lycopersici* is regionally important in the Venezuelan, Peruvian and Ecuadorian Andes, from where reports of over 60% leaf destruction have been made [6,7]. Indeed, "septoriosis" has long been known to have serious effects [8]. In Bolivia, where it is known as khasahui, it is not a devastating disease, although it is of moderate importance and can cause early leaf death [3].

The main production areas of the Andean highlands in the Departments of La Paz (>4000 m) and Cochabamba (>3500 m) are permanently cloudy, very damp, and cold - typical of the transition zone between the lowlands and the Andean Region or Cordillera Real (Figure 1). In recent years, *S. lycopersici* has caused damage to potato crops in these areas, especially in the native *imillas* and *Solanum×juzepczukii* forms [5]. The distributions of this and other forms of leaf spot disease are now expanding, and types of leaf spot new to the highlands of Cochabamba have been recorded [9]. The present work describes the main forms of leaf spot in the highland regions of the Departments of La Paz and Cochabamba, and in a new, milder, lowland potato production area of Bolivia.

Materials and Methods

In January and February of 2012, 2013 and 2014, survey expeditions

were made to record the leaf spot diseases affecting potato plants in Bolivia's traditional highland growing areas of the Departments of La Paz (around 4350 m) and Cochabamba (2900-3300 m), and in the

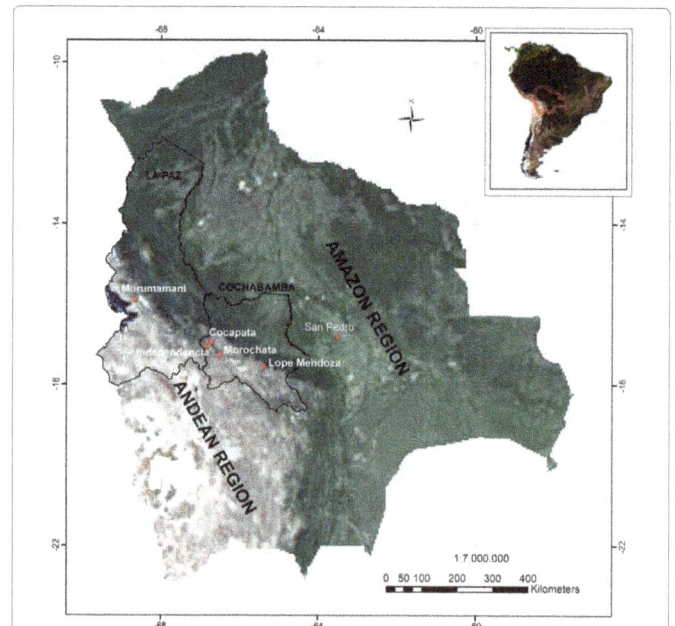

Figure 1: Sampling sites during the field surveys (red spots). (Map after CISTEL, *Facultad de Ciencias Agrícolas y Pecuarias*, UMSS).

Corresponding author: Coca Morante M, Faculty of Agricultural Sciences, Livestock, Forestry and Veterinary, Plant Science and Plant Production Department, Cochabamba, Bolivia, E-mail: agr.mcm10@gmail.com

Department	Province	Area of influence	Locality	Altitude	Latitude	Longitude
				(m)	S	W
La Paz	Omasuyos	Murumamani	Murumamani	4300	15°54'0.00"	68°39'0.00"
			Humanata	4250	15°52'41.15"	68°38'49.37"
	Camacho	Humanata	Humanata	4150	15°28'28.82"	69° 7'20.24"
Cochabamba	Carrasco	Lope Mendoza	Escalante	2950	17°31'55.49"	65°22'55.32"
			Chullchungani	3050	17°33'39.15"	65°20'6.75"
			Phuyuhuasi	3150	17°34'14.46"	65°20'26.32"
			Montepunku	2950	17°35'10.76"	65°18'27.14"
			Laimetoro	3100	17°39'4.00"	65° 6'30.52"
	Ayopaya	Morochata	Piucilla	3320	17° 13' 34"	66o 30' 17"
	Ayopaya	Independencia	Sensei	3200	17° 4'49.18"	66°50'11.56"
		Cocapata	Choro	4100	16°58'42.23"	66°42'54.20"
	Chapare	Coranipampa	Coranipampa	2650	17° 7'20.68"	65°56'2.58"
			Pairumani	2600	17° 7'54.35"	65°56'8.47"
Santa Cruz	Cordillera	Chane	San Pedro	500	16°49'37.61"	63°28'52.86"

Table 1: Locations where leaf samples were taken.

Locality	Variety	Potato species	Common name of disease	Causal agent
Murumamani	Waych'a	*Solanum tuberosum* subsp. *andigena*	Khasahui (septoriosis)	*Septoria lycopersici*
Humanata	Rosita	*Solanum tuberosum* subsp. *andigena*	Khasahui (septoriosis)	*Septoria lycopersici*
Humanata	Waych'a	*Solanum tuberosum* subsp. *andigena*	Khasahui (septoriosis)	*Septoria lycopersici*
Escalante	Desireé	*Solanum tuberosum* subsp. *tuberosum*	EBLS	*Alternaria solani*
Chullchungani	Desireé	*Solanum tuberosum* subsp. *tuberosum*	Mancha negra (EBLS)	*Alternaria solani*
Phuyuhuasi	Desireé	*Solanum tuberosum* subsp. *tuberosum*	Mancha negra (EBLS)	*Alternaria solani*
Montepunku	Desireé	*Solanum tuberosum* subsp. *tuberosum*	Mancha negra (EBLS)	*Alternaria solani*
Laimetoro	Puka (*)	*Solanum andigena* x *tuberosum*	Manchón (leaf blotch)	*Cercospora spp.*
Piucilla	Waych'a	*Solanum tuberosum* subsp. *andigena*	Mancha plateada (CLB)	*Passalora concors*
Sensei	Waych'a	*Solanum tuberosum* subsp. *andigena*	Mancha negra (EBLS)	*Alternaria solani*
Choro	Waych'a	*Solanum tuberosum* subsp. *andigena*	Khasahui (septoriosis)	*Septoria lycopersici*
Coranipampa	Phureja	*Solanum phureja*	Gray mold	*Botrytis sp.*
San Pedro	Desireé	*Solanum tuberosum* subsp. *tuberosum*	Mancha negra (EBLS)	*Alternaria solani*
San Pedro	Desireé	*Solanum tuberosum* subsp. *tuberosum*	Mancha marrón (BLS)	*Stagonospora sp.*

(*)=Puka Toralapa, an improved variety that shows inherent resistance, developed by the *Estación Experimental Toralapa, Instituto Boliviano de Tecnología Agropecuaria* in 1980-1982. EBLS=Early Blight Leaf Spot. CLB=Cercospora Leaf Blotch. BLS=Brown Leaf Spot.

Table 2: Types of leaf spot diseases detected, where they were detected, and potato varieties affected.

north of the Department of Santa Cruz, the country's new, lowland growing region (San Pedro locality, around 500 m) (Figure 1) (Table 1). In the Department of La Paz, leaf samples were collected in the communities of Murumamani, Humanata (both in the Province of Omasuyos) and Humanata (in the Province of Camacho) (Table 1). Given their proximity to the Andean peaks of the Cordillera Real mountain range, these communities are under permanent cloud, damp, and cold. In the Department of Cochabamba, samples were collected in the communities within the areas of influence of Lope Mendoza (in the Province of Carrasco), Morochata and Independencia (Province of Ayopaya), and Coranipampa (Province of Chapare) (Table 1). These areas have the same kind of climate as described above, a consequence of their proximity to the transition zones to the Andean region to Amazon region (Figure 1). In the Department of Santa Cruz, samples were collected in the area of Chané to the north of Santa Cruz (Province of Cordillera) (Table 1). Soybean (*Glycine soja*) and sugar cane (*Saccharum officinarum*) are extensively grown here; potatoes are grown during the winter (March to August). The climate at this time is mild to cold and damp, a consequence of cold winter winds arriving from Argentina.

In the highlands of the Departments of La Paz and Cochabamba, leaves showing signs of leaf spot disease (total N=10-20 depending on the year) were collected from different plants growing in small holdings sown with native varieties such as Waych'a (a variety of *imilla*) (*S. tuberosum* subsp. *andigena*), and phureja blanca (a variety of phureja) (*S. phureja*), and bitter potato (*S. x juzepzukii*). In the lowlands of Santa Cruz, similar samples (total N=10-20) were collected from plants growing in extensive plots sown with variety Desireé (*S. tuberosum* subsp. *tuberosum*).

The collected samples were taken to the laboratory in plastic bags inside preservation boxes within two days, and the fungal reproductive structures stained with lactophenol methylene blue [10] and examined under a stereomicroscope. The fungi causing the leaf spot symptoms were identified using the taxonomic keys and indicators of different authors as indicated below.

Results

Six types of leaf spot were found affecting the native *S. t.* subsp. *andigena* and introduced *S. t.* subsp. *tuberosum* (Table 2).

Early blight leaf spot caused by *Alternaria solani* sorauer

This was detected in the highlands of Cochabamba near Escalante, Chullchungani, Phuyuhuasi and Montepunku (in the Lope Mendoza area, Province of Carrasco). It was also detected in the lowland area of the Department of Santa Cruz at San Pedro (in the Chane area) (Table 2). In both agroecosystems the climate during the growing period is damp and largely cold (8-18°C). In the highlands it affected the native variety Waych'a (*S. tuberosum* subsp. *andigena*), while in the lowlands if was found on variety Desireé (*S. tuberosum* subsp. *tuberosum*). In all the places it was detected alongside (i.e., on the same plant or even the same leaf) late blight (caused by *P. infestans*). Incidence was generalized in both agroecosystems, and the disease more severe in cv. Desireé.

The characteristic symptom was the formation of dark brown leaf spots surrounded by a halo that becomes more obvious with time (Figures 2A and 2B). In Waych'a they were more angular and irregular, limited by the veins and the interior of the spots showed discontinuous, irregular rings (Figure 2A). In cv. Desireé, the spots were more circular and regular, larger, and the inner rings more notable (Figure 2B).

Within the spots, small, dark brown, superficially growing structures were seen, formed from small clusters of conidiophores, straights and, septate (Figure 2C). Dark conidia formed at the apices of the latter, solitary, slightly flexuous with the body of the conidium ellipsoidal tapering to a beak which is same length as longer than the body, pale or dark brown, smooth, with 9 or more transverse and 0 longitudinal or oblique septa; beak is flexuous (Figure 2D). Using the keys of Ellis [11], the fungus was identified as *A. solani*.

Septoria leaf spot (khasahui) caused by *Septoria malagutii* ciccar. and Boerema ex E.T. Cline, sp. nov. (Syn. *Septoria lycopersici speg.*)

This was identified in samples from the Departments of La Paz and Cochabamba collected at an altitude of around 4100 m (Table 2). In both of these highland agroecosystems, the conditions are damp and cold (7-15°C) during the growing season (November to May). In the Department of La Paz, it was detected in the Murumamani area in the communities of Murumamani, Paconi and Umanata, where it affected the native varieties Waych'a and the bitter potato (variety Luk'ys) (*Solanum x juzepzukii*) (Table 2). In the Department of Cochabamba it was identified in the areas of Cocapata and Choro (Province of Ayopaya), and in the community of Phuyuhuasi within the Lope Mendoza area (Province of Carrasco), where it affected the variety Waych'a (Table 2). In all these areas the disease appeared alongside (i.e., on the same plant or leaf) late blight. In Murumamani (Dept. of

Figure 2: Leaf spot caused by *Alternaria solani* on the native and introduced potatoes varieties. A: Angular form of lesion on the variety Waych'a (*S. tuberosum* subsp. *andigena*). B: Circular to irregular lesions on variety Desireé (*S. tuberosum* subsp. *tuberosum*). C: Formation of irregular rings in the lesions (Mag. 75X). D: Conidiophores in small clusters, straights and conidia are ellipsoidal tapering to a beak same length as longer than the body (Mag. 400X).

Figure 3: Leaf spot caused by *Septoria lycopersici* in the native variety Waych'a (*S. tuberosum* subsp. *andigena*). A: Small circular lesions on the leaf. B: Formation of irregular, concentric rings in the lesion (Mag. 50X). C: Formation of pycnidia (Mag. 100X). D: Pycnidia immersed in the tissue releasing elongated, hyaline-colored conidia through the ostiole (Mag. 400X). The conidia are hyaline, filiform, or occasionally straight.

La Paz) and Choro (Dept. of Cochabamba), the disease was sometimes destructive.

The spots were small (1-5 mm), circular to irregular, dark brown, and with concentric, irregular rings (Figures 3A and 3B). At first these spots appear isolated from one another, but over time coalesce, giving rise to lesions similar to those seen in late blight. Small (although visible with the naked eye) blackish pycnidia were seen at the center of the spots (Figure 3C). The pycnidial conidiomata were epigenous, solitary, scattered with several per lesion, immersed in the epidermis, and just visible (<100 μm in diameter). The ostiole was brown-black and some 45-70 μm diameter. The conidia were hyaline, filiform, slightly curved, occasionally straight or sigmoid, sharply pointed at both ends, and sometimes slightly rounded at the base. The apex was rounded when the conidia were still attached (90 ± 4.4 [50-122] μm long×1.9 ± 0.03 [1.0-2.0] μm wide). There were 4-7 septa with no constrictions (Figure 3D). Using the keys of Cline and Rossman [1], the fungus was identified as *Septoria malagutii*.

Leaf blotch caused by *Cercospora spp.* Atk.

This was detected in plants in the Laimetoro area (Province of Carrasco) in the Department of Cochabamba (Table 2), affecting (in isolation) the native varieties Waych'a and Chaucha blanca (*Solanum chaucha*) and the improved variety Puka toralapa (Hybrid *Solanum tubersoum* variety with R genes). Cloud cover is permanent in this area, the humidity is high, and the temperatures between 5 and 15°C. Where the humidity is very high the disease can become destructive.

The characteristic symptoms included damp-looking angular and circular-to-irregular spots, sometimes coalescing to form blotches similar to those caused by late blight (Figure 4A). The disease affected both the leaves and shoots. The spots themselves showed just-visible, irregular, concentric rings. The center of the leaf on both sides had a whitish felt-like covering (Figures 4B and 4C) formed by the fruiting structures of the fungus, i.e., clusters of conidiophores (Figure 4D) and acicular conidia (hyaline in color) (Figure 4D). The conidiophores emerged through the stomata, were unbranched, subcylindrical to geniculate-sinuous, appeared in small, dense fascicles, and showed dimensions of up to 70×5.5 μ. The conidia were slightly curved, pale olivaceous, smooth, with 6-8 septa, and with dimensions of up to 82×5.5 μ. The keys of Chupp [12], Ellis [11] and Braun and Crous [13] identified the causal agent as *Cercospora* spp.

Grey mold caused by *Botrytis cinerea* Pers.:Fr.

This was identified in leaves (only) of plants in the Coranipampa area (Province of Chapare, Municipality of Colomi) in the Department of Cochabamba (Table 2). The climate here is temperate to warm temperate (foothill climate). The disease affected the phureja blanca variety (*S. phureja*), but with just a few spots, and appeared alongside late blight. The disease appeared as isolated, damp-looking, blackish spots, circular to irregular in shape. The tissue bordering the spots was slightly chlorotic (Figures 5A and 5B). On the underside of the leaves, a lead-gray woolly carpet was seen (Figure 5B), made up of sporangiophore clusters and spores (Figures 5C and 5D). The conidiophores were 1.5-2.2 mm long, branched, with a stipe, and had a rather open head. They are smooth, clear brown below, paler near the apex, and with the ends of the branches quite colorless. The conidia were ellipsoid or obovoid in shape, with a slightly protuberant hilum, colorless to pale brown, smooth, and of dimensions 8-12×6-10 μ. using the keys of Ellis [11], the fungus was identified as *Botrytis cinerea* Pers.

Cercospora leaf blotch caused by *Passalora concors* (Casp.) Sacc. (Syn. *Cercospora concors* (Casp.) Sacc.

This disease affected the leaves of native Waych'a Paceña plants (*S. tuberosum subsp. andigena*), commonly alongside late blight. The severity of infection ranged from 10-20% (determined using Image Analysis software). The lower to middle leaves were preferentially affected, with symptoms starting as yellowish-green, circular to irregular blotches on the upper leaf surface (Figure 1A). Gradually the centers of these blotches become grey to black with a soft yellow halo (Figure 1B). A fluffy grey layer of conidiophores and conidia eventually forms on the abaxial surface of the leaves, (Figures 1C and 1D). Hyphae were observed growing over the leaf hairs (Figures 1D and 1E). The conidiophores occurred in dense fascicles above the stomata; these were irregular in width, grayish, and highly branched (Figure 1F). The conidia were variable in size, catenate and slightly curved, obclavate, pale olivaceous, smooth, non-septate, and 12-16 μ in length×3-6 μ in width (Figure 1F). Using the keys of Ellis [11] and Crous and Braun [14], the fungus was identified as *Passalora concors*, a synonym of *Cercospora concors* (Casp.) (Crous). This is the first report of this pathogen causing Cercospora leaf blotch in the Andean highlands of Bolivia.

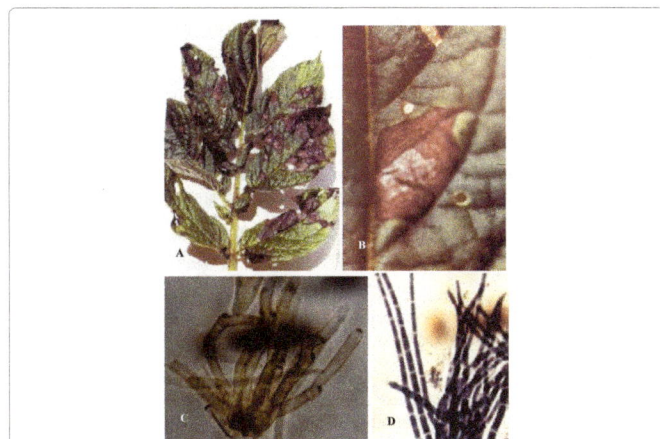

Figure 4: Leaf spot caused by *Cercospora solanicola* in the variety *Puka toralapa* (an improved variety that shows resistance to *P. infestans*). A: Circular to irregular lesions of damp, blackish appearance (Mag. 50X). B: Lesion showing irregular rings and the growth of a felt-like, whitish carpet in the central area. C: Conidiophore in cluster formation (Mag. 400X). D: Elongated, septate conidia stained with methylene blue (Mag. 400X).

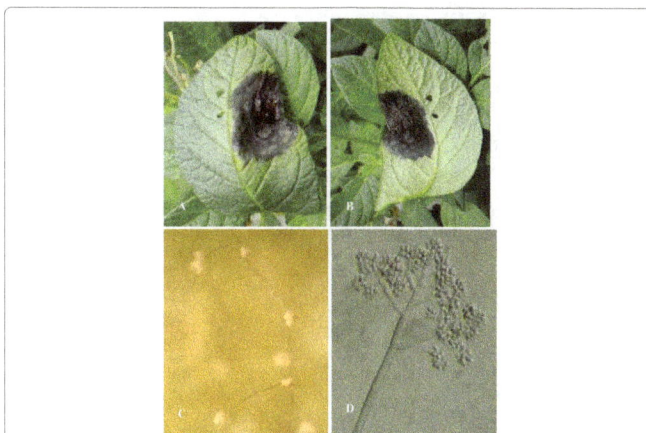

Figure 5: Leaf spot caused by *Botrytis* spp. in the native variety phureja blanca (*Solanum phureja*). A: Dark lesion on the upper leaf surfauce. B: Lesion on the underside of the leaf showing a gray carpet-like growth. C: Conidiophores showing spore cluster formations (under the stereomicroscope, Mag. 100X); D: Conidiophore with clustering conidia (Mag. 400X).

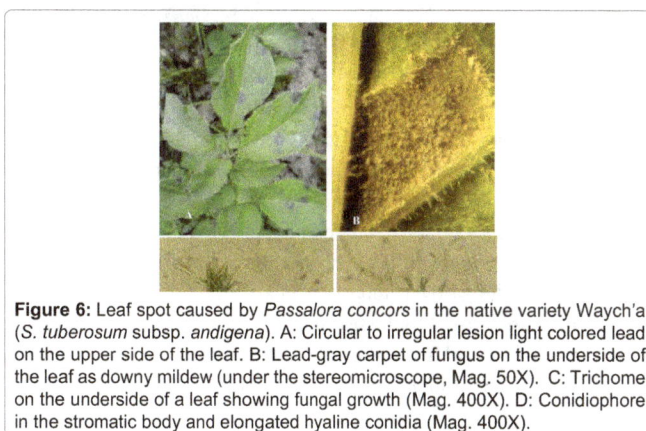

Figure 6: Leaf spot caused by *Passalora concors* in the native variety Waych'a (*S. tuberosum* subsp. *andigena*). A: Circular to irregular lesion light colored lead on the upper side of the leaf. B: Lead-gray carpet of fungus on the underside of the leaf as downy mildew (under the stereemicroscope, Mag. 50X). C: Trichome on the underside of a leaf showing fungal growth (Mag. 400X). D: Conidiophore in the stromatic body and elongated hyaline conidia (Mag. 400X).

Brown leaf spot caused by *Stagonospora* spp. *(Sacc.) Sacc.*

This was detected in cv. Desireé plants growing in the area of Chane (Province of Cordillera) in the north of the Dept. of Santa Cruz (Table 2) during the winter (March-August). The climate in this area is damp and temperate to temperate-cold (10-20°C), a consequence of cold winter winds blowing in from Argentina. Spots were few in number, and appeared alongside those caused by *A. solani*. The characteristic symptom is the formation of dark brown irregular spots with marked brown edges, that become elongated by coalescence (Figure 6A) and which may then acquire a yellowish halo (Figure 6A). Small, light or dark brown fruiting bodies (pycnidia) were visible within the spots (Figure 6B), partially embedded in necrotic tissue (Figure 6B). The conidia, released from the ostiole, were elongated (Figure 6C), hyaline to clear brown in color, and with several transverse septs (Figure 6D) and (Figure 7).

Discussion

The six causal agents of leaf spot diseases reported in this work are *A. solani* [11], *S. malagutii* [1], *Cercospora* spp. [12], Ellis [11], *P. concors* [14]; Ellis [11], *B. cinerea* [15], and *Stagonospora* spp. [16]. All have been reported in other Andean countries and indeed other regions of the world [2-4,8,9,17-20]. *Stagonospora* spp. is a new causal of agent of leaf spot in lowland Bolivia.

Most of the leaf spot diseases affecting Bolivia's potato crops were

Figure 7: Leaf spot caused by *Stagonospora* spp. in the variety Desireé (*S. tuberosum* subsp. *tuberosum*). A: Irregular, clear-brown, elongated lesion on the upper side of the leaf. B: Formation of small pycnidia on the leaf surface (red arrow). C: Microscopic view of the pycnidia (Mag. 400X). D: Conidia with 2-4 septs (Mag. 600X).

found in the damp and cold highland areas of the Departments of La Paz and Cochabamba. This agrees with that reported by Turkensteen [4], Hoopes and Sage [3] and Hooker [2], who indicate them to be appear in Andean Bolivia, Peru, Ecuador and Colombia. In the lowland area examined leaf spot disease was also found, but much less commonly; this agrees with that reported by Hooker [2]. A relationship appears to exist between the appearance of disease and altitude/environmental conditions (dampness and cold temperatures). Note that khasahui, caused by *Septoria malaguti* (a synonym of *S. lycopersici*) was the only disease at altitudes of over 4000 m. Coca-Morante [9] indicates that this disease can cause 40% losses of production in high areas of the Department of La Paz. In addition, Torres [20] indicates that the spots caused by this disease appear in cold, wet areas at altitudes of 3800-400 m. However, Hooker [2], indicates that in Venezuela the disease occurs at 1600-2500 m, while in Ecuador it is reported to be seen mostly at over 3000 m [7].

The diseases caused by *A. solani, Cercospora* spp., *P. concors* and *B. cinerea* appeared in highland areas between 2600 and 3300 m in the Departments of La Paz and Cochabamba. That caused by *Stagonospora* spp. appeared at 250 m in the lowlands of Santa Cruz. Early blight caused by *A. solani* is found worldwide wherever potatoes are grown [2]. Neither leaf blotch caused by *Cercospora spp.*, nor gray mold caused by *B. cinerea*, have been previously reported in the Bolivian Andes. Grey mold is usually considered of minor economic importance in the region [2].

In the present work, leaf blotch caused by *P. concors* (Casp.) Sacc (Syn. *C. concors* [Casp.] Sacc.) was destructive under high humidity and cool temperatures. It has been previously reported in the highlands of Cochabamba [9]. It is also known in cool areas of Europe, the former Soviet Union, North America, in some restricted areas of Africa and Asia, and has recently been reported from China [21]. This disease is considered of minor importance, and may occur simultaneously with late blight (caused by *P. infestans*) or early blight (caused by *A. solani*).

Until now, *Stagonospora* spp. has never been reported to cause disease in Bolivia. For the time being it is of minor importance, and normally occurs simultaneously with other potato leaf disease such as late blight.

The present results suggest that leaf spot diseases in Bolivia should be monitored in order to prevent possible losses of production.

Acknowledgements

The author is grateful to ASDI-DICYT-UMSS for financial support of the Project "Evaluación del impacto de las variaciones climáticas en el manejo de la enfermedad tizón tardío de la papa en zonas endémicas altoandinas del departamento de Cochabamba, Bolivia "over 2010-2012. Thanks are also owed to Javier Burgos Villegas, principal investigator at the Centro de investigación en Sistemas y Teledetecion (CISTEL), Facultad de Ciencias Agricolas y Pecurias "Dr. Martin Cárdenas", Universidad Mayor de San Simón, for map production and the collection of samples in the field. The authors also thank the farmers of the Piucilla-Morochata, Lope Mendoza, Laimetoro, Choro-Cocapata and Chane areas for the help received during this work, and Adrian Burton for editorial assistance and valuable comments and suggestions.

References

1. Erica TC, Amy YR (2006) Septoria malagutii sp. nov., cause of annular leaf spot of potato. Mycotaxon 98: 125-135.

2. Hooker WJ (1990) (Ed.) Compendium of potato diseases. The American Phytopathological Society. St. Paul, Minnesota, EUA p: 60.

3. Hoopes RW, Sage C (1982) Factores que restringen la producción de papa en Bolivia perspectivas para su mejoramiento. Consorcio para el Desarrollo Internacional/Departamento de Geografía Universidad de Dirham Reino Unido. Cochabamba, Bolivia.

4. Turkensteen LJ (1978) Tizón Foliar de la Papa en el Perú: I Especies de Phoma Asociadas. Fitopatología 13: 67-69.

5. Coca-Morante M, Castillo-Plata W, Méndez Á (2014) Control químico de septoriosis (Septoria lycopersici Speg.) de la papa (Solanum tuberosum subsp. andigena) en zonas altoandinas de altura de Bolivia. Revista Latinoamericana de la Papa 18: 116-121.

6. Carrera J, Orellana H (1978) Estudio de la Mancha Foliar de la Papa Septoria lycopersici Sub-grupo A. en el Ecuador. Fitopatología 13: 51-57.

7. Cabi E (1984) Data sheets on quarantine organisms N° 142, Septoria lycopersici var. malagutii. Bulletin OEPP/EPPO 14: 43-49.

8. Jiménez AT, French ER (1972) Mancha Anular Foliar (Septoria lycopersici Subgrupo A) de la papa. Fitopatología 5: 15-20.

9. Coca Morante MC (2014) First Report of Passalora concors (Casp.) Causing Cercospora Leaf Blotch in the Andean Region of Cochabamba, Bolivia. J Plant Pathol Microb 5: 221.

10. French ER, Hebbert TT (1980) Métodos de Investigación Fitopatológica. Instituto Interamericano de Ciencias Agrícolas. San José, Costa Rica p: 289.

11. Ellis MB (1971) Dematiaceous Hyphomycetes. Commonwealth Mycological Institute. Kew, Surrey, England p: 608.

12. Chupp C (1954) A monograph of the fungus genus Cercospora. Ithaca, New York p: 667.

13. Braun U, Crous PW (2007) The diversity of cercosporoid hyphomycetes – new species, combinations, names and nomenclatural clarifications. Fungal Diversity 26: 55-72.

14. Crous PW, Braun U (2003) Mycosphaerella and its anamorphs: 1. Names published in Cercospora and Passalora. CBS Biodiversity Series 1: 1-571.

15. Elad Y, Williamson B, Tudzynski P, Denle N (2007) Botrytis: Biology, Pathology and Control. Kluwer Academic Publisher. P.O. Box 17, 3300 AA Dordrecht, The Netherlands p: 403.

16. Sutton BC (1980) The Coelomycetes: Fungi Imperfecti with Pycnidia, Acervuli and Stromata. Commonweath Mycological Institute. Kew, Surrey, England p: 696.

17. Gandarillas A, Ortuño NY (2009) Compendio de enfermedades, insectos, nematodos y factores abióticos que afectan al cultivo de la papa en Bolivia. Fundación PROINPA. Cochabamba, Bolivia p: 181.

18. Otazu V, Brown WM, Quitón MH (1984) Enfermedades de las plantas en Bolivia. Ministerio de Asuntos Campesinos y Agropecuarios, Instituto de Tecnología Agropecuaria, Consorcio Internacional para el Desarrollo. Cochabamba, Bolivia p: 31.

19. Turkensteen (1981) Compendium of Potato Diseases: Cercospora leaf blotches or Cercospora leaf spots. American Phytopathological Society p: 65.

20. Torres H (2002) Otras enfermedades foliares. En: Manual de las enfermedades más importantes de la papa en el Perú. Centro Internacional de la Papa. Lima, Perú p: 86.

21. Tian SM, Ma P, Liu DQ, Zou MQ (2008) First Report of Cercosporaconcors Causing Cercospora Leaf Blotch of Potato in Inner Mongolia, North China. Plant Disease 92: 4-654.

3

Biocontrol Potentiality of Active Ingredients from Endophytic *Bacillus subtilis* Isolated from *Alhagi pseudalhagi* Desv on Maize Spot Diseases

Mahmutjan Dawut[1,3], Abdukadir Abliz[1,2], Wenera Ghopur[1], Pezilet Behti[1], Burabiyem Obulkhasim[1] and Ghopur Mijit[1*]

[1]College of Life Science and Technology, Xinjiang University, China
[2]Media Institute of Urumqi Vocational University, China
[3]Graduate School of Science, Hokkaido University, Japan

Abstract

Two endophytic bacterial strains, XJAS-AB-13 and XJAS-AB-11, with broad antifungal activity against maize spot pathogens *Exserohilum turcicum* and *Bipolaris maydis*, were screened from the *Alhagi pseudalhagi* Desv, by using agar diffusion method. Both of them were identified as *Bacillus subtilis* according to the physiological and biochemical properties, as well as the molecular analysis based on 16S rDNA sequence. The inhibitory activity of broth cultured XJAS-AB-13 and XJAS-AB-11, measured *in vivo*, and the result showed that the diseases inhibition efficiency of XJAS-AB-13 and XJAS-AB-11 broth on *E. turcicum* and *B. maydis* reached 63.33%, 45.0% and 23.33%, 58.34% respectively. Infection by different diseases caused an increase in total amount of the protein in general, where the increment generated by the *Exserohilum turcicum* was greater than *Bipolaris maydis*. Defense enzymes activity of Superoxide dismutase (SOD), however, declined sharply compare to the control group, while the others, catalase (CAT), peroxidase (POD) and ascorbate peroxidase (APX), were increasing. Through extraction and silica-gel chromatography, two different monomers, XJAS-B and XJAS-G, were isolated from the ethyl acetate extraction of the culture broth of XJAS-AB-11, and identified according to ^{1}HNMR and ^{13}CNMR spectral data as well as ESI-MS molecular weight analysis as cyclo-(D-leucyl-trans-4-hydroxy-L-proline) [(3R,7R,8aS)-7-hydroxy-3-isobutylhexahydropyrrolo [1,2-a] pyrazine-1,4-dione] and 2-(3,4-dihydroxyphenyl)-3,5,7-trihydroxy-4H-chromen-4-one. The preliminary comparative analysis on XJAS-AB-11 Naringenin-chalcone synthase (*CHS*) gene also indicated their high homology with some prokaryotic and eukaryotic *CHS* amino acid sequences. In addition, the amino acid sequence of XJAS-AB-11 *CHS* gene has showed similar amino acid composition with those of *Streptomyces griseus*, *Arachis hypogaea* and *Gerbera hybrida* at the active site.

Keywords: Antagonistic endobacteria; *In vivo* disease control test; Active monomers; Structure identification; Chalcone synthase

Introduction

Maize is one of the most essential cereal crops used in the human diet in most parts of the world and it is vital constituent for animal feeds. Among cereal crops, maize remains third after wheat and rice in total area and production in the world. Despite the huge production and consumption ratio of the maize, diseases caused by different pathogens were becoming worse recently. The damage of main cereal crops around the world is up to 10% to 15%, even if number of different precautions had taken, where 70% to 80% of which caused by pathogenic fungi [1,2]. Traditionally, diseases are prevented using chemical disinfectants, but this practice is considered unsustainable and harmful due to their potential threat towards natural environment and human health [3,4]. Thus, developing a new bio-pesticide, which is non-toxic and non-polluting, is attaining increasing interests among researchers [5-8].

An 'endophyte', a concept which was put forward by Barry in 1866, is a microorganism living in the tissue of a healthy plant, but does not cause any harm to the host [9]. Endophytes include fungi, actinomycetes, and bacteria; these can be found in any different parts of host plants, such as in roots, flowers, fruits, and seeds [10,11]. The secondary metabolites produced by endophytes, especially by those living in the medicinal plants could be the abundant source of new chemical compounds which is essential not only to human health but also for the prevention and treatment of different diseases among the animals and plants [12-15].

The traditional medicinal plant *Alhagi pseudalhagi* Desv is mainly distributed in the flat saline land of Badain Jaran Desert in inner Mongolia, Xinjiang, and Gansu province in China. *Alhagi pseudalhagi* Desv have not been only used as a sand fixing plant in Xinjiang but is also been used as an effective natural traditional medicine particular in curing joint pain, detoxification [16]. Although, there are several studies on screening of active secondary metabolites of *A. pseudalghi*, and resulted to the isolation and identification of 15 different compounds [17]. However, there is no report on screening and testing of active secondary metabolites of endophytic microorganisms isolated from Xinjiang *A. pseudalghi in vivo* level. For the first time, we report herein the result of our study on isolating secondary metabolites from an *A. pseudalghi*-endophytic microorganism and the compound's activity against pathogenic organisms associated with maize spot disease.

Materials and Methods

Strains and cultures

The medicinal plant *Alhagi pseudalhagi* Desv was collected from Xinjiang Turpan, in 2010.

The plant pathogens, *Fusarium oxysporum f.sp. cucumerinum*, *Fusarium oxysporum f.sp. vasinfectum*, *Fusarium oxysporumf.sp. niveum*, *Alternaria mali*, *Exserochilum turcicum*, *Bipolaris maydis*,

*Corresponding author:** Ghopur Mijit, Plant Cell Biology Laboratory of The College of Life Science and Technology, Xinjiang University, Xinjiang 830046, China, E-mail: gmijit@163.com

Bacillus subtilis were purchased from the culture collection center of the institute of agricultural resources and regional planning of Chinese academy of agricultural sciences (CAAS); the animal pathogens, *Staphylococcus aureus* and *Escherichia coli*, were kindly provided by microbiology lab of the college of life science and technology, Xinjiang University.

Escherichia coli DH5α was purchased from TransGene Biotech Company. The traditional maize seed without any genetic modification (No. SC-704) was purchased from Xinjiang academy of agricultural science.

The culture medium LB (Luria Broth), MEA (Malt Extract Agar), and PDA (Potato Dextrose Agar) used in this study were prepared according to Qian et al. [18].

Screening and identification

After washing those fresh leaves and stems, which are obtained from the newly collected medicinal plant *Alhagi pseudalhagi* Desv with tap water, the surface was sterilized according to Geris dos Santos [19]; and then grinned it in sterilized mortar and filtered with gauze [20]. The filtrate was evenly coated on the different culture medium to grow. Endophytic microorganisms grown in their culture mediums were purified according to their colony color and morphological treats, gradually.

The fermentation broth of isolated endophytic fungi and bacteria were prepared under the condition of 28°C (fungi), 37°C (bacteria), 150 r/min agitation, for 96 h and 72 h, respectively. The inhibitory activity of isolated microorganism`s broth on the different plant and animal pathogens were measured via agar diffusion method [21], and the inhibition zone diameter was gauged. Then, the morphological characteristics of endophytes, which are showed remarkable effects on pathogens described according to Buchanan and Dong et al. [22,23]. Plus, transmission electron microscope (TEM; H-600, Japan) was also used for more detailed morphological analyses.

For the molecular analyses, the total genomic DNA of selected endophytes was extracted by UNIQ-10 column bacterial genomic DNA extraction Kit (Sangon Biotech) and 16S rDNA region was sequenced using the bacterial universal primer 27F and 1492R.

Disease prevention *in vivo* test

The test plant seeds, SC-704, was planted in the nutrition bottle which contains of 3:1 sill/perlite, and irrigated with 1/2, 1/4 fold of MS nutrition solution; the cultivation room temperature was controlled at 25°C.

Since two leaf stage, maize seedlings were separated into seven groups (each group contains 30 maize seedling, with a total of 60 leaves per group): one blank control group (no any treatment); two negative control groups (only treated with *E. turcicum* and *B. maydis* spore suspension, 50 spore/40 uL); two treatment groups (after treated with fermentation broth, next day treated with *E. turcicum* and *B. maydis* spore suspension, respectively); two positive control groups (first day, treated with 0.1% Carbendazol (broadly used pesticide), next day treated with *E. turcicum* and *B. maydis*). Observation of the result and calculation of disease prevention efficiency were done 10 days after.

Just after finishing up the observation of disease prevention efficiency, the same seven group subjected to measurement of total protein content, and variation on the activity of 4 different disease control related defense enzymes, superoxide dismutase (SOD), catalase (CAT), peroxidase (POD) and ascorbate peroxidase (APX), according

to Chen et al. [24]. One-way ANOVA analysis, at α=0.05 and α=0.01, was applied for further statistical analysis.

Isolation and structural analysis of XJAS-AB-11 metabolites

Fifth day of culturing the strain XJAS-AB-11 in LB medium at 37°C, collected the broth supernatant by centrifuging, then used it for extraction by four different chemical solutions, isopyknic petroleum ether, normal butyl alcohol, chloroform, ethyl acetate. This extraction proceeded until the extracts became colorless. Then, the bacteriostatic effect of those extracts against different plant/animal pathogens were examined by adopting the same method mentions earlier and the aqueous methanol (5%) as a negative control. After determined the active site of the culture broth of XJAS-AB-11, the bulk fermentation and its extraction took placed.

Thin layer chromatographic (TLC) pre-test was applied for the purpose of choosing the optimal developing solvent for the use in the silica gel column chromatography. Took a rectangular silica plate (type GF254, 5 × 10 cm) draw two horizontal lines at the 0.5 cm from the bottom and the top, used one of which (bottom) as the spotting line or the baseline, others as end line. After dissolving the extract in methanol, inhaled it into 0.33 mm capillary tube and gently put a dot on the spotting line. Total of three developing solvent with distinct polarity were used, which are petroleum ether/acetone, Petroleum ether/ethyl acetate, chloroform/methanol, as developer. After pulled the plate out from the urn, dried it naturally at the room temperature and visualized it by sulfuric acid-ethanol solution, then observed it under UV-analyzer, at 254 nm and 365 nm wavelength, and the retention value (R_f=distance from baseline travelled by solute/distance from baseline travelled by solvent) was calculated (considered the value between 0.2-0.8 is the best).

The silica gel column chromatography of elected extract was done through gradient eluting by different proportion of chloroform: methanol (100:1, 70:1, 50:1, 30:1, 10:1, 7:1, 5:1, 1:1). The eluents were collected into test tubes in the order of from weak polarity to strong. TCL test was applied on each test tube, and concentrated them in vacuum. The ^{1}HNMR and ^{13}CNMR, API-ES mass spectrometer and infrared spectrogram were applied to identify the structure of the isolated pure monomers.

Preliminary study on the Naringenin-Chalcone synthase (*CHS*) gene

Obtaining *CHS* partial sequence: after extracting the genomic DNA using the purchased UNIQ-10 bacterial DNA extraction kit, the *CHS* partial gene sequence was amplified by using two pairs of primers: *B. subtilis* bcsA-ypbQ operon *CHS* gene primer (BpsA-1 and BpsA-2) [25] and the newly designed primer by our laboratory according to *B. subtilis* naringenin-chalcone synthase gene (Bsn5_01495) sequence (accession No. YP_004203960.1) Bsn-1: 5'-AGCGGCTGTTCGTGAGTGCC-3' and Bsn-2: 5'-CCCGGCCCTAATGCGCCAAT-3'. The PCR amplification was conducted in 50 μm reaction system, which includes distilled water (31 μm), PCR buffer (with Mg^{2+}, 5 μm), Deoxynucleotide (dNTP, 4 μm), each primer (2 μm), template DNA (5 μm) and Taq enzyme (1.0 μm). The reaction was done with 35 cycle (denatured at 94°C for 60s, annealing at (Bsn/58.4, BpsA/66.0) for 90s, extension at 72°C for 2 min), which was initialized at 94 for 5 min, and elongated 72 for 10 min, finally terminated at 4°C. After obtaining the target sequences, the sequence amplified by *B. subtilis* bcsA-ypbQ operon *CHS* gene primer was ligated to the Invitrogen TA cloning vector, and then transformed it into *E. coli* competent cell, and sequenced the vector. DNAMAN was applied for the purpose of multiple comparisons between the DNA sequences of Bsn and BpsA (the *B. subtilis* BSn5 naringenin-chalcone synthase was selected as standard sequence).

Contrastive analysis of XJAS-AB-11 CHS partial sequence: GenBank blast tools (http://www.ncbi.nlm.nih.gov/) were used to compare similar CHS gene and amino acid sequences. The construction of the secondary and three-dimensional structure of enzyme was done by SWISS-model molecular modeling system (http://swissmodel. expasy.org/), and NRPS-PKS database (http://nrps.igs.umaryland. edu/) was applied to searching of active site amino acid sequence of CHS enzyme from different species. The CHS amino acid sequences comparison analysis was done through online ClustalW 2 program (http://www.ebi.ac.uk/Tools/msa/clustalw2/).

Results

Morphology and identity of two antagonistic bacteria

Total of 14 endophytic bacteria and 6 endophytic fungi strains were isolated from *A. pseudalhagi* Desv. Among which the bacteria XJAS-AB-13 and XJAS-AB-11 showed much stronger and stable inhibitory effect especially on the *B. maydis* (Tables 1 and 2). The inhibition zone average diameters of XJAS-AB-13 on the *E. turcicum* and *B. maydis* up to 24.6 mm and 32.8 mm, respectively. As mentioned in our previous work, XJAS-AB-11 had showed strong inhibitory effect on both *E. turcicum* and *Alternaria mali*, the effect on the former was much stronger (31.5 mm) than the latter [26]. Both bacterial strains, XJAS-AB-13 and XJAS-AB-11, forms circular colony, and cells were rhabditiform, colonies were slimy and jagged, has ivory gloss on surface, and easy to pick up (Table S1 and Figure S1).

The phylogenetic tree inferred from 16S rDNA sequence (Figure S2) demonstrated the high homology between the XJAS-AB-13 and *Bacillus subtilis* BS-HOT1 (99%). So, based on the molecular and morphological characteristics, XJAS-AB-13 identified as *B. subtilis* and registered into GeneBank (registration number is JF826131) while strain XJAS-AB-11was identified as a subspecies of *B. subtilis* in our previous work [26].

In vivo test and determination of defense enzyme activity

For XJAS-AB-13, all leaves were growth normally without any

infection in the blank control group. In the negative control group, however, almost all leaves were infected except 9 of them, the infection rate of *E. turcicum* and *B. maydis* up to 85% and 76.67%. Yet there was no significant difference between the infection rate of *E. turcicum* and *B. maydis*. But the significant difference (P<0.01) was occurred only between the infection rate of the treatment group and its relative negative control group (Table 1 and Figure S3). This also explains that the culture broth of XJAS-AB-13 is more efficient on preventing *E. turcicum* than *B. maydis*.

As shown in the Table 2 and Figures S4-S6, all leaves were growth healthy in blank control group, whereas the 85% (51 leaves) and 76.67% (46 leaves) of the leaves were infected in the negative control group. The statistical significant difference (P<0.01) between the treatment group and its relative negative control group of *B. maydis* showed the strong inhibitory efficiency of the XJAS-AB-11 culture broth on *B. maydis*, while its control rate only up to 23.33% on *E. turcicum*.

In general, the infection by diseases increased the production of different proteins in both cases (AB-11and AB-13). As shown in the Figure 1, infection without any preventing protocols caused increase in different proteins (Figure 1, NC 36260 and NC36265). However, application of AB-13 culture broth decreased the total protein content of the leaves sharply (Figure 1, T 36260 and T 36265). The multiple comparison result showed significant differences (α=0.01) between each pair of group, except between the blank control and negative control group.

On the contrary, there was no significant difference between negative control group and treatment group of *E. turcicum*, although there was a significant difference (α=0.01) between negative control group and treatment group of *B. maydis* (Figure 2).

The changes among the enzyme activity of four different enzymes were varies greatly. As shown in the Figure 3A, disease infection by both *E. turcicum* and *B. maydis* distinctly ceased the enzyme activity of SOD, especially of the negative control group. However, the exposure to the culture broth of XJAS-AB-13 kept the activity of SOD almost the same with blank control group, and there is no significant difference between them (P<0.01) (Figure 3A). Enzyme activity of POD and CAT

Pathogen inoculated	Treating type	Inoculated	Infected	Infection rate	Normal	Control rate
Exserohilum turcicum	Treated	60	13	21.67%	47	63.33%**
Exserohilum turcicum	Negative control	60	51	85%	9	—
Exserohilum turcicum	Positive control	60	4	6.67%	56	78.33%
No pathogen inoculated	Blank	0	0	0	60	—
Bipolaris maydis	Negative control	60	46	76.67%	14	—
Bipolaris maydis	Positive control	60	4	6.67%	56	70%
Bipolaris maydis	Treated	60	19	31.67%	41	45%**

Note: In the table, "**" means that between that treatment group and its relevant negative control group has significant difference (P<0.01)

Table 1: Controlling efficiency of XJAS-AB-13 culture broth on *Exserochilum turcicum* and *Bipolaris maydis*.

Pathogen inoculated	Treating type	Inoculated	Infected	Infection rate	Normal	Control rate
Exserohilum turcicum	Treated	60	37	61.67%	23	23.33%
Exserohilum turcicum	Negative control	60	51	85%	9	—
Exserohilum turcicum	Positive control	60	4	6.67%	56	78.33%
No pathogen inoculated	Blank	0	0	0	60	—
Bipolaris maydis	Negative control	60	46	76.67%	14	—
Bipolaris maydis	Positive control	60	4	6.67%	56	70%
Bipolaris maydis	Treated	60	11	18.33%	49	58.34%**

Note: in the table, "**" means that between that treatment group and relevant negative control group has significant difference (P<0.01)

Table 2: Controlling efficiency of XJAS-AB-11 culture broth on *Exserochilum turcicum* and *Bipolaris maydi*.

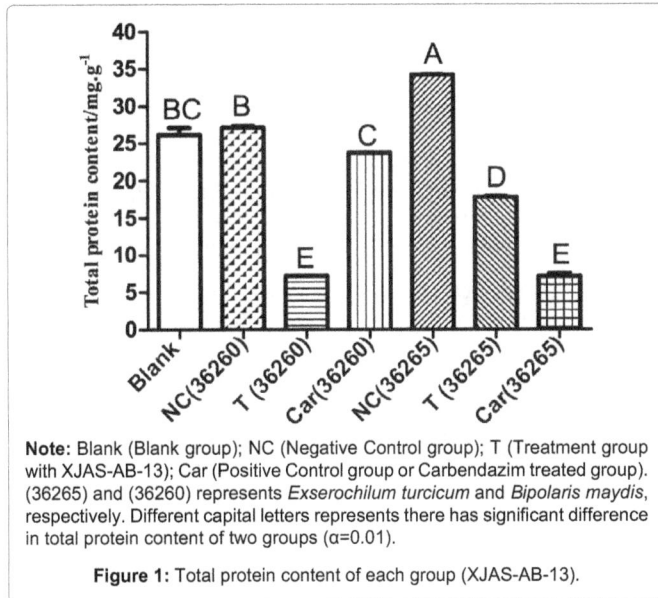

Note: Blank (Blank group); NC (Negative Control group); T (Treatment group with XJAS-AB-13); Car (Positive Control group or Carbendazim treated group). (36265) and (36260) represents *Exserochilum turcicum* and *Bipolaris maydis*, respectively. Different capital letters represents there has significant difference in total protein content of two groups ($\alpha=0.01$).

Figure 1: Total protein content of each group (XJAS-AB-13).

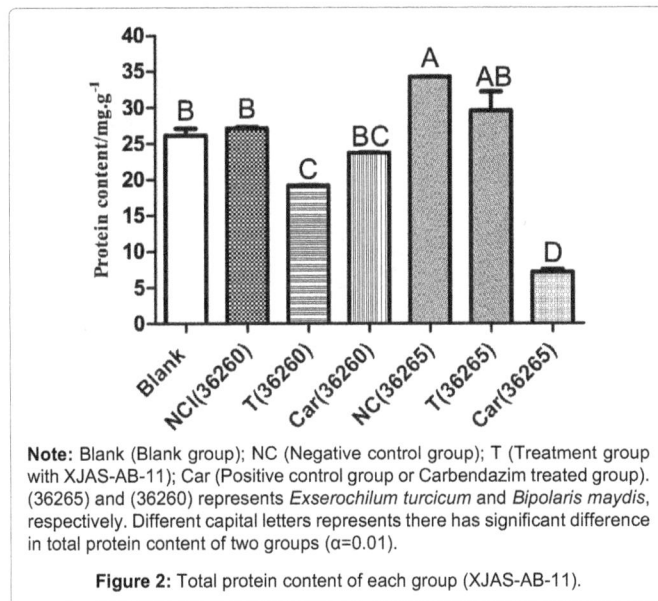

Note: Blank (Blank group); NC (Negative control group); T (Treatment group with XJAS-AB-11); Car (Positive control group or Carbendazim treated group). (36265) and (36260) represents *Exserochilum turcicum* and *Bipolaris maydis*, respectively. Different capital letters represents there has significant difference in total protein content of two groups ($\alpha=0.01$).

Figure 2: Total protein content of each group (XJAS-AB-11).

as a mobile phase (developing solvent) for the silica-gel column chromatography. As a final result for the silica-gel chromatography, two different monomers were obtained, which were XJAS-B (colorless acicular crystal) and XJAS-G (yellow powder).

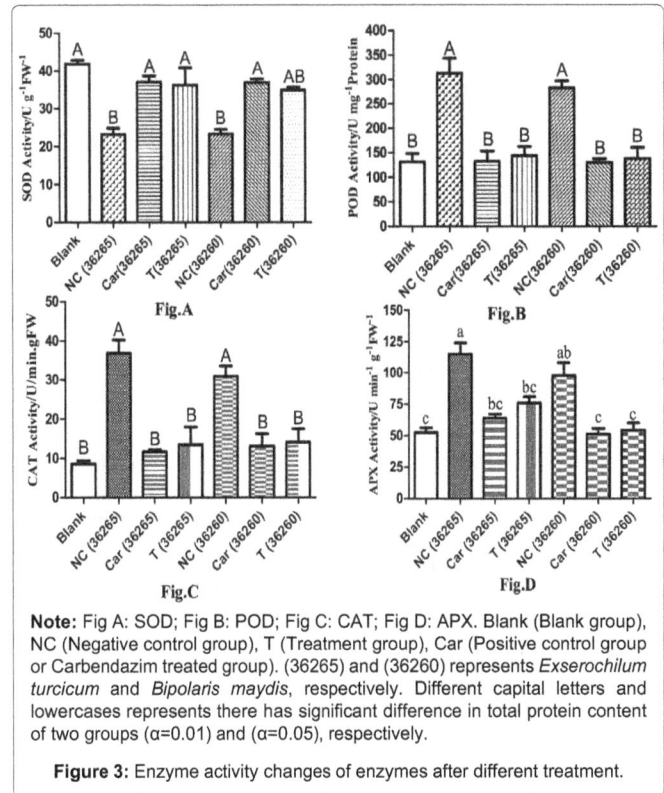

Note: Fig A: SOD; Fig B: POD; Fig C: CAT; Fig D: APX. Blank (Blank group), NC (Negative control group), T (Treatment group), Car (Positive control group or Carbendazim treated group). (36265) and (36260) represents *Exserochilum turcicum* and *Bipolaris maydis*, respectively. Different capital letters and lowercases represents there has significant difference in total protein content of two groups ($\alpha=0.01$) and ($\alpha=0.05$), respectively.

Figure 3: Enzyme activity changes of enzymes after different treatment.

among the treatment group (with AB-13) and positive control group was significantly lower than negative control group, approximately 2.162 and 2.737 time lower that negative control group (Figure 3). For XJAS-AB-11 (Figure 4), a similar SOD activity change with XJAS-AB-13 was observed. However, after exposure to pathogens and AB-11 fermentation broth, in general, all of CAT, POD and APX showed increasing tendency.

Structural analysis of metabolic active ingredients of XJAS-AB-11

Through extracting the fermentation broth of XJAS-AB-11, by isopyknic petroleum ether, normal butyl alcohol, chloroform, ethyl acetate, four brown yellow compounds were obtained, in weight of 2.717 g, 6.875 g, 28.354 g, 1.038 g respectively. Among which, the ethyl acetate extract showed the highest inhibition efficiency against all plant and animal pathogens, while the chloroform extract has no obvious effects. Hence, the ethyl acetate extract applied in the next experiment. Through TLC pre-test the chloroform/methanol (6:1) system chosen

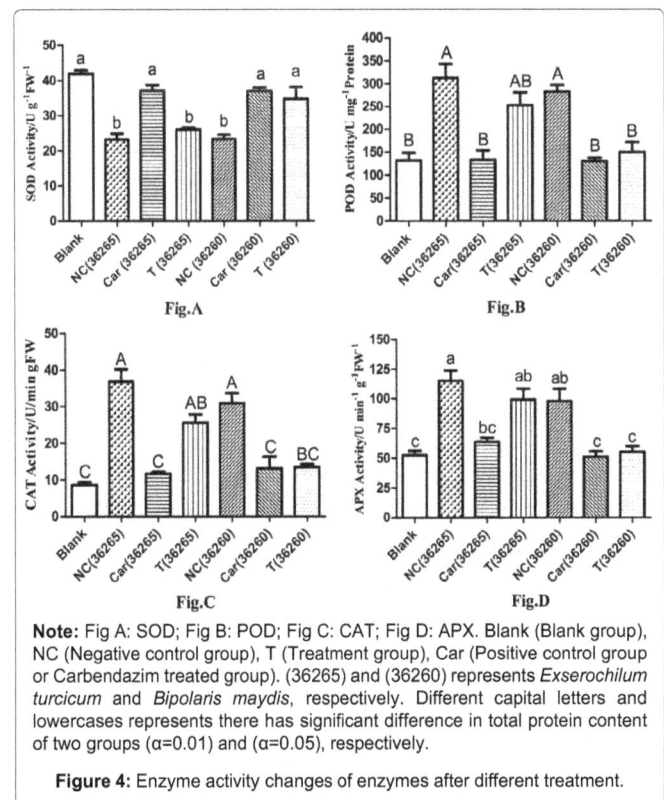

Note: Fig A: SOD; Fig B: POD; Fig C: CAT; Fig D: APX. Blank (Blank group), NC (Negative control group), T (Treatment group), Car (Positive control group or Carbendazim treated group). (36265) and (36260) represents *Exserochilum turcicum* and *Bipolaris maydis*, respectively. Different capital letters and lowercases represents there has significant difference in total protein content of two groups ($\alpha=0.01$) and ($\alpha=0.05$), respectively.

Figure 4: Enzyme activity changes of enzymes after different treatment.

According to ¹HNMR (Figures S7-S14), ¹³CNMR (Figure S8 and Figure S11) and API-MS data set (Figure S6 and Figure S9), the monomer XJAS-B, with a molecular weight of 226.1, identified as Cyclo-(D-leucyl-trans-4-hydroxy-L-proline) [(3R,7R,8aS)-7-hydroxy-3-isobutyl hexahydro-pyrrolo[1,2-a] pyrazine-1,4-dione] (Figure 5), while XJAS-G, with a molecular weight of 302.0, identified as 2-(3,4-dihydroxyphenyl)-3,5,7-trihydroxy-4H-chromen-4-one (Figure 6).

Naringenin-chalcone synthase (*CHS*) gene

A total of 823 bp and 1014 bp of XJAS-AB-11 *CHS* partial gene was amplified by Bsn and BpsA primers, respectively. Both sequence data showed high similarity with *Bacillus subtilis* BSn5 naringenin-chalcone synthase gene (99% and 97%, respectively). Based on the similarity and aligned gene length, the *CHS* partial sequence amplified by BpsA primer after TA cloning was selected for further analysis. Firstly, the selected *CHS* gene was translated into amino acid sequence with a total of 337 amino acids via ORF Finder system, and then the similar amino acid sequences were found through searching SWISS database. The phylogenetic tree (Figure S14) inferred from the amino acid sequence showed high homology with some experimentally verified prokaryotes and eukaryotes CHS amino acid sequence, such as those of *Streptomyces griseus* CHS, *Medicago sativa* CHS malonyl-CoA binding subunit, *Arachis hypogaea* stilbene synthase, and *Gerbera hybeid* CHS.

The secondary and three-dimensional structures of the *CHS* gene (Figure S12 and S13) showed the richness of α-helix and random coil structure in the XJAS-AB-11 *CHS* protein secondary structure. In addition, the spatial structure of this gene was highly conserved despite the considerable difference in the gene sequence between prokaryotes and higher plants, and their distant relationship. Moreover, high homology of *CHS* active site amino acid was also observed with *Arachis hypogaea* stilbene synthase, *Streptomyces griseus* CHS, and *Gerbera hybeid* CHS (Table 3).

Furthermore, the comparative analysis between XJAS-AB-11 CHS protein sequence and *Streptomyces griseus* CHS showed high homology between their amino acid sequences (Figure 7). Moreover, both of them also have similar amino acid sequences such as CMIYP and poly-G, which play a conclusive role in constructing of the spatial structure. Likewise, the catalytic site also contains the same amino acids, Cysteine (Cys) and Proline (Pro).

Discussion

Although *Bacillus subtilis* is a common bacteria, but not all of them possess the potentiality to inhibit the growth of plant/animal pathogens, as those isolated from medical plant *Alhagi pseudalhagi* Desv collected from Xinjiang. Despite the high inhibitory effect on different pathogens, these two bacteria, XJAS-AB-11 and XJAS-AB-13, were only effective on plant pathogens; this might be explained by one of these two reasons: (1) both of them could not produce any secondary metabolites that which is efficient on inhibiting animal pathogens; (2) they can produce some metabolites but not enough to observe the effects due to low production.

When plants are infected, they instantly produce and accumulate huge amount of reactive oxygen--called oxidative burst phenomenon; this is a most important early response, also one of the fastest response for initiating of defense mechanism against pathogens [27,28]. Thus, the activity changes of reactive oxygen related defense enzymes, SOD, POD, CAT, and APX, could be used as important parameters to indicate plant infection level. In this study, these defense enzymes showed increased activity, but those of SOD activity decreased. This is an interesting response, which needs further study to be explained.

The cyclic dipeptide XJAS-B isolated from XJAS-AB-11 is a derivative of cyclic dipeptide leucine-proline. Although there are several reports related to dipeptides isolated different organisms, for example, marine bacteria [29,30], however, no reports about finding of cyclic dipeptide from any medical plants or in any living bacterial strain. Thus, this is the first report about a cyclic dipeptide, (i.e., XJAS-B (cyclo-[D-leucyl-trans-4-hydroxy-L-proline])), isolated from a bacterial endophyte of medical plant *Alhagi pseudalhagi* Desv. In the other hand, XJAS-G is belonging to Quercetin. Quercetin and its derivatives are widely distributed among the plant world and it has been known to have anti-tumor, anti-inflammation, anti-platelet aggregation, anti-oxygen free radicals and vascular dilation properties [31]. The fermentation broth of XJAS-AB-11 showed inhibitory effect only on *Exserohilum turcicum* and *Alternaria mali*, but showed no apparent effects on animal pathogens. Interestingly, the monomers XJAS-B and XJAS-G were only effective on controlling animal pathogens. This might be due to (1) other active compounds in the fermentation broth of XJAS-AB-11 worked together with the two

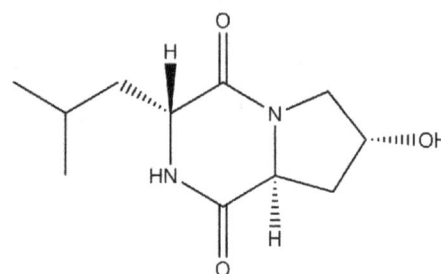

Figure 5: Chemical structure of compound XJAS-B.

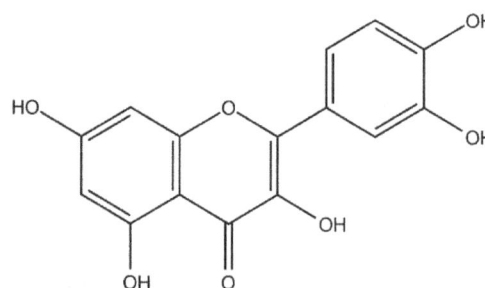

Figure 6: Chemical structure of compound XJAS-G.

	A	B	C	D	E	F	G	Substrate	Extender	Product
1	THN-RPPA	RNR	FF	CHN	CMIYP	TECCA	PGGLDGAGGGGGG	M-CoA	Mal	THNap
2	Stilbene	KRRA	FF	CHN	TLIGP	SETTS	PGGLDGPGGGGGA	pCoum	Mal	RESVR
3	Chalcone	KRKA	FF	CHN	TMIGP	SETTS	PGGLDGPGGGGGG	pCoum	Mal	Chalc

Note: A=protein name: THN-RPPA is *Streptomyces griseus* CHS; Stilbene is *Arachis hypogaea* Styrene synthetase; Chalcone is *Gerbera hybrid* CHS. B=CoA Tunnel Residues, C=Gatekeepers, D=Catalytic Triad, E=Cyclization Pocket, F=Coumaroyl Pocket, G=Geometry Shaping residues

Table 3: Comparison of *Streptomyces griseus* (1), *Arachis hypogaea* (2), *Gerbera hybrid* (3) type III polyketide synthase` active site amino acid sequences.

```
Streptomyces    MATLCRPAIAVPEHVITMQQTLDLARETHAG-HPQRDLVLRLIQNTGVQTRHLVQPIEKT 59
XJAS-AB-11      ------------AYNVNQEKAAEFARYMFQHSFKDIDRLLSSFKNGQIYSRQFVKPIEWY 48
Active residue  ----------------------------------------------------------

Streptomyces    LAHPGFEVRNQVYEAEAKTRVPEVVRRALAN----AETEPSE-IDLIVYVSCTGFMMPSL 114
XJAS-AB-11      KEGHSFEEKNQIYIEETLKHSRAAVRECLSHPEFFQEAIPYEKVEAVFFVSSTGLSTPSI 108
Active residue  ----------------------------------------------------------

Streptomyces    TAWIINSMGFRPETRQLPIAQLGCAAGGAAINRAHDFCVAYPDSNVLIVSCEFCSLCYQP 174
XJAS-AB-11      EARLMNELPFSPYTKRIPIWGLGCAGGASGLARAAEYCKAYPEVFGLVIAAELCSLTFIP 168
Active residue  ----RNRF-F-----------CHN-----------CMIYP--------TECC----AP  19
                    * : *               *           *  **        ** *      *

Streptomyces    TDIGVGSLLSNGLFGDALSAAVVKGQGGTG--------MRLERNGSHLVPDTEDWISYAV 226
XJAS-AB-11      QHDTTSKRIDTSQFDDHLAAALLGGEKADRRMSKLKLAPKIMDAQSVLMKQSEDVMGWDF 228
Active residue  G----G--LDG----------AGGGGGG------------------------------ 31
                    .    :.              *   .

Streptomyces    RDTGFHFQLDKRVPG-TMEMLAPVLLDLVDLHGWSVPNMDFFIVILAGGPRILDDLCHFLD 285
XJAS-AB-11      TDQGFKVIFSRDIPTLVEKWLKTNVQIFLDKHKLSFHDISVFLAHPGGKKVIDAYIKSLG 288
Active residue  ----------------------------------------------------------

Streptomyces    LPPEMFRYSRATLTERGNIASSVVFDALARLFDDGGAAESAQGLIAGFGPGITAEVAVGS 345
XJAS-AB-11      LSSEKLSSAQSILQKHGNMSSATILYVIKDHLQNGHKKEAERGMIGALG----------- 337
Active residue  ----------------------------------------------------------
```

Figure 7: Cluatal W comparison of amino acid sequence of XJAS-AB-11 CHS and *Streptomyces griseus* CHS.

monomers, which becomes more potent against plant pathogens, and (2) fermentation dilutes the concentration of the monomer, making it less effective against animal pathogens.

In the last decade, as a result of the widespread use of PCR and DNA sequencing, 16S rDNA sequencing has played a pivotal role in the accurate identification of bacterial isolates and the discovery of novel bacteria in clinical microbiology laboratories. particularly important in the case of bacteria with unusual phenotypic profiles, rare bacteria, slow-growing bacteria, uncultivable bacteria and culture-negative infections. Not only has it provided insights into aetiologies of infectious disease, but it also helps clinicians in choosing antibiotics and in determining the duration of treatment and infection control procedures.

For bacterial identification, 16S rDNA sequencing has played a pivotal role in the accurate identification of bacterial isolates and the discovery of novel bacteria in clinical microbiology laboratories. Using 16S rDNA sequences, numerous bacterial genera and species have been reclassified and renamed, classification of uncultivable bacteria has been made possible, phylogenetic relationships have been determined, and the discovery and classification of novel bacterial species has been facilitated. In the last decade, sequencing of various bacterial genomes and comparison between genome and 16S rDNA gene phylogeny has confirmed the representativeness of the 16S rDNA gene in bacterial phylogeny [32-35].

Endophytes of medical plants could produce the same or similar metabolites with host plant. For example, plants in the Yew genus produce anticancer chemical, paclitaxel; similarly, endophytes of this plant were also able to produce the same chemical [36]; endophytes of Podophyllum produce podophyllotoxin [37], of Camptotheca, produce analogue of camptothecin [38], of Ginko produce flavonoids compounds [39]. This phenomenon may be due to horizontal transfer of genetic materials between host plant and its endophytes, which might be mutual exchange [14,40-42]; Prokaryotic cell genome sequencing pointed out that horizontal transfer of genetic materials is the most important mechanism of evolution [43-45]. Nevertheless, phylogenetic relationship between the species is always different despite the genetic material's transfer. During the horizontal gene transfer from

eukaryotes to prokaryotes, gene(s) could undergo accelerated evolution and lost; as a result, the genetic relationship between the transferred gene(s) would be far. Moreover, the species misassignment caused by DNA contamination during gene transfer, may also phylogenetic relationships, even after the transfer of same genetic material [46].

In this study, the comparison result of *CHS* active site amino acid sequence showed that despite of differences of nucleic acid sequences in different species, the translated product (i.e., CHS amino acid sequence), is highly similar to those of *Actinomycetes*, as well as higher plant *Medicago*; thus, it is also possible that the endophytes of *Alhagi pseudalhagi* Desv, XJAS-AB-11, has its independent *CHS* gene that similar to higher plant.

In conclusion, the mechanism of gene horizontal transfer has been not clear, but this study further supported the probability of functional gene horizontal transfer between endophyte and its host plant. Moreover, this study provided useful data for development and utilization of new eco-friendly pesticide.

Acknowledgment

This study funded by the National Natural Science Foundation of China (grant №: 31160015).

Compliance with ethical standards

Conflict of Interest

All authors declare that they have no conflict of interest.

Ethical Approval

This article does not contain any study with human participants or animals performed by any of the authors.

References

1. FAOSTAT (2012) Statistical database of the food and agriculture of the united nations.

2. Kang ZS (2010) Current status and development strategy for research on plant fungal disease in China. Plant Protection 36: 9-12.

3. Yan MH (2004) The applied research of endophytic bacteria in biological control of plant disease. Biotechnol Infor 3: 8-10.

4. Shi JY, Chen WX, Liu AY (2006) Advance in the study of endophytes and their effects on control of plant disease. Acta Ecological Sinica 26: 2395-2401.

5. Liang JG, Zhu LH, Wu JA, Sang JL, Yao HL, et al. (2007) Control efficiency of biocontrol strain B-3 against pepper wilt and its identification. Acta Phytophylacica Sinica 34: 529-533.

6. Yi YJ, Xiao LT, Wang RZ, Bo LY, Liu EM, et al. (2007) Growth effect of endophyte B-001 (*Bacillus subtilis*) on tobacco seedling and its fluctuation. Acta Phytophylacica Sinica 34: 619-623.

7. Strobel GA (2003) Endophytes as sources of bioactive products. Microbes Infect 5: 535-544.

8. Bian JC, Feng YH, Yang J, Ding YW, Jing H (2010) Research Progress in Traditional Chinese Medicine Endophytes. GuangMing J Chinese Med (CJGMCM) 25: 164-165.

9. Bary A (1866) Morphologie und physiologie der pilze, flechten und myxomyceten. W. Engelmann.

10. Chen JH, Liu JJ, Zang GG, Li YJ, Zhuo LN (2004) Screening of Taxol-producing Endophytic Fungi and Regulation of Fermentation Condition. J Central South University (Sciences and Technology) 35: 65-69.

11. Zhang LQ, Guo B, Li HY, Zeng SR (2000) Preliminary study on the isolation of endophytic fungus of catharanthus roseus and its fermentation to produce products of therapeutic value. Chinese Traditional and Herbal Drugs 31: 805-807.

12. Stone JK, Bacon CW, White JF (2000) An overview of endophytic microbes: Endophytism defined. Microbial endophytes 3: 29-33.

13. Sturz AV, Christie BR, Nowak J (2000) Bacterial endophytes: Potential role in developing sustainable systems of crop production. Crc Cr Rev Plant Sci 19: 1-30.

14. Stierle A, Strobel G, Stierle D (1993) Taxol and taxane production by Taxomyces andreanae, an endophytic fungus of Pacific yew. Science 260: 214-216.

15. Strobel GA, Miller RV, Martinez-Miller C, Condron MM, Teplow DB, et al. (1999) Cryptocandin, a potent antimycotic from the endophytic fungus Cryptosporiopsis cf. quercina. Microbiology 145: 1919-1926.

16. Yang XW, Jiang YM, Li JS, Lou ZC (1996) A New Flavonal Glucoside from Aerial Parts of Manaplant Alhagi (Alhagi pseudoalhagi). Chinese Traditional and Herbal Drugs 27: 707-711.

17. Ailat S, Dilnur T, Ablimit A, Stiwaldi H (2005) Studies on Flavones Compounds from Alhagi pseudoalhagi. Natural Product Research and Development 17: 1-2.

18. Qian CR, Huang YX (2003) Laboratory Experiments in Microbiology. Beijing: Peking University Press p: 215.

19. Geris dos Santos RM, Rodrigues-Fo E, Caldas Rocha W, et al. (2003) Endophytic fungi from Melia azedarach. World J Microbiol Biotechnol 19: 767-770.

20. Conn VM, Franco CMM (2004) Analysis of the endophytic actinobacterial population in the roots of Wheat (*Triticum aestivum* L.) by terminal restriction fragment length polymorphism and sequencing of 16S rRNA Clones. Appl Environ Microbio 70: 1787-1794.

21. Li HY, Qing C, Zhang YL, Zhao ZW (2005) Screening for endophytic fungi with antitumor and antifungal activities from chinese medicinal plants. World J Microbiol Biotechnol 21: 1515-1519.

22. Buchanan RE, Gibbons NE (1984) Bergey's manual of determinative bacteriology. (18thedn.), Beijing: Science Press.

23. Dong XZ, Cai MY (2001) Common Bacterial Systems Identification Manual. Beijing: Science Press.

24. Chen JX, Wang XF (2006) Experimental Guide for Plant Physiology. (2ndedn.), Guangzhou: South China University of Technology Press.

25. Nakano C, Ozawa H, Akanuma G, Funa N, Horinouchi S (2009) Biosynthesis of aliphatic polyketides by type III polyketide synthase and methyltransferase in *Bacillus subtilis*. J Bacteriol 191: 4916-4923.

26. Aynazar Z, Ghopur M, Abdukadir A, Alimjan A, Gulzira Z (2012) Isolation of Endophytic Bacteria from Alhagi pseudalhagi Desv. and Analysis of Active Metabolites. Acta Botanica Boreali-Occidentalia Sinica 32: 1466-1473.

27. Song FM, Zheng Z, Ge XC (1996) Role of Active Oxygen and Membrane Lipid Peroxidation in Plant-Pathogen Interactions. Plant Physiol Commun 32: 377-385.

28. Wojtaszek P (1997) Oxidative burst: An early plant response to pathogen infection. Biochem J 322: 681-692.

29. Ai F, Xu QZ, Yang Y, Liu XY, Shi XQ, et al. (2006) Bioactive cyclodipeptides extracted from marine microbes in east China Sea. Academic Journal of Second Military Medical University 27: 22-24.

30. Yao Y, Tian L, Li J, Pei YH (2007) Cyclo-dipeptide Metabolites from The Broth of Marine Bacterium Bacillus sp. Chinese J Med Chem 17: 310.

31. Wang YF, Wang XH, Zhu YT (2003) Advancement of Researches in Quercetin. Nat Prod Res Dev 15: 171.

32. Fox GE, Magrum LJ, Balch WE, Wolfe RS, Woese CR (1977) Classification of methanogenic bacteria by 16S ribosomal RNA characterization. Proc Natl Acad Sci USA 74: 4537-4541.

33. Gupta R, Lanter JM, Woese CR (1983) Sequence of the 16S Ribosomal RNA from Halobacterium volcanii, an Archaebacterium. Science 221: 656-659.

34. Snel B, Bork P, Huynen MA (1999) Genome phylogeny based on gene content. Nat Genet 21: 108-110.

35. Woo PCY, Lau SKP, Teng JLL, Tse H, Yuen K (2008) Then and Now: Use of 16S rDNA Gene Sequencing for Bacterial Identification and Discovery of Novel Bacteria in Clinical Microbiology Laboratories. Clin Microbiol Infec 14: 908-934.

36. Wang JF, Lv HY, Su WJ (2000) Study on taxol production of plant endophytic fungi. Microbiology 27: 58-60.

37. Puri SC, Nazir A, Chawla R, Arora R, Riyaz-Ul-Hasan S, et al. (2006) The endophytic fungus Trametes hirsuta as a novel alternative source of podophyllotoxin and related aryl tetralin lignans. J Biotechnol 122: 494-510.

38. Liu JH, Yu BY (2004) Isolation of endophytic fungi from camptotheca acuminala and the screening method of antitumor secondary metabolite produced by the fungi. J Plant Resou Env 13: 6-10.

39. Zhao QY, Fan MT, Shi JL (2007) Isolation and identification of flavones-forming endophytes from Ginkgo bilob L. Acta Agriculture Boreali-occidentalis Sinica 16: 169-173.

40. Mower JP, Stefanovic S, Young GJ, Palmer JD (2004) Plant genetics: Gene transfer from parasitic to host plants. Nature 432: 165-166.

41. Lawrence JG (2001) Catalyzing bacterial speciation: correlating lateral transfer with genetic headroom. Syst Biol 50: 479-496.

42. Boucher Y, Doolittle WF (2000) The role of lateral gene transfer in the evolution of isoprenoid biosynthesis pathways. Mol Microbiol 37: 703-716.

43. Nara T, Hshimoto T, Aoki T (2000) Evolutionary implications of the mosaic pyrimidine-biosynthetic pathway in eukaryotes. Gene 257: 209-222.

44. Ochman H (2001) Lateral and oblique gene transfer. Curr Opin Genet Dev 11: 616-619.

45. Dujon B, Sherman D, Fischer G, Durrens P, Casaregola S, et al. (2004) Genome evolution in yeasts. Nature 430: 35-44.

46. Hall C, Brachat S, Dietrich FS (2005) Contribution of horizontal gene transfer to the evolution of Saccharomyces cerevisiae. Eukaryot Cell 4: 1102-1115.

4

Determination of Suitability of Deoiled Cakes of Neem and Jatropha for Mass Multiplication of *Pseudomonas fluorescens*

Ajay Tomer*, Ramji Singh and Manoj Maurya

Department of Plant Pathology, Sardar Vallabhbhi Patel University of Agriculture & Technology, Meerut (UP), India

Abstract

Jatropha cake was found to be best substrate for supporting the highest population dynamics and longevity of *P. flourescens* in vitro. Both these cakes supported the population of *P. flourescens* up to 120 days at 28 ± 2°C. Highest population of *P. flourescens* was noticed on jatrofa cake after 45 days of inoculation and incubation at 28 ± 2°C when maintained with 15% moisture, while on neem cake, highest population of *P. flourescens* was noticed after 60 days of inoculation when maintained with 25% moisture. Jatropha cake was better than neem cake for supporting population and longer shelf life of *P. flourescens* in vitro. Result of this study may help for using deoiled cakes of different TBOs (tree born oil seeds) in mass multiplication of different bio agent's especially *P. flourescens*.

Keywords: Longevity; Survival; Deoiled cakes; Neem; Jatropha; *Cfus*; *Pseudomonas flourescens*

Introduction

Various factors act as impediments which are known to affect the commercialization of microbial bio-control agent such as *Pseudomonas fluorescens* (an essential step for improving efficiency of bio-control agents). One of these impediments is the lack of technology for cost-effective mass production of bio-control agents (BCAs) and their insufficient longevity during storage and transportation with a sufficient level of effective population (cfus). It is being noticed that *Pseudomonas fluorescens* are often effective in the laboratory, but the level of disease control, achieved in the field is sometimes disappointing and unpredictable. Some of these failures can be attributed to inadequate establishment and survival of this microorganism in soil [1]. Since they are well adapted in soil, that's why *P. fluorescens* strains are being investigated extensively for use in bio-control of plant pathogens in agriculture [2]. It is known to enhance plant growth along with yield enhancement and to reduce severity of many diseases [3].

Many agro-industrial byproducts such as deoiled cakes of tree born oils seeds (TBOs) like Neem and Jatropha which are either going waste or being used as a less profitable and less usable products since quite long time. Deoiled cakes of these trees, either remain unexploited or poorly exploited, generally as low value soil amendments or organic manure. These deoiled cakes contains good amount of N, P, K, Ca, Mg, S, Fe, Mn, Zn and Cu [4]. In addition they also contain carbohydrates, proteins, fatty acids, and many more biochemical constituents. With these qualities, these deoiled cakes may be exploited as substrate for mass multiplication of bacterial bio-control agents such as *Pseudomonas flourescens*. Mass multiplication of *Pseudomonas flourescens* will not only leads to value added products development from deoiled cakes of Neem and Jatropha; rather it will prevent huge wastage and inappropriate use of these by-products.

The mass cultures made at industrial scale are generally talc based, with no nutritional background to support the life of BCAs during storage, transportation and other stress. Deoiled cakes of TBOs may serve as source of diversified nutrition for BCAs when used as substrate for mass multiplication of antagonists.

Materials and Methods

Sources and maintenance of culture

Pseudomonas fluorescens culture was isolated from tomato rhizospheric soil, collected from crop research centre (CRC) of SVPUAT Meerut. For isolation of microorganism, Ten gm of soil sample adhered to roots and rootlets of tomato were collected and placed in 250 ml conical flasks containing 100 ml of sterilized distilled water (SDW) and mixed thoroughly. Different dilutions of working samples were prepared by serially diluting the stock solution upto 10^{-8}. One ml of last serial dilution *i.e.*, 10^{-8} was spread on *Pseudomonas fluorescens* selective king's B Medium [5] for isolation of *Pseudomonas fluorescens*. The plates were incubated for 2 days at 28 ± 2°C and after two days of incubation, pure culture was maintained in PDA slants. Conformity of culture was done on the basis of color of bacterial colony which was initially yellow but turned yellow green as pigmentation were produced (Bonds 1957). Further culture was again reconfirmed by molecular conformity test at National Beauro of Agriculturally Important Microbes (ICAR) Mau (UP) India. The culture thus obtained was stored in refrigerator at 5°C for further studies and was sub cultured periodically.

Determination of population dynamics and longevity of *pseudomonas fluorescens* on deoiled cakes of neem and jatropha

Deoiled cakes of two tree born oilseeds *i.e.*, Neem and Jatropha were collected from agricultural product-processing units situated in eastern Distt. Mau of Uttar Pradesh (UP) and Raipur in Chhattisgarh. Before using the cakes, they were grounded in a metallic pastel and mortar to prepare fine powder and three different level of moisture *i.e.*, 15%, 25% and 35% (w/v) were maintained by adding sufficient amount of sterilized distilled water. Before inoculation of *Pseudomonas fluorescens*, cakes containing different level of moisture were placed in

*Corresponding author: Ajay Tomer, Department of Plant Pathology, Sardar Vallabhbhi Patel University of Agriculture & Technology, Meerut- 250110 (UP), India, E-mail: ajaytomer1489@gmail.com

Moisture level	15 Days	30 Days	45 Days	60 Days	75 Days	90 Days	105 Days	120 Days
15 %	122.00	157.33	187.00	203.00	183.67	125.00	84.67	54.33
25%	129.33	162.00	192.00	208.00	157.67	132.7	48.00	9.33
35%	139.00	168.00	196.67	176.67	130.07	98.00	48.00	10.33

CD @ 5%   Moisture % = 7.0655
Days = 11.5378
MxD = 19.984

Table 1: CFUs of *Pseudomonas fluorescens* at different moisture level on sterilized Neem cake.

Moisture level	15 Days	30 Days	45 Days	60 Days	75 Days	90 Days	105 Days	120 Days
15%	127.33	188.00	271.00	211.3	141.33	97.67	52.00	16.67
25%	140.00	212.33	262.67	197.00	131.00	92.00	43.00	14.00
35%	148.00	236.67	254.00	214.00	171.00	103.00	46.67	17.67

C.D @ moisture% = 1.2504
Days = 2.0419
MxD = 3.537

Table 2: CFUs of *Pseudomonas fluorescens* at different moisture level on sterilized Jatropha cake.

250 ml capacity conical flasks (@75 gram/flask), plugged tightly with cotton plugs, wrapped with butter paper and autoclaved at 121.6°C (1.1 kg/cm^2) for 20 minutes. After inoculation, flasks were allowed to cool overnight at room temperature prior to inoculation. Flasks containing sterilized deoiled cakes with different level of moisture were inoculated with 3-4 days old actively growing culture of *Pseudomonas fluorescens* (2-3 bits of 5 mm size from the culture grown on PDA in Petri plates) under aseptic conditions in laminar flow. For each moisture level and each set of duration (Tables 1 and 2) three replicates were maintained. Flasks inoculated with *Pseudomonas fluorescens* were incubated at 28 ± 2°C in a BOD incubator and shaken thoroughly once a day to provide sufficient substrate to grow properly.

Monitoring of population dynamics in deoiled cakes

Population of *Pseudomonas fluorescens* was monitored from the deoiled cakes of Jatrofa and Neem maintained with different level of moisture (15%, 25% and 35% respectively) after each 15 days interval upto 120 days. For this purpose, 1 gm of each cakes where *Pseudomonas flourescens* was inoculated was taken from each flasks maintained for different duration i.e., 15 to 120 days and CFUs were counted using PDA through dilution plate technique/Simplified agar plate method for quantifying viable bacteria [6].

Results

Screening of deoiled cakes of Neem and Jatropha for mass multiplication of *Pseudomonas fluorescens*

Deoiled cakes of two Tree Born Oilseeds (TBOs) i.e., Neem and Jatropha Were tested for their suitability to support the population dynamics and longevity of *Pseudomonas fluorescens* at three different level of moisture i.e 15%, 25% and 35%. Results obtained have been presented in (Tables 1 and 2).

Neem cakes

Neem cake with 35% moisture resulted in 139.00×10^8 number of CFUs of *Pseudomonas fluorescens* at 15 days of inoculation. At 30 days of inoculation neem cake with 35% moisture resulted in 168.00×10^8 CFUs of *P .fluorescens*, whereas at 45 days of inoculation the population of *P .fluorescens* increased to 196.67×10^8 and at 60 days of inoculation, it further goes down to the level of 176.67×10^8. At 75 days of inoculation Neem cake with 35% moisture showed further reduction in the *Pseudomonas* population and accordingly CFUs decreased to 137.00×10^8. At 90 days of inoculation, the Neem cake with

35% moisture resulted in 98.00×10^8 number of CFUs of *P .fluorescens*. At 105 days the number of CFUs of *P .fluorescens*, recovered from the cakes were 48.00×10^8, whereas at 120 days, the CFUs of *P. Fluorescens* recovered, were 10.33×10^8. In the case of neem cake the population of *P. Fluorescens* observed at each 15 days interval differed significantly from each other while different level of moisture hardly had significant effect on the increase or decrease of population.

Jatropha cake

Jatropha cake containing 35% moisture resulted in 148.00×10^8 level of CFUs of *Pseudomonas fluorescens* at 15 days of inoculation. At 30 days of inoculation, the number of CFUs increased to the level of 236.67×10^8, whereas at 45 days the population further increased to the level of 254.00×10^8 CFUs of *Pseudomonas fluorescens*. At 60 days onward there was declining trend and population declined to the level of 214.00×10^8 and at 75 days it was 171.00×10^8. At 90 days of inoculation the Jatropha cake containing 35% moisture, resulted in 103.00×10^8 level of population of *P .fluorescens*. At 105 days the *P .fluorescens* population decreased to the level of 46.67×10^8 and at 120 days population further declined to the level of 17.67×10^8. Level of CFUs recorded after each 15 days interval and each level of moisture were significantly different from each other.

In the comparison of two deoiled cakes it was found that jatropha cake was quite superior over neem cakes in supporting the population dynamics of *Pseudomonas fluorescens*. In case of neem cake, increasing the level of moisture didn't have any significant effect on population dynamics of *Pseudomonas fluorescens*, whereas in case of jatropha cakes, with increasing in the level of moisture had resulted in significant increase of population dynamics of *Pseudomonas fluorescens*.

Discussion

Cakes of neem and jatropha

With a purpose to find out a suitable substrate for mass multiplication and also for a longer shelf life of *Pseudomonas fluorescens*, an experiment was conducted to test the suitability of two de-oiled cakes of Neem and Jatropha with three moisture level i.e. 15%, 25% and 35%. Results indicated that Jatropha cake was found to be comparatively better than neem cake for enhancing population of Pseudomonas *fluorescens* with a highest level after 45 days of inoculation with 15% moisture. It was also noticed that Jatropha and neem cake both could support the population and longevity upto 120 days with×10^8 level of population. In case of neem cake it was observed that on neem cake population of *Pseudomonas fluorescens* was found to be increasing upto 60 days after inoculation with a highest at 60 days after inoculation with 25% moisture, while on Jatropha cake population of *Pseudomonas fluorescens* was found to be increasing upto 45 days only and after that there was a trend of decline in the population of *Pseudomonas fluorescens* . In case of neem cake upto 60 days increase in the moisture level resulted with increase of population but after 60 days onward increase in the moisture resulted in decrease of population. In case of Jatropha cake upto 45 days only increase in the moisture level resulted with increase in population but after 45 days onward increase in the moisture resulted in decrease in population of *Pseudomonas fluorescens.*

Reason behind higher population dynamics of *Pseudomonas fluorescens* on the de-oiled cakes of two tree born oilseeds (TBO's) may be because of their richness and sufficiency of different type of nutrients, minerals and other constituents which are required and may be supportive for multiplication of *Pseudomonas fluorescens* . Reason behind decline of population dynamics after 45/60 days during present investigation may be that, at the initial level there may be plenty of nutrition available to be utilized for multiplication of *Pseudomonas fluorescens* which later get declined, because they might have been exhausted day by day due to utilization and exploitation by growing *Pseudomonas fluorescens* in the substrate itself and resulted in poor supply after 45/60 days and thereby lower population dynamics with prolonging duration of storage.

Nilkamal et al. [7] assessed Deoiled Jatropha cake as substrate for enzyme production by solid-state fermentation (SSF). Solvent tolerant *Pseudomonas aeruginosa* PseA strain was used for fermentation. The seed cake supported good bacterial growth and enzyme production (protease, 1818 μg/g of substrate and lipase, 625 μg /g of substrate). Maximum protease and lipase production was observed at 50% substrate moisture, at a period of 72 and 120 h, pH of 6.0 and 7.0, respectively. Murugalakshmi et al. [8] reported that Agricultural residues rich in carbohydrates can be utilized in fermentation process to produce microbial protein which in turn can be used to determine the factors influencing cell biomass production. *Pseudomonas fluorescens* was cultivated using banana peel out, watermelon skin, and Cane molasses showed that the strain was capable of meeting its components required for growth. The organism was capable of growth at 28°C, when supplemented with agricultural wastes in different concentration mixed with agar. The number of colony forming units was more when compared with nutrient agar. Abhinav et al. [9] evaluated PGPR strain of *Pseudomonas fluorescens PS1* to formulate carrier based bioformulations. The viability of *Pseudomonas fluorescens PS1* was monitored at different time intervals during the period of storage at room temperature in different carriers such as soil, charcoal, sawdust and sawdust-soil. Sawdust-soil was found to be the most efficient carrier material for *P. fluorescens PS1* followed by other carriers. Sangeetha et al. [10] studied the survival of PGPR isolates by using different carrier materials. The carrier based PGPR consortium with four selected strains *viz., Azospirillum lipoferum* VAZS-18, *Azotobacter chroococcum* VAZB-6, *Bacillus megaterium* VBA-2, *Pseudomonas fluorescens* VPS-19 was prepared and the shelf life for each inoculants was studied upto six months of storage. The surviving population in the lignite based consortium was 1.64×10^8 cfu g^{-1} for *Azospirillum lipoferum* VAZS-18, 1.46×10^8 cfu g^{-1} for *Azotobacter chrococcum,* VAZB-6, 1.22×10^8 cfu g^{-1} for *Bacillus megaterium* VBA-2 and 2.01×10^8 cfu g^{-1} for *Pseudomonas fluorescens* VPS-19 after six month of storage. The surviving population in vermiculite based consortium was 4.32×10^8 cfu g^{-1} for *Azospirillum lipoferum* VAZS-18, 1.98×10^8 cfu g^{-1} for *Azotobacter*

chroococcum VAZB-6, 1.14×10^8 cfu g^{-1} for *Bacillus megaterium* VBA-2 and 3.32×10^8 cfu g^{-1} for *Pseudomonas fluorescens* VPS-19 after six months of storage. In the pressmud based consortium, the surviving population was 3.25×10^8 cfu g^{-1} for *Azospirillum lipoferum* VAZS-18, 3.00×10^8 cfu g^{-1} for *Azotobacter chroococcum* VAZB-6, 2.14×10^8 cfu g^{-1} for *Bacillus megaterium* VBA-2 and 3.42×10^8 cfu g^{-1} for *Pseudomonas fluorescens* VPS-19 after six months of storage. In the alginate bead based consortium the surviving population was 64.61×10^8 cfu g^{-1} for *Azospirillum lipoferum* VAZS-18, 56.81×10^8 cfu g^{-1} for *Azotobacter chroococcum* VAZB-6, 47.83×10^8 cfu g^{-1} for *Bacillus megaterium* VBA-2 and 63.89×10^8 cfu g^{-1} for *Pseudomonas fluorescens* VPS-19 after six months of storage. Although scanty literatures are available regarding use of deoiled cakes for mass multiplication of *Pseudomonas fluorescens* but after thorough scanning of literature it is clear that the carriers rich in organic substances and carbohydrate are highly supportive of multiplication of *Pseudomonas fluorescens* thus the findings of present studies are well supported by previous findings as mentioned above [11].

Based on the findings reported by all the groups mentioned above it is clear that the material rich in either protein or carbohydrates have been found to be comparatively better carrier for mass multiplication of *Pseudomonas fluorescens* as compared to those substrates which or either nutrient less like agar based or having less nutrient. Thus the findings of these workers are in conformity with the present findings [12].

Conclusion

De-oiled cakes of two trees born oilseeds (TBOs) viz., Neem and Jatropha were tested as solid substrate for their suitability for mass multiplication of *Pseudomonas flourescens* Among two deoiled cakes, Jatropha cake was found to be best substrate in supporting the population dynamics and longevity as well of *Pseudomonas flourescens in vitro*. Neem cake was next in order to support the population of *Pseudomonas flourescens* rather closely followed the Jatropha cake in supporting the population and longevity of *Pseudomonas flourescens in vitro*.

Both these cakes (Neem and Jatropha) supported the population of *Pseudomonas flourescens* up to 120 days with a considerable level of viable counts of *Pseudomonas flourescens*.

Highest population of *Pseudomonas flourescens* was noticed on jatrofa cake after 45 days of inoculation when maintained with 15% moisture, while on neem cake highest population of *Pseudomonas flourescens* was noticed after 60 days of inoculation when maintained with 25% moisture. Jatropha cake was better than neem cake for supporting population and longer shelf life of *Pseudomonas flourescens* in-vitro.

Jatropha cake helped to increase the population of *Pseudomonas flourescens* up to 45 days and thereafter it was decreased, whereas neem cake helped to increase the population of *Pseudomonas flourescens* up to 60 days and thereafter it was decreased.

Increasing the moisture level resulted in enhancement of total viable count of *Pseudomonas flourescens* on both deoiled cakes i.e. neem cake and Jatropha cake. After duration of 120 days also the viable counts of *Pseudomonas flourescens* were$\times 10^8$.

Both these cakes supported the population of *Pseudomonas flourescens* up to duration of 120 days with a considerable level of viable counts of *Pseudomonas flourescens*. Based on the findings of present studies these deoiled cakes i.e. Neem and jatropha both can be used as a carrier for mass multiplication of *Pseudomonas flourescens* and can be applied to crop/soil.

Acknowledgments

Help and support received from SVPUAT Meerut and NOVOD Board during the course of investigation is dully acknowledged.

References

1. Elliot LF, Lynch JM (1995) The International Workshop on the Establishment of Microbial Inocula in soils. Cooperative research project on biological resource management of the Organisation for Economic Cooperation and Development (OECD). Am J Altern Agric 10: 50-73.

2. Ganeshan G, Kumar MA (2006) Pseudomonas fluorescens a potential bacterial antagonist to control plant diseases. J Plant Interactions 3: 123-134.

3. Hoffland E, Halilinen J, Van Pelt JA (1996) Comparison of systemic resistance induced by avirulent and non-pathogenic Pseudomonas sp. Phytopathology 86: 757-762.

4. Patolia (2007) Preparing Status Reports on Themes Related to Technical and Scientific Aspects of Biofuels Utilization by Winrock International India (WII), New Delhi India

5. King EO, Ward MK, Raney DE (1954) two simple media for the demonstration of Pyocyanin and fluorescin. Journal of Laboratory and Clinical Medicine 36: 100-102.

6. Jett BD, Hatter KL, Huycke MM, Gilmore MS (1997) Simplified agar plate method for quantifying viable bacteria. Biotechniques 23: 648-650.

7. Neelkamal M, Gaur R, Gupta A, Khare SK (2008) Purification and characterization of lipase from solvent tolerant Pseudomonas aeruginosa PseA. Biological Sciences 43: 1040-1046.

8. Murugalakshmi CN, Sudha SS (2010) The Efficacy of agro waste on cultivation of Pseudomonas fluorescens. A potential biocontrol agent International Journal of Biological Technology 1: 32-34.

9. Abhinav A, Dubey RC, Maheshwari D K, Pandey P, Bajpai K, et al. (2011) PGPR strain Pseudomonas fluorescens PS1 was evaluated to formulate carrier based bioformulations. European Journal of Plant Pathology I: 81-93.

10. Sangeetha D (2012) Survival of plant growth promoting bacterial inoculants in different carrier materials. International Journal of Pharmaceutical & Biological Archive 3: 221-243.

11. Ambardar VK, Sood AK (2010) Suitability of different growth substrate for mass multiplication of bacterial antagonists. Indian Phytopathology 63: pp.4

12. Bonde GJ, Jensen CE, Thamsen J (1957) Studies on a water soluble fluoresceing bacterial pigment which depolymerizes hyaluronic acid. Acta Pharmacol Toxicol (Copenh) 13: 184-193.

5

Effect of Climate Change Resilience Strategies on Common Bacterial Blight of Common Bean (*Phaseolus vulgaris* L.) in Semi-Arid Agro-Ecology of Eastern Ethiopia

Negash Hailu[1]*, Chemeda Fininsa[1], Tamado Tana[1] and Girma Mamo[2]

[1]*School of Plant Sciences, Haramaya University, P.O.Box 138, Dire Dawa, Ethiopia*
[2]*Melkasa Agricultural Research Institute, P.O. Box 436, Adama, Ethiopia*

Abstract

Common bacterial blight (CBB) caused by Xanthomonas axonopodis pv. phaseoli is the most important biotic production constraint to common bean in eastern Ethiopia. Climate change could have an impact on the disease epidemiology by influencing both common bean growth and the pathogen reproduction. The effects of climate change needs to be mitigated using climate change resilience strategies. Field experiments were conducted in the 2012 and 2013 cropping seasons at Haramaya and Babile research stations in eastern Ethiopia to assess the effects of integrating climate change resilience strategies on CBB of common bean. Gofta (G2816) and Mexican 142(11239) common bean varieties were used. Eight climate change resilience strategies used were compost application, row intercropping and furrow planting alone and in combination. Factorial combinations of two common bean varieties and eight climate change resilience strategies totally 16 treatment combinations were studied in randomized complete block design (RCBD) with three replications and repeated once. Disease severity data were recorded from 10 randomly tagged plants from four central rows per plot. Disease severity, area under disease progress curve (AUDPC) and disease progress rate were significantly different among climate change resilience strategies, between varieties, cropping seasons and locations. Disease severities, AUDPC and disease progress rate were consistently less on row intercropping + compost application + furrow planting and row intercropping + compost application compared to singly applied climate change resilience strategies and sole planting plots in both locations and seasons. The disease epidemic was relatively higher on Mexican 142 than Gofta and during 2012 than 2013 at Babile than Haramaya. Integrated climate resilience strategies reduced CBB epidemics and could be applied as a component in management of CBB in eastern Ethiopia and in areas with similar agro-ecological zones.

Keywords: *Climate Change Resilience, Phaseolus vulgaris, Xanthomonas axonopodis pv. phaseoli*

Introduction

There are several biotic and abiotic production constraints on common bean (*Phaseolus vulgaris* L.) in semi-arid agro-ecologies of eastern Ethiopia. Diseases, insect pests, low soil fertility and periodic water stress are the major constraints [1,2]. The major diseases of common bean in the tropical regions, including Ethiopia that should be targeted for management are common bacterial blight (CBB) caused by *Xanthomonas axonopodis* pv. *phaseoli*, halo blight caused by *Pseudomonas syringae* pv. *phaseolicola,* bacterial brown spot caused by *Pseudomonas syringae* pv. *syringae,* rust caused by *Uromyces appendiculatus,* anthracnose caused by *Colletotrichum lindemuthianum* and other viral and root rot diseases [3-6]. These diseases are frequently occurring and widely distributed in Ethiopia and are destructive agents of common bean production causing heavy yield loss and decreasing seed quality [6-9].

Common Bacterial Blight (CBB) is one of the major diseases and the most important constraint to common bean production in eastern Ethiopia. When environmental conditions are favourable for the pathogen during long periods of warm and humid weather causing reductions CBB becomes the most destructive in both yield and seed quality [10]. Common bacterial blight incidence (53%) and severity (63%) were recorded in sole cropping system of common bean in eastern Ethiopia [8] and the relative yield loss of 22-40% was found in pure stand cropping system.

Change in rainfall pattern, soil moisture, soil temperature, and soil fertility has direct impact on the disease epidemiology by influencing host plant growth and susceptibility; pathogen reproduction, spread,

survival, activity and host-pathogen interaction [11]. According to International Panel for Climate Change, the average global surface temperature will increase by 2.8°C ranging from 1.8 to 4.0°C during 2050 assuming no emission control policies. Climate models suggest that Ethiopia will see further warming in all seasons of +2.2 (1.4°C-2.9°C) by the 2050's [12]. It is likely that this warming will be associated with higher heat waves and higher evapotranspiration rate. Nowadays, the doubt on climate change is not on its occurrence and effects, but rather on how to mitigate the ever-happening effects of climate change to manage properly crop production and protection to ensure food security of the ever-growing population and the proper functioning of the natural ecosystem. This needs to set resilience strategies that can mitigate the existing and ever happening effects of climate change.

The effect of increased temperature, rainfall variability and decreased soil moisture on plant diseases depend on the nature of their effects on both the host and the pathogens [13,14]. Those variables

*Corresponding author: Negash Hailu, School of Plant Sciences, Haramaya University, P.O.Box 138, Dire Dawa, Ethiopia
E-mail: negashhailu76@yahoo.com

could first affect disease directly by changing the encounter rate between pathogens and hosts.

Because disease and environment are closely related, climate change may create favourable conditions for pathogens and alter the spatial and temporal distribution of disease epidemics [15]. The host plant agro-climate will be altered and pathogens will be affected negatively or positively [13]. To this effect, new diseases may arise in certain regions, and other diseases may cease to be economically important, especially when the host plant is introduced into new areas [15]. Pathogens tend to follow the host plant in their geographical distribution, but the rate at which pathogens become established in the new environment is a function of the mechanism of pathogen dispersal, suitability of the environment for dispersal, over seasoning, and physiological and ecological changes in the host plant [15]. Climate variability has the potential to modify host physiology and disease resistance and to change the stages and rates of pathogens development. The most likely impacts would be shifts in the spatial distribution of host and pathogen, changes in the physiology of host-pathogen interaction, alteration in crop loss and changes in the efficacy of management strategies [16,17]. Climate variability itself may be an important factor of pathogen selection pressure [13].

Inspite of crop diseases reducing crop productivity and food supply, there has been limited field based empirical research to assess the potential effect of climate change on plant disease [11,17,18]. Most research on how climate change influences plant disease has concentrated on the effects of one or two of the changing climatic factors on the host, pathogen, or the interaction of the two under controlled conditions that are very different from those in the real field. Other situational studies are based on modelling of data from controlled experiments [16].

Research in climate change related issues could result in improved understanding and management of crop diseases in the face of current and future climate extremes [13,16]. Understanding the effect of climate change resilience strategies such as plant nutrients through compost application [19,20], soil and water conservation [20,21], and species mixture combinations [8] management practices on disease intensities will assist identification of the most important variables and focus efforts in developing integrated management packages. The epidemics of CBB needs to be assessed under sole and integrated field based climate change resilience strategies such as intercropping, compost application and furrow planting. The objective of this study, therefore, was to assess the effects of climate change resilience strategies against CBB epidemiology of common bean in eastern Ethiopia.

Materials and Methods

Experimental location

Field experiments on CBB of common bean were conducted at Haramaya University experimental field stations at Babile and Haramaya, eastern Ethiopia during 2012 and 2013 main cropping seasons (June to November each year). Haramaya is located at 09° 26′ N and 42° 3′ E. The altitude of the area is 1980 meters above sea level with average annual rainfall of 786.8 mm, mean minimum temperature of 10.4°C, maximum temperature of 23.4°C and mean temperature of 16.8°C. The soil of Haramaya is alluvial type with organic matter content of 6.8%, total nitrogen of 0.34%, available phosphorus of 2.2 mg kg soil[-1], pH of 7.13 with percent sand, silt and clay content of 62.9, 19.6 and 17.5, respectively. Babile is situated at 09°13′ N and 42°19′ E, with altitude of 1655 meters above sea level with annual total rainfall of 719.2 mm. The mean minimum and maximum temperatures are

15.4 and 28.3°C, respectively, with an average temperature of 21.83°C. The soil of Babile is characterized with organic matter content of 2.95%, total nitrogen of 0.1%, available phosphorus of 1.28 mg kg soil[-1,] pH of 6.6 with percent sand 45%, silt 24% and clay 31%.

Both locations represent important common bean growing areas and are conducive for CBB disease development of common bean. Babile is a representative of semi-arid agro ecology whereas Haramaya is a representative of midland agro ecology. The weather variables (mean maximum and minimum monthly temperature, monthly total rainfall) for 2012 and 2013 for both locations during experimental months are presented in Figure 1.

Experimental design and management

Three field based climate change resilience strategies such as compost application, row intercropping, and furrow planting were used solely and in combination (Table 1) in two common bean varieties Gofta (G2816) and Mexican 142 (G11239). Gofta is moderately resistant while Mexican 142 is susceptible to CBB. Sorghum variety, Teshale (3442-2OP) developed for the semi-arid environment as moisture stress and striga weed tolerant was used.

Totally 16 treatment combinations were arranged in a randomized complete block design in three replications on a plot size of 3.4 m × 3.6 m (12.24 m²). Compost was applied two weeks before sowing at a rate of 8 tons per hectare, about half the rate recommended for cereals [22] for both crops. Compost was produced from the maize straw, teff straw and khat remaining (3:1:2 V/V) ratio in well-prepared composting pits. The selection of proportions was based on the availability of crops remaining in the study area. The chops of those materials were watered and allowed to decompose for four months through intermittent mixing up. Its chemical properties were also analyzed.

Close-ended furrows were prepared by digging 20 centimeters deep rows two weeks before sowing in order to harvest the rainwater. Sorghum (variety Teshale) seeds were sown on 21 June 2012 and 28 June 2013 at Babile, 20 June 2012 and 02 July 2013 at Haramaya by drilling seeds at the seed rate of 5 Kg ha[-1]. Common bean seeds were sown on 11 July 2012 and 05 July 2013 at Babile, 07 July 2012 and 09 July 2013 at Haramaya manually by planting two seeds per hill. A standard intra and inter-row spacing for both crops were used according to the recommended spacing. Sorghum was planted in 0.8 m inter-row and 0.25 m intra-plant spacing. In row intercropping, a row of common bean was planted in the center of sorghum rows at 0.1 m intra-plant and 0.4 m inter-row spacing. Simultaneous planting was used in row intercropping. Similarly, in sole planting of common bean 0.4 m inter-row and 0.1 m inter-plant spacing with 9 rows per plot were used. Spacing between blocks was 1.2 m and between plots was 1 m [23].

After emergence and establishment of seedlings, the rows were thinned to one plant per hill. Recommended agronomic practices such as hand weeding and hoeing were used for both crops [23]. Fertilizer application and artificial inoculation were not applied for common bean. Plants were hand weeded three times at Haramaya and two times at Babile and cultivated once during the growth periods at both locations and seasons.

Disease data

All disease data were collected from central four rows. Disease severity (leaf area showing characteristic CBB symptom) was assessed six times at an interval of seven days during the experimental periods beginning from 48-51 days after planting (DAP) until physiological maturity. Disease severity rating was performed on 10 randomly pre-

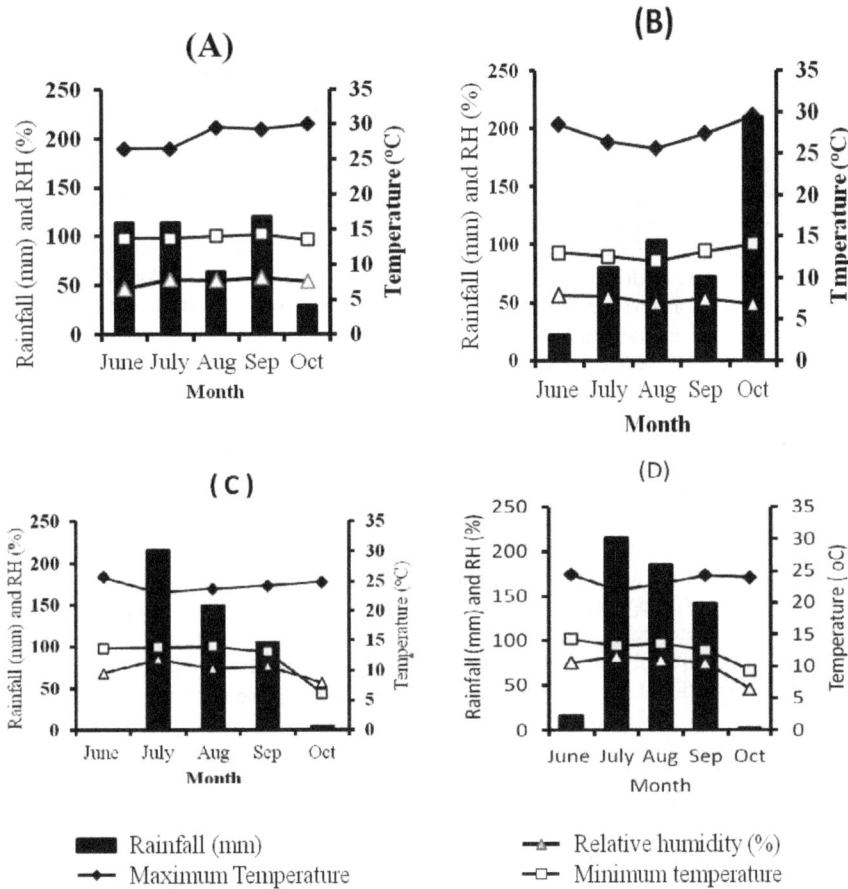

Figure 1: The weather variables (mean maximum and mean minimum temperature (°C), mean relative humidity (%) and total monthly rainfall (mm) at Babile in 2012. (A), in 2013. (B), at Haramaya in 2012. (C) and in 2013. (D).

S. No	Treatment	Treatment combination description
1	SP	sole planting
2	CA	compost application
3	FP	furrow planting
4	RI	row intercropping
5	CA+FP	row intercropping + furrow planting
6	RI+CA	row intercropping + compost application
7	RI+FP	row intercropping + furrow planting
8	RI+CA+FP	row intercropping + compost application + furrow planting

Table 1: Climate change resilience treatments for management of common bean common bacterial blight for both Gofta and Mexican 142 varieties during 2012 and 2013 at Babile and Haramaya, eastern Ethiopia.

tagged plants per plot at both locations and seasons. Severity was rated using standard scales of 1-9 [24-26] where 1=no visible symptom and 9=disease covering more than 25% of the foliar tissue and the severity grades were converted into percentage severity index (PSI) for analysis using:

$$PSI = \frac{\text{Sum of numerical ratings} \times 100}{\text{Number of plants scord} \times \text{maximum score on scale}}$$

Area under disease progress curve (AUDPC) and disease progress rate (r) were calculated from the severity data. AUDPC was computed from PSI data calculated on each date of assessment as described by Madden [27].

$$AUDPC = \sum_{i=1}^{n-1} 0.5(x_i + 1 + x_i)(t_i + 1 - t_i)$$

Where n is the total number of assessments, t_i is the time of the ith assessment in days from the first assessment date, x_i is the percentage of disease severity at i^{th} assessment. AUDPC was expressed in percent-days because the severity (x) was expressed in percent and time (t) in days.

Weather data

Weather variables of minimum temperature (°C) at 09:00 h and maximum daily air temperature (°C) at 18:00 h, relative humidity (%) at 06:00 h, 09:00 h, 12:00 h, 15:00 h and 18:00 h, and daily rainfall (mm) were collected from the meteorological stations of both location and the averages were calculated and presented in Figure 1.

Data analysis

Mean disease severity and yield data from each variety and treatment were examined and used for data analysis. Disease severity at different DAP and AUDPC were subjected to analysis of variance using the PROC GLM procedure of AUDPC or SAS version 9.1 [28] to determine the treatment effects. Homogeneity of variances was tested using the procedure described by Gomez and Gomez [29] and as the test showed heterogeneity of variances, separate analysis of the two-year and two location data was performed. Differences among treatment means were compared using the Fisher's least significant difference (LSD) test at 5% level of significance.

Results

Disease severity

Common bacterial blight epidemics on both varieties during both cropping seasons were varied significantly among the climate change resilience strategies. Disease severity was consistently less on the most integrated plots than sole planted and less integrated plots. During both cropping seasons, the disease severity was significantly different (P<0.01) among climate change resilience strategies throughout the whole disease recording dates while significant difference (P<0.01) of disease severity between the two varieties were recorded at late disease recording days. A higher disease severity was observed in the year 2012 compared with the 2013 cropping season. The interaction between and among the climate change resilience strategies, varieties and location was not significant.

In 2012, the highest initial disease severity (18.5%) at 48 DAP was calculated from sole planted plots of Gofta variety, while lower was recorded in row intercropping + compost application plots of common bean varieties at Babile (Figure 2). Similarly, higher final disease severity at 83 DAP was recorded from sole planting on Mexican 142 (54%) and on Gofta (53.8%) whereas lower final disease severity was recorded from row intercropping + compost application + furrow plating on Gofta (40%) and from row intercropping + compost application on Mexican 142 (40.7%) (Figure 2). During the 2013 cropping season, higher final disease severities at 85 DAP on Gofta (43%) and on Mexican 142 (50.4%) were obtained from sole planting while lower final disease severity on Gofta (27.4%) was obtained from row intercropping + compost application and on Mexican 142 (38.2%) from row intercropping + compost application + furrow planting at Babile (Figure 2). The other climate change resilience strategies reduced CBB severity intermediately and showed similar trend of disease reduction form the first to last date of disease recording and the data were not presented in the Figure 2 for clarity.

At Haramaya, the highest initial disease severity (20%) at 50 DAP was calculated from sole planted plots of Gofta variety, while lower was recorded in row intercropping + compost application + furrow planting plots of Gofta and in row intercropping of Mexican 142 in 2012 cropping season (Figure 3). Similarly, higher final disease severity at 85 DAP was recorded from sole planting on Mexican 142 (56%) and on Gofta (48%) whereas lower final disease severity was recorded from row intercropping + compost application on Gofta (40%) and from row intercropping + compost application + furrow planting on Mexican 142 (42.9%). During the 2013 cropping season, higher final disease severities at 86 DAP on Gofta (43%) and on Mexican 142 (54.8%) were obtained from sole planting followed by furrow planting while lower final disease severity on Gofta (28%) and on Mexican 142 (40%) were obtained from row intercropping + compost application + furrow planting at Haramaya (Figure 3). The other climate change resilience strategies reduced CBB severity intermediately and showed similar trend of disease reduction form the first to last date of disease recording and the data were not presented in the Figure for clarity.

With respect to mean initial and final disease severity of the varieties during 2012 cropping season, the highest mean initial disease severity was 18.2% at Babile and 19.6% at Haramaya in sole planting (Table 2). The row intercropping + compost application had the lowest 13.3% at Babile and row intercropping + furrow planting had the lowest (14.1 %) mean initial disease severity at Haramaya (Table 2). The highest mean final disease severity (53.9%) was recorded in sole planting at Babile and 52.4% at Haramaya (Table 2). The lowest mean final disease severity was from row intercropping + compost application + furrow planting (40.7%) at Babile (Table 2) and 41.8% at Haramaya during 2012 (Table 2).

During 2013 cropping season, the highest mean initial disease severity at 50 DAP was 16.3% in sole planting at Babile and at 51 DAP was 17.8% in sole planting at Haramaya, while the row intercropping

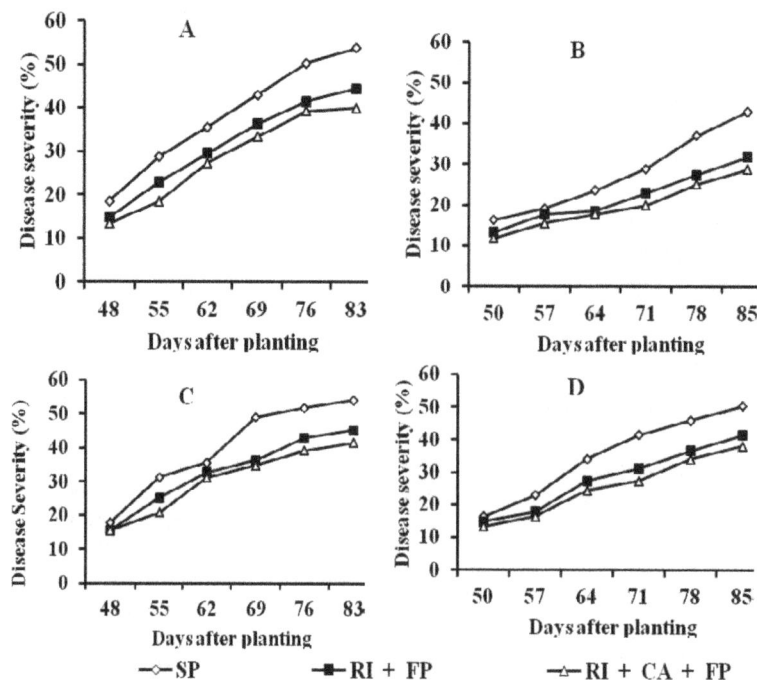

Figure 2: Disease progress curve of common bean common bacterial blight on (A) Gofta in 2012 (B) Gofta in 2013, (C) Mexican 142 in 2012 and (D) Mexican 142 in 2013 at Babile. RI + CA + FP, row intercropping + compost application + furrow planting; RI + FP, row intercropping + furrow planting; SP, sole planting.

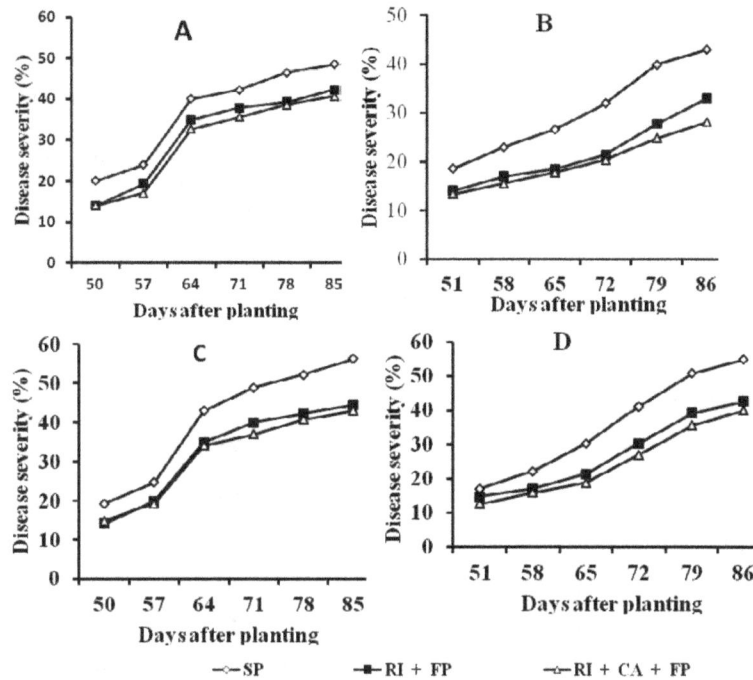

Figure 3: Disease progress curve of common bean common bacterial blight on (A) Gofta in 2012 (B) Gofta in 2013, (C) Mexican 142 in 2012 and (D) Mexican 142 in 2013 at Haramaya. RI + CA + FP, row intercropping + compost application + furrow planting; RI + FP, row intercropping + furrow planting; SP, sole planting.

Location	Babile				Haramaya			
Year	2012		2013		2012		2013	
PSI	Initial	Final	Initial	Final	Initial	Final	Initial	Final
Variety								
Mexican 142	15.8a	47.5a	15.3a	42.7a	15.8a	48.7a	15.2a	47.2a
Gofta	15.1a	46.8a	13.9b	34.1b	15.8a	43.4b	15.1a	34.3b
LSD (0.05)	1.32	1.59	1.04	2.26	1.28	2.42	1	2.1
Strategy								
SP	18.2a	53.9a	16.3a	46.7a	19.6a	52.4a	17.8a	48.9a
CA	14.8c	51.0ab	15.6ab	39.3bc	17.4ab	52.8a	16.3ab	46.3ab
FP	17.8ab	51.1ab	15.9ab	42.6ab	16.7bc	44.8bc	15.7b	43.3bc
RI	15.2bc	44.6cd	14.4abcd	36.7cd	14.4cd	44.1bc	14.4bc	38.0de
CA+FP	14.8c	49.7b	14.8abc	38.5bc	15.bcd	47.0b	15.9ab	41.3cd
RI+CA	13.3c	41.5de	12.9cd	33.2d	14.8cd	41.8c	13.7c	36.1e
RI+FP	15.2bc	44.8c	14.1bcd	36.7cd	14.1d	43.3bc	14.4bc	37.8de
RI+CA+FP	14.4c	40.7e	12.6d	33.5d	14.4cd	41.8c	13.0c	34.1e
LSD (0.05)	2.65	3.18	2.07	4.52	2.55	4.83	2	4.2
CV (%)	14.5	5.71	12.05	9.99	13.68	8.91	11.2	8.7

PSI= percentage severity index, LSD=Least Significant difference, CV= Coefficient of Variance, CV= Coefficient of variation, CA=compost application, FP=furrow planting, RI=row intercropping, CA + FP=compost application + furrow planting, RI + CA=row intercropping + compost apllication , RI + FP=row-intercropping + furrow planting, RI + CA + FP=row intercropping + compost application + furrow planting, SP= sole planting.

Table 2: Effects of integrated climate change resilience strategies on CBB disease severity at Babile and Haramaya during 2012 and 2013 cropping seasons.

+ compost application + furrow planting had the lowest mean initial disease severity 12.6% at Babile and 13% at Haramaya (Table 2). The highest mean final disease severity at 85 DAP (46.7%) was recorded in sole planting at Babile and at 86 DAP was 48.9% at Haramaya while the lowest mean final disease severity was from row intercropping + compost application + furrow planting (33.5%) at Babile and 34.1% at Haramaya (Table 2).

In both cropping seasons while CBB occurred on both varieties it did not vary at early stage. However, significant variations in the disease severity between varieties started from 55 to 69 DAP at Babile and from 64 to 83 DAP at Haramaya during 2012, whereas 50 to 85 DAP at Babile and 65 to 86 DAP at Haramaya during 2013 (Figure 3). At both locations and during both cropping seasons significantly higher disease severity was recorded from Mexican 142 than Gofta variety. Likewise, higher mean initial disease severity (15.8%, 15.3%) and final disease severity (45.5%, 42.7%) were recorded on Mexican 142, during 2012 and 2013 respectively than on Gofta variety at Babile (Table 2).

Considering the range of disease severity and percentage of

disease severity reduction, the solely applied climate change resilience strategies had higher disease severity and lower reduction compared to most integrated and integrated ones. The most integrated climate change resilience strategy i.e. row intercropping + compost application + furrow planting and row intercropping + compost application caused higher CBB severity reduction. The resilience strategies reduced the mean final disease severity of both varieties from 5.2-24.5% (mean 14.3%) during 2012 at Babile (Table 2). Similarly, the resilience strategies reduced the mean final disease severity from 12.1-34.5% (mean 20.3%) on Gofta (Figure 3B) and from 10.8-27% (mean 13.9%) on Mexican 142 (Figure 3D) at Haramaya during 2013.

Area under disease progress curve

There were significant differences (P<0.001) among the climate change resilience strategies in both locations and seasons for mean AUDPC (Figures 4 and 5). There was significant difference (P<0.01) between the two varieties of common bean, with the mean AUDPC value (1230%-days) higher on the Mexican 142 variety than Gofta (1158%-days) during 2012 cropping season and mean AUDPC value (1041%-days) higher on the Mexican 142 variety than Gofta (807%-days) during 2013 at Babile. At Haramaya, the mean AUDPC value (1253%-days) was higher on Mexican 142 variety than (1153%-days) on Gofta during 2012 cropping season and mean AUDPC value was higher (1057%-days) on Mexican 142 variety than (822%-days) on Gofta during 2013.

At Babile, highest AUDPC The values were obtained from sole planting of Gofta and Mexican 142, respectively. The lowest AUDPC value was computed from row intercropping + compost application + furrow planting on Gofta and from row intercropping + compost application on Mexican 142, during 2012 (Figure 4A) .The row intercropping + compost application + furrow planting reduced the AUDPC value by 25% on Gofta and row intercropping + compost application reduced AUDPC value by 26.3% on Mexican 142. During 2013 cropping season, on Mexican 142 variety maximum AUDPC value (1244.5%-days) was computed from the sole planting whereas the minimum value (895.9%-days) was from row intercropping + compost application + furrow planting plots followed by row intercropping + compost application (908.7%-days) at Babile (Figure 5A). The resilience strategies reduced AUDPC values from 6.4-30.8% (mean 16.8%) on

Gofta and from 8.7-28% (mean 16.3%) on Mexican 142 compared to sole planting during 2013 cropping season at Babile (Figure 5A).

At Haramaya during 2012, the maximum AUDPC value was computed from the sole planting on Mexican 142 and the minimum value from row intercropping + compost application + furrow planting followed by row intercropping + compost application (Figure 4B). In the same year, on Gofta variety, row intercropping + compost application + furrow planting reduced AUDPC by 19.2%, row intercropping + compost application by 19.1% compared to sole planting whereas row intercropping + compost application + furrow planting reduced AUDPC by 22.6%, row intercropping + compost application by 21.7% and row intercropping + furrow planting by 19.5% compared to sole planting on Mexican 142. The climate change resilience strategies reduced AUDPC value by 6-19% (mean 11.9%) on Gofta and 10.6-22.6% (mean 13.4%) on Mexican 142 compared to sole planting during 2012 cropping season at Haramaya.

Higher AUDPC Values were recorded from sole planting of Gofta and Mexican 142 respectively, while lowest AUDPC value on Mexican 142 form row intercropping + compost application + furrow planting during 2013 (Figure 5B). Therefore, row intercropping + compost application + furrow planting reduced the AUDPC values by 34.8% on Gofta and by 31.4% on Mexican 142. All of climate change resilience strategies reduced AUDPC by 16.8-34.8% (mean 22.9%) on Gofta and 11.4-31.4% (mean 16.3%) on Mexican 142 compared to sole planting during 2013 cropping season at Haramaya.

Disease progress rate

Comparisons of disease progress rates were made among treatments based on the Logistic model by fitting disease severity data with dates of assessment. The rates of disease progress were varied among treatments, between locations and seasons. During 2012 cropping season, the highest disease progress rate (0.055-logit day[-1]) was computed from Mexican 142 variety at Haramaya while the lowest epidemic rate (0.027 logit day[-1]) was from Gofta variety during 2013 cropping season at Haramaya. The disease progress rates calculated for varieties, climate change resilience strategies, years and locations were different and presented in Table 3 for 2012 and Table 4 for 2013 seasons. During both cropping seasons, the reduction of disease progress rate

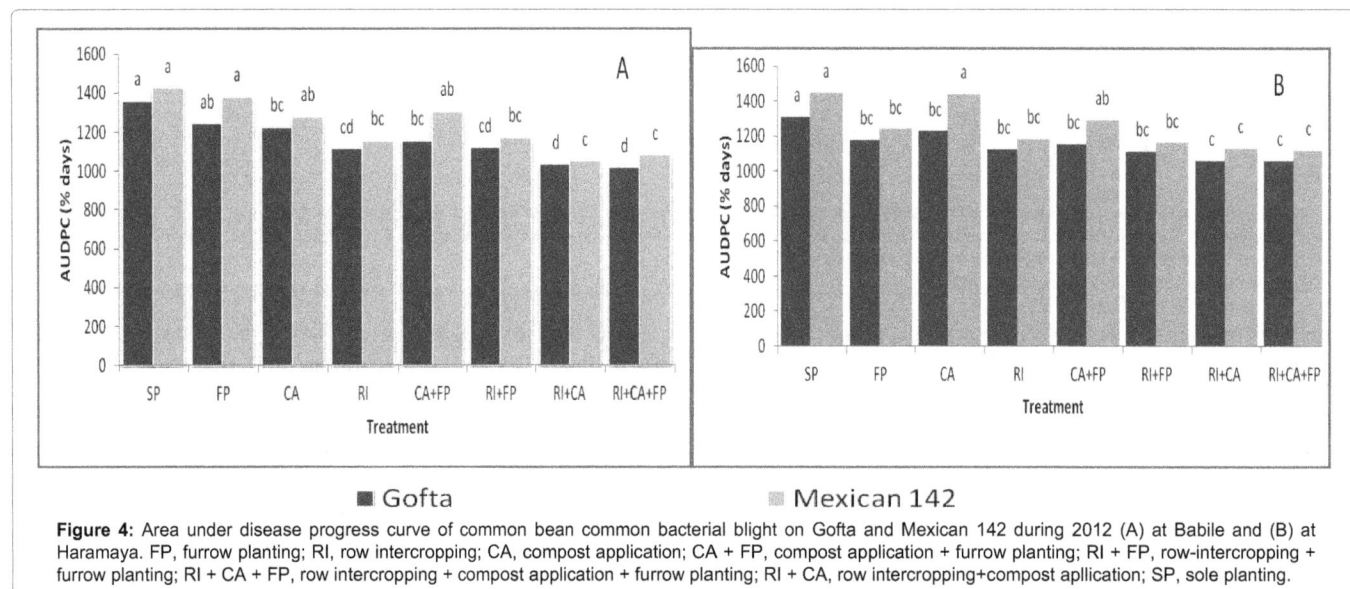

Figure 4: Area under disease progress curve of common bean common bacterial blight on Gofta and Mexican 142 during 2012 (A) at Babile and (B) at Haramaya. FP, furrow planting; RI, row intercropping; CA, compost application; CA + FP, compost application + furrow planting; RI + FP, row-intercropping + furrow planting; RI + CA + FP, row intercropping + compost application + furrow planting; RI + CA, row intercropping+compost apllication; SP, sole planting.

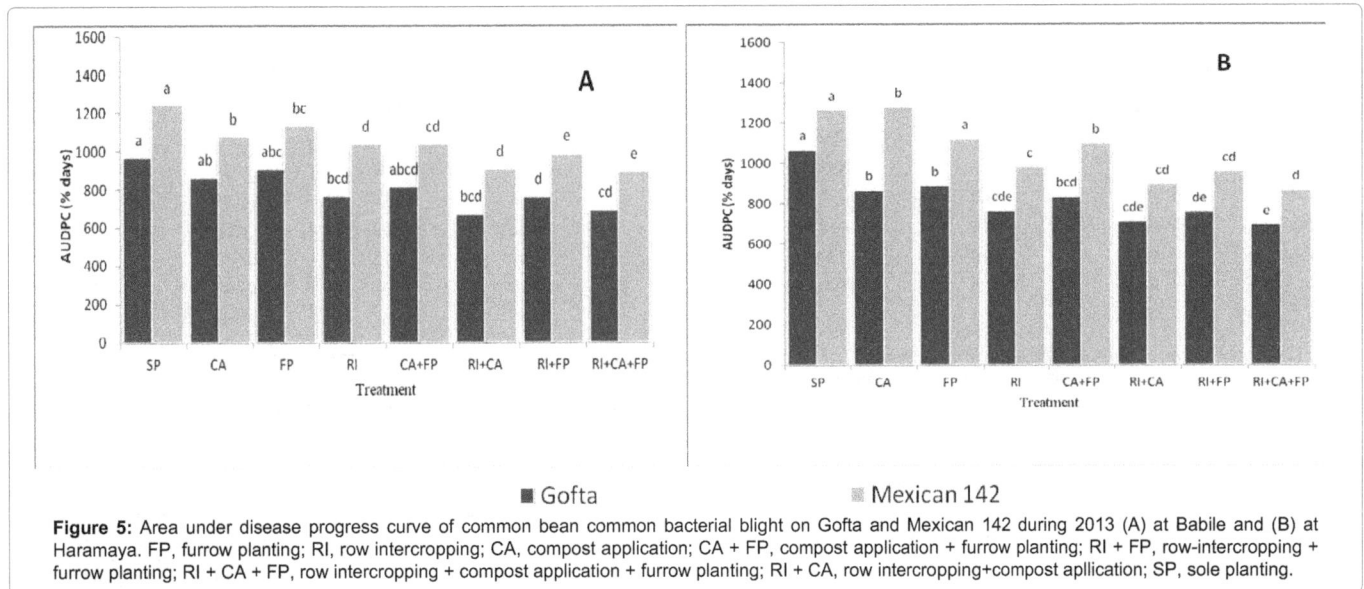

Figure 5: Area under disease progress curve of common bean common bacterial blight on Gofta and Mexican 142 during 2013 (A) at Babile and (B) at Haramaya. FP, furrow planting; RI, row intercropping; CA, compost application; CA + FP, compost application + furrow planting; RI + FP, row-intercropping + furrow planting; RI + CA + FP, row intercropping + compost application + furrow planting; RI + CA, row intercropping+compost apllication; SP, sole planting.

achieved through application of climate change resilience strategies have shown similar trend to the disease severity and area under disease progress curve in both varieties and locations.

In the 2012 cropping season, the disease progress rate was higher on the Gofta (0.053 logit day^{-1}) treated with compost application, on Mexican 142 (0.050 logit day^{-1}) treated with compost application + furrow planting at Babile and on Mexican (0.055 logit day^{-1}) treated with compost application, at Haramaya (Table 3). Application of row intercropping + furrow planting reduced disease progress rate by 6.5% on Gofta and application of row intercropping + compost application + furrow planting reduced disease progress rate by 18.7% on Mexican 142 compared to sole planting at Babile during 2012 cropping season (Table 3). Lower rate of disease progress was obtained from row intercropping + compost application on Gofta (0.040logit day^{-1}) and from row intercropping + compost application + furrow planting on Mexican 142 (0.44 logit day^{-1}). Generally, application of row intercropping + compost application + furrow planting reduced the rate of disease progress by 12% on Mexican 142 variety at Haramaya during 2012.

In 2013 cropping season, slower disease progress rate was calculated on Gofta (0.029 logit day-^{1}) and on Mexican 142 (0.037 logit day^{-1}) treated with row intercropping + compost application at Babile and on Gofta (0.027 logit day^{-1}) and on Mexican 142 (0.046 logit day^{-1}) treated with row intercropping + compost application + furrow planting at Haramaya compared to sole planting. In 2013 the disease progress rate from compost application was faster (0.055 logit day^{-1}) than the remaining climate change resilience strategies on Mexican 142 (Table 4), and sole planting had faster disease progress rate (0.04 logit day^{-1}) on Gofta and (0.048 logit day^{-1}) on Mexican 142 at Babile. The disease progress rate in compost application was highest in both cropping seasons on Mexican 142 at Haramaya. Application of the most integrated climate change resilience strategy: row intercropping + compost application + furrow planting reduced the rate of disease progress by 22.9% on Gofta variety and by 14.8% on Mexican variety during 2013 cropping season at Haramaya (Table 4).

Discussion

Common bacterial blight epidemics significantly varied among and between climate change resilience strategies, between common bean

varieties, locations and cropping seasons. Variety Mexican 142 had higher disease severity and higher area under disease progress curve (AUDPC) than variety Gofta might be due to the higher resistance level of Gofta than Mexican 142. The result of this study agreed with the findings of Fininsa and Tefera [9] that described Gofta as moderately resistant to common bacterial blight and Mexican 142 as susceptible variety. Higher disease epidemic was found during 2012 cropping season than 2013 cropping season at both locations. In both location higher relative humidity, higher minimum and maximum temperatures (Figure 1) recorded in August and September months during 2012 cropping season could have created conducive environment for increased common bacterial blight severity and progress rate than 2013 cropping season. Similarly, lower relative humidity, higher minimum and maximum temperatures were recorded at Babile than at Haramaya during both cropping seasons (Figure 1), were suitable for common bacterial blight disease development and early disease appearance common bacterial blight. At Haramaya, lower disease severity and lower AUDPC were computed than at Babile during both cropping seasons. The result was similar with the findings of Fininsa [8] as high temperature and low relative humidity favor higher common bacterial blight severity.

Climate change resilience strategies had lower final disease severity (8.6-36.2%) and lower AUDPC than the sole planting on both varieties of common bean during 2012 and 2013 cropping seasons at both locations. The variation of final disease severity was based on the type of climate change resilience strategy, resistance level of common bean, common bacterial blight conduciveness of location and weather variables of two cropping seasons.

Intercropping common bean with sorghum has significantly lowered the severity level of common bacterial blight compared with sole planting. Row intercropping + compost application + furrow planting and row intercropping + compost application showed significantly lower common bacterial blight severity than the sole plantings during both cropping seasons and locations. All types of common bean–sorghum intercropping systems (row intercropping, row intercropping + furrow planting, row intercropping + compost application, row intercropping + compost application + furrow planting) significantly reduced the final disease severity (13-36.2%)

Resilience strategy	Babile				Haramaya			
	Gofta		Mexican		Gofta		Mexican	
	Rate (r)	R²	Rate (r)	R²	Rate (r)	R²	Rate (r)	R²
SP	0.046	90.8	0.048	88.4	0.04	83.7	0.05	83.1
CA	0.053	92.7	0.048	89.9	0.043	81.7	0.055	86.9
FP	0.046	89.7	0.046	87	0.04	79	0.044	82.1
RI	0.046	89.3	0.042	86.9	0.042	82	0.049	84.2
CA + FP	0.05	93.9	0.05	84.6	0.042	79.5	0.052	85.1
RI + CA	0.048	87.8	0.044	84.1	0.04	81.2	0.045	83.8
RI + FP	0.043	87.6	0.041	86.5	0.043	79	0.046	84.2
RI + CA + FP	0.044	86.7	0.039	86	0.044	80.8	0.044	82.7

CA=compost application, FP=furrow planting, RI=row intercropping, CA + FP=compost application + furrow planting, RI + CA=row intercropping + compost apllication , RI + FP=row-intercropping + furrow planting, RI + CA + FP=row intercropping + compost application + furrow planting, SP= sole planting.

Table 3: Common bean common bacterial blight disease progress rate (r) in logit per day and adjusted coefficient of determination (R²) on Gofta and Mexican 142 varieties at Babile and Haramaya during 2012 cropping season.

Resilience strategy	Babile				Haramaya			
	Gofta		Mexican		Gofta		Mexican	
	Rate (r)	R²	Rate (r)	R²	Rate (r)	R²	Rate (r)	R²
SP	0.04	90.3	0.048	90.1	0.035	91	0.054	93.3
CA	0.034	83.5	0.041	86.2	0.033	80.1	0.055	95.3
FP	0.037	87.2	0.044	88.5	0.036	84.7	0.048	93.4
RI	0.031	77.9	0.041	89.7	0.029	82.1	0.047	90.4
CA + FP	0.036	83.6	0.039	88.2	0.033	85.9	0.048	89.2
RI + CA	0.029	87.6	0.037	69	0.028	78.5	0.048	88.4
RI + FP	0.03	77.6	0.042	92.9	0.031	82.9	0.046	88.8
RI + CA + FP	0.031	86.1	0.041	90.7	0.027	82.9	0.046	91.1

CA=compost application, FP=furrow planting, RI=row intercropping, CA + FP=compost application + furrow planting, RI + CA=row intercropping + compost apllication , RI + FP=row-intercropping + furrow planting, RI + CA + FP=row intercropping + compost application + furrow planting, SP= sole planting.

Table 4: Common bean common bacterial blight disease progress rate (r) in logit per day and coefficient of determination (R²) on Gofta and Mexican 142 varieties at Babile and Haramaya during 2013 cropping season.

and AUDPC compared to sole planted plots at both locations and seasons. Similarly, Fininsa [8] reported reduction of 17-40% common bacterial blight severity in bean-maize intercropping. In sorghum-common bean intercropping, common bacterial blight disease epidemics were reduced because the sorghum may be used as physical barrier against bacterial inoculum from reaching to common bean. Changes in microclimate such as temperature and wind velocity reduction may disfavor the pathogen and cause reduction in disease. Higher temperature and relative humidity are favorable conditions for common bacterial blight disease epidemics and a decrease in temperature in the intercropping system might have lowered the severity of common bacterial blight. A cool microclimate in row intercropping + compost application + furrow planting disfavours common bacterial blight pathogen infection that could be initiated from infected seeds, infested soil or infested debris in the soil. The microclimate may also retard proliferation and spread of the bacteria between plants as the result of non-host nature of component crop sorghum for the bacteria.

Intercropping common bean with sorghum therefore could have been reduced the common bacterial blight severity and AUDPC compared to sole planting, during 2012 and 2013 cropping seasons at both location. In addition to disfavor common bacterial blight severity, intercropping can maintain soil fertility and balanced nutrition that might be the case for physiological and morphological fitness of the crop to build resistance to common bacterial blight. The result of this study is in line with the report of Matusso, et al. [30] that the principal reasons for intercropping are soil conservation and improvement of soil fertility, diseases control and balanced nutrition.

Furrow planting reduced the final disease severity during both cropping seasons at both locations when applied singly (2.7-18.4%) or in combination (5.7-34.5%) with other resilience strategies such as compost application and intercropping. Furrow planting could conserve soil moisture and create sufficient water availability in the root system of common bean that might have enabled common bean to reduce disease epidemics due to creation suitable condition for plants might favour the host than the pathogen. The result of this study is in agreement with the findings of Aydinalp and Cresser [21] revealed periodic low soil moisture make plant easily susceptible to diseases. Furrow planting had promising potential of mitigating climate change and reducing disease severity when integrated with intercropping and compost application. Furrow planting was effective when the soil had good water holding capacity similar to soil of Haramaya and less effective in sand soils with high drainage capacity for example soil of Babile.

Compost application reduced the final common bacterial blight disease severity (3.9-17.3) when applied singly and (14.2-36%) of disease severity reduction when combined with other resilience strategies during 2013 cropping season. The result of this experiment is in agreement with the results of Vallad et al. [31] and Abdasi et al. [32] conducted on foliar plant diseases. Vallad et al. [31] revealed that compost had 34-65% of disease symptom reduction of bacterial speck of *Arabidopsis thaliana* caused by *Pseudomonas syringae* pv. tomato compared with non-amended soil. Abbasi et al. [32] found that application of compost resulted in reduced bacterial spot incidence on tomato fruit by 28-33%, compared with non-amended soil. Disease control with compost has been attributed to successful competition for nutrients by beneficial microorganisms, antibiotic production by beneficial microorganisms and activation of disease-resistant genes in plants. Using compost not only could supply plant nutrients, but also can increase tolerance and resistance to diseases and retains soil moisture [33]. Composts' contribution to nutrient fertility must also be taken into account because nutrient effects may influence the severity of pathogens and consequently reduce greenhouse gas emission largely carbon dioxide emission as the result of carbon sequestration [19].

When the climate change strategies are integrated their synergetic effect significantly reduced disease severity, AUDPC and disease progress rate. Row intercropping + compost application and row intercropping + compost application + furrow planting have shown significant difference in disease severity reduction compared to less integrated climate change resilience strategies and sole planted plots at Babile and Haramaya. Compost application aggravated the diseases severity on the susceptible variety of Mexican 142 when applied solely, at Haramaya during both cropping seasons. This was because of compost application could have enhanced the growth of Mexican 142 variety at fast rate and created closed canopy earlier, consequently increased temperature and humidity, which sequentially could create favorable condition for common bacterial blight development and spread. Generally, there were higher disease progress rates on Mexican 142 than on Gofta during both seasons and both locations. When composting and furrowing are integrated to intercropping, they are highly effective in reducing disease severity, disease progress rate and the AUDPC as the result of synergetic effect.

Comparisons of disease progress rates among climate change resilience strategies, between varieties, locations and cropping seasons have shown different trends when compared to disease severity and AUDPC. The resilience strategies having lower disease severity exhibited higher disease progress rate. High disease rates were observed in some of resilience strategies that had lower disease severity. This could be due to high density of initial inoculum from the infected seeds, infested debris or infested soil that might have increased the initial disease severity. The disease progress rate with higher initial disease severity was higher even though there was lower final disease severity. Experimental studies have shown that the number of initial inoculum [34] considerably influenced the rates of disease increase. In an experiment with southern blight of processing carrot, the rate of disease increase generally increased as the number of initial foci increased [35].

Generally, common bacterial blight severity was reduced due to the reduction of inoculum dispersal and inhibition of inoculum proliferation and creating disfavoring conditions for the bacteria by climate change resilience strategies. These climate change resilience strategies of disease management therefore are cheaper, sustainable and could be easily adopted by small-scale farmers in eastern Ethiopia. The results obtained from this study suggested the importance of climate change resilience strategies applied singly and in combination as management option of common bacterial blight and other common bean diseases in the eastern Ethiopia and in areas with similar agro-ecological conditions. The furrow preparation and compost application rate of common bean should also need further investigation in the study area.

Conclusions and Recommendations

Application of integrated climate change resilience strategies in field experiments enhanced productivity of common bean and reduced disease severity and AUDPC of common bacterial blight compared to solely applied climate change strategies and sole planting common beans over locations and seasons. In addition, the most integrated strategy: row intercropping + compost application + furrow planting, has shown promising results in maintaining soil temperature and moisture. Thus, it could be concluded that farmers in eastern Ethiopia should design a strategy to promote common bean production through the application of row intercropping + compost application + furrow planting and row intercropping + compost application to improve the physico-chemical properties of soil and sustain enhanced production and productivity of common bean. It is strongly believed that integrated climate change strategies through reducing CBB disease epidemics as ecofriendly disease management option and enhancing soil fertility management, contribute substantially to the efforts of increasing food production and household food security in the study area.

Competing interests

The authors declare that they have no competing interests

Acknowledgments

The authors are grateful to Masresha Lebie, Woinishet Feleke, Birhanu Asfaw, Azeb Tegenu and the late Cholamu Faltamo, field assistants in the School of Plant Sciences of Haramaya University, for their assistance in data collection and fieldwork. We are grateful to Abraham Negash for his help in field data collection during both seasons. The study was financed by Haramaya University Research office.

References

1. Tana T, Fininsa, C, Worku, W (2007) Agronomic performance and productivity of common bean (*Phaseolus vulgaris* L.) varieties in double intercropping with maize (*Zea mays* L.) in eastern Ethiopia. Asian J Plant Sci 6: 749-756.

2. Katungi E, Farrow A, Chianu J, Sperling L, Beebe S (2009) Base line research Report on Common bean in Eastern and Southern Africa: a situation and outlook analysis of targeting breeding and delivery efforts to improve the livelihoods of the poor in Drought prone areas. ICRISAT, Kampala, Uganda.

3. Amin M, Fitsum S, Selvaraj T, Mulugeta N (2014) Field Management of Anthracnose (*Colletotrichum lindemuthianum*) in Common Bean through Fungicides and Bioagents. Adv Crop Sci & Techno 2: 124-129.

4. Fininsa C, Tefera T (2002) Inoculum sources of bean anthracnose and their effect on bean epidemics and yield. Trop Sci 42: 30-34.

5. Fininsa C, Yuen J (2002) Temporal progression of bean common bacterial blight (*Xanthomonas campestris pv. phaseoli*) in sole and intercropping systems. Eur J Plant Patho 108: 485-495.

6. Lemessa F, Sari W, Wakjira M (2011) Association between angular leaf spot (*Phaeoisariopsis griseola* (Sacco) Ferraris) and common bean (*Phaseolus vulgaris* L.) Yield Loss at Jimma, Southwestern Ethiopia. Plant Patho J 110: 57-65.

7. Fininsa C, Yuen J (2001) Association of bean rust and common bacterial blight epidemics with cropping systems in Hararghe highlands, eastern Ethiopia. Inter J Pest Mang 47: 211-219.

8. Fininsa C (2003) Relationship between common bacterial blight severity and bean yield loss in pure stand and bean-maize intercropping systems. Inter J Pest Mang 49: 177-185.

9. Fininsa C, Tefera T (2006) Multiple disease resistance in common bean genotypes and their agronomic performance in eastern Ethiopia. Inter J Pest Mang 52: 291-296.

10. Abo-Elyousr KA (2006) Induction of systemic acquired resistance against common blight of bean (*Phaseolus vulgaris*) caused by *Xanthomonas campestris pv. Phaseoli*. Egyptian J Phytopathol 34: 41-50.

11. Ghini R, Hamada E, Bettiol W (2008) Review on climate change and plant diseases. Scie & Agri 65: 98-107.

12. IPCC (Inter-governmental Panel on Climate Change) (2007) Climate change: Mitigation. Contribution of working group iii to the fourth assessment report of the intergovernmental panel on climate change. Cambridge University Press, New York, USA.EPA (Environmental Protection Authority) (2010) Ethiopia's Vision for a Climate Resilient Green Economy. EPA, Addis Ababa. Ethiopia.

13. Garrett KA, Dendy SP, Frank EE, Rouse MN, Travers SE (2006) Climate change effects on plant disease: genomes to ecosystems. Annu Rev Phytopathol 44: 489-509.

14. Newton AC, Johnson SN, Gregory P (2011) Implications of climate change for diseases, crop yields and food security. Euphytica 179: 3-18.

15. Chakraborty S, Newton AC (2011) Climate change, plant diseases and food security. Plant Pathology 60: 2-14.

16. Doll JE, Baranski M (2011) Climate Change and Agriculture. Fact Sheet Series E3149 on Field Crop Agriculture and Climate change. Michigan State University, East Lancing, USA.

17. Gautam RH, Bhardwaj ML, Kumor R (2013) Climate change and its impact on plant diseases. Current Sci 105: 1685-1691.

18. Boland GJ, Melzer MS, Hopkin A, Higgins V, Nassuth A (2004) Climate change and plant diseases in Ontario. Cana J Plant Pathol 26: 335-350.

19. Luske B (2010) Reduced greenhouse gas emissions due to compost production and compost use in Egypt Comparing two scenarios. Louis Bolk Institute, Amestardem, Netherlands.

20. Sullivan P (2004) Soil System Guide on Sustainable Management of Soilborne Plant Diseases with Compost and Organic Amendments. Appropriate Technology Transfer for Rural Areas (ATTRA), California, USA.

21. Aydinalp C, Cresser MS (2008) The effects of global climate change on agriculture. Ameri-Eurasian J Agri & Environ Sci 3: 672-676.

22. Toulmin C (2011) Prospering Despite Climate Change: New Directions for Smallholder Agriculture. PP 1-25. Paper presented at the international fund for agricultural development (IFAD) conference. 24-25 January 2011. IFAD, Rome. Italy.

23. EARO (Ethiopian Agricultural Research Organization) (2004) Directory of Released Varieties and Their Recommended Cultural practices. EARO, Addis Ababa, Ethiopia.

24. Stutz J, Donahue S, Mintzer E, Cotter A (2003) Technical report on compost in landscaping applications. Tellus Institute, Boston, USA.

25. Buruchara R, Mukankusi C, Ampofo K (2010) Bean disease and pest identification and management. pp.1-67. the handbooks for small-scale seed producers. International Centre for Tropical Agriculture (CIAT). Kampala, Uganda.

26. CIAT (Centro Internacional De Agricultura Tropical) (1987) Standard System for the Evaluation of Bean Germplasm. CIAT, Cali, Colombia.

27. Madden LV (2006) Botanical epidemiology: some key advances and its continuing role in disease management. Eur J Plant pathol 115: 3-23.

28. SAS (Statistical Analysis System) (2003) SAS/STAT Guide for Personal Computers, Version 9.1 edition. SAS Institute Inc., Cary, NC.

29. Gomez KA, Gomez AA (1984) Statistical Procedures for Agricultural Research. 2nd Edition. John Willey and Sons, Inc, New York, USA.

30. Matusso JMM, Mugwe JN, Mucheru-Muna M (2014) Potential role of cereal-legume intercropping systems in integrated soil fertility management in smallholder farming systems of Sub-Saharan Africa. Research J Agri & Enviro. Mang 3: 162-174.

31. Vallad GE, Cooperband L, Goodman RM (2003) Plant foliar disease suppression mediated by composted forms of paper mill residuals exhibits molecular features of induced resistance. Physio & Mole Plant Pathol 63: 65-77.

32. Abbasi PA, Al-Dahmani J, Sahin F, Hoitink HAJ, Miller SA (2002) Effect of compost amendments on disease severity and yield of tomato in conventional and organic production systems. Plant Disease 86: 156-161.

33. Barker AV, Bryson GM (2006) Comparisons of composts with low or high nutrient status for growth of plants in containers. Soil Sci & Plant Analy 37: 1303-1319.

34. Jeger MJ, Termorshuizen AJ, Nagtzaam MPM, Bosch FV (2004) The effects of spatial distributions of mycoparasites on biocontrol efficacy: a modelling approach. Bioco Sci & Techno 14: 359-373.

35. Smith VL, Campbell CL, Jankins SF, Benson DM (1998) Effects of host density and number of disease foci on epidemics of southern blight of processing carrot. Phytopathol 78: 595-600.

Antimicrobial Behavior of Intracellular Proteins from Two Moderately Halophilic Bacteria: Strain J31 of *Terribacillus halophilus* and Strain *M3-23* of *Virgibacillus marismortui*

Badiaa Essghaier*, Cyrine Dhieb, Hanene Rebib, Saida Ayari, Awatef Rezgui Abdellatif Boudabous and Najla Sadfi-Zouaoui

Microorganisms and Active Biomolecules Laboratory, Faculty of Sciences of Tunis, University Campus, Tunisia

Abstract

In the present study, we firstly aimed to determine the ability of halophilic bacteria to improve tomato growth. as well as to detect the antimicrobial activities from two moderately halophilic bacteria strain M3-23 of *Virgibacillus marismortui* and strain J31 of *Terribacillus halophilus* exhibited by their intracellular proteins. The results showed that both bacteria were able to improve stem tomato growth by comparison of untreated tomato. The halophilic bacteria were also able to produce intracellular antifungal enzyme: glucanase produced by *V. marismortui* (1.74U/mg) and chitinase (39.39U/mg) produced by *T. halophilus*. Both chitinases were halotolerants (active in the presence of (0% to 30% NaCl (w/v)). Chitinase produced by strain J31 was alcaline (pH optimum 12), but chitinase from strain M3-23 was acidic (pH 4 optimum) more than 90% and 80% of activities were retained in the presence of pH value from 4 to 12, respectively for strain J31 and M3-23. Both enzymes were thermotolerants; optimum temperature was 80°C and 90°C respectively for strain J31 and strain M3-23. Both strains have lysozyme activity and value ranging from 6.6 U/ml to 6.8 U/ml respectively for strain J31 and strain M3-23. On the whole, the most potent *in vitro* antifungal effect was demonstrated by intracellular compound produced by strain J31 compared to strain M3-23. This study, was the first showing the antimicrobial efficiency of moderately halophilic bacteria by means of their intracellular compounds, by means of the spore germination reduction and the destroy of mycelial growth of *Botrytis cinerea*, *in vitro*. The distinguishable characteristics of their intracellular halotolerant and thermotolerant chitinases make them as good candidates for biotechnological applications.

Keywords: Moderately halophilic bacteria; Intracellular enzymes; Chitinase; Antimicrobial activity

Introduction

Biological control, using microorganisms to repress plant diseases, offers an environmentally friendly strategy to control agricultural phytopathogens. It has been studied for more than 70 years and is becoming a realistic alternative to chemical treatments that are still widely used to control diseases caused by plant pathogens. The emergence of fungicide-resistant strains and deregistration of fungicides may provide non-chemical methods especially biocontrol methods. Moderately halophilic bacteria are a group of halophilic micro-organisms able to grow optimally in media containing a wide range of NaCl concentrations (3-15% NaCl) [1]. They constitute a heterogeneous group of micro-organisms and have been studied for their ecology, physiology, biochemistry and genetics [1]. However their biotechnological possibilities have not been extensively exploited. Furthermore, halophilic bacteria produce a wide range of extracellular salt-tolerant hydrolytic enzymes such as amylases, proteases, nucleases, phosphatases, lipases, DNases, pullulanases and xylanases that have diverse potential usages in different areas such as food technology, feed additives and chemical industries [2,3]. Moreover, extremophilic micro-organisms are adapted to survive in ecological niches such as at high temperature, extremes of pH, high salt concentrations and high pressure. These micro-organisms produce unique biocatalysts that function under extreme conditions comparable to those prevailing in various industrial processes [2].

An important bacterial trait for biological control of plant disease is the production of hydrolytic enzymes which degrade fungal cell walls, especially chitinases and β-1,3-glucanases, these cell wall degrading enzymes are among the key factors involved in the suppression of pathogenic fungi by biocontrol agents [4]. Our laboratory was the pioneer to take an interest in the study of antifungal enzyme from moderately halophilic bacteria such as *Virgibacillus marismortui*, *Terribacillus halophilus*, *Salinococcus roseus* and *Planococcus rifitoensis*. Previously, these moderately halophilic bacteria were isolated from shallow salt lakes in Tunisia and selected as strong antagonists of *B. cinerea* the causal agent of grey mold disease on strawberries and tomatoes under the commercial standard conditions applied in Tunisia [5,6]. Previous research also reported the characterization of extracellular antifungal enzymes such as chitinase, and protease produced by moderately halophilic bacteria [7,8]. From our knowledge, there are no reports focusing on the investigation of intracellular antimicrobial activity produced by halophilic bacteria.

In previous work, we have reported the efficiency of the moderately halophilic strain M3-23 of *V. marismortui* and stain J31 of *T. halophilus*, in biological control, and their high production of extracellular antifungal enzymes. In this paper, we firstly aimed to evaluate the antimicrobial ability of intracellular enzymes such as chitinase, glucanase and protease produced by these selected

*Corresponding author: Badiaa Essghaier, Microorganisms and Active Biomolecules Laboratory, Faculty of Sciences of Tunis, University Campus, Tunisia
E-mail: badiaaessghaier@gmail.com

moderately halophilic bacteria and tested *in vitro* their ability to inhibit spore germination and mycelial growth of *Botrytis cinerea*.

Material and Methods

Microorganisms

Bacterial strain: Two moderately halophilic bacteria strain M3-23 of *V. marismortui* and strain J31 of *T. halophilus* were isolated from a Tunisian Sebkha. The morphological, physiological and molecular characteristics of these strains were previously described and their 16S rDNA sequences have been deposited in the GenBank database under the accession number GQ2825501 for strain M3-23 and EU435359 for strain J31 as previously described by us [5].

Pathogen strains: To determine the antimicrobial activity of the intracellular protein from two halophilic bacteria described in this paper; gram-positive pathogen bacterium (*Staphylococcus aureus*), and a *phytopathogenic fungus; Botrytis cinerea*, were employed and taken from the culture collection of the laboratory of Microorganisms and Biomolecules Actives, Faculty of Sciences in Tunisia.

Effect on tomato stems growth

Tomato plants (hybrid F1 Maria) used in this study were grown in 25 cm diameter pots in a plastic house. Two moderately halophilic isolates, *Terribacillus halophilus* (J31) and *Virgibacillus marismortui* (M3-23) were used for this study. The bacterial antagonists were grown for 48 h on TSA (Tryptic Soy Agar, Difco) supplemented with 5% NaCl. After 48 h, the cells were then scraped from the surface of the Petri dishes and diluted in sterilised saline solution (1% NaCl). Bacterial suspensions were adjusted to 10^8 CFU/ml. Bacterial treatments were applied as spray at 10, 20, 30 and 40 days. Each treatment had 10 replicate plants. All plants were grown in the greenhouse at about 22 C, 75% relative humidity. The statistical analysis was made with test of multiple comparisons (Dunnett test). Analysis were performed using logicial XLSTAT, test ANOVA. Data are reported in the text and figures as mean values for all replicated experiments ± standard error of the mean.

Media composition and Intracellular protein fraction preparation

The investigation of the antifungal activities production by strain M3-23 of *V. marismortui* and strain J31 of *T. halophilus* was carried out in a colloidal chitin medium containing 5 g/l tryptone, 5 g/l yeast extract, 1 g/l K_2HPO_4 at pH 7 and 0.1% (w/v) colloidal chitin. The incubation was maintained for 3 days at 37°C, 180 rev/min. After incubation, each bacterial culture broth was centrifuged at 9000 rpm for 15 min, pellet were collected and rinsed with 1% NaCl, centrifuged at 12000 rpm for 15 min. 800 μl of extraction protein buffer TEP (25 mM Tris-HCl pH 7.5, 100 mM KCl, 5% glycerol) were added and taken for sonication in ice bath (Am25, pulse 0.5-0.5, time 1 min). After a centrifugation at 12000 rpm for 30 min, supernatant containing intracellular protein was obtained and sterilized by filtration through a 0.22 μm pore size filter (Life Sciences, PALL, Ann Arbor, MI, Acrodisc 32 mm syringe filter with 0.2 μm Supor membrane) and stored at -20°C until further use.

Chitinase assay: Chitinase was determined according to the method of Gomez Ramirez et al. [9] as previously detailed by Essghaier et al. [5]. The mixture volume per volume (v/v) of intracellular protein extracted and colloidal chitin suspension (10%) was incubated for 1 h at 50°C. The reaction was stopped by adding 1 ml of 1% NaOH and shaking. The product was determined by 3,5-dinitrosalicylic acid assay (DNS) and the absorbance was measured at 535 nm. The chitinase

activity was defined as the amount of enzyme required to produce 1 μmol of N-acetylglucosamine (NAG, Sigma) per h per mg of protein [10].

Glucanase assay: The standard assay was performed according to the method of Leelasuphakul et al. [11]. The amount of reducing sugar from Laminarin was measured. The standard assay contained 10 μl of the intracellular protein and 90 μl Laminarin at 5 mg/ml in 0.1 M Sodium acetate buffer pH5. After incubation at 40°C for 10 min, the reaction was stopped by boiling for 5 min then 0.2 ml of 1% 3,5-dinitrosalicylic acid (DNS) and 0.2 ml of sodium acetate buffer were added and boiled for another 5 min, then placed in an ice bath; finally 0.9 ml distilled water was added. The optical absorbance was measured at 540 nm. The amount of reducing sugars released was calculated from a standard curve prepared with glucose and the glucanase activity was expressed in units (μmol glucose equivalent per min).

Protease assay: This was determined by incubating 500 μl of 0.5% Azocasein (sigma) in 0.1 M Tris-HCl buffer pH 8 with 100 μl of intracellular protein for 1 h at 37°C. The reaction was stopped by adding 500 μl of 15% TCA (trichroloroacetic acid) and shaking. This was left for 15min and centrifuged at 7000 rpm at 4°C for 10 min. 1 ml of supernatant was added to 1 ml of 1 M NaOH and absorption was measured at 440 nm. One unit (U) of protease activity was defined as the amount of enzyme that liberated 1μmol of tyrosine per min at 37°C [12].

Lysozyme activity

After culture of *Staphylococcus aureus* for 48 h at 30°C, bacterial cell was centrifuged, twice washed with distilled water, suspended in 50mM sodium phosphate buffer pH 6.5. After that, volume per volume (v/v) bacterial suspension and intracellular proteins from each halophilic bacterium were incubated at 37°C for 30 min. The optical density was measured at 660nm. One unit of lysozyme activity was determined as the decrease of OD at 660 nm with 0.01 per min compared to control tube contained volume (v) pathogen bacteria and volume (v) buffer without the intracellular complex.

Spore germination Inhibition

B. cinerea was grown at 25°C on potato dextrose agar in 9 cm diameter Petri dishes for 10 to 15 days. Sterile water (20 ml) was added to each plate, and the surface was scraped gently with a sterile loop to release the spores. The resulting spore suspension was filtered through a sterile 30 μm filter to remove any mycelial fragments and diluted with sterile water to the desired concentration of about 106 spores/ml. Conidial suspension of fungus was adjusted to 10^6 spores/ml by counting with a haemocytometer. To investigate the effects of the intracellular proteins from both strain M3-23 and J31, on spores germination; 20 μl of conidial suspensions adjusted at 10^6 spores/ml and 20 μl of the intracellular proteins were pipetted into an Eppendorf tube containing 1 ml of sterile distilled water with 5% glucose, the mix was then incubated at 21°C for 24 h. Control tubes were inoculated only with fungal spores. The percentage of spore germination inhibition (**I%**) was determined by microscopic examination of spores in the presence of the intracellular proteins from each bacterium (**E**), compared to control tube containing only the spore suspensions. The percentage of spores germination was determined as; **I (%): (C-E)/C X 100**. Three replicates were used for each treatment [13].

Mycelial hyphae destruction

Fungal culture was rinsed with distilled sterile water. After centrifugation at 9000 rpm for 10 min pellet (mycelium) was replaced in an Eppendorf tube containing an appropriate Tris-HCl buffer (0.01 M, pH 8), in order to obtain the same concentration of mycelial solution (expressed in mg/ml) for each intracellular bacterial fraction tested. A volume of each intracellular fraction was added. The mixture was incubated at 37°C for 14 h. After that, optical density was measured at 540 nm. Increase of OD compared to control tube (containing only mycelial suspension), make destruction of fungal hyphae by intracellular fraction.

Intracellular chitinase characterization

The optimum temperature for intracellular chitinases was determined by monitoring each activity at pH8 at various temperatures ranging from 40 to 100°C. The heat stability was analyzed by measuring the residual activity after pre-incubation of the enzymes for 30 min at various temperatures 40 to 100°C [14].

The pH optimum of the enzymes was determined by applying a substrate solution at different pH from 4 to 12 and was measured at optimum temperature (90°C for strain M3-23 and 80°C for strain J31). The pH was adjusted by using the following buffers: 0.1 M phosphate buffer (pH 4.0 to 6.0), 0.1 M Tris-HCl buffer (pH 7.0 to 8.0), 0.1 M H_3BO_3-NaOH buffer (pH 9.0 to 10) and 0.1M Na_2HPO_4-NaOH (pH 12). The pH stability was examined by incubating enzymes in the mentioned buffers for 1h at 4°C before adding the substrate. The remaining activities (%) were subsequently determined [15]. The experiments were repeated three times and mean values were taken.

The optimum salinity

To determine salt optimum for enzymatic activity, each buffer with pH optimum was added with NaCl to give a gradient of salt concentration varied from 0 to 30% NaCl (w/v). Enzymatic reactions were maintained at temperatures and pH optimum for each bacterium (80°C, pH 4 for strain J31 of *T. halophilus* and 90°C, pH 12 for strain M3-23 of *V. marismortui*).

Results and Discussion

Previous studies have shown that halophilic bacteria from shallow salt lakes located in Tunisia were successful for pre- and postharvest

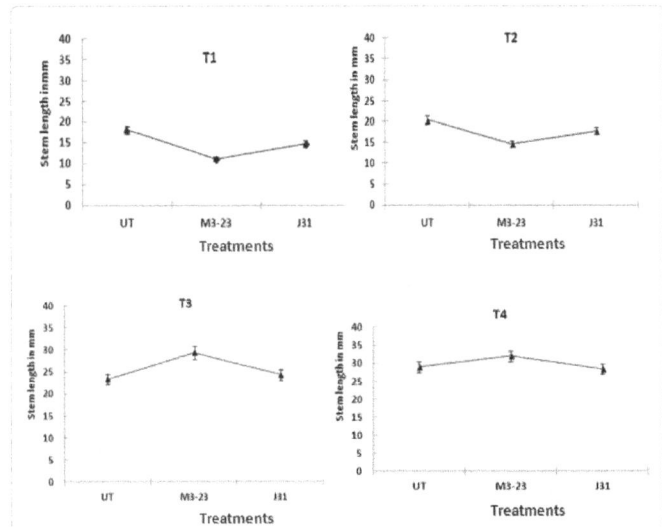

Figure 1B: Analysis of the averages of length differences between the untreated (UT) and the treated plant by biological solutions adjusted to 10^8 CFU/ml of strain M3-23 of *Virgibacillus marismortui* (M3-23) and strain J31 of *Terribacillus halophilus* (J31) under four treatments (T1, T2, T3 and T4) according to Dunnett test.
NB: T1, T2, T3 and T4 mean respectively treatment after 10, 20, 30 and 40 days from the first biological treatment.

treatments of tomato and strawberry crop in Tunisia [5,6,16]. In addition, on one hand, previously, we were the pioneer to evaluate the ability of halophilic bacteria to protect plant diseases from phytopathogenic fungi. On the other hand, these halophilic bacteria were able to produce extracellular antifungal enzymes such as chitinase, glucanase and protease. In earlier work, these extracellular antifungal enzymes exhibited potential properties (tolerance and stability in the presence of extreme conditions (pH, temperature and salt) compared to others published bacterial antifungal enzymes. Therefore, the biotechnological exploitation of these enzymes should be of great importance in agro-industries.

As an extension of our studies, in this work, we extracted the intracellular proteins from two halophilic bacteria and tested them *in vitro* for their potential antimicrobial properties such as on cell wall degradation, spore germination and mycelial destruction of the phytopathogen (*Botrytis cinerea*). Previously, we have demonstrated the ability of halophilic bacteria to reduce greymould disease on tomato plant. There are no reports on the use of halophilic bacteria to induce plant growth, so in this work, we are screening two antagonistic halophilic bacteria strain J31 of *T. halophilus* and strain M3-23 of *V. marismortui* to investigate their effect on growth plant. The results presented in Figure 1A, had shown that both strains used were able to improve stem growth by comparison with control plant (untreated plant) with the most efficiency of strain M3-23 compared to strain J31 specially in treatments T3 and T4 (Figure 1B). The performance of strain M3-23 to improve stem tomato growth was markedly evaluated and confirmed by statistical analysis according to Dunnett test, based on the comparison of the difference between untreated plant and the biological treated (Figure 1B). The highest effect of growth improve have been obtained by strain M3-23 after 30 and 40 days of treatment.

Fungal Cell wall degrading enzymes

The ability of each halophilic bacterium, to produce antifungal intracellular enzyme was mentioned in Figure 2. The results shown

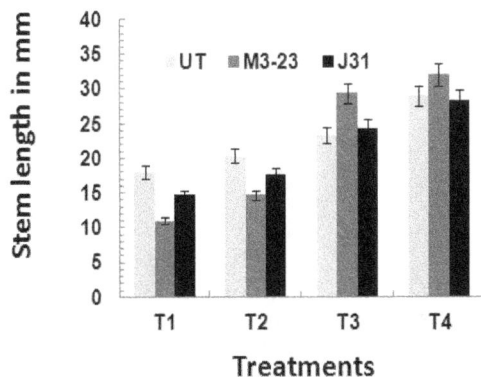

Figure 1A: Effect of biological treatment by means of strain solutions spray on tomato plant growth by measuring stem length, and four repetitions were applied at period of about 10 days.
NB: T1, T2, T3 and T4 mean respectively treatment after 10, 20, 30 and 40 days from the first biological treatment. UT, M3-23 and J31 means respectively untreated plant, biological solution adjusted to 10^8 CFU/ml of strain M3-23 and strain J31. Bars indicate standard deviation.

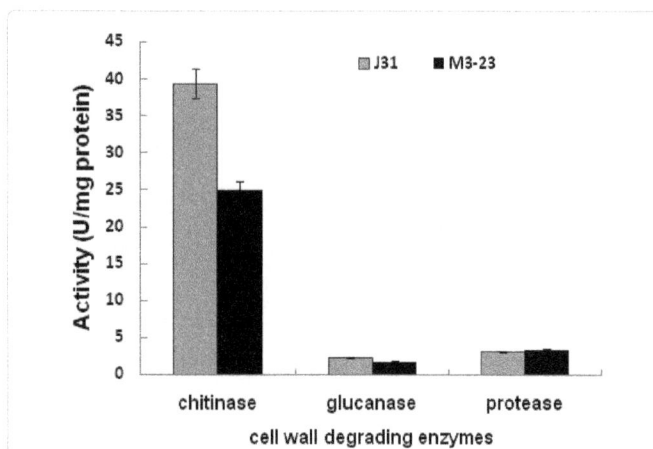

Figure 2: Quantification of cell wall degrading enzymes given in the intracellular proteins extracted from two moderately halophilic bacteria stain M3-23 of *Virgibacillus marismortui* and strain J31 of *Terribacillus halophilus*. Values shown are the average from triplicate experiments. Error bars Standard deviation.

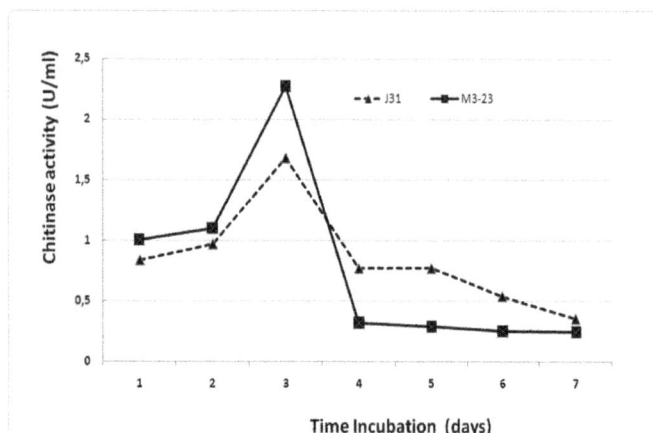

Figure 3: Kinetic production of intracellular chitinases by both moderately halophilic bacteria strain M3-23 of *V. marismortui* and strain J31 of *T. halophilus*.

that in the conditions used here, both moderately halophilic bacteria were able to produce intracellular antifungal enzymes: chitinase, glucanase and protease. But the production of intracellular chitinase was highly induced than the production of intracellular glucanase and protease. Since the value of chitinase production range from 39.39U/mg of protein for strain J31 to 24.87 U/mg of protein for strain M3-23. But values for others intracellular enzyme production were less than 3.27 U/mg of protein for intracellular protease from strain M3-23 and 2.3 U/mg of protein for intracellular glucanase produced by strain J31. For that reason, we have selected the intracellular chitinases from both strains J31 and M3-23 for biochemical characterization. In consideration of the quantification of enzyme production, the results shown that both moderately halophilic bacteria, were able to produce high intracellular chitinases by comparison with others intracellular antifungal enzymes:Glucanase and Protease. Similar results published by us shown the high extracellular chitinase production by halophilic bacteria compared to extracellular glucanase and protease [5]. Both strains were able to produce a maximum of intracellular chitinases after three days of incubation in medium containing colloidal chitin as sole carbon source (Figure 3).

Chitinase bacterial induction by chitin from crustacean shells in the marine environment has been extensively demonstrated. Bacteria produce chitinase to utilize chitin producing C and N sources to meet their nutritional requirements [17]. Some Tunisian sebkha (e.g El Melah), which was the origin of our bacteria were connected to the sea since the sea water pours in the sebkha could take with him by example chitin. So similar chitinase induction from our bacteria isolated from Tunisian sebkha could also be induced by chitin from crustacean shells.

One bacterium usually produces several chitinases probably due to efficient hydrolysis of chitin or hydrolyzes different forms of chitin found in nature [17,18]. As chitin is probably the most abundant biopolymer in the marine environment, marine chitinase producing bacteria play quite an important role in chitin degradation. There are many chitinase genes cloned, sequenced and studied from marine bacteria [19].

Intracellular physicochemical chitinase characterization

Chitin, an insoluble b-1,4-linked polymer of N-acetylglucosamine, is the second most abundant polysaccharide in nature and a major constituent of the cell walls of many fungi, insect exoskeletons, and crustacean shells (eg [20]. Degradation of chitin is essentially catalyzed by chitinases [21,22]. Chitinases (E.C. 3.2.1.14) are found in bacteria, fungi, virus, and higher plants. Furthermore, bacterial chitinases are considered primarily to digest and utilize chitin as a carbon and nitrogen nutrient. Several *Bacillus* species produces enzymes that degrade chitin. Recently, our laboratory showed the production of extracellular chitinase enzymes inducibles by the pathogen *in vitro* by antimicrobial halotolerant to moderately halophilic bacteria [5,7,8].

Effect of salinity on intracellular chitinases production was presented in Figure 4. The results shown that the intracellular chitinases from both moderately halophilic bacteria, were significantly produced in the absence and at high salt concentration (30%NaCl (w/v) which clearly indicated the highly salt-tolerant nature of the enzymes. Optimal activity (100% of original activity) was obtained at 0% NaCl. By comparison with few studies on halo-tolerance of chitinases, same results of salt tolerant chitinase has been proved with others chitinases produced from halophilic bacteria such as *Salinivibrio costicola* 5SM-1 [23], *Planococcus rifitoensis* M2-26 [7]. As well as by the marine bacteria *Vibrio harveyi* and *Alteromonas sp.* strain O-7 [18,24]. Chitinases with these characteristics may have interesting applications in biotechnological sectors. Especially in biological agricultural process,

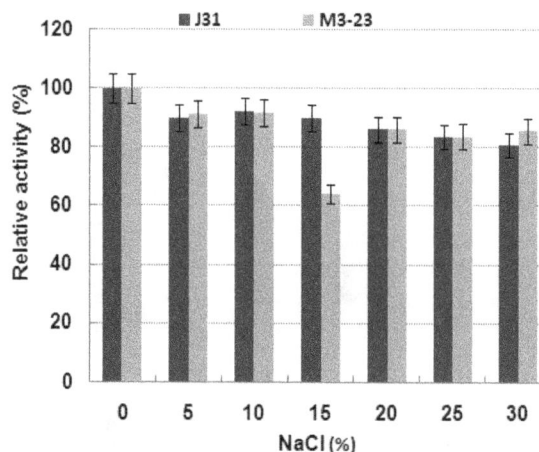

Figure 4: Effects of salt on intracellular chitinases activities of *V. marismortui* strain M3-23 and *T. halophilus* strain J31. The values shown are percentages of the maximum activity of the enzyme which is taken as 100%. Value is the average from triplicate experiments. Error bars represent the standard deviation.

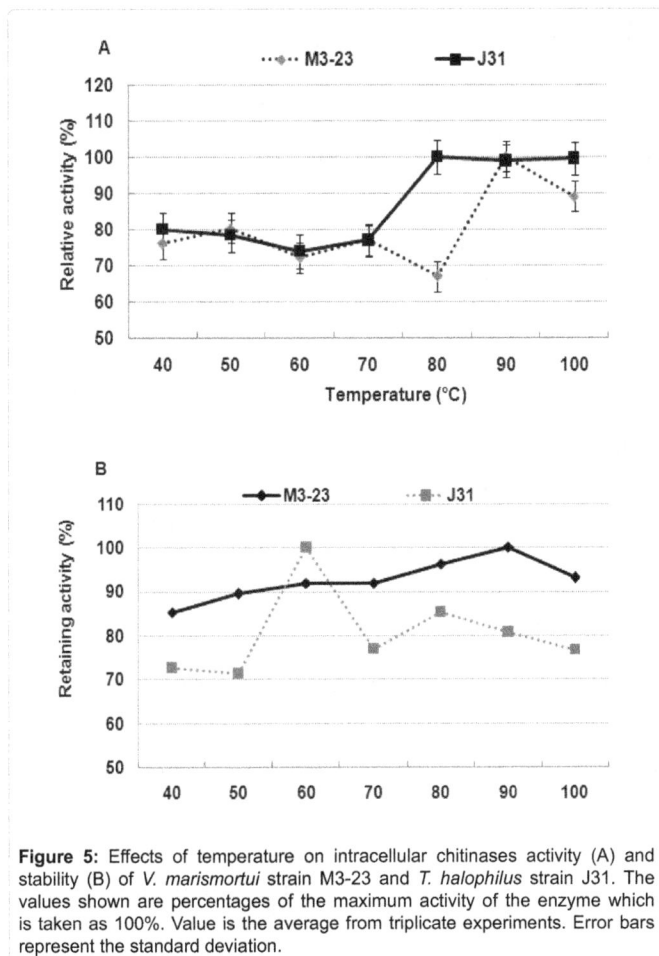

Figure 5: Effects of temperature on intracellular chitinases activity (A) and stability (B) of *V. marismortui* strain M3-23 and *T. halophilus* strain J31. The values shown are percentages of the maximum activity of the enzyme which is taken as 100%. Value is the average from triplicate experiments. Error bars represent the standard deviation.

12.0 and was active over a broad pH range between 4 and 12, retaining greater than 90% of its original activity. Strain M3-23 produces an alkaline chitinase; similar observations were reported for *Alcaligenes xylosoxydans* (pH 8.0) [29], *Streptomyces thermoviolaceus* OPC-520 (pH 8.0-10) [27] and *Planococcus rifitoensis* (pH 8.0) [7], *Bacillus atrophaeus* SC081 (pH8) [30]. The effects of pH on the activity and stability of J31 intracellular chitinase was examined at optimum temperature 80°C. Maximum activity was observed at pH 4.0. The enzyme was active and retained greater than 90% of its activity within a pH range of 4.0 to 12.0. Unlike the alcaline intracellular chitinase from strain M3-23, an enzyme from strain J31 was acidic. Few reports showed acidic chitinase from bacteria such as chitinase (ChiA) from *Bacillus licheniformis* [31].

Antimicrobial potentiality of intracellular compounds

Botrytis cinerea is the causal agent of grey mold disease on a variety of fruits, vegetables, and field crops. In this report we describe novel assays for inhibition of spore germination and growth hyphae in *B. cinerea*, which are based on the use of the intracellular proteins from halophilic bacteria. For investigation of the antimicrobial behavior of intracellular proteins produced by both strain J31 and M3-23, we have tested the ability of lysozyme activity, spores germination inhibition and mycelial destruction of *B. cinerea*. The results of lysozyme activity produced by intracellular proteins have demonstrated that both bacteria

applied in saline soil such almost Tunisian soil due to the expansive use of a heavy dressing of manure and the irrigation with brackish waters characterized by the presence of high NaCl concentrations of about 3 to 4 g/l [25].

The effect of temperature on the activity and stability of the intracellular chitinases is presented in Figure 5. Maximal chitinase activities was found respectively at 90°C and at 80°C for strain M3-23 and strain J31. However, the results of thermostability showed that intracellular chitinases from strain M3-23 was more stable over a broad range of temperatures between 40 and 100°C retaining 85.27% of its activity compared to only 71.75% for chitinase stability from strain J31. Although a large number of chitin hydrolyzing enzymes have been isolated, only a few thermostable chitin-hydrolyzing enzymes are known to be active at above 70°C [2]. The optimum temperature of thermostable chitinases from *B. stearothermophilus* CH-4 (75°C) [26]. *Streptomyces thermoviolaceus* OPC-520 (70-80°C) [27], and *Planococcus rifitoensis* [7]. These characteristics of thermo-tolerant and halotolerant enzyme are of great importance in metabolic production, and for the biocontrol ability against fungi and for protecting plants under biocontrol saline conditions as well as in future applications of these enzymes in any biotechnological process that depends on high salinity or osmotic pressure for long incubation periods [28].

The effect of pH on the activity and stability of the chitinase are illustrated in Figure 6 which shows that the intracellular chitinase from strain M3-23 of *V. marismortui* at 90°C, has the highest activity at pH

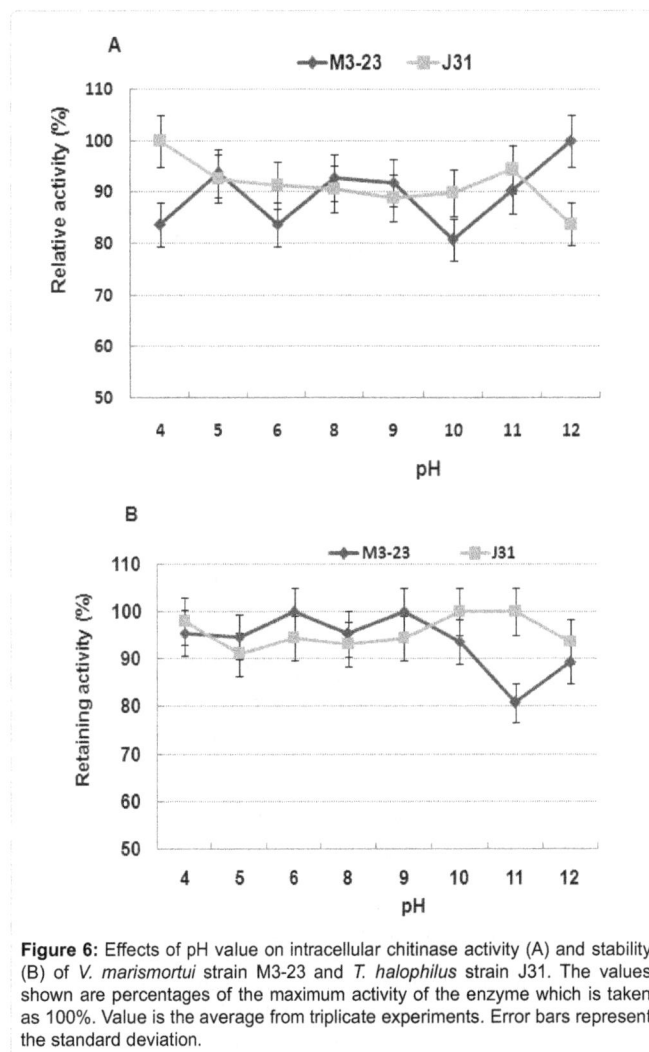

Figure 6: Effects of pH value on intracellular chitinase activity (A) and stability (B) of *V. marismortui* strain M3-23 and *T. halophilus* strain J31. The values shown are percentages of the maximum activity of the enzyme which is taken as 100%. Value is the average from triplicate experiments. Error bars represent the standard deviation.

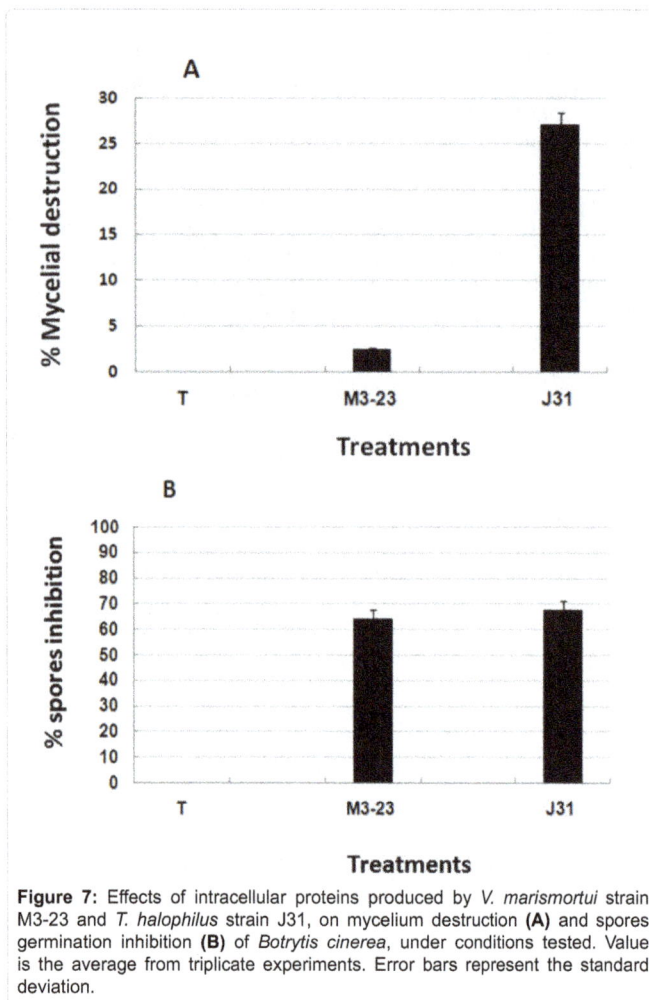

Figure 7: Effects of intracellular proteins produced by *V. marismortui* strain M3-23 and *T. halophilus* strain J31, on mycelium destruction **(A)** and spores germination inhibition **(B)** of *Botrytis cinerea*, under conditions tested. Value is the average from triplicate experiments. Error bars represent the standard deviation.

grey-mould disease in number crop. Nowadays, consumers are aware of the health concerns regarding food additives; the health benefits of "natural" and "traditional" foods, processed without any addition of chemical preservatives, are becoming more attractive. Thus, because of recent consumer demand for higher quality and biodegradable bioactive compound, researchers all over the world are looking for new bioactive metabolites. In previous work, we have been reporting on the efficiency of halophilic bacteria in agriculture as a biological agent. Furthermore, in this study, we have mentioned the ability of the moderately halophilic bacteria to produce not only intracellular antimicrobial compounds but also stable at extreme conditions (salt, pH and temperature). Characterization of biochemical properties suggested that the intracellular chitinases were salt thermo and pH-stable, suitable for various biotechnological applications, including bioconversion of colloidal chitin and chitobiose into NAG or the juice industry.

Acknowledgments

This work was supported by funds from the Ministry of Higher Education and Scientific Research (Tunisia).

References

1. Ventosa A, Nieto JJ, Oren A (1998) Biology of moderately halophilic aerobic bacteria. Microbiol Mol Biol Rev 62: 504-544.

2. Niehaus F, Bertoldo C, Kähler M, Antranikian G (1999) Extremophiles as a source of novel enzymes for industrial application. Appl Microbiol Biotechnol 51: 711-729.

3. Pandey A, Soccol CR, Mitchell DA (2000) New developments in solid-state fermentation I—bioprocesses and products. Proc Biochem 1153-1169.

4. Ordentlich A, Elad Y, Chet I (1988) The role of chitinase of *Serratia marcescens* in biocontrol of *Sclerotium rolfsii*. Phytopathol 78: 84-88.

5. Essghaier B, Fardeau ML, Cayol JL, Hajlaoui MR, Boudabous A, et al. (2009) Biological control of grey mould in strawberry fruits by halophilic bacteria. J Appl Microbiol 106: 833-846.

6. Sadfi-Zouaoui N, Essghaier B, Hajlaoui MR, Fardeau ML, Cayol JL, et al. (2008) Ability of moderately halophilic bacteria to control grey mould disease on tomato fruits. J Phytopathol 156: 42-52.

7. Essghaier B, Rouaissi M, Boudabous A, Jijakli H, Sadfi-Zouaoui N (2010) Production and partial characterization of chitinase from a halotolerant *Planococcus rifitoensis* strain M2-26. World J Microbiol Biotechnol 26: 977-984.

8. Essghaier B, Hedi A, Bejji M, Jijakli H, Boudabous A, et al. (2012) Characterization of a novel chitinase from a moderately halophilic bacterium *Virgibacillus marismortui* strain M3-23. Ann Microbiol 62: 835-841.

9. Gomez Ramirez M, Rojas Avelizapa LI, Rojas Avelizapa NG, Cruz Camarillo R (2004) Colloidal chitin stained with Remazol Brillant Blue R, a useful substrate to select chitinolytic microorganisms and to evaluate chitinases. J Microbiol Methods 56: 213-219.

10. Roja Avelizapa LI, Cruz Camarillo R, Guerro MI, Rodriguez Vazquez R, Ibarra JE (1999) Selection and characterization of a proteo-chitinolytic strain of *Bacillus thuringiensis*, able to grow in shrimp waste media. World J Microb Biotech 15: 299-308.

11. Leelasuphakul W, Sivanunsakul P, Phongpaichit S (2006) Purification, characterization and synergistic activity of β-1-3-glucanase and antibiotic extract from an antagonistic *Bacillus subtilis* NSRS89-24 against rice blast and sheath blight. Enzyme Microbiol Tech 38: 990-997.

12. Olajuyigbe FM, Ajele JO (2005) Production dynamics of extracellular protease from Bacillus species. Afr J Biotech 4: 776-779.

13. Sarangi N, Athukorala P, Dilantha Fernando WG, Rashid KY, Kievit TD (2010) The role of volatile and non-volatile antibiotics produced by *Pseudomonas chlororaphis* strain PA23 in its root colonization and control of *Sclerotinia sclerotiorum*. Biocontrol Sci Technol 20: 875-890.

14. Issam SM, Mohamed G, Farid L, Sami F, Thierry M, et al. (2003) Production, purification, and biochemical characterization of two beta-glucosidases from *Sclerotinia sclerotiorum*. Appl Biochem Biotechnol 111: 29-40.

have lysozyme activity presented in their intracellular protein fraction with respectively 6.6 U/ml and 6.8 U/ml for strain J31 of *T. halophilus* and strain M3-23 of *V. marismortui*. It is worth pinpointing that, the combination of lysozyme activity to antifungal enzymes (chitinase, protease and glucanase) were rare and offer great potential use of such antimicrobial compounds in biotechnological area for example to assure food safety and consumer health by producing good quality wines [32].

The intracellular fraction produced by each bacterium was able to destroy mycelium of *B. cinerea* (Figure 7A) and to inhibit its spore germination (Figure 7B). On the whole, the most potent *in vitro* antifungal effect was demonstrated by intracellular compound produced by a strain J31 of *T. halophilus* with respectively 67.58% compared to 64.25% spores inhibition given by strain M3-23 of *V. marismortui*. As regards only the intracellular proteins produced by strain J31 of *T. halophilus* have a strong effect on the mycelial destruction with 27% compared to only 2.55% for strain M3-23 of *V. marismortui*. Numerous studies have demonstrated the ability of bacteria to inhibit spore germination, but this work was the first addresses the ability of halophilic bacteria to produce intracellular proteins inhibiting spore germination and mycelial growth of fungus (Figures 7A and 7B).

Recently, in the course of our research on the biologically active compound, in this work, we investigated the efficiency of intracellular enzymes from halophilic bacteria to inhibit *B. cinerea* agent causal of

15. Ellouze O, Mejri M, Smaali I, Limam F, Marzouki MN (2007) Induction, properties and application of xylmanase activity from *Slerotinia sclerotiorum* S2 Fungus. J Food Biochem 31: 96-107.

16. Sadfi-Zouaoui N, Hannachi I, Andurand D, Essghaier B, Boudabous A, et al. (2008) Biological control of *Botrytis cinerea* on stem wounds with moderately halophilic bacteria. World J Microbiol Biotech 24: 2871-2877.

17. Clarke PH, Tracey MV (1956) The occurrence of chitinase in some bacteria. J Gen Microbiol 14: 188-196.

18. Svitil AL, Chadhain S, Moore JA, Kirchman DL (1997) Chitin Degradation Proteins Produced by the Marine Bacterium *Vibrio harveyi* Growing on Different Forms of Chitin. Appl Environ Microbiol 63: 408-413.

19. Suginta W, Robertson PA, Austin B, Fry SC, Fothergill-Gilmore LA (2000) Chitinases from Vibrio: activity screening and purification of chiA from *Vibrio carchariae*. J Appl Microbiol 89: 76-84.

20. Felse PA, Panda T (1999) Regulation and cloning of microbial chitinase genes. Appl Microbiol Biotechnol 51: 141-151.

21. Flach J, Pilet PE, Jollès P (1992) What's new in chitinase research? Experientia 48: 701-716.

22. Cohen-Kupiec R, Chet I (1998) The molecular biology of chitin digestion. Curr Opin Biotechnol 9: 270-277.

23. Aunpad R, Panbangred W (2003) Cloning and characterization of the constitutively expressed chitinase C gene from a marine bacterium, *Salinivibrio costicola* strain 5SM-1. J Biosci Bioeng 96: 529-536.

24. Tsujibo H, Kondo N, Tanaka K, Miyamoto K, Baba N, et al. (1999) Molecular analysis of the gene encoding a novel transglycosylative enzyme from Alteromonas sp. strain O-7 and its physiological role in the chitinolytic system. J Bacteriol 181: 5461-5466.

25. Messaï A, Hannachi C, Zid E (2006) *In vitro* Regeneration of NaCl-adapted Tomato Plants (Lycopersicon esculentum Mill.) Tropicultura 24: 221-228.

26. Sakai K, Narihara M, Kasama Y, Wakayama M, Moriguchi M (1994) Purification and characterization of thermostable beta-N-acetylhexosaminidase of *Bacillus stearothermophilus* CH-4 isolated from chitin-containing compost. Appl Environ Microbiol 60: 2911-2915.

27. Tsujibo H, Minoura K, Miyamoto K, Endo H, Moriwaki M, et al. (1993) Purification and properties of a thermostable chitinase from *Streptomyces thermoviolaceus* OPC-520. Appl Environ Microbiol 59: 620-622.

28. Margesin R, Schinner F (2001) Potential of halotolerant and halophilic microorganisms for biotechnology. Extremophiles 5: 73-83.

29. Vaidya RJ, Shah IM, Vyas PR, Chhatpar HS (2001) Production of chitinase and optimization from a novel isolate Alcaligenes xylosoxydans: potential in antifungal biocontrol. World J Microbiol Biotech 17: 691-696.

30. Cho EK, Choi IS, Choi YJ (2011) Overexpression and characterization of thermostable chitinase from *Bacillus atrophaeus* SC081 in *Escherichia coli*. BMB Rep 44: 193-198.

31. Songsiriritthigul C, Lapboonrueng S, Pechsrichuang P, Pesatcha P, Yamabhai M (2010) Expression and characterization of *Bacillus licheniformis* chitinase (ChiA), suitable for bioconversion of chitin waste. Bioresour Technol 101: 4096-4103.

32. Visan L, Margarit G, Groposila D, Varga M (2009) Lysosyme enzymatic activity changes in control of malolactic fermentation used in Romanian wines. Lucrări Ştiinţifice 438-444.

7

Enzymatic Responses of Ginger Plants to *Pythium* Infection after SAR Induction

Rajyasri Ghosh*

Mycology and Plant Pathology Laboratory, Department of Botany, Scottish Church College, Kolkata, India

Abstract

A variety of enzymatic responses of ginger plants to *Pythium* infection after induction of SAR (systemic acquired resistance) have been investigated. Results of pathogenicity test of *P. aphanidermatum* on a susceptible ginger cultivar showed that disease intensity increased with time up to 28 days but Polyphenol oxidase (PPO), Lipoxygenase (LOX) and Phenyl alanine ammonia lyase (PAL) activities increased up to 14 days following inoculation and then declined whereas Peroxidase (PO) activity reached their peaks on 21st day after inoculation and then decreased sharply. To induce SAR, rhizome seeds were soaked separately in salicylic acid (SA-5 mM) and *Acalypha* leaf extract (ALE – 10%) for 1 hour prior to sowing. Significant disease reduction was observed in both SA and ALE treated plants. SA and ALE treatment enhanced activities of all four defence related enzymes in ginger leaves but the rate of increase was higher in untreated inoculated and treated non-inoculated plants in relation to their respective controls. Treated inoculated plants exhibited maximum activity for all four enzymes. SA stimulated PO and PAL more than that of ALE. Results suggest that a correlation exists between reduction of disease intensity due to SAR induction and greater stimulation of specific enzymatic activities in ginger plants although not all four enzymes are equally responsive to a defence activator.

Keywords: Ginger; *Pythium aphanidermatum*; Salicylic acid; *Acalypha* leaf extract; Defence related enzyme; SAR

Introduction

Induced resistance is defined as an enhancement of the plant defensive capacity against a broad spectrum of pathogens and pests that is acquired after appropriate stimulation. The resulting elevated resistance due to an inducing agent upon infection by a pathogen is called ISR or SAR [1]. Induction of resistance in plants to a broad spectrum of microorganisms by different plant defence activators and involvement of defence related proteins and enzymes have been discussed earlier in detail [2-5]. Various biotic and abiotic inducers can also enhance the activities of defence related enzymes, Two different biotic inducers [*Pseudomonas fluorescens* and *Pseudomonas putida*] and three different abiotic inducers [copper sulphate, indole butyric acid and potassium chloride] were tested for their efficacy in inducing resistance in lupin plants against *Fusarium* wilt disease caused by *Fusarium oxysporum* f. sp. *Lupine*. A time course of defence-related enzymes showed substantial increases in enzyme activities in induced infected seedlings compared with untreated healthy plants or infected controls [6]. Induction of defense-related marker enzyme activity, namely, peroxidase, polyphenol oxidase, β-1,3 glucanase, chitinase, and phenolics was observed in banana (Grand Naine variety) plants when interacting with dead or live pathogen, *Fusarium oxysporum f. sp. cubense*, a causative agent of Panama disease [7]. These defence related enzymes are associated with the biosynthesis of lignin, phenolic compounds and phytoalexins, which are considered as important plant, defence components. Therefore, their role in induced resistance is significant. Peroxidase (PO) oxidizes phenolics to quinones and generates H_2O_2. The latter not only is antimicrobial in itself, but it also releases highly reactive free radicals and in that way further increases the rate of polymerization of phenolic compounds into lignin like compounds. These substances are deposited in cell walls and papillae and inhibited further growth and development of pathogen. Increased activity of polyphenol oxidase (PPO) also results in accumulation of higher concentrations of toxic products of oxidation and therefore greater degree of resistance to infection occurs. In contrast to peroxidase and polyphenol oxidase, lipoxygenases (LOX) use molecular oxygen to oxygenate unsaturated fatty acids such as linoleic acid, linolenic acid producing fatty acid hydroperoxide [8,9]. Lipoxygenases are primarily

induced by pathogen attack and to a lesser extent by wounding and herbivore damage [9]. This enzyme has several important roles in defence, for example, LOX are required for the synthesis of jasmonates that has a major role in defence. Phenylalanine ammonia lyase (PAL), one of the key enzymes in the phenyl propanoid pathway, has a role in phytoalexin, phenolic compound and salicylic acid synthesis. Other defense related enzymes include pathogenesis related proteins (PRs) such as β-1,3-glucanase (PR2) and chitinase (PR1) which degrade fungal cell wall and cause cell lysis.

Recent studies on host pathogen interactions revealed the existence of SAR in non-vascular plants. Andersson et al. [10] reported SA-dependent defense pathway in *Physcomitrella patens* against *Erwinia ceratovora*. Ponce de Leon et al. demonstrated that treatment of whole plant of *P. patens* with elicitors or cell-free culture filtrates of the bacterium *Erwinia carotovora* or inoculation with spores of the fungus *Botrytis cinerea* altered expression of the genes *PR-1*, *CHS*, *PAL*, and *LOX*, which are all up-regulated upon pathogen attack in vascular plants. Oliver et al. [11] showed that *P. patens* activated multiple and similar responses against *Pythium irregulare* and *Pythium debaryanum*, including the reinforcement of the cell wall, induction of the defense genes CHS, LOX and PAL, and accumulation of the signaling molecules jasmonic acid (JA) and its precursor 12-oxo-phytodienoic acid (OPDA). This study indicated that in *P. patens*, *Pythium* infection activates common responses to those previously characterized in flowering plants. Winter et al. [12] demonstrated moss species *Amblystegium serpens* could initiate SAR like reactions upon inoculation with *Pythium irregulare*. Results of their investigation,

---

*Corresponding author: Rajyasri Ghosh, Assistant professor, Mycology and Plant Pathology laboratory, Department of Botany, Scottish Church College, Kolkata, India, E-mail: rajyasri_5@rediffmail.com together with previous studies, suggest that SAR arose prior to the divergence of vascular and non-vascular plants.

Ginger occupies an important place in plantation crops and is grown in tropical and subtropical regions of the country for its aromatic rhizomes, which are used both as a spice and as a medicine. Rhizome rot caused by *Pythium aphanidermatum* is a serious disease of ginger, which causes considerable loss in yield every year. In the present investigation an attempt has been made to study the specific enzymatic responses of ginger plants to *Pythium aphanidermatum* after induction of resistance by two plant defence activators. The purpose of this study is to determine how far defence related enzymes such as PO, PPO, LOX and PAL activities are affected by SAR induction in ginger.

Materials and Methods

Source of ginger and fungal culture

Rhizome seeds (cultivar Suprabha) were collected from Amtala seed stores, South 24 Parganas, West Bengal. The culture of *Pythium aphanidermatum* (Edson) Fitz, a causal organism of rhizome rot or soft rot disease of ginger was supplied by the Indian Institute of Spices Research, Calicut, Kerala.

Pathogenicity test

Healthy rhizome seeds of ginger were disinfected and sown in earthenware pots (one rhizome seed/pot of 20 cm diameter) containing non-infected sandy soil (sand: soil: 1:3). The plants were kept under ordinary conditions of daylight, temperature (28-34°C) and humidity (77-85%) of the experimental garden of the Department of Botany. Water supply was maintained once a day except during rainy days. The top soil in pots was inoculated with *P. aphanidermatum* (in sand maize meal) after 28 days of sowing rhizome seeds following the method of Karmakar et al. [13]. Disease intensity was assessed usually 4 weeks after inoculation and 8 weeks after sowing rhizome seeds. Yellowing of leaves is one of the important symptoms of rhizome rot disease of ginger and hence yellowing index (Y.I.)/plant was calculated as follows:

$$\text{Y.I. / Plant} = \frac{\text{Total number of leaves showing yellowing symptom}}{\text{Total number of test plants / treatment}}$$

Apart from Y.I./Plant, percentage loss in fresh weight of rhizomes due to putrifaction caused by the pathogen was also estimated in relation to non-inoculated control.

Method of application of plant defence activators

To induce resistance in ginger plants rhizome seeds are soaked separately in salicylic acid (SA) and 10% leaf extract of *Acalypha indica* (ALE) for 1 hour prior to sowing. Control seeds were soaked in distilled water for a similar period. The leaf extract was prepared following the method of Doubrava et al. [14]. The leaves were homogenized with distilled water (50 g fresh weight of leaves/100 ml) in an electric blender. The homogenate was strained through two layers of muslin and the filtrate centrifuged at 5,000 x g for 15 minutes at 4°C. Usually, 100 ml of supernatant (50%) were added to 500 ml distilled water (1: 5), mixed thoroughly and used for treating rhizome seeds by soaking method.

Extraction and assay of enzymes

PO: Extraction and assay of PO from ginger leaves were carried out following the method of Sadashivam and Manickam [15] with modifications. PO was extracted from ginger leaves with 0.1 M phosphate buffer, pH 7.0 (2 ml g^{-1} fresh tissue) by grinding the leaves in a prechilled mortar and pestle with sea sand. The extract was strained through muslim and centrifuged at 15,000 rpm. for 15 min at 4°C. The supernatant was used as enzyme source.

For enzyme assay, 200 μl of enzyme extract was added to 3.5 ml of phosphate buffer (pH 6.5) and to this mixture 0.1 ml freshly prepared o-dianisidine solution (1 mg ml^{-1} methanol) was added. The assay mixture was brought to 28-30°C and the reaction started after addition of 0.2 ml (0.2 M) H$_2$O$_2$ to the mixture. The initial OD at 436 nm was noted and then readings were taken at an interval of 30 sec up to 3 min. A water blank was also included in the assay. The enzyme activity was expressed as change in OD (optical density) at 436 nm per unit time per mg protein.

PPO: PPO activity was measured following the method of Sadashivam and Manickam [15] with modifications. For enzyme extraction, leaf tissues of ginger were homogenized in 50 mM Tris-HCl buffer, pH 7.2 (2 ml g^{-1} tissue). The homogenate was strained through muslin, centrifuged at 15,000 rpm for 10 min at 4°C and the supernatant was used as enzyme extract.

For enzyme assay 0.6 ml catechol solution (0.01 M) was first added to 5 ml phosphate buffer (0.1 M, pH 6.5) and then 1 ml enzyme extract was supplemented to this mixture. The change in optical density was recorded at 490 nm for every 30 seconds up to 3 min. The enzyme activity was expressed as change in OD at 490 nm per unit time per mg protein.

LOX: For extraction and assay of LOX activity the method of Vick and Zimmermann [16] was adopted with modifications. LOX was extracted from ginger leaves using potassium phosphate buffer (0.05 M, pH 6.0). The homogenate was filtered and centrifuged at 15,000 rpm for 15 min at 4°C and the supernatant was used for LOX assay. LOX activity was measured spectrophotometrically using linoleic acid as substrate. Hundred microlitres of crude enzyme extract was added to 2.9 ml potassium phosphate buffer (0.05 M, pH 6.0). The reaction was initiated by 20 μl of linoleic acid (8 mM) substrate at room temperature (30°C). LOX activity was measured by conjugated diene absorption of the hydroperoxide at 234 nm. The enzyme activity was expressed as μmol conjugated diene produced per unit time per mg protein.

PAL: For extraction and assay of PAL, the method described by Smith et al. [17] was followed. Ginger leaf tissues were homogenized in 50 mM Tris-HCl buffer (pH 8.6) containing 10 mM ascorbic acid (3 ml g^{-1} fresh weight) with a mortar and pestle. The homogenate was filtered, centrifuged at 15,000 r.p.m for 20 min at 4°C and the supernatant dialysed against 50 mM Tris-HCl (pH 8.6) for 16 h at 4°C.

The dialysate was assayed spectrophotometrically for PAL activity after addition of 1 ml of 30 mM L-phenylalanine (dissolved in appropriate volume of 50 mM Tris-HCl, pH 8.6) and 2 ml of dialysate (protein content 1 mg/ml). Absorbance was measured at 290 nm in Shimadzu UV-160A spectrophotometer after 2 h incubation at 30°C. A control was maintained for a similar period using 10 mM D-phenylalanine instead of L-isomer. The activity was expressed as the amount of cinnamic acid (μmol) produced per unit time per mg protein.

Result and Discussion

Pathogenicity test

Disease severity was assessed after 7, 14, 21 and 28 days of inoculation. Initially, symptoms developed on leaves as slight fading of green colour followed by yellowing of leaf tips (chlorosis) that extended downwards resulting in withering and drying of leaves. Rhizomes became soft and gradually the internal tissues rot completely (putrefaction), plants showed wilting symptoms and finally collapsed. It appears from the results that the intensity of disease increased with time up to 28 days following inoculation. About 34% loss in fresh weight of rhizomes was noted. Yellowing index/plant also increased with time up to 28 days. Significant difference in percentage weight loss was recorded between 14 and 21 days (P<0.01) and also between 21 and 28 days (P<0.05) of inoculation. Difference was not significant between 7 and 14 days in terms of weight loss. Statistical analysis of data shows a positive correlation (P<0.05) between intensity of disease and duration of incubation time up to 28 days following inoculation (Table 1 and Figure 1).

Effect of incubation period on enzyme activity

Defence related enzymes (PO, PPO, LOX and PAL) were extracted from leaves of ginger plants after 7, 14, 21 and 28 days of inoculation in all cases after assessment of disease intensity. The enzyme activities were assayed as described and results are given in Figure 2A-2D.

PO activity increased in leaves of untreated inoculated plants up to 21 days of inoculation and then declined sharply (Figure 2A). Whereas PPO activity increased and reached its peak on the 14th day of inoculation and then declined gradually in leaves of untreated inoculated plants (Figure 2B). Significant differences (P<0.01) in enzyme activities were noted between 7 and 14, 14 and 21 and also between 21 and 28 days of inoculation in case of both PO and PPO. The LOX activity also increased in leaves of untreated inoculated plants up to 14 days following inoculation and then declined (Figure 2C). Differences in enzyme activities were significant (P<0.01) between 14 and 21 and also between 21 and 28 days but not between 7 and 14 days. The PAL activity also increased up to 14 days and then declined in leaves of untreated inoculated plants. There is a significant (P<0.01) decrease in enzyme activity between 14 and 21 days but not between 21 and 28 days (Figure 2D). Activities of all four enzymes increased in leaves of untreated inoculated plants as follows: PO (4-fold), PPO (4.2-fold), LOX (3.2-fold), and PAL (5-fold).

Results revealed that incubation period has significant effect on disease development and enzyme activities. But there is no direct correlation between the rate of enzyme activity and intensity of disease.

Effects of plant defence activators (SA and ALE) on disease development

Effects of two plant defence activators (SA and ALE) were tested on disease development and enzyme (PO, PPO, LOX and PAL) activities in ginger. Healthy rhizome seeds were soaked separately in 5 mM SA and 10% ALE for 1 h prior to sowing. Plants were inoculated as usual after 28 days of sowing and disease severity was assessed after 14 and 28

days of inoculation. Results indicated that both SA and ALE reduced disease significantly (P<0.01). The difference between SA and ALE treatments in terms of percentage loss in fresh weight of rhizomes, even after 28 days of inoculation was not significant (Table 2).

Effects of plant defence activators (SA and ALE) on enzyme (PO, PPO, LOX and PAL) activities

After assessment of disease intensity on 14th and 28th day of inoculation, enzyme were extracted from leaves of untreated and treated non-inoculated and inoculated ginger plants and assayed. The results are given in Table 3.

It is evident from the results that both SA and ALE enhanced PO activity in leaves of non-inoculated and inoculated ginger plants but markedly in treated inoculated ones. Enzyme activity decreased on 28th day of inoculation in all cases except untreated non-inoculated plants. The difference in PO activity between SA and ALE-treated plants was significant (P<0.01) on 14th day but not significant on 28th day of inoculation. Like PO, PPO activity also increased in leaves of SA and ALE-treated non-inoculated and inoculated plants up to 14 days of inoculation. The rate of increase was significantly higher (P<0.05)

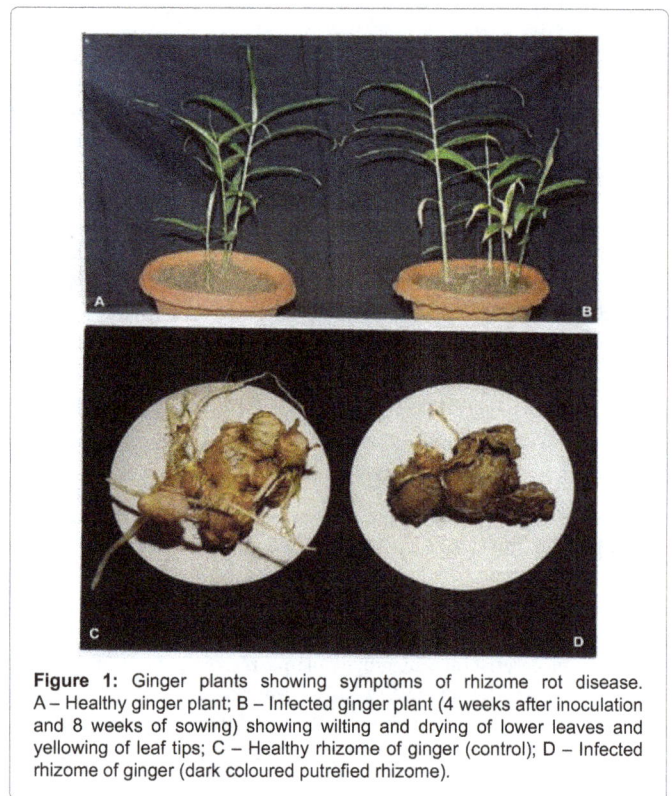

Figure 1: Ginger plants showing symptoms of rhizome rot disease. A – Healthy ginger plant; B – Infected ginger plant (4 weeks after inoculation and 8 weeks of sowing) showing wilting and drying of lower leaves and yellowing of leaf tips; C – Healthy rhizome of ginger (control); D – Infected rhizome of ginger (dark coloured putrefied rhizome).

Incubation time (days)*	Yellowing index/plant		**Average fresh wt (g) of rhizomes		% loss in fresh weight of rhizome	Correlation coefficient
	Non inoculated	Inoculated	Non inoculated	Inoculated		
7	0	0	39.40 ± 0.57	38.00 ± 1.08	3.55	r=0.992
14	0	1.00	44.20 ± 1.15	38.95 ± 1.61	11.87	
21	0	1.25	50.00 ± 0.77	37.50 ± 1.65	25.00	P<0.05
28	0	2.66	55.00 ± 1.29	36.15 ± 3.23	34.27	
CD at 5%					8.73	
CD at 1%					12.25	

*Days after inoculation
**Average of 4 rhizome seeds/treatment

Table 1: Effect of incubation period on the development of rhizome rot disease of ginger.

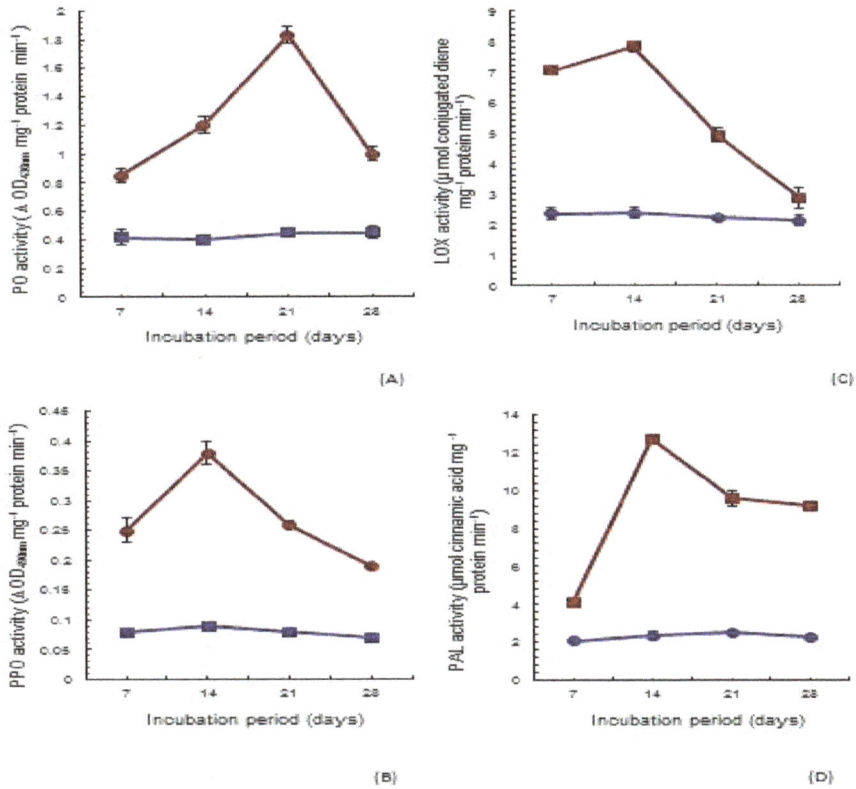

Figure 2: Activities of defence related enzymes in leaves of ginger after different periods of incubation with P. aphanidermatum .Red square line denoted untreated inoculated and blue circle line denoted ntreated noninoculated. A – PO; B – PPO; C- LOX; D : PAL.

Treatment	Yellowing index/plant		**Average fresh wt (g) of rhizomes		% loss in fresh wt of rhizomes	
	14 days*	28 days	14 days	28 days	14 days	28 days
Untreated						
Non inoculated	0	0	34.50 ± 1.55	46.00 ± 1.29		
Inoculated	1.16	2.20	30.85 ± 1.17	31.00 ± 0.91	10.57 ± 0.68	32.60 ± 0.30
Treated with SA (5 mM)						
Non inoculated	0	0	32.00 ± 1.47	42.87 ± 1.16		
Inoculated	0.50	0.83	30.67 ± 1.34	39.35 ± 2.11	4.15 ± 1.92	8.21 ± 2.72
Treated with ALE (10%)						
Non inoculated	0	0	33.20 ± 1.00	44.00 ± 1.58		
Inoculated	0.28	0.71	32.35 ± 1.96	41.32 ± 0.75	2.56 ± 1.40	6.09 ± 1.71
CD at 5%					4.10	6.52
CD at 1%					5.76	9.37

SA: Salicylic acid; ALE: *Acalypha* leaf extract
*Days after inoculation
**Average of 4 replicate rhizome seeds/treatment

Table 2: Effect of selected plant defence activators (SA and ALE) on disease development.

in SA-treated inoculated plants. The difference in PPO activity was not significant between SA- and ALE-treated non-inoculated plants. LOX activity increased in both SA- and ALE-treated plants up to 14 days of inoculation but markedly in SA-treated plants. PAL activity also increased in leaves of untreated inoculated as well as treated (SA or ALE) non-inoculated and inoculated plants. Activity was always higher in SA-treated than in ALE-treated plants (both non-inoculated and inoculated). Results showed that enzyme activities increased in ginger plants as a result of both treatment and inoculation. Highest activity was noted in treated inoculated plants. The rate of increase in treated

inoculated plants were significantly higher in comparison to untreated inoculated plants (P<0.05) (Figure 3A-D).

In the present investigation results of pathogenicity test of *P. aphanidermatum* on a susceptible ginger cultivar showed that disease intensity increased with time up to 28 days but PPO, LOX and PAL activities increased up to 14 days following inoculation and then declined whereas PO activity reached their peaks on 21st day after inoculation and then decreased sharply. Plant defence activators SA and ALE not only reduced rhizome rot disease significantly in ginger, but they also

Treatment	PO activity (ΔOD_{436nm} mg^{-1} protein min^{-1})		PPO activity (ΔOD_{490nm} mg^{-1} protein min^{-1})		LOX activity (μ mol conjugated diene mg^{-1} protein min^{-1})		PAL activity (μ mol cinnamic acid mg^{-1} protein min^{-1})	
	14 days*	28 days	14 days*	28 days	14 days*	28 days	14 days*	28 days
Untreated								
Non- inoculated	***0.38 ± 0.02	0.42 ± 0.03	0.07 ± 0.006	0.05 ± 0.007	2.35 ± 0.27	2.03 ± 0.12	2.28 ± 0.12	2.22 ± 0.06
Inoculated	1.35 ± 0.03	1.12 ± 0.06	0.31 ± 0.008	0.22 ± 0.002	7.71 ± 0.24	2.67 ± 0.14	11.04 ± 0.57	10.08 ± 0.48
Treated with SA (5 mM)								
Non-inoculated	0.82 ± 0.06	0.80 ± 0.04	0.15 ± 0.005	0.13 ± 0.005	4.04 ± 0.13	3.83 ± 0.16	10.98 ± 0.54	9.54 ± 0.88
Inoculated	2.05 ± 0.03	1.77 ± 0.01	0.85 ± 0.14	0.56 ± 0.003	9.45 ± 0.22	4.28 ± 0.21	18.42 ± 0.78	16.56 ± 0.50
Treated with ALE (10%)								
Non -inoculated	0.72 ± 0.05	0.67 ± 0.06	0.14 ± 0.011	0.12 ± 0.005	3.64 ± 0.18	3.49 ± 0.09	9.24 ± 0.77	8.28 ± 0.52
Inoculated	1.86 ± 0.03*	1.65 ± 0.11	0.67 ± 0.015	0.42 ± 0.02	8.53 ± 0.14	3.85 ± 0.30	14.70 ± 0.92	12.84 ± 1.53
CD at 5%	0.124	0.174	0.156	0.035	0.612	0.623	2.83	2.44
CD at 1%	0.174	0.259	0.223	0.048	0.858	0.873	3.97	3.42

SA: Salicylic acid; ALE: *Acalypha* leaf extract
*after inoculation
**Average of 4 replicate rhizome seeds/treatment
***3 replicates/treatment

Table 3: Effect of plant defence activators (SA and ALE) on disease development and enzyme activities in ginger leaves.

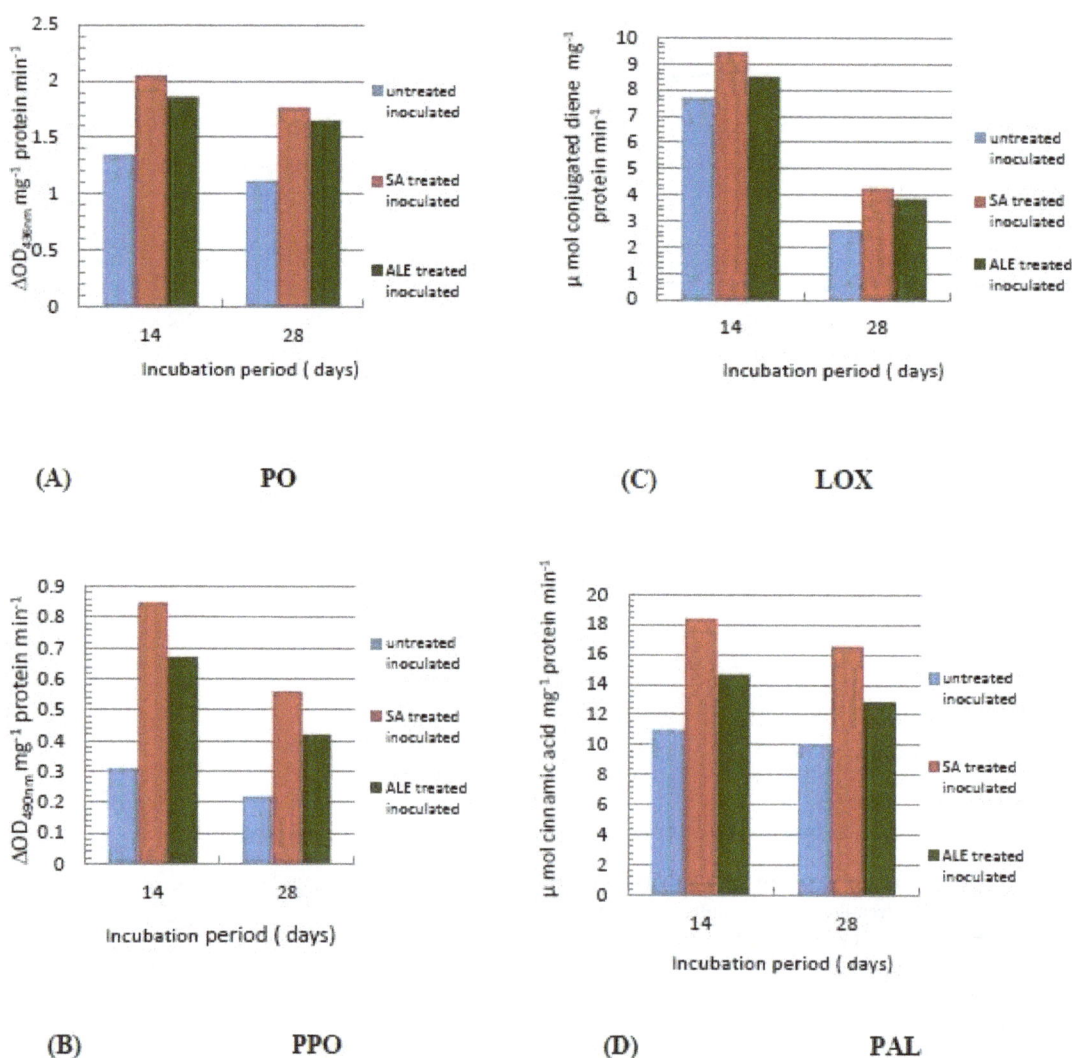

(A)   PO

(C)   LOX

(B)   PPO

(D)   PAL

Figure 3: Activities of defence related enzymes in leaves of untreated inoculated and treated inoculated ginger plants.

enhanced activities of all four defence related enzymes in ginger leaves although the rate of increase was higher in untreated inoculated and treated non-inoculated plants in relation to their respective controls. The test enzymes preferentially responded to different kinds of defence activators. Systemic protection against *P. aphanidermatum* was induced in ginger by chemicals or specific herbal extracts. Among 12 plant defence activators tested, jasmonic acid (5 mM) and 10% leaf extract of *Acalypha indica* (ALE) reduced disease significantly with concomitant increase of defence related proteins [18]. In this study, treated inoculated plants exhibited maximum activity for all four enzymes. SA stimulated all the four defence related enzymes than that of ALE. It was reported earlier that nonpathogenic rhizobacteria and a fungal pathogen *P. aphanidermatum* stimulated PO, PPO and PAL activities in cucumber roots [19]. Schneider [20] also recorded high enzyme (PO, PPO, PAL) activities in cucumber and tobacco after 2-3 days of treatment with various biotic and abiotic inducers including salicylic acid. Aqueous leaf extract of neem provided protection to barley against leaf stripe pathogen *Drechslera graminis* and enhanced PAL and TAL (Tyrosine ammonia lyase) activities along with rapid accumulation of fungitoxic phenolic compounds [21]. Induction of some defense related enzymes and phenolics in roots and shoots of two different genotypes of chickpea cultivars which were susceptible (L550) and resistant (ICCV10) to wilt disease treated with salicylic acid, spermine (Spm), SA+Spm and *Fusarium oxysporum* f. sp. *ciceri* was investigated. Higher levels of polyphenol oxidase (PPO), phenylalanine ammonia-lyase (PAL), β-1,3-glucanase (PR-2) and phenolics were observed in roots and shoots of resistant cultivar than that of susceptible cultivar on treatment with elicitors and pathogen [22]. Involvement of LOX in induced resistance was also reported by a number of workers. Bohland et al. [23] observed that LOX activity increased in wheat leaves when treated with either glycopeptide elicitor from germtube of *P. graminis* f. sp. *tritici* or methyl jasmonate. Similarly, arachidonic acid (10^{-8} µM) also activated LOX activity in tuber tissues and enhanced resistance in potato against late blight [24]. Mitchell and Walters [25] reported that systemic resistance of barley to powdery mildew could be induced by potassium phosphate (25 mM). Treatment of first leaves of barley with potassium phosphate led to significant increase in activities of PAL, PO and LOX in second leaves. Activities of PAL and PO increased further when second leaves of phosphate treated plants were inoculated with powdery mildew (*Blumeria graminis* f. sp. *hordei* Marchal).

The utilization of plants owns defence mechanism is the area of major interests in the management of pests and diseases. Result of present investigation clearly indicated that defence activators SA and ALE reduced disease significantly and also induced PO, PPO, LOX and PAL activities in ginger. Although enzyme activity increased in all cases, the rate of activity varied with the nature of enzymes and defence activators. In conclusion, it could be suggested that activation of defence related enzymes are quite likely to govern some of the basic mechanisms of biochemical resistance induced by SA and ALE in ginger plants against rhizome rot disease.

Acknowledgement

Financial assistance received from Department of Biotechnology, Government of India is gratefully acknowledged.

References

1. Hammerschmidt R, KuÄ J (1995) Induced resistance to disease in plants. Kluwer, Dordrecht.

2. Kessmann H, Staub T, Hofmann C, Maetzke T, Herzog J, et al. (1994) Induction of systemic acquired disease resistance in plants by chemicals. Annu Rev Phytopathol 32: 439-459.

3. Sticher L, Mauch-Mani B, Métraux JP (1997) Systemic acquired resistance. Annu Rev Phytopathol 35: 235-270.

4. Dow M, Newman MA, von Roepenack E (2000) The induction and modulation of plant defense responses by bacterial lipopolysaccharides. Annu Rev Phytopathol 38: 241-261.

5. Durrant WE, Dong X (2004) Systemic acquired resistance. Annu Rev Phytopathol 42: 185-209.

6. Abd El-Rahman SSA, Mazen MM, Mohamed HI and Mahmoud NM (2012) Induction of defence related enzymes and phenolic compounds in lupin (Lupinus albus L.) and their effects on host resistance against Fusarium wilt. European Journal of Plant Pathology 134: 105-116

7. Thakker JK, Patel A, Dhandhukia PK (2013) Induction of Defense-Related Enzymes in Banana Plants: Effect of Live and Dead Pathogenic Strain of Fusarium oxysporum f. sp. cubense. ISRN Biotechnology.

8. Galliard T, Chan HWS (1980) Lipoxygenases: The Biochemistry of Plants, Academic Press, New York.

9. Siedow JN (1991) Plant lipoxygenase: structure and function. Annu Rev Plant Physiol Plant Mol Biol 42: 145-188.

10. Andersson RA, Akita M, Pirhonen M, Gammelgard E, Valkonen JPT (2005) Moss-Erwinia pathosystem reveals possible similarities in pathogenesis and pathogen defense in vascular and nonvascular plants. J Gen Plant Pathol 71: 23-28.

11. Oliver JP, Castro A, Gaggero C, Cascon T, Schmelz EA, et al. (2009) Pythium infection activates conserved plant defense responses in mosses. Planta 230: 569-579

12. Winter PS, Bowman CE, Villani PJ, Dolan TE, Hauck NR (2014) Systemic acquired resistance in moss: further evidence for conserved defense mechanisms in plants. PLoS One 9: e101880.

13. Karmakar NC, Ghosh R, Purkayastha RP (2003) Plant defence activators inducing systemic resistance in Zingiber officinale Rosc. against Pythium aphanidermatum (Edson) Fitz. Indian Journal of Biotechnology 2: 591-595

14. Doubrava NS, Ralph AD and KuÄ J (1988) Induction of systemic resistance to anthrachose caused by Colletotrichum lagenarium in cucumber by oxalate and extracts from spinach. Physiol Mol Plant Pathol 33: 69-79.

15. Sadashivam S, Manickam A (1996) Biochemical Method, (2ndedn) New Age International, New Delhi.

16. Vick BA, Zimmerman DC (1976) Lipoxygenase and hydroperoxide lyase in germinating watermelon seedlings. Plant Physiol 57: 780-788.

17. Smith CJ, Milton JM, Williams JM (1995) Induction of phytoalexin synthesis in Medicago sativa (Lucerne) - Verticillium albo-atrum interaction. Handbook of phytoalexin metabolism and action. Marcel Dekker, Inc, New York.

18. Ghosh R, Purkayastha RP (2003) Molecular diagnosis and induced systemic protection against rhizome rot disease of ginger caused by Pythium aphanidermatum. Current Science 85: 1782-1787.

19. Chen C, Bélanger RR, Benhamou N, Paulitz TC (2000) Defense enzymes induced in cucumber roots by treatment with plant growth promoting rhizobacteria (PGPR) and Pythium aphanidermatum. Physiol Mol Plant Pathol 56: 13-23.

20. Schneider S (1994) Differential induction of resistance and enhanced enzyme activities in cucumber and tobacco caused by treatment with various abiotic and biotic inducers. Physiol Mol Plant Pathol 45: 291-304.

21. Paul PK, Sharma PD (2002) Azadirachta indica leaf extract induces resistance in barley against leaf stripe disease. Physiol Mol Plant Pathol 61: 3-13.

22. Raju S, Jaylakshmi SK, Sreeramulu K (2008) Comparative study on the induction of defense related enzymes in two different cultivars of chickpea (Cicer arietinum L) genotypes by salicylic acid, spermine and Fusarium oxysporum f. sp. Ciceri. Australian Journal of Crop Science 2: 121-140.

23. Bohland C, Balkenhohl T, Loers G, Feussner I, Grambow HJ (1997) Differential Induction of Lipoxygenase Isoforms in Wheat upon Treatment with Rust Fungus Elicitor, Chitin Oligosaccharides, Chitosan, and Methyl Jasmonate. Plant Physiol 114: 679-685.

24. Ilinskaya L, Perekhod EA, Chalenko GI, Gerasimova NG, Romanenko EN, et al. (2000) Lipoxygenase activity in plants with induced resistance to disease. Russian Journal of Plant Physiology 47: 449-455

25. Mitchell AF, Walters DR (2004) Potassium phosphate induces systemic protection in barley to powdery mildew infection. Pest Manag. Sci 60: 126-134

8

Distribution and Importance of Maize Grey Leaf Spot *Cercospora zeae-maydis* (Tehon and Daniels) in South and Southwest Ethiopia

Alemu Nega*, Fikre Lemessa and Gezahegn Berecha

Department of Horticulture and Plant Sciences Jimma University P.O. Box 307 Jimma, Ethiopia

Abstract

Maize (*Zea mays* L.) is one of the most important cereal crops used in the human diet in large parts of the world as well as in Ethiopia and its production are limited by diseases such as grey leaf spot caused by a fungus *Cercospora zeae-maydis*. Currently grey leaf spot has become the most important threat to maize production in maize belt areas of Ethiopia causing significant yield loss. The objective of this study was therefore, to assess the distribution and importance of maize grey leaf spot in south and southwest Ethiopia. The field assessments were conducted during the year of 2014 cropping seasons by sampling 110 farmer fields in 11 districts selected from two zones of Oromia region and two zones of Southern Nation, Nationalities and People region (SNNPR). The result revealed that the disease occurs in the entire assessed districts having different agro-ecological zones. Maize grey leaf spot was prevalent in all surveyed farms of south and southwest Ethiopia, 74% of maize fields were infected by grey leaf spot. However, the mean incidence and severity of grey leaf spot on maize was significantly varied from district to district. The highest grey leaf spot incidence (71.2%) and severity (46.2%) was recorded in Boricha district, while Damote Gale had the lowest mean incidence (51.8) and severity (33.5%). Among the surveyed four zones in two regions, the highest incidence was found in Sidama (65.6%) and Illubabore (63.1%) followed by Jimma (62.5%) and Wolaita (57.6%). The highest mean severity of maize grey leaf spot was observed in Sidama (44.5%) followed by Illubabore (43.7%) and Jimma (42.63%) while the lowest severity was recorded in Wolaita (36.4%) zone. The disease was more severe in intermediate/humid areas with intermediate annual rainfall. The current study revealed maize grey leaf spot pressure in maize farms of south and southwest Ethiopia and the need to design an efficient, inexpensive and sustainable management approaches against the pests.

Keywords: *Cercospora zeae-maydis*; Grey leaf spot; Incidence; Prevalence; Severity; Zea mays

Introduction

Maize (*Zea mays* L.) is one of the most essential cereal crops used in the human diet in most parts of the world and it is an imperative feed constituent for domestic animals. Among cereal crops, maize has the highest average yield per hectares and remains third after wheat and rice in total area and production in the world [1]. The estimated area under maize production in the world is about 144 million hectares, and has an annual production of about 700 million tons [2], with about 60% produced in the developed countries, mainly by the United States, produces nearly one half of the total world production. The next largest maize producing countries are China and Brazil [2]. About 70% of the world maize production area is found in developing countries. However, these countries contribute to only 49% of the world's maize production [1]. In sub-Saharan Africa maize is produced in an estimated area of about 26 million hectares with an average yield of 460 million tons. In east Africa, maize occupies about 11 million hectares production area with an average yield of 16 million tons [2].

Maize is one of the most important cereal crops cultivated in Ethiopia ranking second after teff in area coverage and first in total production. The recent post-harvest crop production survey of 2011/12 indicated that total land areas of about 12 million ha are covered by grain crops. Out of the total grain crop areas, 79.34% (about 9 million ha) is under cereals. Of this, maize covered 17% (about 2 million ha) and gave about 6.1 million tons of grain yields [3]. The major maize producing regions of Ethiopia are southern, western and southwestern and in some northern, northwestern and eastern parts where over 90% of the maize produced are used as food among the low income groups [4].

Despite the large area under maize, the national average yield of maize is about 2.95 tons/ha [3]. This is by far below the world's average yield, which is about 5.21 tons/ha [2]. The low productivity of maize is attributed among other, to frequent occurrence of drought [5], declining of soil fertility, poor agronomic practice, limited use of input, lack of credit facilities, poor seed quality [6], diseases [7,8], insect pests and weed [9].

Among these constraints diseases play a major role and a foliar disease particularly grey leaf spot (GLS) is among the important constraints limiting tropical maize production [10]. Grey leaf spot has been reported as one of the major yield-limiting diseases of maize in sub-Saharan Africa [11]. It also poses a serious threat to maize production in central, eastern, southern, and western Africa. Grey leaf spot is apparent on plants as small spots first on lower leaves of plants at tassel initiation. The disease moves upwards and spots change into long characteristics lesions within a month turning plants into a diseased field. The disease is significant since it rapidly destroys foliage when the plant is near at grain maturity. This disease is most rigorous and injurious when comprehensive periods of high relative humidity occur, resulting from slow-drying dews and prolonged late-season fogs [12]. Increased incidence of grey leaf spot in Africa has been associated with cultural practices such as reduced tillage, continuous cultivation of maize, and use of susceptible maize cultivars [13]. Conservation

*Corresponding author: Alemu Nega, Department of Horticulture and Plant Sciences Jimma University P.O. Box 307 Jimma, Ethiopia
E-mail: alemunega531@gmail.com

Distribution and Importance of Maize Grey Leaf Spot Cercospora zeae-maydis (Tehon and Daniels)...

45

tillage leaves infested residue from previous crop on the soil surface that increases initial inoculums of the disease.

Documented yield losses of maize attributed to grey leaf spot vary from 11 to 69% [11], with estimated losses as high as 100% when severe epidemics contributed to loss of photosynthetic area, severe lodging also can adversely affect mechanical harvesting and results in further grain loss due to a reduction in harvestable grain yield [14]. The yield losses caused by the disease were estimated to reach 50% for moderately resistant and 65% for susceptible hybrid maize in South Africa [11].

Grey leaf spot was first reported in Ethiopia in 1997 in the border of west Wellega and Ilubabor zones, of western Ethiopia [15,16]. The survey report of [15] showed increased prevalence of grey leaf spot in the major maize producing regions of Western, Southern and Northwestern parts of Ethiopia. According to the report, grey leaf spot has become the principal maize disease since 1998 in Ethiopia. In Ethiopia, [17] reported that yield losses due to grey leaf spot on resistant, moderately resistant, and susceptible varieties were between 0-14.9%, 13.7-18.3% and 20.8-36.9%, respectively during 2003/2004 cropping seasons in Bako and its nearby areas. The severity of grey leaf spot was high in the warm humid maize belt areas of the country which adversely affecting farmers who live with limited resources. The disease has been observed spreading over the years in the maize belt areas of the country. Despite the observed high grey leaf spot incidence and severity in the farmers' fields, little empirical information is available regarding the distribution and importance of grey leaf spot (GLS) in maize growing areas of south and southwest Ethiopia. Therefore, the objective of this study was to assess the distribution and importance of maize grey leaf spot in four major maize growing zones in south and southwest Ethiopia.

Materials and Methods

Survey area

The survey was conducted in different agro-ecological Zones of Oromia Regional States (Jimma and Illubabore) and SNNPR (Wolaita and Sidama) during the 2014 cropping season (Table 1). The survey route followed major roads to towns and localities in each regional state. A total of 110 farmers' fields in 11 districts of 4 administrative zones within 2 regional states were assessed for the occurrence and intensity of maize grey leaf spot (Figure 1).

Assessment of grey leaf spot prevalence, incidence and severity

The survey programs covered the most important maize growing zones in southern and southwestern Ethiopia with frequent stopping at different intervals depending on the variability of fields in terms of altitude and cropping systems. Size of the survey site and availability and accessibility of maize field was given due consideration in deciding where to stop on survey routes. Disease prevalence was assessed by determining the number of fields where grey leaf spot was recorded in relation to the number of fields sampled in different origins of south and southwest Ethiopia. Disease assessment was made in 30 randomly selected spots in each field. Grey leaf spot incidence was assessed as:

$$\text{Disease incidence} = \frac{\text{Number of diseased plants}}{\text{Total number of plants inspected}} \times 100$$

Disease severity: Grey leaf spot severity was rated according to Maroof [18], using 1-5 scales: 1 = no symptoms; 2 = moderate lesion development below the leaf subtending the ear; 3 = heavy lesion development on and below the leaf subtending the ear with a few

lesions above it; 4 = severe lesion development on all but the uppermost leaves, which may have a few lesions; and 5 = all leaves dead. The survey was conducted at tasseling and physiological maturity stage to assess the average leaf area of maize covered by grey leaf spot symptom for 30 randomly selected diseased plants per field. The numerical rating was converted to percentage severity index (PSI) using the following equation suggested by Wheeler [11]:

$$\text{PSI} = \frac{\text{Sum of all numerical ratings} \times 100}{\text{Total number of rated} \times \text{Maximum disease score on scale}}$$

Data analysis

Data was first checked for various ANOVA assumptions. The mean incidence and severity data were calculated for each district. The field survey data for maize grey leaf spot (incidence and severity) was analyzed using three stage nested design. Mean grey leaf spot incidence and severity of each zone were used to make comparison between the surveyed zones within two regions. The same process was followed to determine grey leaf spot incidence and severity for the different agro-climatic districts within zones. All the statistical analyses were carried out using SASV 9.2 and the Least Significant Difference (LSD) test was used for mean comparison.

Result and Discussion

Distribution and importance of grey leaf spot

Maize grey leaf spot was prevalent in all maize producing districts of south and southwest Ethiopia included in this survey. The disease prevalence of maize gray leaf spot ranged from (54%-98%) (Figure 2). This indicated the disease was the most destructive diseases during the season. The highest prevalence of gray leaf spot (98%) was recorded in Boricha district followed by Urumu, Bedele and Seka chekorsa which had 95%, 82% and 76% gray leaf spot prevalence, respectively, while the lowest prevalence of grey leaf spot (54%) was recorded in Damote Gale district (Table 2). In general, grey leaf spot was (74%) prevalent in all 110 surveyed farms of the different eleven districts of south and southwest Ethiopia in 2014. The widely distribution and the high prevalence of grey leaf spot during the survey season in all assessed district could be attributed to the favorable environmental conditions coupled with cultivation of susceptible maize cultivars worsened the problem to maximum. A grey leaf spot epidemic is favored by high rainfall and relative humidity, warm temperatures, and the presence of large amounts of inoculums. The most possible reason could be due to the inconsistency in environmental condition, production systems and practices and the variety grown. As a result, producers can have an impact on the prevalence and intensity of this disease by deciding from a number of management strategies including host plant resistance, cultural practices (adjusting planting date, crop rotation, tillage practices) and careful use of foliar-applied fungicides [19]. The previous study of Dagne et al. [15] reported that increased prevalence of gray leaf spot in the major maize producing regions of Western, Southern and Northwestern parts of Ethiopia. According to the report, GLS has become the principal maize disease since 1998 in Ethiopia.

Sr.No.	Regional States	No of Zones	No of District	No of Farms surveyed	Total
I	Oromia	Jimma	2	10	20
		Illubabore	3	10	30
II	SNNPR	Wolaita	3	10	30
		Sidama	3	10	30
Sub Total	2	4	11	40	110

Table 1: Description of surveyed areas and number of farms surveyed in 2014.

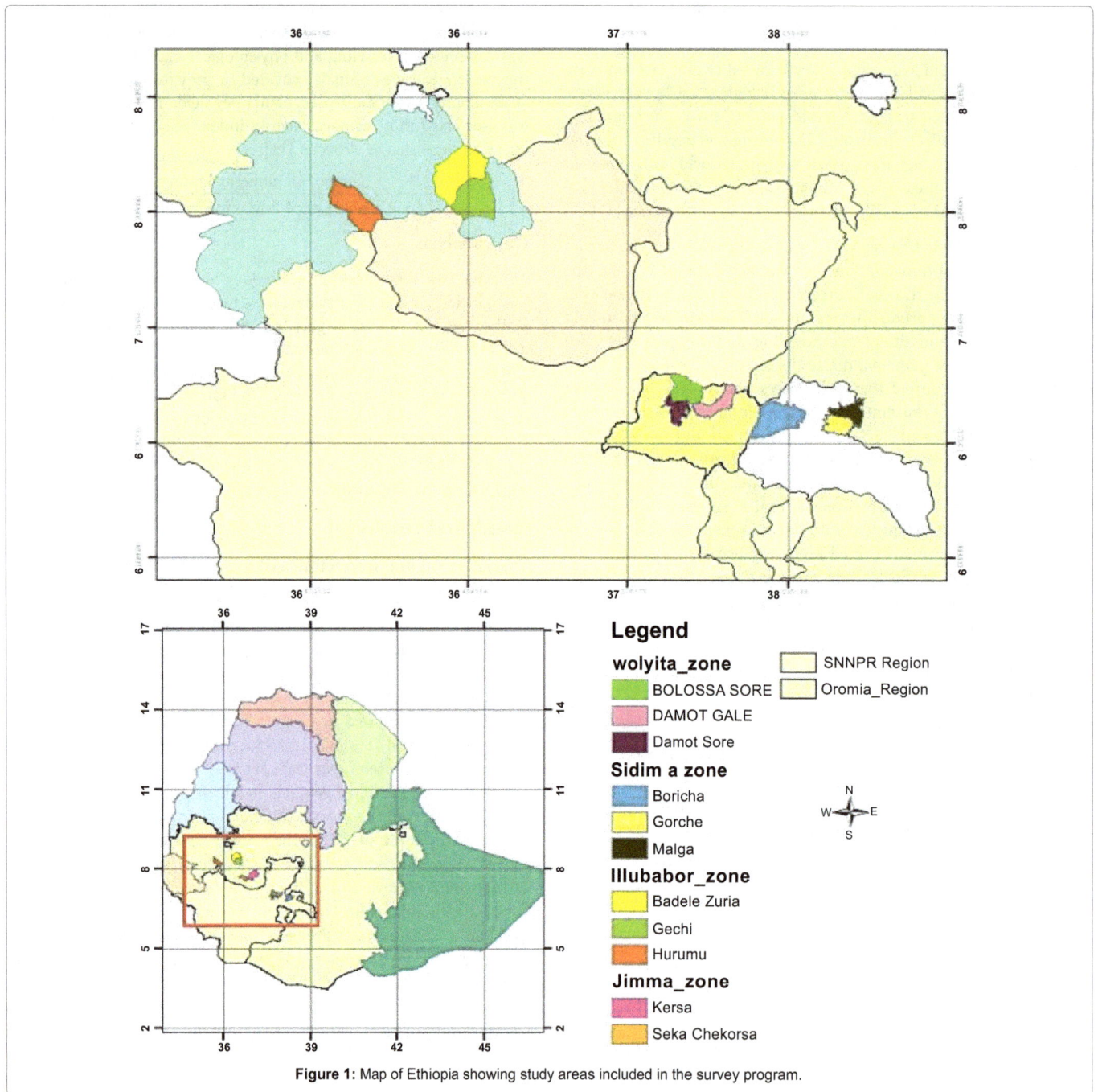

Figure 1: Map of Ethiopia showing study areas included in the survey program.

Incidence and Severity of grey leaf spot across zone within regions

A total of four zones (Sidama, Wolaita, Jimma and Illubabore) were included in the survey program and maize grey leaf spot was prevalent in the entire surveyed zone/areas but with varying intensity. The mean incidence and severity of maize grey leaf spot was significantly varied from zone to zone (Tables 3 and 4). The mean incidence of maize grey leaf spot in the different zones ranged from (57.6-65.6%) (Table 3). There was a statistically significant (p<0.001) difference among zones in terms of maize grey leaf spot incidence within two regions. We found higher incidence of maize grey leaf spot in Sidama (65.6%) and Illubabore (63.1%) followed by Jimma (62.5%) and Wolaita (57.6%). The widely distribution and the high incidence of grey leaf

spot in Sidama and Illubabore followed by Jimma and Wolaita could be attributed to intermediate rainfall occurrence, variation in altitude and germplasm planted by farmers and lack of crop rotation in the zone that are favorable for stimulating the growth of the fungus [13]. Showed that increased incidence of grey leaf spot in Africa has been associated with cultural practices such as reduced tillage, continuous cultivation of maize, and use of susceptible maize cultivars. Conservation tillage leaves infested residue from previous crop on the soil surface that increases initial inoculums of the disease. Moreover, increasing crop residue on the soil proportionally increases the amount of primary inoculums. Although Anderson [12] reported that the beneficial effects of stubble tillage on soil and water conservation are widely recognized, these benefits are frequently offset by the increased crop damage by

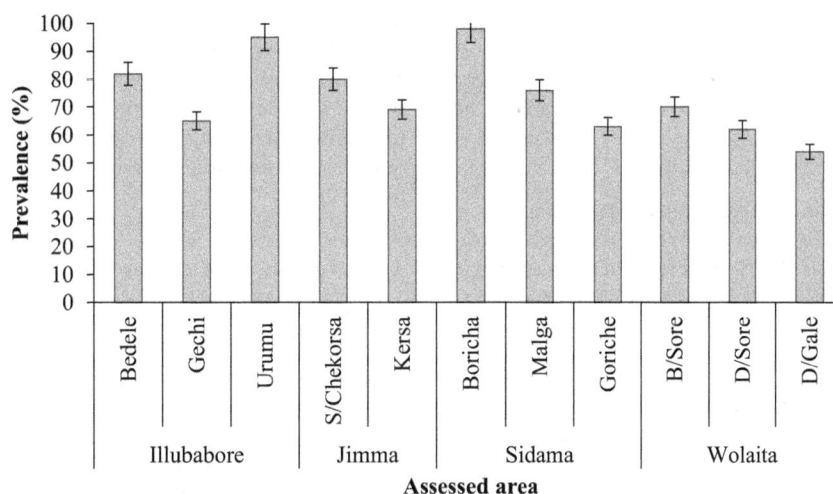

Figure 2: Percentage prevalence of grey leaf spot of maize in south and southwest Ethiopia in 2014.

Region	Zone	District	Location	Altitude (m.a.s.l.)	Annual rainfall (mm)	Mean Temp. (°C)	
						Max	**Min**
SNNPR	Sidama	Boricha	06°56'30.8" N and 03°825'07.4" E	1866	1317	27.9	14.5
		Goriche	06°51'00.3" N and 038°18'04.6" E	1950	1319	28	14.3
		Malga	7°4'N and 38° 31'E	1073	1700	26.3	14.1
	Wolaita	D/Gale	077°03'06" N and 036°16'30"E	1612-2964	900-1400	24	12
		B/Sore	7°04'N-37°42'E	1830	101.3-158.1	26.2	11.4
		D/Sore	37°20'80''N	1900-2010	850-1200	25	20
Oromia	Jimma	Kersa	7°35'- 8° 00'N and 36°46'- 37° 14'E	1600-2400	1587	15	10
		Seka Chekorsa	7°17'- 7° 44'N and 17'- 36°52' E	1580-2560	1800-2300	25	15
	Illubabore	Hurumu	8°33'N 35°67' E	1780	1015	27	18
		Bedele	8°46'-8°56' N and 36°01'-36°15' E	1500-2300	144-173	21.9	18.5
		Gechi	8° 8'58" N - 8° 24' 30" N and 36°17' 30" E - 36°31'45"E	1300 -2270	2214-2300	18	9

Table 2: Location and the climatic characteristics of the study area in 2014.

Level of Zone		Number of farms surveyed	Disease Incidence	
			Mean	**SD**
Sidama	SNNPR	30	65.60[a]	±5.30
Wolaita	SNNPR	30	57.60[c]	±6.28
Jimma	Oromia	20	62.50[b]	±3.44
Illubabore	Oromia	30	63.10[b]	±3.63
LSD(0.05)			2.06	
CV (%)			6.12	

Mean within a column followed by the same letters are not significantly different according to LSD at 5% probability level. SD = Standard Deviation, LSD = Least Significant Difference, CV = Coefficient of Variation.

Table 3: Incidence of maize grey leaf spot in different Zones of South and Southwest Ethiopia in 2014.

Level of Zone		Number of farms surveyed	Disease Severity	
			Mean	**SD**
Sidama	SNNPR	30	44.50[a]	±2.05
Wolaita	SNNPR	30	36.41[c]	±4.38
Jimma	Oromia	20	42.63[b]	±3.20
Illubabore	Oromia	30	43.70[ab]	±2.81
LSD(0.05)			1.60	
CV (%)			7.08	

Mean within a column followed by the same letters are not significantly different according to LSD at 5% probability level. SD = Standard Deviation, LSD = Least Significant Difference, CV = Coefficient of Variation.

Table 4: Severity of maize grey leaf spot in different Zones of South and Southwest Ethiopia in 2014.

fungal pathogens that survived in the previous season's debris. This indicated that variations in grey leaf spot severity within a zone may be due to differences in weather and maize variety during the survey year. Reportedly, in the U.S.A. (Iowa) the outbreak of grey leaf spot based on empirical observations, moderate to high temperatures and prolonged periods of high relative humidity generally are accepted as being favorable for the development of this disease [11].

Statistically there was also significant (p<0.001) difference in severity of maize grey leaf spot in the surveyed zones of maize belt areas of South and Southwest Ethiopia. The mean severity of grey leaf spot in the different zone ranged from (35.4 - 44.5%) (Table 5). Severity of maize grey leaf spot could also be higher in the Sidama (44.5%) and Illubabore (43.7%) followed by Jimma (42.63%) whereas the lowest severity was recorded in Wolaita (36.41%), south and southwest Ethiopia in 2014. Our result revealed that high variation in disease incidence and severity were recorded in Sidama and Illubabore followed by Jimma and Wolaita (Table 4). This suggests that variation in existing weather condition, cultural practices, lack of crop rotation and difference in genetic background of maize genotype planted by farmers in the zone encourage the development of the fungus. Therefore, farmers can have a contact on the variation in incidence and severity of grey leaf spot by choosing from a number of management techniques including cultural practices, which include crop rotations, soil cultivation and removal of crop residues, resistant host genotypes, use of short-season maize varieties, regulating planting time and use of fungicides [14,20].

Incidence and severity of grey leaf spot across districts within zones

The survey result on the distribution and importance of grey leaf spot in maize growing districts of separate zone of South and Southwest Ethiopia are presented in Table 5 and 6. Maize grey leaf spot was prevalent in the all assessed areas of south and southwest Ethiopia, both incidence and severity of the disease varied significantly across the districts with in four zones of different agro-ecological condition (Tables 5 and 6). The mean incidence of maize grey leaf spot in the different districts ranged from (51.8%-71.2%). Statistically there was highly significant (p<0.001) difference among districts in terms of maize grey leaf spot incidence and severity (Tables 5 and 6). The highest grey leaf spot incidence (71.2%) was recorded in Boricha district followed by Urumu, Gechi and Seka chekorsa which had 65.7%, 64.1% and 62.8% grey leaf spot incidence, respectively, while the lowest (51.8%) was recorded in Damote Gale district (Table 6). The mean severity values of maize grey leaf spot in the different districts of farmer field ranged from (33.5 - 46.2%) (Table 6). The severity of maize grey leaf spot was highest in Boricha (46.2%) followed by Urumu, Kersa and Goriche (45.13%, 44.06% and 43.2%), respectively. Although the lowest disease severity (33.5%) was recorded in Damote Gale district (Table 7). The seriousness of grey leaf spot was elucidated by the damage (severity) demonstrated on the plant. Generally, the overall severity of the disease was 41.64% (Table 6). As a result, farmers are frustrated by the nature and epidemic of the grey leaf spot disease during the season. In view of the fact that, all of the surveyed areas, maize grey leaf spot was recorded in all districts, thus grey leaf spot was present in all areas of the assessed districts. Earlier results of various surveys conducted in most maize growing regions of Ethiopia indicated that the disease has a wide distribution and significant impact on maize yield reduction on both local and improved varieties [21]. According to field visit report [22] of all released hybrid maize, only BH-660 and PHB-30H83 were found relatively tolerant to grey leaf spot. The disease has also been observed in maize seed multiplication sites at East Shewa and Sidama in most released maize varieties. Reportedly, in the U.S.A. (Iowa) the

epidemic of grey leaf spot is severe under monoculture maize with no rotation practices and minimum tillage practices [23]. It was apparent that maize is grown over a large area of south and southwest Ethiopia. The current survey in different parts of south and southwest Ethiopia showed great variations in grey leaf spot ranging from mild to severe infections. All of the surveyed districts had moderate to severe grey leaf spot infection indicating the potential of the disease in hindering maize productivity.

Level of District	Level of Zones	No of farms surveyed	Disease Incidence	
			Mean	SD
Bedele	Illubabore	10	62.10cd	±3.47
Gechi	Illubabore	10	64.10bc	±3.07
Urumu	Illubabore	10	65.70b	±3.49
S/Chekorsa	Jimma	10	62.80bc	±3.79
Kersa	Jimma	10	62.10cd	±3.21
Boricha	Sidama	10	71.20a	±4.16
Malga	Sidama	10	61.40cd	±2.55
Goriche	Sidama	10	61.50cd	±2.67
B/Sore	Wolaita	10	59.40d	±4.78
D/Sore	Wolaita	10	61.50cd	±6.46
D/Gale	Wolaita	10	51.80e	±2.17
LSD(0.05)			3.37	
CV (%)			6.12	

Mean in a column followed by the same letters are not significantly different according to LSD at 5% probability level. SD = Standard Deviation, LSD = Least Significant Difference, CV = Coefficient of Variation.

Table 5: Incidence of maize grey leaf spot in different district of South and Southwest Ethiopia in 2014.

Level of District	Level of Zones	No of farms surveyed	Disease Severity	
			Mean	SD
Bedele	Illubabore	10	42.90b	±2.88
Gechi	Illubabore	10	43.06b	±3.02
Urumu	Illubabore	10	45.13ab	±2.13
S/Chekorsa	Jimma	10	42.56b	±3.32
Kersa	Jimma	10	44.06ab	±2.04
Boricha	Sidama	10	46.20a	±1.69
Malga	Sidama	10	42.70b	±3.36
Goriche	Sidama	10	42.20b	±1.12
B/Sore	Wolaita	10	39.06c	±4.35
D/Sore	Wolaita	10	36.70c	±3.57
D/Gale	Wolaita	10	33.50d	±3.55
LSD(0.05)			2.64	
CV (%)			7.08	

Mean within a column followed by the same letters are not significantly different according to LSD at 5% probability level. SD = Standard Deviation, LSD = Least Significant Difference, CV = Coefficient of Variation.

Table 6: Severity of maize grey leaf spot in different district of South and Southwest Ethiopia in 2014.

Variables	Incidence	Severity	Altitude	Rainfall
Incidence	1.00			
Severity	0.76**	1.00		
Altitude	0.02	-0.06	1.00	
Rainfall	0.22	0.41	-0.10	1.00
Temperature	0.17	-0.21	-0.23	-0.39

Altitude, Rainfall and Temperature represent the actual weather of each location. Significant level: NS = Non-significant, * p ≤ 0.1, ** p ≤0.01 and *** p ≤0.001

Table 7: Pearson correlation between weather condition and grey leaf spot incidence and severity.

Relationship between weather variables, grey leaf spot incidence and severity

The relationship between disease development and weather variables was established using correlation analysis to look for the effect of individual variable on disease development. Correlations between weather variables and grey leaf spot (GLS) incidence and severity were determined in Table 7. The result revealed that grey leaf spot incidence and severity had negative and insignificant correlation with altitude (r = 0.02 and r = -0.06, respectively). There was strong positive correlation between rain fall and mean grey leaf spot incidence and severity (r = 0.22 and r = 0.41, respectively). Rainfall during the actual maize production season of each surveyed area of 2014 correlated also strongly and insignificantly with both grey leaf spot incidence and severity. On the other hand temperature had insignificant correlation with both disease incidence and severity. This indicated the strong influence of weather condition particularly that of rain rainfall on grey leaf spot development. A strong positive and highly significant correlation (r = 0.76) was also found between disease incidence and severity.

It was evident that maize is grown over a large area in many parts of the country, particularly in southern and southwestern Ethiopia. In a current survey conducted in southern and southwestern parts of Ethiopia showed great variation in grey leaf spot intensity ranging from mild to severe infections. In this study, all of the surveyed areas with moderate to severe grey leaf spot infection, which usually have, humid or warm temperature and high rainfall. Information on the relative humidity of the assessed areas was not on hand for all of the assessed areas but it is understandable that comparatively warm temperature and high rainfall could give rise to high relative humidity. Reportedly, warm temperature, well distributed rainfall and high relative humidity are weather conditions favoring this disease development [11].

Cropping methods such as mono or inter-cropping and use of cultivar mixture are also recognized to supply to disease pressure in positive or negative ways [19]. It was observable from the present study that farmers in surveyed maize belt areas of Ethiopia do not apply any specific management practice to contest grey leaf spot at least intentionally. Furthermore, crop rotations are another means of reducing initial inoculums and even a single year of crop rotation will significantly reduce the initial level of *C. zeae-maydis* inoculums. However, it normally takes several years of rotation to reduce the inoculums to levels achieved by deep ploughing [24]. Most of the surveyed fields were continuously planted to maize while some were rotated with bean/sorghum/teff and this might have contributed to the high inoculums buildup of *Cercospora zeae-maydis* especially in areas with conducive environmental conditions.

Conclusion

Our results revealed that gray leaf spot of maize was widely distributed in all surveyed areas. Gray leaf spot incidence and severity varies from location to location within two regional states of South and Southwest Ethiopia. Maize grey leaf spot was prevalent in all maize producing zones and districts of south and southwest Ethiopia included in this survey. Grey leaf spot incidence and severity showed a noticeable variation between zones. Sidama zone had the highest grey leaf spot level and it was followed by the Illubabore, Jimma and Wolaita zones of south and southwest Ethiopia. Variations in grey leaf spot incidence and severity within a zones and districts may be due to differences in weather and maize variety planted by farmers during the survey year. The current study revealed that maize farms surveyed in south and southwest Ethiopia are under maize grey leaf spot pressure and consequently, management should be required to control the disease in an effective, affordable and sustainable approach. Furthermore, in order to get full depiction of the distribution and importance of maize grey leaf spot disease and to design appropriate control technique, it is worthwhile to conduct similar assessments in different maize belt areas of the country.

Acknowledgment

We would like to thank Jimma University College of Agriculture and veterinary medicine for financial support for this work.

References

1. FAOSTAT (2012) Statistical Database of the Food and Agriculture of the United Nations.

2. FAO (Food and Agriculture organization of the United Nations) (2011) FAOSTAT online database.

3. CSA (Central Statistical Agency) (2012) Agricultural Sample survey: report on area and production of major crops (private peasant holdings, Meher season). Statistical Bulletin, Addis Abeba, Ethiopia.

4. Ferdu Z, Demissew K, Birhene A (2001) Maize insect pests of maize and their management in Ethiopia: a review, In: Mandefro N., Tanner D. and Twumass-Afriyie S. eds. Enhancing the contribution of maize to food security in Ethiopia. EARO and CIMMYT, AddisAbeba, Ethiopia pp: 97-105

5. Gezahegn B, Dagne W, Lealem T, Deseta G (2012) Maize improvement for low moisture stress areas of Ethiopia: Achievements and progress in the last decade. In: Worku, M., Twumasi-Afriyie, S., Wolde, L., Tadesse, B., Demisie, G., Bogale, G., Wegary, D. and Prasanna, B.M. (Eds.). Meeting the Chalenges of Global Climate Change and Food Security through Innovative Maize Research. Proceedings of the 3rd National Maize Workshop of Ethiopia. 18-20 April 2011, Addis Ababa, Ethiopia pp: 193-201.

6. CIMMYT (International Maize and Wheat Improvement Center) (2004) Second Semi-Annual Progress Report for the Quality Protein Maize Development Project for the Horn and East Africa (XP 31519). July 1- December 31, 2003.

7. Temesgen D, Wondimu F, Kasahun Z, Wogayehu W, Takele N, et al. (2012) Weed management research on maize in Ethiopia. In: Worku, M., Twumasi-Afriyie, S., Wolde, L., Tadesse, B., Demisie, G., Bogale, G., Wegary, D. and Prasanna, B.M. (Eds.). Meeting the Chalenges of Global Climate Change and Food Security through Innovative Maize Research. Proceedings of the 3rd National Maize Workshop of Ethiopia. 18-20 April 2011, Addis Ababa, Ethiopia pp: 128-133.

8. Ward JMJ, Laing MD, Cairns ALP (1997) Management practices to reduce grey leaf Spot of maize. Journal of Crop Science 37: 1257-1262.

9. Tewabech T, Dagne W, Girma D, Meseret N, Solomon A, et al. (2012) Maize pathology research in Ethiopia in the 2000s: A review. In: Worku, M., Twumasi-Afriyie, S., Wolde, L., Tadesse, B., Demisie, G., Bogale, G., Wegary, D. and Prasanna, B.M. (Eds.). Meeting the Challenges of Global Climate Change and Food Security through Innovative Maize 23. Research. Proceedings of the 3rd National Maize Workshop of Ethiopia. 18-20 April 2011, Addis Ababa, Ethiopia. pp: 193-202.

10. Simmonds NW, Smartt J (1999) Principles of Crop Improvement. Blackwell Science, Oxford, UK.

11. Wheeler BEJ (1969) An Introduction to Plant Diseases, John Wiley and Sons Limited, London p: 301.

12. Anderson BM (1995) Grey leaf spot resistance. Proceedings of the 31st Annual Conference of AL Illinois Corn Breeders School, University of Illinois, 6-7 March, 1995, USA.

13. Gevers HO, Lake JK (1994) GLS1 a major gene for resistance to grey leaf spot in maize. South African Journal of Science 90: 377-379.

14. Patricia C (2000) Studies on *Cercospora zeae-maydis*, the cause of grey leaf spot of maize in Kwazulu-natal. PhD thesis. University of Natal Pietermaritzburg Republic of South Africa.

15. Dagne W, Fekede A, Legesse W, Gemechu K (2001) Grey leaf spot disease: A potential threat to maize in Ethiopia. In: Proceedings of the Ninth Annual Crop Science Conference of the Crop Science Society of Ethiopia (CSSE), 22- 23 June 1999. Sebil 9: 147-157.

16. Aschalew S, Fekede A, Kedir W (2012) Influence of cultural practices on the

development of grey leaf spot (GLS) on maize at bako, western Ethiopia. Plant Pathology journal 01: 19-26.

17. Dagne W, Demisew A, Girma D (2004) Assessments of losses in yield and yield components of maize varieties due to grey leaf spot. Pest Management Journal of Ethiopia 8: 59-69.

18. Maroof S, Van MA, Scoyoc SW, Yu YG (1993) Grey leaf spot disease of maize: Rating methodology and Inbred line evaluation. Plant Disease 77: 583-587.

19. Agrios GN (2005) Plant Pathology. 5th Edition, Academic Press, New York, USA p: 952.

20. Ward JMJ, Stromberg EL, Nowell DC, Nutter FWJ (1999) Grey Leaf Spot: A disease of global importance in maize production. Plant Disease 83: 884-895.

21. Tadesse A (2008) Increasing Crop Production through Improved Plant Protection -Volume I. Plant Protection Society of Ethiopia (PPSE), 19-22 December 2006. Addis Ababa, Ethiopia. PPSE and EIAR, Addis Ababa, Ethiopia p: 598.

22. Tefferi A (1999) Survey of maize diseases in western and northwestern Ethiopia. Pp. 121-124. In: Maize Production Technology for the Future: Challenges and Opportunities. Proceedings of the 6th Eastern and Southern Africa Regional Maize Conference. CIMMYT/ EARO.

23. Perkins JM, Smith DR, Kinsey JG, Dowden DL (1995) Prevalence and control of grey leaf spot. 31st Annual Conference of the Illinosis Corn Breeders' School. University of Illinois, USA.

24. CSA (Central Statistical Agency) (2011) Report on area and production of crops: Agricultural sample survey on private peasant holdings of 2010/2011 Meher season. Central Statistic Authority, Addis Ababa, Ethiopia.

9

Evaluation of Rice Germplasm against Bacterial Leaf Streak Disease Reveals Sources of Resistance in African Varieties

Issa Wonni[1,2]*, Gustave Djedatin[3], Léonard Ouédraogo[1,2] and Valérie Verdier[3]

[1]Institut de l'Environnement et de Recherches Agricoles (INERA), 01 BP 910 Bobo Dioulasso, Burkina Faso
[2]Laboratoire Mixte Internationnal, observatoire des agents phytopathogènes en Afrique de l'Ouest, Biosécurité et Biodiversité (LMI Patho-Bios), 01 BP 910 Bobo Dioulasso, Burkina Faso
[3]Institut de Recherche pour le Développement, UMR IPME, IRD-CIRAD-UM2, 911 Avenue Agropolis BP 64501, 34394 Montpellier Cedex 5, France

Abstract

Bacterial leaf streak caused by *Xanthomonas oryzae* pv. *oryzicola* (*Xoc*) is a rice disease emerging in West Africa. Its emergence is correlated with the recent expansion of rice cultivation and the introduction of new rice varieties. Our goal is to identify resistance sources to control BLS in rice. We evaluated six *Oryza sativa* and two *Oryza glaberrima* accessions for resistance to bacterial leaf streak under greenhouse conditions. Three week-old plants were inoculated with different *Xoc* strains originated from Mali and the Philippines. Two *Oryza sativa* accessions (FKR14 and ITA306) show a high level of resistance to African *Xoc* while are susceptible to the Philippines one. The others accessions tested are susceptible to all *Xoc* strains tested. We identify new resistance sources to *X. oryzae* pv. *oryzicola* that could be used by breeders, thus improving yield of rice crops in West Africa.

Keywords: *Xanthomonas oryzae* pv. Oryzicola; Resistance; Rice; Africa

Introduction

Bacterial leaf streak (BLS) is an important disease of rice and is caused by *Xanthomonas oryzae* pv. *oryzicola* (*Xoc*). *Xoc* is occuring in the tropical and subtropical areas of Asia, and Australia [1]. In Africa the disease was reported in Madagascar, Senegal and Nigeria in the 1980's [2]. More recently BLS was observed in Mali and Burkina Faso [3,4]. Recently increase of BLS disease was observed in Asia and Africa likely due to the planting of susceptible varieties [3-5]. The increase of BLS disease may also be related to climate changes occuring in Sub-Saharan Africa [6].

BLS develops in the field at any growth stage of rice. Initial symptoms are water-soaked, interveinal streaks along the leaf. *Xoc* is an intercellular pathogen that enters plants through wounds or by invading open stomata [1]. *Xoc* multiplies in the substomatal chamber and colonizes the apoplast of the mesophyll cells [7,8]. *Xoc* oozes from natural openings in strands or strings on the leaf surface and exudates can spread the disease directly from plant to plant by contact, or indirectly via irrigation water and by windblown rain [9]. *Xoc* is a seedborne and a seed-transmitted pathogen [10]. Yield losses due to this disease depend on the rice variety cultivated and climatic conditions but typically range from 10-20% [1]. Effective quarantine and deployment of resistant germplasm are key factors to control BLS disease. Accurate detection of *Xoc* is critical for diagnostic and regulatory purposes. A multiplex PCR assay was developed to simultaneously detect and distinguish the different pathovars of *X. oryzae* infecting rice [11]. The multiplex is currently used to identify *Xoc* strains in West Africa [4,12].

Planting of resistant cultivars is the most effective method for controlling BLS. However, most rice germplasm cultivated in Asia is susceptible to BLS. Raymundo et al. [13] reported that the African variety Morobereckan (upland japonica) is one of the most resistant to Asian *Xoc* strains. The wild rice *O. meyeriana* and *O. officinalis* showed a high level of resistance to Chinese *Xoc* strains [14].

Single rice resistant genes have not been found to control BLS resistance and no strategy has been pursued for controlling the disease in Africa so far. *Rxo1*, a gene from maize, confers resistance to *Xoc* strains in transgenic rice when the corresponding effector gene *avrRxo1* (also named *xopAJ*) is present in the pathogen [15]. *avrRxo1* is present in all the Philippines *Xoc* strains [10] while absent in most African *Xoc* strains isolated in 2003 and 2009 [3,4]. More recently it has been shown that a larger number of *Xoc* strains isolated in Mali and Burkina Faso harbor a functional *avrRxo1* [12]; Dao et al. personal communication). According to Han et al. [16], *AvrRxo1*-ORF1 is adjacent to *AvrRxo1*-ORF2 gene, which was predicted to encode a molecular chaperone of *AvrRxo1*-ORF1. They found that *AvrRxo1*-ORF1 contributes to *X. oryzae* proliferation on rice plants, and strongly suppresses bacterial growth in the absence of *AvrRxo1*-ORF2.

While *Rxo1* may be a useful gene for combatting BLS in Asia, it is not clear yet if that will be the case in West Africa. Also the genetic diversity observed among *Xoc* strains require to identify an efficient breeding strategy for BLS in Africa [3]. Comparative mapping of BLS resistance to Asian *Xoc* strains has led to the identification of Quantitative Trait Loci (QTL) [17]. Recently a recessive *R* gene, *bls1*, conferring resistance on *Xoc* was localized to chromosome 6 from a rice line DP3 derived from *Oryza rufipogon* [18].

Given the severity and extent of BLS epidemics in recent years in West Africa, identifying resistance against *Xoc* is an important goal for breeding programs. The objective of this study was to evaluate rice varieties for resistance to *Xoc* strains.

*Corresponding author: Issa Wonni, Institut de l'Environnement et de Recherches Agricoles (INERA), 01 BP 910 Bobo Dioulasso, Burkina Faso
E-mail: wonniissa@yahoo.fr

Materials and Methods

Germplasm evaluated

Six *O. sativa. susbp. indica* accessions Curinga, FKR14, ITA306, PaDckono, TN1, and IR64; two accessions of *O. glaberrima* TOG6767 and TOG5672 and the transgenic line Kitaake-*Rxo1* were evaluated for resistance to *Xoc* strains (Table 1). Most of these accessions are grown in West Africa and possess good agronomic characteristics such as drought tolerance and resistance to other diseases. TN1 is known to be susceptible to BLS [19] and was used in this study as a susceptible control. IR64 and ITA306 were identified as highly resistant to African *Xanthomonas oryzae* pv. *oryzae*, the causal agent of Bacterial Blight (BLB) while TOG6767 and TOG5672 are moderately resistant [20]. FKR14 is grown in main rice areas in Burkina Faso. Originated from India, FKR14 was introduced in Burkina Faso in 1976 [21]. It is a highly productive, plastic and suitable variety for upland and irrigated rice growing regions. It is sensitive to iron toxicity and susceptible to Rice Yellow Mottle Virus and Bacterial Blight diseases [21,5]. All plants were grown in the greenhouse under controlled conditions at 28°C, 12h/12 day/night with 80% humidity in IRD Montpellier.

Bacterial strains and plant inoculation

Three *Xoc* strains were used in this study (Table 2). *Xoc* MAI3 (CFBP7326) and MAI10 (CFBP7331) were isolated in Mali in 2003 [3]. Both strains exhibit different virulence level on the rice cultivar Nipponbare and possess different Transcription Activator like (TAL) genes [3,12]. *Xoc* BLS256 originated from the Philippines was isolated in 1985. Except *xopAJ* and *xopW* genes, African *Xoc* strains share the same type III effectors genes with *Xoc* strain BLS256 [12]. Additional 50 *Xoc* strains collected in Mali and Burkina Faso in 2009 representative of the genetic diversity of *Xoc* in West Africa [12] were tested on the FKR14 variety. All strains were stored in 15% glycerol at-80°C. *X. oryzae* strains were grown on PSA medium (peptone 10 g, sucrose 10 g, glutamic acid 1 g, agar 8 g, pH 7.0) for 24 h at 30°C, then resuspended in sterile water at 0,2 OD_{600} (approximately 10^8 cfu ml^{-1}).

Accessions	Subspecies	IRGC accession number-ID	R gene to BB or BLS	Origin
Curinga	*indica*	-	ND	Brazil
FKR14	*indica*	-	ND	Burkina Faso
IR64	*indica*	66970	ND	IRRI
ITA306	*indica*	-	ND	Nigeria
PaDcKono	*indica*	-	ND	Sierra Leone
TN1	*indica*	105	*Xa14*	Taiwan
TOG 6767	*glaberrima*	-	ND	Liberia
TOG5672	*glaberrima*	-	ND	Nigeria
Kitaake *Rxo1*	*Transgenic line*	-	*Rxo1*	IRRI

Table 1: Accessions used in this study.

Strains	MAI3	MAI10	BLS256
Origin	Mali	Mali	Philippines
virulence classe	5	na	5
[a]**xopAJ**	+	-	+
[b]**xopW**	+	>	+
Reference	Wonni et al. [20]	Wonni et al. [20]	Wonni et al. [20]

[a]All the *Xoc* strains that are positive for xopAJ induced a hypersensitive response 72 hours after inoculation while the strains that are negative for xopAJ induce watersoaking lesions on Kitaake-*Rxo1*.
[b]Presence or absence of xopW; > indicates that a DNA arrangement (insertion element) was identified within the *xopW* gene.

Table 2: Strains used in this study to screen cultivars for resistance to *Xoc.*

Rice leaves from 3-week-old plants were inoculated by leaf infiltration as described previously [22]. Leaf reactions were observed 72 hr after inoculation (hai), and lesions were measured 12 days after inoculation (dai). At least six infiltrations were done per leaf with two leaves per plant and two to three plants per strain. A scale was established for BLS disease according to the size of the lesion length (LL) induced by *Xoc* strains: Resistant (R), 0<LL ≤ 1 mm; Moderately Resistant (MR), 1<LL ≤ 10 mm; Moderately Susceptible (MS), 10<LL ≤ 30 mm; Susceptible (S), LL>30 mm. The entire experiment was repeated three times.

Colonization of leaf tissue by *Xoc*

Four rice varieties TOG5672, TOG6767, FKR14 and TN1 were selected and further inoculated as described above. Multiplication of *Xoc* strains was measured *in planta* at three time points (0, 7 and 15 dai). Each leaf was cut into 5 cm section below and above the leaf-infiltrated area. The leaf pieces were ground in 1 ml of sterile water. Bacterial numbers were assessed in serial dilutions that were spread onto PSA agar plates. The plates were incubated at 28°C until single colonies could be counted. The experiment was repeated four times.

Results

Xoc virulence vary depending on rice variety

To determine whether *Xoc* strains vary in their virulence, eight rice varieties were leaf infiltrated with *Xoc* strains BLS256, MAI3 and MAI10. *Xoc* BLS256 caused leaf streak symptoms on all varieties (unless KitaakeRxo1) with lesion length varying from 7,6 to 59,5 mm depending the variety (Table 2). Lesion length induced by MAI3 and MAI10 vary from 0 to 30,8 mm.

The reaction induced by *Xoc* strains varied according to the variety. For example, on varieties TN1, ITA306, *Xoc* strain BLS256 caused large lesions, while on others (TOG5672, TOG6767) lesions were small (Figure 1). Both strains MAI3 and MAI10 induced very similar lesions on all the varieties tested but one.

While BLS256 induces small and large lesions on FKR14 and ITA306 respectively, MAI3 and MAI10 induced a resistant reaction. On FKR14, a hypersensitive reaction (HR) was observed 48 h to 72 h post inoculation with strains MAI3 and MAI10 (Figure 2). FKR14 was then tested with 50 African *Xoc* with an HR consistently observed (data not shown). Out of eight varieties, two (FRK14 and ITA306) showed a strong resistance response to African *Xoc* strains. The HR response observed with FKR14, suggests the resistance to be mediated by African *Xoc* effectors. It is noted that MAI3 and BLS256 harboring *xopAJ* gene induce HR on Kitaake-*Rxo1*. ITA306 is susceptible when challenged with strain BLS256. On the opposite, ITA306 is immune to African *Xoc* strains (Table 3). Variety TN1 is highly susceptible to all *Xoc* strains tested.

Resistance to BLS is associated with low multiplication of *Xoc* strains

Xoc strain BLS256 caused substantial lesions on rice varieties FKR14 and TN1 when compared to MAI3 and MAI10. To determine the effect of plant on bacterial population growth, bacterial numbers were determined in infiltrated leaves.

There was no significant difference in population growth between MAI3 and MAI10 on all varieties tested except Kitaake-*Rxo1* (Figure 3). In TOG6767 and TOG5672, no difference in population growth

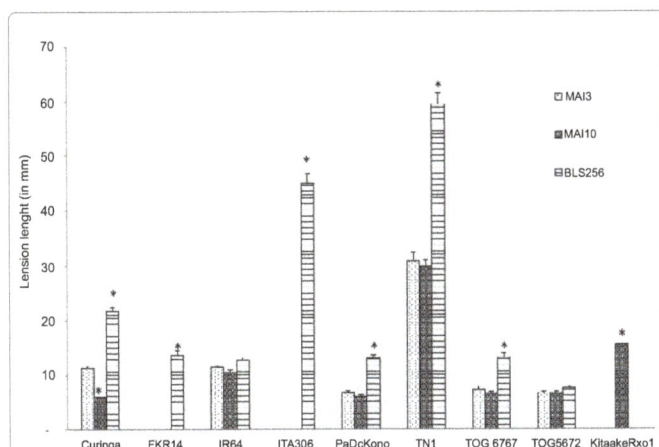

Figure 1: Lengths of lesions caused by *X. oryzae* pv. *oryzicola* strains BLS256, MAI10 and MAI3 on six rice accessions of *O. sativa* sp. *Indica*, two accessions of *O. glaberrima* (TOG6767, TOG5672) and trangnic indica line Kitaake-*Rxo1*. Lesions were measured 15 days after infiltration. An asterisk denotes a significant difference between BLS256 and African *Xoc* (*, P<0.05). The experiment was repeated three times with similar results.

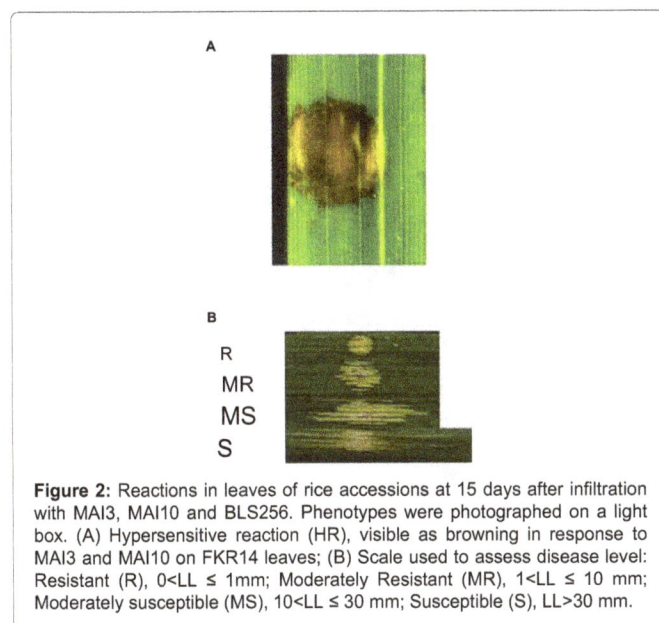

Figure 2: Reactions in leaves of rice accessions at 15 days after infiltration with MAI3, MAI10 and BLS256. Phenotypes were photographed on a light box. (A) Hypersensitive reaction (HR), visible as browning in response to MAI3 and MAI10 on FKR14 leaves; (B) Scale used to assess disease level: Resistant (R), 0<LL ≤ 1mm; Moderately Resistant (MR), 1<LL ≤ 10 mm; Moderately susceptible (MS), 10<LL ≤ 30 mm; Susceptible (S), LL>30 mm.

Discussion

Eight rice accessions were tested for their resistance to African and Asian *Xoc* strains. Among these, two *O. sativa* FKR14 and ITA306 show a resistance response when challenged with African *Xoc* strains while are moderately to highly susceptible to the Philippines strain *Xoc* BLS256. The reaction induced by African *Xoc* strains on FKR14 corresponds to a hypersensitive reaction with a dark browning area appearing 48 to 72 h after inoculation. Resistance and susceptibility to *Xoc* in rice are correlated with quantitative differences in the bacterial population. The hypersensitive reaction is associated with lower levels of bacterial population in leaves when compared to a susceptible reaction.

No single resistance gene to BLS disease has been characterized in rice. The transgenic rice lines with *Rxo1* gene exhibit an HR symptom when inoculated with *Xoc* harboring the corresponding effector gene *avrRxo1* [3,4]. The HR phenotype observed in African cultivars (FKR14) occurs independently of the presence or absence of the *avrRxo1* gene in *Xoc* strains. This suggests that the resistance observed in FKR14 to African *Xoc* strains differs from the one induced by *Rxo1*.

Zhou et al. [23] showed that upon infection with *Xoc* a larger number of Differentially Regulated Genes (DRG) are expressed in transgenic rice line expressing *Rxo1* gene when compared to wild type. We need to characterize genes that are induced in resistant cultivars such as FKR14 upon infection with African *Xoc* strains.

Also, TOG5672 a glaberrima accession exhibits a moderately resistant reaction correlated with low bacterial numbers when challenged with African and Asian *Xoc*. To confirm TOG5672 as a broad source of resistance to *Xoc*, we need to test a larger number of strains.

We identify novel and broad-spectrum resistance sources to contsrol BLS in rice in Africa. Characterization of genes and markers underlying BLS resistance mechanisms will be mandatory for future use in breeding program.

Crossing of FKR14 with TN1 that is highly susceptible to *Xoc* strains and/or the use of Multi-Parent Advanced Generation Inter-Cross lines (MAGIC) can be achieved to identify markers associated to BLS resistance [24]. Meanwhile, resistance in FKR may be introgressed in others rice accessions by classical breeding to further manage BLS disease in Africa.

was observed between plants infected with BLS256, MAI3 and MAI10. Unlike FRK14 no significant increase of bacterial growth of *Xoc* strains was observed with MAI3 and MAI10 between 7 and 15 days after inoculation (Figure 3).

On TN1 although *Xoc* BLS256 induced larger lesions compared to that of MAI3 and MAI10 (Figure 1), there was no significant difference between the population growth of *Xoc* strains (Figure 3). A significant increase of the bacterial population was observed 15 days after inoculation with all the strains tested (Figure 3).

In FKR14, BLS256 numbers were significantly greater than MAI3 and MAI10 at 7 and 15 days after inoculation. Both African strains were detected at low levels in leaf tissues. Together these results show that bacterial population increases in a susceptible cultivar (TN1) and remains stable in resistant accessions FKR14.

	Strains used					
	MAI3		**MAI10**		**BLS256**	
Accession	LL + SE	DR	LL+ SE	DR	LL + SE	DR
Curinga	11.3 ± 0.4	MS	6	MR	21.7 ± 0.8	S
FKR14	0	R	0	R	13.67 ± 0.9	MS
IR64	11.4 ± 0.4	MS	10.5 ± 0.5	MS	12.7 ± 0.5	MS
ITA306	0	R	0	R	45 ± 1.8	S
PaDcKono	6.8 ± 0.3	MR	6 ± 0.4	MR	13 ± 0.6	MS
TN1	30.8 ± 1.5	S	29.71 ± 1.2	S	59.4 ± 2	S
TOG 6767	7.3 ± 0.5	MR	6.7 ± 0.3	MR	13.2 ± 0.7	MS
TOG5672	6.7 ± 0.2	MR	6.7 ± 0.2	MR	7.6 ± 0.3	MR
Kitaake-*Rxo1*	0	R	15.5 ± 0.11	MS	0	R

LL: Length lesion induces by *Xoc* strains upon infiltration 15 days after inoculation, P <0.005; SD: Standard Deviation, DR: Disease Reaction, Resistant (R), 0 < LL ≤ 1mm; Moderately Resistant (MR), 1 < LL ≤ 10mm; Moderately susceptible (MS), 10 < LL ≤ 30mm; Susceptible (S), LL > 30mm.

Table 3: Length lesion induced by *Xoc* strains on rice varieties.

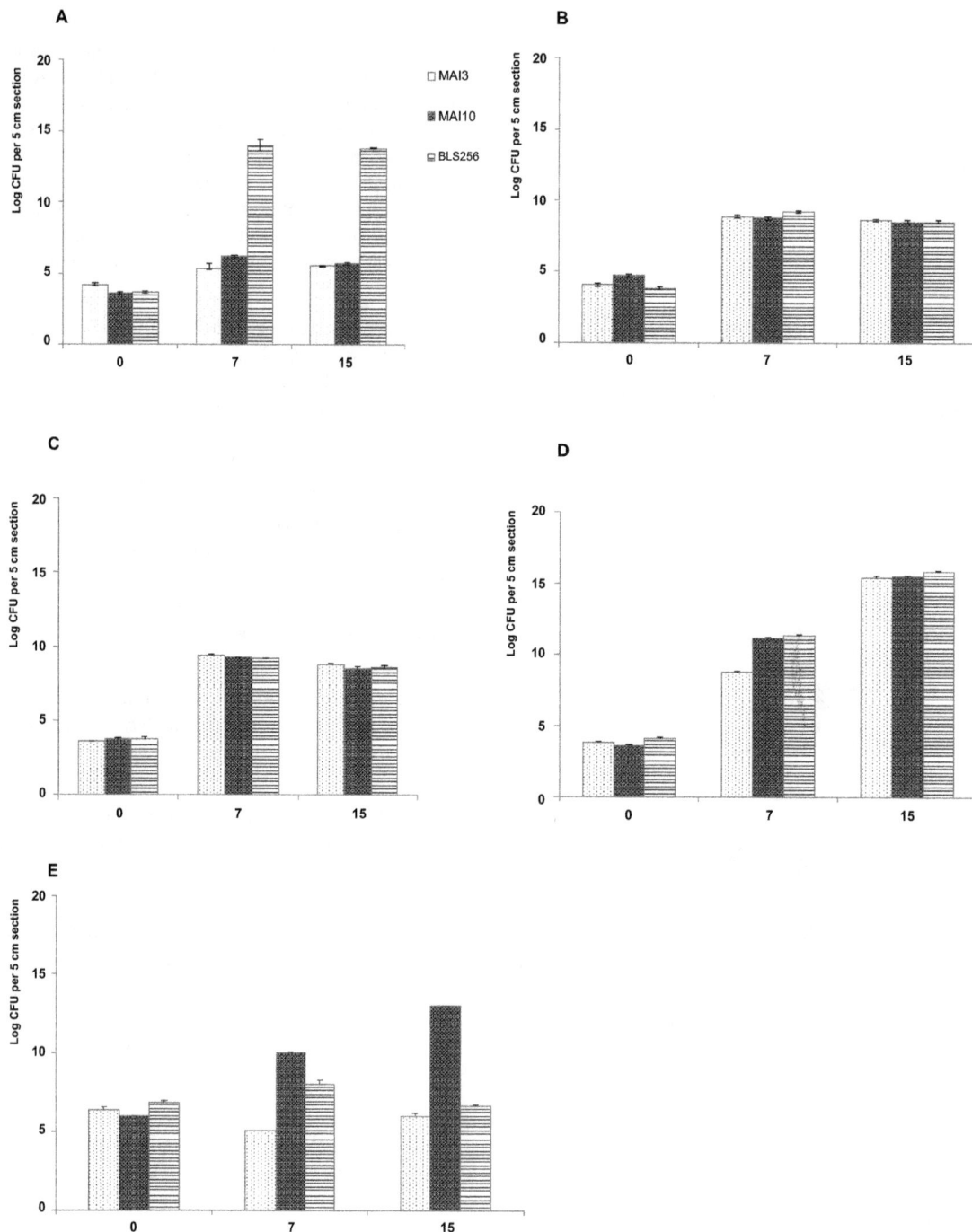

Figure 3: Bacterial growth in 3-week-old leaves of FKR14 (A), TOG5672 (B), TOG6767 (C), TN1 (D) rice accessions and trangenic line Kitaake-*Rxo1* (E). Bacterial population was measured in a 5 cm leaf segment around the infiltration point at 0, 7 and 15 days after inoculation with *Xoc* strains MAI3, MAI10 and BLS256. The experiment was repeated three times. Error bars represent standard error.

Acknowledgements

I. Wonni is currently supported by an IRD DPF (Direction des Programmes de recherche et de la Formation au Sud) fellowship. This work was partially supported by the INERA (Burkina Faso), IRD UMR RPB, LBMA (Mali), and the International Foundation For Science (IFS) grant awarded to I. Wonni (C/4813-1 and C/4813-2).

References

1. Ou SH (1985) Rice Diseases. 2nd Ed. Commonwealth Mycological. Instt. Kew. England pp: 247-256.

2. Awoderu VA, Bangura N, John VT (1991) Incidence, distribution and severity of bacterial diseases on rice in West Africa. Trop Pest Manag 37: 113-117.

3. Gonzalez C, Szurek B, Manceau C, Mathieu T, Séré Y, et al. (2007) Molecular and pathotypic characterization of new *Xanthomonas oryzae* strains from West Africa. Mol Plant Microbe Interact 20: 534-546.

4. Wonni I, Ouedraogo L, Verdier V (2011) First report of bacterial leaf streak caused by *Xanthomonas oryzae* pv *oryzicola* on Rice in Burkina Faso. Plant Dis Rep 1: 72.

Evaluation of Rice Germplasm against Bacterial Leaf Streak Disease Reveals Sources of Resistance...

55

5. Ouedraogo SL, Somda I, Wonni I, Seré Y (2007) Résistance au flétrissement bactérien de lignés inter- et intraspécifiques de riz de bas fonds en conditions d'infestation artificielles. African Crop Science Journal 15: 191-199.

6. Müller C, Cramer W, Hare WL, Lotze-Campen H (2011) Climate change risks for African agriculture. Proc Natl Acad Sci U S A 108: 4313-4315.

7. Mew TW, Vera Cruz CM (1986) Colonization of host and non-host plants by epiphytic phytopathogenic bacteria. Microbiology of the Phyllosphere. Eds. N.J. Fokkema, J. Van Den Heuvel, New York: Cambridge Univ. P. 269-282.

8. Niño-Liu DO, Ronald PC, Bogdanove AJ (2006) *Xanthomonas oryzae* pathovars: model pathogens of a model crop. Mol Plant Pathol 7: 303-324.

9. Mew TW, Alvarez AM, Leach JE, Swings J (1993) Focus on bacterial blight of rice. Plant Dis Rep 77: 5-12.

10. Xie GL, Mew TW (1998) A leaf inoculation method for detection of *Xanthomonas oryzae* pv. *oryzicola* from rice seed. Plant Dis 82: 1007-1011.

11. Lang JM, Hamilton JP, Diaz MGQ, Van Sluys MA, Burgos MAG, et al. (2010) Genomics-based diagnostic marker development for *Xanthomonas oryzae* pv. *oryzae* and *X. oryzae* pv. *oryzicola*. Plant Dis 93: 311-319.

12. Wonni I, Cottyn B, Detemmerman L, Dao S, Ouedraogo L, et al. (2014) Analysis of *Xanthomonas oryzae* pv. *oryzicola* population in Mali and Burkina Faso reveals a high level of genetic and pathogenic diversity. Phytopathology 104: 520-531.

13. Raymundo AK, Briones AMJ, Ardales EY, Perez MT, Fernandez LC, et al. (1999). Analysis of DNA polymorphism and virulence in Philippine strains of *Xanthomonas oryzae* pv. *oryzicola*. Plant Dis 83: 434-440.

14. Peng SQ, Wei ZS, Mao CX (1982) Identification of multi resistance of *O. meyeriana*, *O. officinalis* and *O. sativa f. spantanea* growing in Yunan Province. Acta Phytopathologica Sinica 12: 58-60.

15. Zhao B, Lin X, Poland J, Trick H, Leach J, et al. (2005) A maize resistance gene functions against bacterial streak disease in rice. Proc Natl Acad Sci U S A 102: 15383-15388.

16. Han Q, Zhou C, Wu S, Liu Y, Triplett L, et al. (2015) Crystal Structure of *Xanthomonas AvrRxo1*-ORF1, a Type III Effector with a Polynucleotide Kinase Domain, and Its Interactor *AvrRxo1*-ORF2. Structure 23: 1900-1909.

17. Wang C, Wen G, Lin X, Liu X, Zhang D (2008) Identification and fine mapping of the new bacterial blight resistance gene, *Xa31(t)*, in rice. Eur J Plant Pathol.

18. He WA, Huang DH, Li RB, Qiu YF, Song JD, et al. (2012) Identification of a resistance gene *bls1* to bacterial leaf streak in wild rice *Oryza rufipogon* Griff. J Integr Agr 11: 962-969.

19. Verdier V, Vera Cruz C, Leach JE (2012) Controlling rice bacterial blight in Africa: needs and prospects. J Biotechnol 159: 320-328.

20. Djedatin G, Ndjiondjop MN, Mathieu T, Vera Cruz CM, Sanni A, et al. (2011) Evaluation of African cultivated rice *Oryza glaberrima* for resistance to bacterial blight. Plant Dis 95: 441-447.

21. INERA (2000) Fiches descriptive des variéties de riz. INERA, Burkina Faso.

22. Reimers PJ, Leach JE (1991) Race-specific resistance to *Xanthomonas oryzae* pv. *oryzae* conferred by bacterial blight resistance gene *Xa10* in rice (*Oryza sativa*) involves accumulation of a lignin-like substance in host tissues. Physiological and Molecular Plant Pathology 38: 39-55.

23. Zhou YL, Xu MR, Zhao MF, Xie XW, Zhu LH, et al. (2010) Genome-wide gene responses in a transgenic rice line carrying the maize resistance gene *Rxo1* to the rice bacterial streak pathogen, *Xanthomonas oryzae* pv. *oryzicola*. BMC Genomics 11: 78.

24. Zhao B, Ardales EY, Raymundo A, Bai J, Trick HN, et al. (2004) The avrRxo1 gene from the rice pathogen *Xanthomonas oryzae* pv. *oryzicola* confers a nonhost defense reaction on maize with resistance gene *Rxo1*. Mol Plant Microbe Interact 17: 771-779.

Biosurfactant-Mediated Biocontrol of *Macrophomina phaseolina* causing Charcoal Rot in *Vigna mungo* by a Plant Growth Promoting *Enterococcus* sp. BS13

Sumit Kumar*, Dubey RC and Maheshwari DK

Department of Botany and Microbiology, Gurukula Kangri University, Haridwar – 249404, India

Abstract

A potential bacterial isolate *Enterococcus* sp. BS13 screened from rhizospheric soil of *Vigna mungo* was identified as *Enterococcus* sp. based on morphological, biochemical and genomic characterization. The biosurfactant producing activity of BS13 was based on many tests such as blood heamolysis test, CTAB agar, emulsification stability (E_{24}) test, oil spreading/displacement assay, drop collapse assay, bacterial adhesion to hydrocarbons (BATH) assay and surface tension (ST) measurement after 72 h of growth. GC-MS and FT-IR analyses established the similarity of biosurfactant with glycolipid type biosurfactant. Furthermore, *Enterococcus* sp. BS13 displayed plant growth-promoting ability, HCN production and antagonistic activity against *Macrophomina phaseolina*. Scanning electron microscopic study of fungal mycelia from zone of inhibition showed hyphal degradation, halo cell formation and mycelial deformities in the pathogen. *M. phaseolina* sclerotia formation and development were arrested towards the zone of interaction; consequently, such mycelia and sclerotia lost the vigour. In pot trials *Enterococcus* sp. BS13 increased the growth of *V. mungo* with considerable diseases reduction. Hence, *Enterococcus* sp. BS13 bears ability of biosurfactant production, plant growth promotion and biocontrol of *M. phaseolina*. Therefore, the exogenous application of BS13 can be a potential strategy to accelerate plant growth promotion and biocontrol of *M. phaseolina*.

Keywords: Biosurfactant; *V. mungo*; *Enterococcus* **sp. BS13**; *Macrophomina phaseolina*; **Biocontrol**

Introduction

The naturally occurring plant growth promoting rhizobacteria (PGPR) aggressively colonize plant roots and benefit plants by providing growth nutrients and hormones [1]. They produce many plant growth-promoting metabolites, solubilize insoluble phosphates, and survive under various stresses environmental conditions. Plants cannot always meet their demand of minerals from natural soils [2]. The environmental enterococci are a taxonomically diverse group of bacteria but the plant-associated enterococci represent the lesser-known bacterial group. A little work has been done on enterococci of rhizospheric soil. The enterococci are temporary resident of agricultural plants and grasses [3]. Ulrich and Muller [4] studied the effect of *Enterococcus faecalis* and *Enterococcus faecium* on plants. The application of non-pathogenic soil bacteria as biofertilizers can be environmentally safe for increasing crop productivity and soil fertility owing to their ability to produce auxin, gibberellins and cytokinin [5]. The European Food Safety Authority recognized some of the strains *E. faecium* as feed additives [6]. The ability of *E. faecium* to produce phytohormones has been reported by Lee et al. [7].

A variety of soil microorganisms demonstrated the biocontrol of various soil-borne phytopathogens. But less work on biocontol potential of enterococci has been done. Enterococci produce several compounds that inhibit the growth of many phytopathogens [8,9]. Biocontrol potential of *Enterococcus faecium* against phytopagens was studied by Fhoula et al. [10].

Several enterococci have been screened during the isolation of biosurfactant producing and plant growth promoting rhizobacteria of *V. mungo*. *V. mungo* is a major Indian pulse crop grown throughout India. Annual production of urdbean in India is about 1.3 million tonnes. It has very high nutritious value possibly due to presence of protein (25 mg/100 g), potassium (983 mg/100 g), calcium (138 mg/100 g), iron (7.57 mg/100 g), niacin (1.447 mg/100 g), Thiamine (0.273 mg/100 g), and riboflavin (0.254 mg/100 g) [11]. *M. phaseolina* is most destructive

phytopathogens. It infects more than 500 plant species [12]. It caused charcoal rot in *V. mungo* and reduces the crop yield. Therefore, the present work was designed to elucidate biosurfactant production, PGP activity and biocontrol of *M. phaseolina* by *Enterococcus* sp. BS13 as well as purified biosurfactant *in vitro*.

Material and Methods

Sample collection

The young plants of *Vigna mungo* were randomly collected from different locations of a farmer's field in Saharanpur (India), put in to a sterile poly bag and brought to the laboratory for further work.

Isolation and selection of bacteria

Rhizospheric soil (1 g) was suspended in 10 ml of 0.85% (w/v) NaCl (saline) water, vortexed vigorously, serially diluted and plated on nutrient agar medium. The plates were incubated at 37°C for 24 h. Colonies appearing with dissimilar morphology on the surface of growth medium were purified and selected for further tests [13].

Screening for biosurfactant production

Blood heamolytic test: Log phase growing culture of isolated bacteria was streaked on blood agar plates and incubated at 37°C for

*Corresponding author: Sumit Kumar, Department of Botany and Microbiology, Gurukula Kangri University, Haridwar – 249404, India
E-mail: Sumitarya360@gmail.com

24-48 h. The plates were then observed for the presence of clear zone around the bacterial colonies. The clear zone indicates the presence of biosurfactant producing microorganisms [14].

Oil displacement method: Distilled water (20 ml) was poured in the Petri plates and 1 ml oil was added in the centre of the plates. Then 20 µl of the culture supernatant of the isolates as added to it. The biosurfactant producing organism displaced the oil and spread in the water [15].

CTAB agar plate method: All the bacterial isolates were initially assayed for biosurfactant production using mineral salt cetyl-trimethyl-ammonium-bromide (CTAB) - methylene blue agar plate method (0.2 mg/ml CTAB and 5 µg/ml methylene blue). Isolates were grown for 24 h in mineral salt medium under appropriate culture conditions. Shallow wells were cut on the surface of the indicator plates. Ten microliters of the appropriate culture was placed into each well and incubated at $30 \pm 1°C$ and growth was observed regularly for 24 - 48 h [16].

Drop collapse method: A drops of culture supernatant are placed on an oil coated solid surface. The polar water molecules repelled from the hydrophobic surface and the drops remained static in the absence of biosurfactants in the liquid. If the liquid contains surfactants, the drops spread or even decadence [17].

Emulsification stability (E_{24}) test: Kerosene (2 ml) was added to 2 ml of supernatant culture in a borosil glass test tube (15 mm diameter) and mixed for 2 minute using a vortex mixer. Kerosene emulsion was formed in the test tube which was then covered with para-film and allowed to stand at room temperature for 24 h. The height of the emulsion in the tube was noted after 1 minute of aging and periodically for about 24 h to monitor the stability of the emulsion. E24 (%) was calculated by using the following formula: [18].

$$E_{24}(\%) = \frac{H1}{H2} \times 100$$

Where, H_1= height of emulsified layer; H_2=height of total solution.

Bacterial adhesion to hydrocarbons (BATH) assay: Cell hydrophobicity was measured by bacterial adherence to hydrocarbons (BATH) assay. The cell pellets collected in the above culture medium and bacterial growth section were washed twice and suspended in a buffer salt solution (g l⁻¹, 16.9 K_2HPO_4 and 7.3 KH_2PO_4) and diluted using the same buffer solution to an optical density (OD) of ~ 0.5 at 610 nm. To the cell suspension (2 ml) in test tubes (10 ml volume with 10 × 100 mm dimension) 100 µl of crude oil was added and vortex-shaken for 3 min. After shaking, crude oil and aqueous phases were allowed to separate for 1 h. Then OD of the aqueous phase was measured at 610 nm in a spectrophotometer (UV VIS 1600 Shimadzu, Japan) and bacterial cells attached to crude oil were calculated using the following formula: [19].

Bacterial cell adherence (%) = [1-(OD shaken with oil/OD original)] × 100

Where, OD with oil = OD of the mixture containing cells and crude oil, OD original = OD of the cell suspension in the buffer solution (before mixing with crude oil).

Surface tension measurement: Measurement of surface tension of cell-free culture broth from isolated strain was determined according to Du Nouy ring method. Deduction in surface tension by cell free culture was compared with a standard biosurfactant CTAB solution (1 mg/ml) [20].

Production and purification of biosurfactant: The isolate BS13 was transferred to nutrient rich (NR) broth containing 1% yeast extract, 1.5% nutrient broth and 1% ammonium sulfate. The culture was incubated at 37°C for 12 h and 120 rpm as seed culture to get optical density of 0.5 at 600 nm. Then, 5 ml suspension was transferred to a 1000 ml Erlenmeyer flask containing 500 ml of LB medium and incubated on a rotary shaker incubator (150 rpm) at 37°C. The bacterial cells were removed by centrifugation at 10,000 g at 4°C for 20 minutes and supernatant was acidified with 6N hydrochloric acid to get the pH 2.0, and incubated overnight at 4°C. The precipitated biosurfactant was collected by centrifugation at 15,000 g for 20 min and precipitate was dissolved in distilled water to get pH 7.0 by using 1N NaOH. Further, it was centrifuged at 10,000 g for 10 min following Sánchez et al. [21].

Chemical characterization of biosurfactant

Fourier transform infrared spectroscopy (FT-IR) spectra of the dried biosurfactant: FT-IR spectra of the dried biosurfactants were recorded on an 8400S, FT-IR spectrometer (Shimadzu, Japan) was equipped with a mercury–cadmium–telluride (MCT) detector and cooled with liquid nitrogen. About 2 mg of dried biomaterial was milled with 200 mg of KBr to form a very fine powder. The powder was compressed into a thin pellet to be analyzed by FT-IR spectra at wave length of 400–4000 cm⁻¹. FT- IR spectra were analyzed by using OPUS 3.1 (Bruker Optics) software.

Gas chromatography-mass spectroscopy (GC-MS) analysis: GC-MS analysis of biosurfactant was performed by using a varian 4000 Mass Spectrometer employing DB5 type capillary column and helium as a carrier gas at a flow rate of 0.5 ml/min. The sample volume was 1µl; temperature was gradually increased from 40°C to 280°C to identify the compound. Total run time was 45 min. The MS transfer line was maintained at a temperature of 280°C. GC-MS analysis was done using electron impact ionization at 70 Ev and data were evaluated using total ion count (TIC) for identification and quantification of compound. A comparative study carried out was between the identified compound spectra and that of known compounds of the GC-MS library NIST.

Biofilm formation *in vitro*: Biofilm formation was tested during bacterial growth in borosilicate glass tubes. Sterile Muller Hinton broth (MHB) (5 ml) was poured in the pre-sterile test tubes inoculated separately with BS13 culture along with proper control, and incubated at 37°C for 24 h. Broth was discarded and BS13 biomass attached to the glass surface was observed by staining with crystal violet to confirm biofilm formation [22].

Biodegradation of polycyclic aromatic hydrocarbon (PAH): Hydrocarbon-utilizing bacteria were screened for their ability to utilize PAH as sole source of carbon and energy using Bushnell-Haas fortified with 2% agar medium. In the present study mainly two PAHs such as kerosene and diesel were used for degradation.

Plant growth promotion (PGP) traits of isolate BS13

Phosphate solubilization: Phosphate solubilization ability of isolate was detected by spotting them separately on Pikovskaya's agar plates [23]. A loopful of bacterial culture was spot inoculated on the plate and incubates at 28°C for 7 days. The presence of clear zone around the bacterial colonies indicates the phosphate solubilization.

Indole-3-acetic acid (IAA) production: IAA production by the isolate BS13 was determined following the methods of Gordon and Weber [24]. BS13 was grown at 37°C for 72 h in LB broth and centrifuged at 8,000 rpm for 30 min. The supernatant (2 ml) was mixed with two drops of ortho-phosphoric acid and 4 ml of the Salkowski reagent (50 ml of 35% of perchloric acid plus 1 ml of 0.5 M $FeCl_3$). Appearance of pink colour confirmed the production of IAA.

Siderophore production: Siderophore production by BS13 was tested using chrome azurol S (CAS) agar plate method of Schwyn and Neilands [25]. Overnight culture (10 μl) was spot inoculated on CAS agar plate and incubated 28 ± 1°C for 48–72 h. Orange to yellow halo formation around the bacterial colonies confirmed siderophore production.

Chitinase production: Chitin plates were prepared using M9 agar medium amended with 1% (w/v) colloidal chitin. The plates were divided into equal sectors; spot inoculated with 10 μl of overnight grown BS13 culture and incubated at 37°C for 48–96 h. Zone of clearance around bacterial colonies indicated chitinase production.

HCN production: HCN production was determined following modified method of Bakker and Schippers [26]. For detection of HCN production, exponentially grown culture of BS13 was streaked on agar plates supplemented with or without 4.4 g glycine l^{-1} with simultaneous addition of filter paper soaked in 0.5% picric acid in 1% Na_2CO_3 in the upper lids of plates along with uninoculated control. The plates were incubated at 28 ± 1°C for 72 h along with a control. The changes in colour from yellow to light brown or moderate brown to strong brown was examined for putative HCN production.

16S rRNA gene sequencing of isolate BS13: Pure culture of BS13 was grown in LB broth until log phase achieved. The genomic DNA was extracted according to Bazzicalupo and Fani [27]. PCR amplification of genomic DNA was carried out following Kumar et al. [28]. The amplification of 16S rRNA gene was done by using universal bacterial primer 1492R primer (5′CGGTTACCTTGTTACGACTT 3′) and 27F (5′AGAGTTTGATCMTGGCTCAG 3′) under the conditions described by Brown and Balkwill [29]. The PCR product was sequenced by Aakar Biotech Pvt. Ltd., Biotech Park, Lucknow (U.P.) India. Sequence search in the EMBL/Gen Bank/DDBJ/PDB data libraries was performed using the BLAST (blastn) [30] search algorithm in order to establish the identity of the strain. Sequences of the closest matched reference strains were retrieved and aligned using ClustalW (multiple and pairwise) with the newly determined sequence. Phylogenetic analysis was performed using MEGA version 6 and phylogenetic tree was constructed by neighbour-joining method with 1500 replicates to produce bootstrap value and to validate the reproducibility of the branching pattern of the tree. 16S rRNA gene sequence was submitted to the NCBI GenBank and to get the accession number.

Antagonistic activity against phytopathogenic fungi: Antagonistic activity of the isolate BS13 against *M. phaseolina* was determined by dual culture technique of Skidmore and Dikinson [31]. A fungal disc of 1 cm diameter was placed in the centre of potato dextrose agar plate. Then overnight grown bacterial culture was spot inoculated on the surface of agar plate 2 cm away from the central fungal disc. The plates were incubated at 28°C for 72 h. Growth inhibition was calculated by using the formula: 100 × (C–T)/C (T, treatment; C, control). Petri plates without bacterial spotted culture served as control.

Post interaction events by scanning electron microscopy (SEM): For preparation of SEM samples, the fungal mycelia disc was cut with the help of a sterile cork borer from the zone of interaction. The discs were fixed overnight using 4% glutaraldehyde in 0.05 M phosphate buffer (pH 7.3) and washed thrice in phosphate buffer (10 min each). Then the samples were dehydrated through 70, 80, 90 and 100% ethanol (5 min each) and re-suspended thrice in 100% ethanol at room temperature. Ethanol was then replaced by liquid CO_2 and the samples were air dried. Then samples were mounted on stubs and coated with gold. These coated specimens were observed at 15 kV in a LEO 485 VP SEM. Photo-micrographs were recorded by the same microscope.

Pot trial

Development of antibiotic-resistant marker in strain BS13: Antibiotic resistant marker strain of BS13 was developed followed by Bhatia et al. [32]. For development of an antibiotic strain marker, *Enterococcus* sp. BS13 was subjected to antibiotic sensitivity test. Antibiotic discs (5 mm diam) of different concentrations were placed at four corners over the surface of *Enterococcus* sp. BS13 seeded plate and incubated at 28 ± 1°C for 24 h to measure the inhibition zone. Resistance marker strains were developed by subjecting the culture successively to low concentration to high concentration of methicillin and fusidic acid.

Preparation of fungal inoculum: Inoculum of *M. phaseolina* was prepared by multiplying the pathogen on potato dextrose broth medium. Fungal culture was filtered after 7 days incubation at 28°C. Filtered mycelial mat and sclerotia were dried at 85°C to evaporate moisture content and crushed to make powder. Fungal powder (500 mg/kg) was mixed in the pre-steriled soil (1 kg) in pots.

Mass production of bacteria and seed bacterization: Antibiotic resistant marker *Enterococcus* sp. BS13 $^{Met+Fus+}$ (100 mg/l) was used for mass production. Starter culture of *Enterococcus* sp. BS13$^{Met+Fus+}$ was prepared by inoculating 50 ml LB broth in flask and incubated at 30°C for 24 h at 150 rpm [33]. Then the culture was separately transferred in 1 L flask containing sterilized broth and incubated at 30°C at 150 rpm. The pH of the medium was maintained at 6.8. Culture broth was periodically excluded to check the cfu (colony forming unit) for detection of any contamination. When cell concentration increased to 1×10^8 cells/ml in pure form, it was used for seed bacterisation and pot trials.

Healthy and viable seeds of *V. mungo* were washed in sterile water and surface-sterilized with 95% alcohol for 30 second followed by 0.1% (w/v) $HgCl_2$ for 1 min. Seeds were washed 5-6 times with sterile distilled water. The bacterial culture introduced as above was centrifuged at 8000 g for 15 min at 4°C. The pellets obtained were suspended in sterile distilled water to obtain a population density of 1×10^8 cfu/ml.

Seed bacterization was carried out following the method of Weller and Cool [34]. The cell suspension was mixed with 1% carboxymethyl cellulose (CMC). Slurry was coated separately on the surface of healthy seeds. The seeds coated with 1% CMC slurry without bacteria served as control [32]. The fungal culture in the form of powder was mixed in sterile sandy loam soil (sand 56%, silt 22%, clay 27%, WHC 41.10%, pH 8.29) put in the earthen pots (25 × 25 × 25 cm). Healthy seed of *V. mungo* inoculated with the *Enterococcus* sp. BS13$^{Met+Fus+}$ strain alone and in combination with *M. phaseolina* separately were examined for their ability to enhance the overall growth of plants. In the pot trial, bacterial inoculated and non-inoculated seed were sown in triplicate along with control. Seed germination was recorded after 15 days of the sowing. Vegetative growth parameters were recorded after 30 days and bacterial root colonization was recorded at an interval of 30 days up to 90 day after sowing (DAS).

Statistical analysis

All data of pot trial statically were analyzed by GraphPad Prism 5. Differences among treatments were assessed using one-way analysis of variance (ANOVA).

Results

The isolate BS13 was found most potent biosurfactant producing and PGP activity bearing isolate. On the basis of morphological, physiological and biochemical characteristics the isolate BS13 was found Gram-positive, coccus, nonspore former, catalase negative, oxidae positive, and producers of white, dry and irregular edged colonies on NAM plates (Table 1). The isolate BS13 was identified as *Enterococcus* sp. based on microscopic and biochemical analysis according to Bergey's manual of determinative bacteriology [35]. Different phenotypic characters of isolate was compared with the standard strains of *Enterococcus faecium* (ATCC 700221) *Enterococcus faecium* (ATCC 51858), *Enterococcus faecium* (ATCC 51559) and *Enterococcus faecium* (ATCC 27270).

Screening of biosurfactant

BS13 was further tested for biosurfactant production employing haemolytic test, oil displacement assay, CTAB agar plate method, drop collapsing, emulsification index, and Surface tension measurement. The isolate BS13 showed β-haemolytic activity and caused the maximum zone of ~1.66 cm. It also showed a zone of displacement in the oil. The flat drop appearance in microtiter plate confirmed the positive result of drop collapse test, proving the use of drop collapse method as a quick and simple method for detection biosurfactant production. A prominent emulsification activity with the emulsification index (E_{24}) of 70.58 was also exhibited by BS13. Production of dark blue halo zone in the CTAB methylene blue agar plate confirmed the presence of anionic biosurfactant. The formation of insoluble ion pair precipitates in the agar plate containing methylene blue exhibited dark blue colour against the light blue background. In BATH assay, the bacterial cells indicated their affinity towards hydrophobic substrate. The isolate caused the maximum reduction in surface tension by 68.974 D/CM.

FTIR analysis

The molecular composition of biosurfactant manifest displayed the most prominent adsorption bands located at 3120 cm^{-1}, 2771 cm^{-1} and 3001 cm^{-1} (C–H stretching bands of CH_2 and CH_3 groups), 1735 cm^{-1} (C = O stretching vibrations of the carbonyl groups), 1278 cm^{-1} (C–O stretching bands; formed between carbon atoms and hydroxyl groups in the chemical structures) and 794 cm^{-1} (CH_2 group) (Figure 1).

GC-MS analysis

The fatty acid composition of biosurfactant was analyzed by GC-MS and compared with the library data. The biosurfactant comprised of long chain fatty acids, mainly C-10 long fatty acids. The major fatty acid of biosurfactant was C-10 decanoic acid. On the basis of FTIR and gas chromatography the structure of biosurfactant produced by *Enterococcus* sp. BS13 was predicted as glycolipid type (Figure 2).

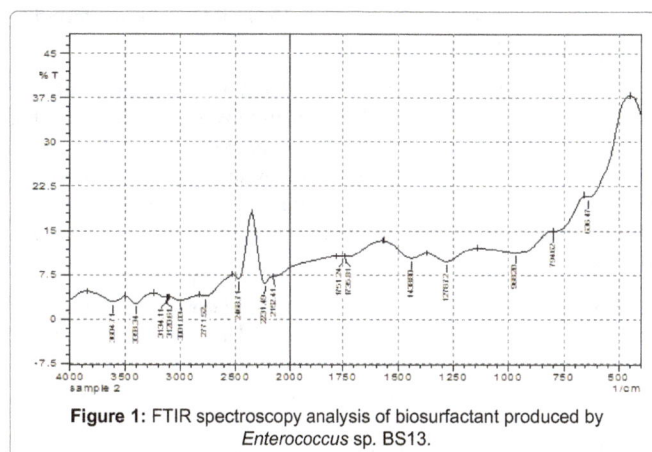

Figure 1: FTIR spectroscopy analysis of biosurfactant produced by *Enterococcus* sp. BS13.

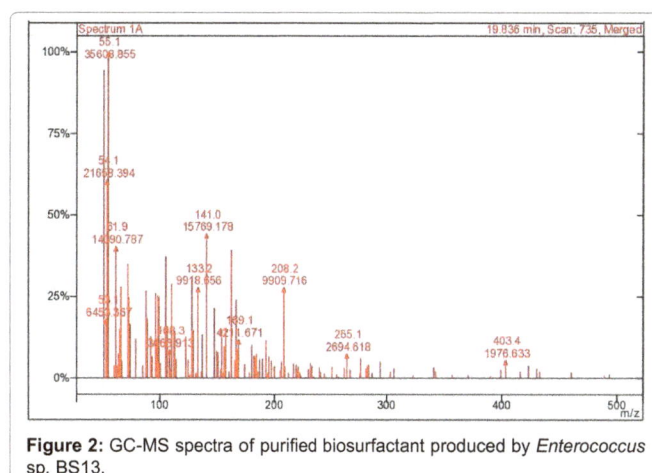

Figure 2: GC-MS spectra of purified biosurfactant produced by *Enterococcus* sp. BS13.

Biofilm forming ability of BS13: BS13 formed biofilm in the test tube assay. Bacterial cells adhered on the internal surface of test tubes which eluded the biofilm production.

Biodegradation of polycyclic aromatic hydrocarbon (PAH): The isolate grew in the Bushnell Haas agar medium containing kerosene and diesel used as carbon source; it developed clear zone in Petri plate showing it's PAH degradation activity.

Plant growth promoting traits of BS13: The isolate BS13 showed IAA production. Similarly BS13 showed marked phosphate solubilization as visualized by a clear zone around the colony after 48 h. The zone of hydrolysis gradually increased with the increase the incubation time reaching to 3.4 mm after 7 days. Change in the color of the chrome-azurol S medium from blue to orange-red confirmed the production of siderophore. The isolate BS13 changed colour of filter paper from yellow to reddish-brown after 2-3 days of inoculation confirming HCN production. BS13 isolate produced a wider zone of chitin hydrolysis in chitin minimal medium plate, indicating the chitinase production.

16S rRNA gene sequencing and phylogenetic analysis: The 16S rRNA gene sequence of the BS13 comprised of 1431 bp (NCBI GenBank accession No. KU99196). It showed 94% sequence similarity with *Enterococcus faecium* L3 (accession No. KJ728981) and *Enterococcus faecium* KCL1 (Accession No. KM497512). The generated 16S rRNA sequences were analyzed with Basic Local Alignment Search Tool (BLAST) available on NCBI website. A multiple sequence was produced followed by a guide-tree generation, phylogram and cladogram

Biosurfactant properties		PGP Properties	
CTAB	++	Siderophore	+
Haemolysis	++	HCN	+
BATH Assay	++	IAA	++
Emulsification assay	+	Chitinase production	+
Drop collapse test	+	Phosphate solubilization	++
Surface Tension measure	+++	Biofilm assay	+
Oil	+	Antagonistic activity against *M. phaseolina*	+++

Note: -: Absence of halo formation or no activity; +: small halo < 0.5 cm, biofilm forming HCN, hydrogen cyanide; Chitinase producing; Chitinase producing; ++: medium halo formation > 0.5 cm wide surrounding colonies; medium IAA production; +++, large holo > 1.0 cm wide surrounding colonies, high surface tension reducing, +++: 51% to 75% inhibition against *Macrophomina phaseolina*.

Table 1: Biosurfactant screening test and plant growth promoting attributes of *Enterococcus* sp. BS13.

60

Principles of Plant Disease Management

formation [36]. Therefore, the isolate BS13 has further been referred to as *Enterococcus* sp. BS13 (Figure 3).

***In vitro* antagonistic activity of BS13 and pure biosurfactant:** The pure culture of *Enterococcus* sp. BS13 and its biosurfactant inhibited the radial growth of *M. phaseolina* by 72.3% and 53.2%, respectively after 7 day of incubation at 28 ± 1°C (Figures 4A and 4B). However, fungal growth inhibition was more pronounced in dual culture as compared to that of pure biosurfactant. Inhibition in fungal growth corresponded with incubation time.

Post-interaction events in mycelia of *M. phaseolina*: *Enterococcus* sp. BS13 in the zone of interaction resulted in halo cell formation, mycelial deformities and hyphal degradation of *M. phaseolina*. Formation and development of sclerotia of the pathogen were also arrested towards the zone of interaction; consequently such mycelia and sclerotia lost their vigour (Figures 5A-5C).

Pot trials: *Enterococcus* sp. BS13$^{Met+Fus+}$ inoculated seeds showed significantly (P<0.01) increased seed germination, plant length and biomass over the control (Table 2). *Enterococcus* sp. BS13$^{Met+Fus+}$ also resulted in reduction of disease caused by *M. phaseolina* after 30, 60 and 90 DAS (66.6%, 37.5%, and 42.5%) (Figure 6). *Enterococcus* sp. BS13$^{Met+Fus+}$strain showed effective root colonization as evidenced by the prominent population recovery from *V. mungo* rhizosphere 90 DAS (Table 3). The strain successfully colonized *V. mungo* roots, alone and in combination with *M. phaseolina*, and increased its population in rhizosphere.

Discussion

The isolate *Enterococcus* sp. BS13 was found more potential to secrete biosurfactant and promote plant growth. The biosurfactant had similarity with glycolipid type surfactant. Sharma et al. [37]

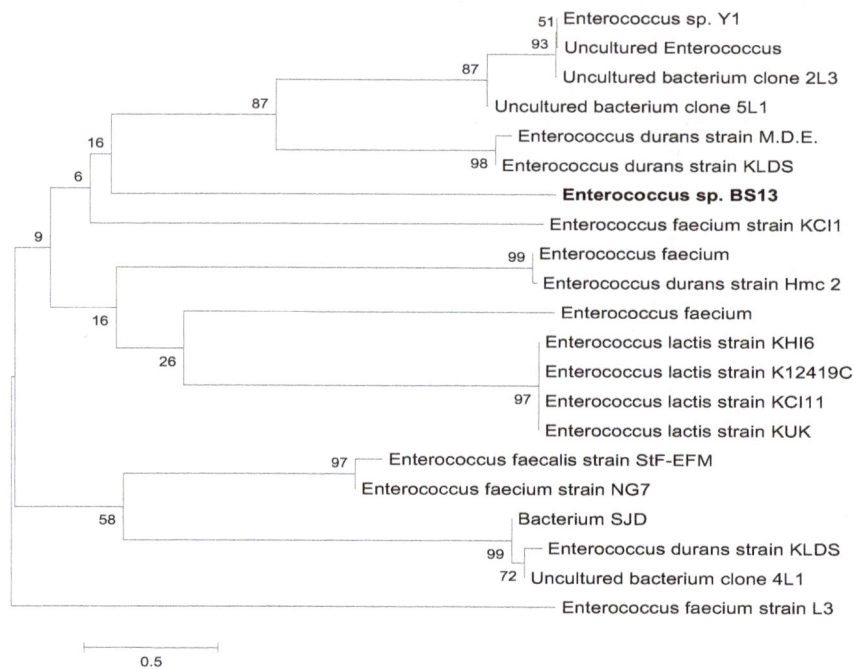

Figure 3: Phylogenetic analysis of rhizobacteria *Enterococcus* sp. BS13 (KU99196) and presenting species based on16S rDNA sequences using cluster algorithm. Distance and clustering with neighbor-joining method was performed by using the MEGA 6.06 software.

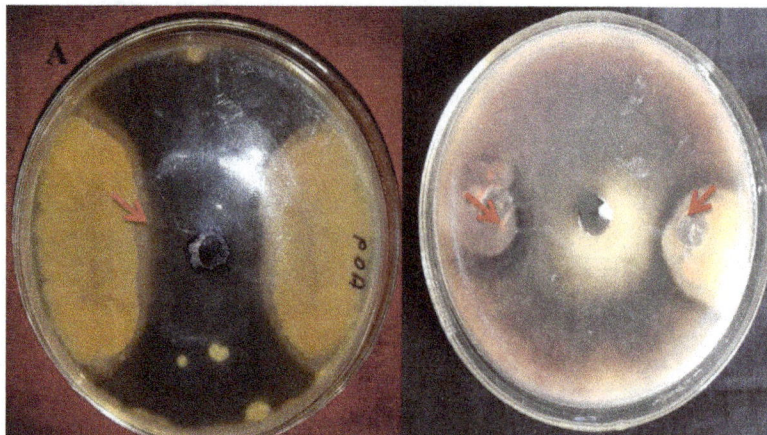

Figure 4: Antagonistic effect of *Enterococcus* sp. BS13 against *M. phaseolina* in dual culture (A); effect of pure biosurfactant on fungal growth (B).

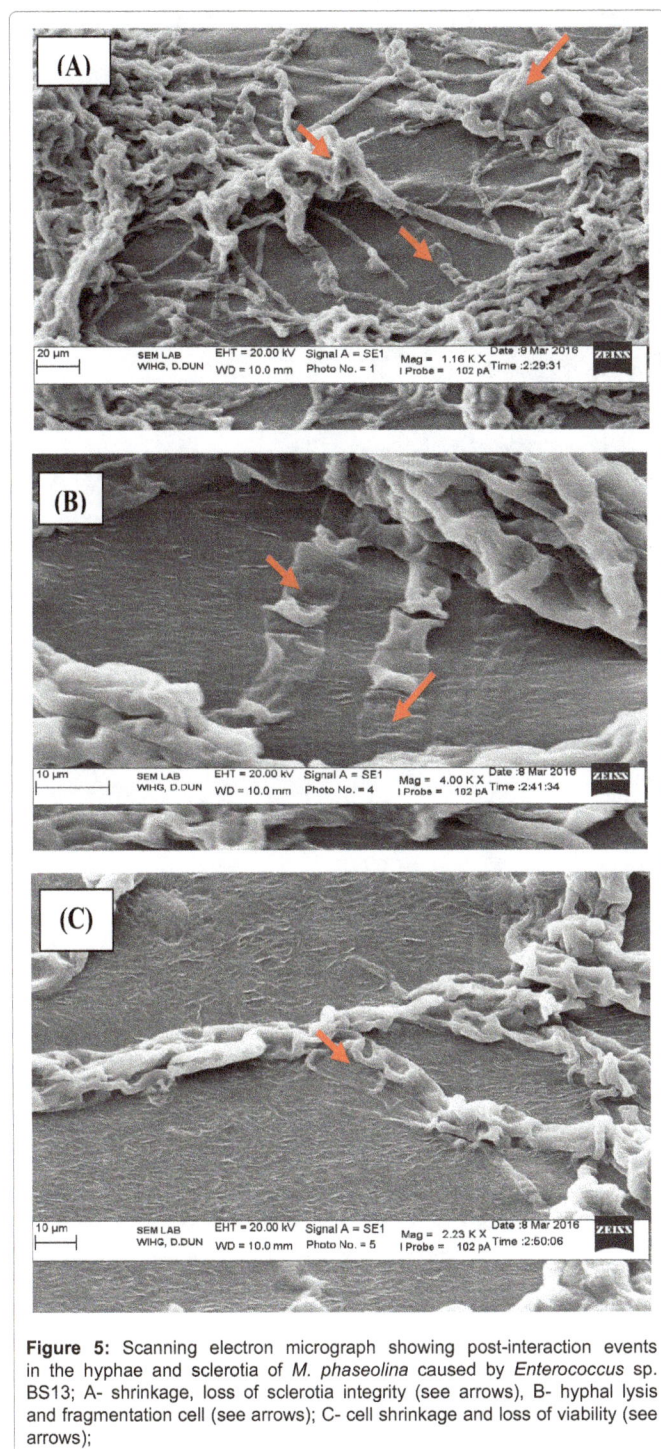

Figure 5: Scanning electron micrograph showing post-interaction events in the hyphae and sclerotia of *M. phaseolina* caused by *Enterococcus* sp. BS13; A- shrinkage, loss of sclerotia integrity (see arrows), B- hyphal lysis and fragmentation cell (see arrows); C- cell shrinkage and loss of viability (see arrows);

have also reported a novel glycolipid biosurfactant producing *Enterococcus faecium*. Besides, the biosurfactant also displayed *in vitro* antifungal activity against *M. phaseolina*. Presence of antifungal activity of glycolipid against phytopathogenic fungi has also been reported by Teichmann et al. [38] and Kulakovskaya et al. [39]. Biofilms are comprehensive colonies of single or multispecies of microbial cells adhered to surfaces or communicate contact with each other, encased in a self-produced matrix of extracellular polymeric substances (EPS). The isolate *Enterococcus* sp. BS13 formed biofilm and produced biosurfactant. The importance of biofilm formation

in PGP rhizobacterial action has been studied by Seneviratne et al. [40]. In the present investigation *Enterococcus* sp. BS13 was found to degraded hydrocarbon. In agricultural soil increased contamination by different hydrocarbons is a serious environmental problem due to their persistence in nature for a long time. The use of biosurfactants is an alternative over the chemical surfactant as the former is a better biodegradable and ecofriendly [41]. Biodegradation of hydrocarbon has also been studied by Zhang et al. [42].

Auxins called indol acetic-acid (IAA) are plant hormones which originate from an amino acid tryptophan. IAA production is widespread among plant associated bacteria. In the present study the *Enterococcus* sp. BS13 was found to produce IAA *in vitro*. IAA production by *Enterococcus* sp. has also been reported by Lee et al. [7]. Phosphorus plays a major role in many biological processes such as cell division, photosynthesis, sugar break down, energy and transfer of nutrient in crop plant. *Enterococcus* sp. BS13 efficiently solubilized phosphate *in vitro*. Ghosh et al. [43] also isolated phosphate solubilizing bacteria from the rhizosphere soil of seagrass. Iron is a necessary nutrient for all living organisms. In the soil it is directly assimilated by microorganisms because ferric iron (Fe III) which predominates in nature is only thriftily soluble and in very low concentration to support microbial growth. *Enterococcus* sp. BS13 was found siderophore producing in CAS medium. Similar study on siderophore production by *Enterococcus faecium* LKE12 has been performed by Lee et al. [7]. *Enterococcus* sp. BS13 showed the HCN production. HCN is a major chemical weapon to depress the growth of phytopathogen. HCN indirectly plays a major role in plant growth promotion by biological control of phytopathogen [44]. Similarly, HCN production has also been carried out by Gupta et al. [45] and Bhatia et al. [46]. Chitin is one of the most natural renewable polysaccharides present in cell wall of fungi, algae, insects and marine invertebrates. *Enterococcus* sp. BS13 showed the most significant chitinolytic activity in chitin minimal medium. Chitin is hydrolysed by chitinase enzyme [47]. Vaaje-Kolstad et al. [48] have also chitinase enzyme production by *Enterococcus* sp.

Several pesticides and herbicide are being used to control of phytopathogens in routine agriculture by farmers. These are not biodegradable and persist for long duration in the soil. Hence, biocontrol is the most vital alternative of this problem to maintain agricultural sustainability. In the present study, *Enterococcus* sp. BS13 displayed strong antagonistic properties and proved as a good biocontrol agent against *M. phaseolina*. Similar work on biocontrol efficiency of *Enterococcus faecium* has been carried out by Fhoula et al. [10].

Bacterial broth culture of *Enterococcus* sp. BS13[Met+Fus+] applied as seed treatments the most effective method for plant growth promotion of *V. mungo* and biocontrol of charcoal rot disease. *Enterococcus* spp. has earlier been reported for the biological control of different fungi including species of *Aspergillus niger*, *Penicillium expansum*, *Botrytis cinerea*, *Vercticillium dahliae* [10]. Plant growth potential of *E. faecium* for enhancing the growth of rice plant has been reported by Lee et al. [7], which significantly enhanced the length and biomass of rice shoots in both normal and dwarf cultivars.

Conclusion

Based on our findings it may be concluded that *Enterococcus* sp. BS13 produces glycolipid type biosurfactant and bear PGP and biocontrol properties which may be exploited to develop an effective commercial biocontrol agent in future.

Acknowledgement

The authors are thankful to the Head, Department of Botany and Microbiology for providing laboratory facility.

Figure 6: Effect of *Enterococcus* sp. BS13[Met+Fus+] growth of *V.mungo* in Pot experiment. T1, *Enterococcus* sp. BS13[Met+Fus+] +*M. phaseolina*; T2, *Enterococcus* sp. BS13[Met+Fus+].

Treatment	Germination	Shoot length	Root length	Shoot Weight		Root Weight	
	(%)	(cm)	(cm)	Fresh	Dry	Fresh	Dry
Enterococcus sp. BS13[Met+Fus]	80.0	48.1**	17.7**	5.433**	0.912**	1.857**	0.087**
Enterococcus sp.BS13[Met+us+]+ *M. phaseolina*	76.66	45.2**	16.5**	5.109**	0.889**	1.819**	0.085**
Control	63.33	38.9	13.9	3.998	0.793	1.173	0.075
*Values are the mean of triplicates; ns-non significant; ** Significant at 1% LSD as compared to control.							

Table 2: Effect of *Enterococcus* sp. BS13[Met+Fus+] on the growth of *V. mungo* under pot assays after 90 days*.

Treatment	Bacterial population (log10 cfu)*		
	30 days	60 days	90 days
Enterococcus sp. BS13[Met+Fus+]	6.38	7.19	7.96
Enterococcus sp. BS13[Met+Fus+]+ *M. phaseolina*	7.20	7.50	7.52
Values are mean of three replicates.			

Table 3: Root colonization of *V. mungo* by *Enterococcus* sp. strain BS13[Met+Fus+] at 30, 60 and 90 days*.

References

1. Saharan BS, Nehra V (2011) Assessment of plant growth promoting attributes of cotton (*Gossypium hirsutum*) rhizosphere isolates and their potential as bio-inoculants. J Environ Res Develo 5: 3.

2. Liang C, Tian J, Liao H (2013) Proteomics dissection of plant responses to mineral nutrient deficiency. Proteomics 13: 624-636.

3. Mundt JO (1961) Occurrence of *Enterococci*: Bud, blossom, and soil studies. Appl Microbiol 9: 541-544.

4. Müller T, Ulrich A, Ott EM, Müller M (2001) Identification of plant-associated *enterococci*. J Appl Microbiol 91: 268-278.

5. Coombs JT, Franco CM (2003) Isolation and identification of actinobacteria from surface-sterilized wheat roots. Appl Environ Microbiol 69: 5603-5608.

6. EFSA (European Food Safety Authority) (2012) Guidance on the assessment of bacterial susceptibility to antimicrobials of human and veterinary importance. Eur Food Safe Authority J 10: 2740.

7. Lee KE, Radhakrishnan R, Kang SM, You YH, Joo GJ, et al. (2015) *Enterococcus faecium* LKE12 Cell-free extract accelerates host plant growth via gibberellin and indole-3-acetic acid secretion. J Microbiol Biotechnol 25: 1467-1475.

8. Gálvez A, Giménez-Gallego G, Maqueda M, Valdivia E (1989) Purification and amino acid composition of peptide antibiotic AS-48 produced by *Streptococcus (Enterococcus) faecalis* subsp. *liquefaciens* S-48. antimicro agents Chemother 33: 437-441.

9. Martínez-Bueno M, Maqueda M, Gálvez A, Samyn B, Van Beeumen J, et al. (1994) Determination of the gene sequence and the molecular structure of the *enterococcal* peptide antibiotic AS-48. J Bacteriol 176: 6334-6339.

10. Fhoula I, Najjari A, Turki Y, Jaballah S, Boudabous A, et al. (2013) Diversity and antimicrobial properties of lactic acid bacteria isolated from rhizosphere of olive trees and desert truffles of Tunisia. Biomed Res Int 2013: 405708.

11. NNDSR (2016) National nutrient database for standard reference: Basic Report 16057: chick peas, (garbanzo beans, bengal gram) mature seeds, cooked, boiled, without salt.

12. Wyllie TD (1988) Charcoal rot of soybean-current status. In Soybean diseases of the north central region. T.D. Wyllie and D.H. Scott, (eds). APS Press, St. Paul, MN.

13. Dubey RC, Maheshwari DK (2012) Practical microbiology. S. Chand and Co., New Delhi.

14. Anandaraj B, Thivakara P (2010) Isolation and production of biosurfactant producing organism from oil spilled soil. J Biosci Tech 1: 120-126.

15. Rodrigues LR, Teixeira JA, Van Der Mei HC, Oliveira R (2006) Physicochemical and functional characterization of a biosurfactant produced by *Lactococcus lactis* 53. Colloids Surf B Biointerfaces 49: 79-86.

16. Siegmund I, Wagner F (1991) New method for detecting rhamnolipids excreted by *Pseudomonas* species during growth on mineral agar. Biotechnol. Technique 5: 265-268.

17. Jain DK, Collins-Thompson DL, Lee H, Trevors TA (1991) Drop collapsing test for screening surfactant-producing microorganisms. J Microbiol Methods 13: 271-279.

18. Sarubbo LA (2006) Production and stability studies of the biosurfactant obtained from a strain of Candida glabrata UCP 1002. J Biotechnol 9: 400-409.

19. Rosenberg M (2006) Microbial adhesion to hydrocarbons: twenty-five years of doing math. FEMS Microbiol Lett 262: 129-134.

20. Lunkenheimer K, Wantke KD (1981) Determination of the surface tension of surfactant solutions applying the method of Lecomte du Noüy (ring tensiometer). Colloid Polym Sci 259: 354-366.

21. Sánchez M, Aranda FJ, Espuny MJ, Marqués A, Teruel JA, et al. (2007) Aggregation behavior of a dirhamnolipid biosurfactant secreted by Pseudomonas aeruginosain aqueous media. J Coll Interface Sci 307: 246.253.

22. Christensen GD, Simpson WA, Bisno AL, Beachey EH (1982) Adherence of slime-producing strains of Staphylococcus epidermidis to smooth surfaces. Infect Immun 37: 318-326.

23. Pikovskaya RI (1948) Mobilization of phosphorus and soil in connection with the vital activity of some microbial species. Microbiol 17: 362-370.

24. Gordon SA, Weber RP (1951) Colorimetric estimation of indole-acetic acid. Plant Physiol 26: 192-195.

25. Schwyn B, Neilands JB (1987) Universal chemical assay for the detection and determination of siderophores. Anal Biochem 160: 47-56.

26. Bakker AW, Schippers B (1987) Microbial cyanide production in the rhizosphere in relation to potato yield reduction and Pseudomonas spp. mediated plant growth-stimulation. Soil Biol Biochem 19: 451-457.

27. Bazzicalupo M, Fani R (1996) The use of RAPD for generating specific DNA probes for microorganisms. Methods Mol Biol 50: 155-175.

28. Kumar H, Dubey RC, Maheshwari DK (2011) Effect of plant growth promoting rhizobia on seed germination, growth promotion and suppression of Fusarium wilt of fenugreek (Trigonella foenum-graecum L.) Crop Protection 30: 1396-1403.

29. Brown MG, Balkwill DL (2009) Antibiotic resistance in bacteria isolated from the deep terrestrial subsurface. Microb Ecol 57: 484-493.

30. Altschul SF, Madden TL, Schäffer AA, Zhang J, Zhang Z, et al. (1997) Gapped BLAST and PSI-BLAST: a new generation of protein database search programs. Nucleic Acids Res 25: 3389-3402.

31. Skidmore AM, Dickinson CH (1976) Colony interaction and hyphal interference between Septoria nodorum and phylloplane fungi. Trans Brit Mycol Soc 66: 57-60.

32. Shweta B, Maheshwari DK, Dubey RC, Arora DS, Bajpai VK, et al. (2008) Beneficial effects of fluorescent pseudomonads on seed germination, growth promotion, and suppression of charcoal rot in groundnut (Arachis hypogea L.). J Microbiol Biotechnol 18: 1578-1583.

33. Singh N, Kumar S, Bajpai VK, Dubey RC, Maheshwari DK, et al. (2010) Biological control of Macrophomina phaseolina by chemotactic fluorescent

34. Weller DM, Cook RJ (1983) Suppressing of take-all of wheat by seed treatments with fluorescent pseudomonads. Phytopathologia 73: 463-469.

35. Holt JG, Krieg NR, Sneath PHA, Staley JT, Williams ST (1994) Bergey's manual of determinative bacteriology. Williams &Wikins Press, Baltimore, USA.

36. Fry NK, Warwick S, Saunders NA, Embley TM (1991) The use of 16S ribosomal RNA analyses to investigate the phylogeny of the family Legionellaceae. Microbiol 137: 1215-1222.

37. Sharma D, Saharan BS, Chauhan N, Procha S, Lal S (2015) Isolation and functional characterization of novel biosurfactant produced by Enterococcus faecium. Springerplus 4: 4.

38. Teichmann B, Linne U, Hewald S, Marahiel MA, Bölker M (2007) A biosynthetic gene cluster for a secreted cellobiose lipid with antifungal activity from Ustilago maydis. Mol Microbiol 66: 525-533.

39. Kulakovskaya T, Shashkov A, Kulakovskaya E, Golubev W, Zinin A, et al. (2009) Extracellular cellobiose lipid from yeast and their analogues: structures and fungicidal activities. J oleo Sci 58: 133-140.

40. Seneviratne G, Weerasekara MLMAW, Seneviratne KACN, Zavahir JS, Kecskés ML, et al. (2010) Importance of biofilm formation in plant growth promoting rhizobacterial action. Plant growth and health promoting bacteria 18: 81-95.

41. Tambekar DH, Gadakh PV (2013) Biochemical and molecular detection of biosurfactant producing bacteria from soil. Int J Life Sci Biotechnol Pharm Res 2: 204-211.

42. Zhang C, Wang S, Yan Y (2011) Isomerization and biodegradation of beta-cypermethrin by Pseudomonas aeruginosa CH7 with biosurfactant production. Bioresour Technol 102: 7139-7146.

43. Ghosh U, Subhashini P, Dilipan E, Raja S, Thangaradjou T, et al. (2012) Isolation and characterization of phosphate-solubilizing bacteria from seagrass rhizosphere soil. J Ocean Uni China 11: 86-92.

44. Schippers B, Bakker AW, Bakker PAHM, Van Peer R (1991) Beneficial and deleterious effects of HCN-producing pseudomonads on rhizosphere interactions. Rhizosphere Plant Growth 211-219.

45. Gupta C, Dubey RC, Maheshwari D (2002) Plant growth enhancement and suppression of Macrophomina phaseolina causing charcoal rot of peanut by fluorescent Pseudomonas. Biol Fert Soils 35: 399-405.

46. Bhatia S, Dubey RC, Maheshwari D K (2005) Enhancement of plant growth and suppression of collar rot of sunflower caused by Sclerotium rolfsii through fluorescent Pseudomonas. Ind Phytopathol 58: 17-24.

47. Patil RS, Ghormade VV, Deshpande MV (2000) Chitinolytic enzymes: An exploration. Enzyme Microb Technol 26: 473-483.

48. Vaaje-Kolstad G, Bøhle LA, Gåseidnes S, Dalhus B, Bjørås M, et al. (2012) Characterization of the chitinolytic machinery of Enterococcus faecalis V583 and high-resolution structure of its oxidative CBM33 enzyme. J Mole Biol 416: 239-254.

Pseudomonas aeruginosa PN1 and its plant growth promotory activity in chirpine. Crop Protection 29: 1142-1147.

11

Antifungal Properties of Phytoextracts of Certain Medicinal Plants against Leaf Spot Disease of Mulberry, *Morus* spp.

Ul-Haq S[1]*, Hasan SS[1], Dhar A[2], Mital V[2] and Sahaf KA[2]

[1]School of Science, Indira Gandhi National Open University, New Delhi, India
[2]Central Sericultural Research and Training Institute, Central Silk Board, Govt. of India, Pampore, (J&K) 192 121, India

Abstract

The fungi toxicant use checks the pathogen, yet its use in the organic farming system is least permitted because of their eco-toxic properties. Hence, the use of plants for their anti-fungal properties which could be used against the pathogen in the organic farming system becomes an area of interest for the eco-friendly mode of disease management. Six commonly available medicinal plants were selected based on their wide application in the state and for their valuable use as ethano-medicines. Cold water extracts of the six plants were tested in vitro against three identified leaf spot pathogens, *Cercospora moricola* Cooke, *Alternaria alternata* and *Cladosporium cladosporiodes* and extracts of plants viz. *Artemisia absanthemum, Allium sativa L, Euphorbia ligularia* Roxb, *Zingiber officinale* and *Datura metel* showed more than 85% conidial inhibition *in vitro* with a 94.56% in *A. absanthemum* and more than 50% decrease in leaf spot disease incidence and severity in field conditions except *A. sativa* L and *E. ligularia* Roxb. at 0.05% decrease in PDI. All the plant extracts screened showed more than 50% percent mycelial inhibition with respect to control (water). The highest mycelial inhibition of more than 70% was found in *E. ligularia* Roxb. Followed by *Z. officinale* in all the three pathogens. So these plants with anti-fungal properties could be utilized against these pathogens, at least to lessen the impact of these pathogens. Similar eco-friendly means of disease control has been appreciated by the present environment conscious generation. Thus exploring new plants for their anti-fungal activity would bring about more resource base for use in eco-friendly and sustainable mode of agriculture especially in organic farming.

Keywords: Antifungal; Eco-organic; Phytoextracts; Leaf spot disease; *Morus* spp.,; Organic farming

Introduction

Mulberry (*Morus* spp.) leaves forms the only food material for the silkworm, *Bombyx mori* L. For the development of silk industry, production of high quality silkworm cocoons is must. To achieve the goal of production of good quality silkworm cocoon crop, certain factors play important role. The most important factor is the mulberry leaf, contributing about 38.2% followed by climate (37.0%), rearing techniques (9.3%), silkworm race (4.2%), silkworm egg (3.1%) and other factors (8.2%) in producing good quality cocoons. Hence, quality of mulberry leaf is one of the basic prerequisite of sericulture and plays a pivotal role for successful silkworm cocoon crop [1]. Healthy mulberry leaves influences the growth, development and quality of cocoons formed and thus decide the superiority of silk to a greater extent. Mulberry is exposed to the ravages of different pests and diseases. All mulberry varieties are attacked by one or more fungi that cause scattered, rather definite, round to oval, angular, or irregularly shaped spots on the leaves. These spots usually become conspicuous from late June through August. Leaf spots are the most common diseases of mulberry plants. The diseases development is favored by cool weather, light and frequent rains, fog or heavy dews, high humidity, and crowded or shady plantings. A few spots on the leaves do little harm to a mulberry plants and are far more unsightly than they are injurious. However, leaf spot infections that start early in the growing season can lead to premature defoliation. If it occurs over two or more successive years, it can seriously weaken a tree, reduce its growth, and increase its susceptibility to bark borers, winter injury, and other diseases. Leaf spots commonly increase in number and size in late summer and early autumn as the leaves begin to senesce. The occurrence of a leaf spot disease late in the growing season generally does not seriously affect the health of a tree. Certain leaf spots with special names, such as anthracnose, black spot, downy spot or white mold, ink spot, spot anthracnose, leaf blister or curl, scab, shot-hole, sooty blotch, and tar spot, all occur in the mulberry plantations by the diversity of pathogens.

Leaf spot disease of mulberry is an important fungal disease of mulberry plantation causing considerable damage to the rearing and ultimately to the cocoon crop parameters. Various methods for the management of the disease have been studied by various workers in other states. There are reports that foliar spray of carbendazim and Mancozeb 70% WP in the ratio of 0.1% and 0.2% were most effective in reducing the disease. However due to residual effect of synthetic fungicides, there is demand for more ecofriendly substances like bio pesticides [2]. As such exploration of plant resources for their antifungal potential against the pathogen is quite inevitable for a sustainable and ecofriendly management of the pathogen. Further these plant extracts could be readily used by the farmers to lessen the impact of the pathogen on their mulberry plantation. Using plant resources for its antifungal activity is an attractive avenue for the development of sustainable mode of moriculture in organic farming system. Hence, new plants especially locally available, need to be explored for their antifungal property. Thus six plants locally used in medicinal purposes were selected based on

*Corresponding author: Ul-Haq S, School of Sciences, Indira Gandhi National Open University, New Delhi-1100068, India
E-mail: sajadulhaqzargar@yahoo.in

their abundant availability during the growing season and for their ethano-economic use [3-5].

Materials and Methods

Pathogen isolation

The pathogen was isolated from the diseased leaf as small, scattered, circular to oval dead areas in the leaves; usually tan, dark brown, yellow, gray, purple, or black. Some spots are raised, shiny, and coal black, others may drop out leaving ragged holes; some are marked with light and dark concentric zones. Numerous spots develop yellow, purple, red, or reddish brown to black margins; and later, in damp weather, increase in size and number and merge into large, angular to irregular dead areas. Dark areas and speck-sized, fungus-fruiting bodies (known as pycnidia, acervuli, and perithecia) commonly form in the dead tissues of many older spots. Heavily infected leaves may turn yellow to brown, wither, and drop early, weakening the tree. Occasionally, some leaf spotting fungi deform or kill leaf tissues, buds, twigs, or even small branches. The pathogen was first isolated on PDA plates at $25 \pm 5°C$ in BOD incubator and transferred on fresh PDA slants at regular interval for pure culturing and for further study.

Plant extracts preparation

Five plants viz. *Artemisia absanthemum*, *Allium sativa* L, *Euphorbia lingularia* Roxb., *Zingiber officinale* and *Datura metel* were selected for the study. Healthy non infected leaves, seeds, rhizomes and bulbs of the six plants were collected from the local area i,e. karewa lands of Panzgam, Pulwama Kashmir and the *zingiber officinale* procured from the local market of Pampore. Extracts were prepared from different plant species (Table 1). Fresh leaves were collected and thoroughly washed in sterilized water before preparing their extract whereas bulbs and rhizomes were used in case of Garlic and Ginger, respectively. The extracts were prepared as per the method of Awuah. The leaves of the selected plants were collected and cleaned with distilled water and dried under shade. Individual samples were ground with the help of mortar

S. No	Botanicals	Pathogen spp.	Treatments #									%I/C
			0.1%			0.2%			0.5%			
			TNO	TNG	% I (χ^2)	TNO	TNG	% I (χ^2)	TNO	TNG	% I (χ^2)	
01	Artemesia absanthemum	C. moricola	63.3 (52.7)	17.7 (24.8)	72.11 {2.28}	72.0 (58.7)	13.3 (21.3)	81.48 {2.41}	68.0 (56.0)	9.67 (18.0)	85.78 {3.43}	94.56
		C. cladosporioides	62.7 (52.6)	15.0 (22.7)	76.06 {2.31}	70.3 (52.6)	13.0 (21.0)	81.52 {3.40}	86.7 (69.0)	15.0 (22.6)	82.69 {3.51}	91.53
		A. alternata	73.3 (58.9)	13.7 (21.6)	81.36 {8.42}	68.3 (58.9)	41.3 (40.0)	39.51 {2.09}	67.7 (56.0)	18.7 (25.5)	72.41 {1.30}	91.24
02	Allium sativa L.	C. moricola	89.0 (70.6)	65.0 (53.7)	26.97 {1.05}	79.0 (70.6)	47.3 (43.4)	40.08 {8.12}	83.0 (65.6)	64.0 (54.6)	22.89 {9.12}	79.6
		C. cladosporioides	85.3 (67.5)	65.3 (53.9)	23.44 {3.04}	86.0 (67.5)	47.3 (43.4)	44.96 {1.15}	83.7 (66.5)	40.0 (39.1)	52.19 {5.20}	86.59
		A. alternata	82.7 (66.3)	72.3 (58.5)	12.50 {0.15}	77.0 (66.3)	53.0 (46.7)	31.17 {4.07}	83.7 (67.9)	49.0 (44.4)	41.43 {1.12}	84.7
03	Euphorbia ligularia Roxb	C. moricola	92.0 (74.5)	21.7 (27.7)	76.45 {3.46}	87.3 (74.5)	17.3 (24.5)	80.15 {2.48}	86.0 (68.0)	10.7 (19.0)	87.60 {6.57}	94.67
		C. cladosporioides	82.7 (65.5)	16.3 (23.8)	80.24 {7.46}	78.0 (65.5)	13.7 (21.6)	82.48 {9.46}	70.3 (57.4)	11.0 (19.3)	84.36 {3.43}	91.7
		A. alternata	89.7 (71.4)	36.3 (37.0)	59.48 {1.27}	74.3 (71.4)	24.7 (29.7)	66.82 {1.28}	58.3 (50.0)	16.7 (24.0)	71.43 {4.26}	91.12
04	Datura metel	C. moricola	78.7 (62.6)	46.7 (43.0)	40.68 {9.12}	89.0 (2.6)	43.0 (40.9)	51.69 {9.21}	74.3 (60.0)	34.3 (35.7)	53.81 {2.19}	91.32
		C. cladosporioides	75.0 (60.3)	24.3 (29.3)	67.56 {12.29}	77.3 (60.3)	21.7 (27.7)	71.98 {2.34}	49.7 (44.8)	16.0 (23.6)	67.79 {1.19}	89.67
		A. alternata	94.0 (76.5)	42.0 (40.3)	55.32 {8.25}	83.7 (76.5)	24.0 (29.3)	71.31 {1.36}	80.0 (63.8)	15.0 (22.6)	81.25 {1.45}	92.2
05	Zingiber officinale	C. moricola	68.3 (56.4)	33.0 (34.8)	51.71 {7.16}	75.7 (56.4)	31.0 (33.8)	59.03 {9.23}	82.3 (65.1)	25.0 (29.9)	69.64 {2.34}	93.29
		C. cladosporioides	9.00 (72.7)	64.7 (53.5)	28.15 {3.06}	92.3 (72.7)	41.0 (39.8)	55.6 {1.24}	71.3 (58.6)	29.0 (32.4)	59.35 {2.22}	88.2
		A. alternata	69.7 (39.5)	58.3 (49.8)	16.26 {0.77}	78.7 (59.5)	48.3 (44.0)	38.56 {3.11}	71.3 (58.2)	30.0 (33.1)	57.94 {1.21}	89.06
06	Control/water	C. moricola	69.0 (55.4)	64.3 (53.1)	4.67 {0.02}							
		C. cladosporioides	87.3 (68.2)	80.3 (60.5)	7.00 {0.15}							
		A. alternata	93.3 (70.1)	87.0 (58.7)	6.34 {0.35}							
SEm ± /F-test			5.01/*	2.14/**		5.45/ns	1.77/**		5.27/**	3.75/**		
Cd at 5%			14.4	6.20		5.75	5.11		15.2	10.8		
CV (%)			13.4	9.70		14.7	9.05		15.0	21.9		
SD ±			13.40	21.16		11.63	14.62		14.33	14.92		

TNO=total no. of spores observed, TNG=total no. of spores germinated, %I/C=Percent inhibition over control (0.5%) and #=avg. of 10 replications and %I/ {χ^2}=percent inhibition and chi test. values in bracket(s) are arc sine values

Table 1: Effect of different Botanicals on the conidial germination of identified leaf spot disease causing pathogens of mulberry.

and pestle or by using the automated grinder followed by addition of sterile distilled water (1.00 ml/g). Then this material was taken in a beaker and boiled at 80°C for ten minutes in a hot water bath. The material was homogenized for five minutes and filtered through muslin cloth. The filtrate was centrifuged at 5000 RPM for fifteen minutes and the clear supernatant was collected [6,7]. This was taken as 100% basic stock solution and further diluted to desired concentrations (2.50, 5.00 and 10.00%) with distilled water before use. Then they were mixed with PDA medium and sterilized as per the method of poisoned food technique. Varying amounts of plant extract were added to PDA to get a final concentration of 5%, 10%, 15%and 20% to access their effect on the mycelial growth of the test pathogen [8-10].

Inoculation

The PDA mixed with the plant extracts were poured in Petri plates and allowed to set. Then, one disc (7 mm) of the test fungus taken from the margin of five days old culture were taken and placed in the reversed orientation at the centre of the Petri plates. Three replications were set up for each treatment. The whole set up is placed in BOD incubator with temperature set at 25°C for five days. Pathogen grown on PDA plates with no plant extracts but with only distilled water acts as control plate. Percent inhibition is calculated as,

The inhibition percent was calculated by the formula given by Vincent.

$$I = \frac{(C-T)}{C} \times 100 \tag{1}$$

Where C=growth in control: T=Growth in treated groups and I=inhibition percent.

Average of four replications of each test is taken for calculations.

In vitro effect of phytoextracts

Spores of three pathogens were taken from 7 day-old cultures on PDA. Spore suspension (103 conidia/ml) were made separately against three different concentrations of the phytoextracts 0.1, 0.2, and 0.5%) when the plates were thoroughly covered with mycelium and spores. The spores were removed and put in triplicate in phytoextracts solution or suspensions in sterile water where different concentrations of each phytoextracts were used. Five ml suspensions of each were taken in small sterilized Petri dishes (65 mm) and kept at 28°C for 30 min. Then a drop of lacto phenol cotton blue was added to conidial suspension on the slides [11]. The slides were finally examined under microscope (×400) for recording the percentage of conidial germination.

Field trials

Field trials were conducted for two consecutive growing seasons (2012 & 2013) at the Central Sericutural Research and Training Institute, Gallander, Pampore, India (TS1) temperate humid, located at an altitude of 1574 meters above mean sea level, 74.93° E longitude and 34.02° N latitude [12]. Mulberry plants raised by following the standard cultural practice. The field shows leaf spot disease incidence during 2011 growing season. Hence experiment is carried out at the natural inoculums potential of the soil. In order to understand the relative distribution of disease and its impact on leaf yield, an experiment was conducted under field conditions for a period of two years (2012 & 2013). Five selected varieties were undertaken by following the paired plot technique with a plant gap of 3×3 ft having three replications. Treatment plots were sprayed with extracts of plants viz. A. absanthemum, A. sativa L, E. ligularia, Z. officinale and D. metel

in 0.2%, 0.1% and 0.05% concentration for the control of leaf spot disease after the initial appearance of disease symptoms. Water sprayed plots were kept as control for comparison. Data were recorded after one week in protected and un-protected plots [13]. Disease severity was calculated at random from 10 plants of each replication. In each plant, all the leaves from three branches, one each at top, middle and bottom position, were counted for recording the disease incidence by using the following grading scale and 0-5 is used for statistical analysis [14-16].

$$Disease\ incience\ (DI) = \frac{No.\ of\ infected\ leaves}{Total\ No.\ of\ leaves\ observed} \times 100 \tag{2}$$

$$Precent\ Disease\ Index\ (PDI) = \frac{Sum\ of\ all\ numerical\ values}{Total\ no.\ of\ leaves\ graded \times Maximum\ grade\ (5)} \times 100 \tag{3}$$

Statistical analysis

The data obtained was analyzed using technique of ANOVA as given by Ronald E Walpole to test the effectiveness of the plant extracts and if there is any significant difference in the antifungal properties of the plant extracts.

Result

In vitro mycelia inhibition

The results as presented in Table 1 shows that the plant extracts were effective in significantly reducing the conidial germination as compared to control where 7%, 6.34% and 4.67% inhibition was observed in C. moricola, C. cladosporiodes and A. alternata respectively. More than 90% conidial germination inhibitions as compared with control plates were observed with 0.5% aqueous extracts of the plants viz. A. absanthemum, A. sativa L., E. ligularia, Z. officinale and D. metel. Again 72.7% mycelial growth inhibition was observed by using followed by Z. officinale, E. ligularia, which showed the 71.6% inhibition. The least mycelial growth inhibition was found in D. metel, which showed 45.5% followed by 48.9% inhibition by A. sativa L. in C. moricola and A. alternata.

Disease incidence

The decrease over percent disease index (PDI) was highest in all the five phytoextracts except A. sativa L. in all the three concentrations (0.05, 0.1 and 0.2%) and E. ligularia at 0.05% (Table 2). The application of the phytoextracts of D. metel showed 76.3% followed by 71.9% in E. ligularia showed decrease in the disease incidence and severity. The finding showed the antifungal effect of the five plants extract against leaf spot disease pathogens (Table 3).

Discussion

Leaf spot disease pathogens are necrotropic foliar pathogens with high competitive saprophytic and biotrophic activity. Hence regular application of the fungicide is needed for the chronic disease of mulberry. Synthetic chemicals might successfully control the disease but their application is against the logic of eco-organic moriculture. Hence exploration of alternative antifungal agents, especially the plant extracts has merits. Plant extracts as potential antifungal substance has been explored against several fungal diseases. In our study, six plants showed 50% or above fungal mycelium inhibition activity against the pathogens in in vitro experiment. These plants have been reported to possess antifungal properties against different fungi. Although none of the plants under study showed 100% mycelia inhibition plant yet most of them showed antifungal activity against leaf spot disease. From the in vitro and field results, it can be safely concluded that the aqueous extracts of the six plants could be used in the organic farming

S. No	Antagonists	Agar Plate method at 0.5% conc.			% Inhibition over control (%)		
		C.m	C.c	A.a	C.m	C.c	A.a
01	Artemesia absanthemum	12.3 (20.5)	12.0 (20.2)	12.0 (20.2)	58.0	61.7	59.1
02	Allium sativa L.	15.0 (22.7)	12.0 (20.2)	15.0 (22.7)	48.9	61.7	48.9
03	Euphorbia ligularia Roxb	8.0 (16.4)	9.33 (17.7)	8.0 (16.0)	72.7	70.2	72.7
04	Datura metel	16.0 (23.5)	15.0 (20.7)	13.3 (21.3)	45.5	52.1	54.5
05	Zingiber officinale	8.33 (16.5)	9.67 (18.0)	11.7 (19.8)	71.6	69.1	60.2
10	Control	29.3 (32.7)	31.3 (34.0)	29.3 (32.7)			
	Cd at 1%	2.92	2.60	4.52			
	Cd at 5%	7.44	6.60	11.4			
	SEm ± /F-test	0.95/**	0.84/**	1.47/**			

Cc=Cerrcospora moricola, Cc=cladosporium cladosporiodes and Aa=alternaria alternate.
Values in Brackets are sine transformed values

Table 2: *In vitro* botanical control on leaf spot disease pathogens at CSR&TI, Pampore in August 2012.

S No	Treatment	Conc.	GSH		KNG		IC N		TR-10		CW	
			Mean PDI	% D/C	Mean PDI	% D/C	Mean PDI	% D/C	Mean PDI	% D/C	Mean PDI	% D/C
01	Artemesia absanthemum	0.20	1.51 (7.03)	70.6	3.31 (10.4)	55.29	3.61 (10.9)	61.9	5.98 (14.1)	53.0	6.95 (15.2)	58.81
		0.10	2.24 (8.60)	56.4	4.03 (11.5)	45.52	4.24 (11.8)	55.3	6.73 (15.0)	47.1	7.65 (16.2)	54.64
		0.05	3.17 (10.2)	38.4	4.48 (12.2)	39.53	5.44 (13.4)	42.5	7.2 (15.5)	43.4	8.51 (16.9)	49.56
02	Allium sativa L.	0.20	3.77 (11.1)	26.7	5.25 (13.2)	29.13	5.01 (12.8)	47.1	7.31 (15.5)	42.5	8.67 (17.0)	48.61
		0.10	4.24 (11.8)	17.6	5.10 (13.0)	31.16	5.15 (13.0)	45.7	7.63 (15.9)	40.0	9.71 (18.0)	42.42
		0.05	4.36 (12.04)	15.2	5.39 (13.3)	27.24	6.38 (14.5)	32.6	9.55 (17.9)	24.9	10.7 (18.9)	36.83
03	Euphorbia ligularia Roxb.	0.20	1.22 (6.33)	76.3	3.08 (10.1)	58.4	4.02 (11.5)	57.6	4.77 (12.2)	62.5	7.03 (15.3)	58.29
		0.10	2.60 (9.28)	49.4	4.06 (11.6)	45.11	4.99 (13.3)	47.4	6.39 (14.6)	49.8	7.63 (16.0)	54.77
		0.05	3.24 (10.3)	36.9	5.29 (13.2)	28.55	6.12 (14.3)	35.4	7.27 (15.6)	42.9	10.9 (19.3)	35.58
04	Datura metel	0.20	1.44 (6.88)	71.9	3.29 (10.4)	55.56	5.12 (13.0)	46.0	5.96 (14.6)	53.1	6.80 (15.0)	59.68
		0.10	2.44 (8.97)	52.6	4.15 (11.7)	43.99	6.52 (14.7)	31.1	7.29 (14.1)	42.7	7.87 (16.2)	53.33
		0.05	2.99 (9.94)	41.9	5.30 (13.3)	28.41	7.62 (15.9)	19.5	7.53 (15.6)	40.8	9.25 (17.7)	45.15
05	Zingiber officinale	0.20	1.91 (7.93)	62.8	3.22 (10.3)	56.55	3.46 (10.7)	63.4	5.82 (15.9)	54.2	6.96 (15.2)	58.75
		0.10	2.62 (9.31)	49.0	4.08 (11.6)	44.89	4.76 (12.5)	49.7	7.55 (13.9)	40.6	6.76 (15.0)	59.93
		0.05	3.30 (10.1)	35.7	5.46 (13.5)	26.25	5.17 (13.1)	45.4	7.94 (14.9)	37.6	7.06 (15.4)	58.13
06	Control/water		5.10 (15.7)		7.40 (17.9)		9.50 (17.6)		13.0 (16.3)		17.0 (24.2)	
Cd at 5%			0.39		0.78		1.57		2.03		2.40	
CV (%)			8.26		10.41		17.30		16.69		16.63	
SEm±			0.13		0.27		0.54		0.70		0.83	

Table 3: Field control of different botanicals on the leaf spot disease of mulberry at CSR&TI, Pampore.

environment to lessen the impact of the pathogens on mulberry leaf crop, although complete control could not be attained. Yet based on their wide availability and ease of application it could be used in wide scale in the moriculture. Even though more useful oils and other components could be extracted through the use of other synthetic solvent and refined techniques yet, their use by the marginal farmers in the organic environment is limited. Hence the use of aqueous extracts has merits and is simple and could be easily followed even by a layman. This study would benefit the farmers who wish to lessen the impact of leaf spot disease and enhance the cocoon crop. More and more plants, locally available need to be explored for a fruitful sustainable moriculture.

Conclusions and Forward Look

From the results some plants under study showed significant inhibitory effect even though none of the plant extracts shows cent percent mycelia inhibition. They are widely available in the state. Hence these plants could be used in the organic farming environment to lessen the impact of the leaf spot disease at global level and mostly in the temperate region of Jammu and Kashmir. More novel plants need to be explored to increase the resource base for use in eco-organic moriculture in a sustainable mode.

Acknowledgements

Authors are highly thankful to Dr. K. A Sahaf, Director CSR&TI, Govt. of India, Pampore, Kashmir for providing all the necessary facilities and encouragement for the present work. We duly acknowledge the contributions of Prof. Vijayshri, Director, School of Sciences, IGNOU, Main Campus, New Delhi and Dr. Irfan Illahi, Scientist-B CSR&TI, Pampore and Prof. Neera Kapoor, (Coordinator School of Sciences, IGNOU), Prof. Amrita Nigam and Prof. Jaswant Sokhi eminent professors of the School of Sciences, IGNOU, Main Campus, Maidan Garhi, New Delhi for their generous and enthusiastic help and constructive suggestions.

References

1. Guttierrez WA, Shew HD, Melton TA (1997) Source of inoculums and management of Rhizoctonia solani causing damping off on tobacco transplants under greenhouse conditions. Plant Diseases 81: 604-608.

2. Adityachaudhury N (1991) Phytochemicals – their potency as fungicides and insecticides and their prospects of manipulating natural production. Biotech. in crop Protection: 203.

3. Mathew KA, Gupta SK (1996) Studies on web blight of French bean caused by Rhizoctonia solani and its management . Journal of Mycology and plant pathology 26: 171-177.

4. Ganesan T (1993) Fungitoxic effect of wild plant extracts. Geobios 20: 264-266.

5. Gopal K, Chhetry N, Mangang HC (2011) Effect of soil amendments on soil borne pathogens of French bean (Phaseolus vulgaris L) in organic farming system of Manipur. Journal of Agricultural science and Technology 1: p68.

6. Srivastava AK, Lal B (1997) Studies on biofungicidal properties of leaf extract of some plants. Indian Phytopathology 50: 408-411.

7. Biswas S, Sen SK, Kumar T (1995) Integrated disease management systems in mulberry. Sericologia 35: 401-415.

8. Sindle GR, Patel RL (2004) Evaluation of plant extracts against Rhizoctonia solani incitant of black scruf disease of potato.Journal of mycologyand plant pathology 32: 284-286.

9. Khare MN, Shukla BN (1998) Utility of plants in crop disease control. Vasundhara 3: 1-15.

10. Sunita Chandel, Manica Tomar (2008) Effectiveness of bioagents and neem formulations against fusarium wilt of carnation, Indian Phytopathology 61: 152-154.

11. Walpole RE (1982) Introduction to Statistics, (3rdedn). Macmillan Publishing Company.

12. Mamatha T, Ravishankar Rai V (2004) Evaluation of Fungicides and plant extracts against Fusarium solani, leaf blight of Terminalia catapa. Journal of Mycology and Plant Pathology 34(20): 306-307.

13. Gopal K, Chhetry N, Belbahri L (2009) Indigenous pest and disease management practices in traditional farming systems in north east India. A review. Journal of plant breeding and crop science 1: 28-38.

14. Sachin SK, Upamanyu G, Shyam KR (2004) Innovative approaches for the management of root rot and web blight (Rhizoctonia soalni) of French bean. Journal of Mycology and plant Pathology 32: 317-331.

15. Nicolls MJ (1970) Antifungal activity in Passiflora Species. Ann Bot 34: 229-237.

16. Satish S, Mohana DC, Ranhavendra MP, Raveesha KA (2007) Antifungal activity of some plant extracts against important seed borne pathogens of Aspergillus sp. Journal of Agricultural Technology 3: 109-119.

12

Genetic Diversity of *Alternaria alternata* Isolates causing Potato Brown Leaf Spot, using ISSR Markers in Iran

Shima Bagherabadi, Doustmorad Zafari* and Mohammad Javad Soleimani

Department of Plant Protection, College of Agriculture, University of Bu Ali Sina, Hamedan, Iran

Abstract

Sampling was carried out from different fields in Hamedan province, Iran during the years 2012 and 2013. Among the 300 obtained isolates of *Alternaria* spp. after morphological identification, it was revealed that *A. alternata* isolates from potato had the highest frequency distribution. Due to the abundance of this species in this region which is considered as a major potato producing province in the country, genetic diversity of its isolates were assessed using ISSR markers. Among *A. alternata* isolates, 11 isolates screened from nine different regions on four potato cultivars were selected. A total of 15 ISSR primers of UBC group were used to investigate the genetic diversity of these isolates. Out of 15 primers used in this study, 5 primers had favorable results and produced a significant number of bands. Based on the results of cluster analysis using the Jaccard's coefficient, isolates were divided into two main groups and there was some correlation between the grouping of isolates regarding their geographic location, pathogenicity and potato cultivars.

Keywords: *Alternaria alternate*; Cluster analysis; Diversity analysis; ISSR markers; Potato

Introduction

Alternaria alternata (Fr.) Keissler is common saprobe found on many plants and other substrate worldwide [1-3] and can cause damage to many plants in different agro ecosystems, including potato brown leaf spot disease. It is mainly found in the soil or on decomposing plant tissues [4]. This species is also an opportunistic pathogen affecting many cultivated plants in the field and during post-harvest storage of fruit and vegetables. This fungus attacks plants including cereals, ornamental plants, magnolia, oilseeds, vegetables such as cauliflower, broccoli, eggplant, pepper, carrot, potato, tomato, bean and fruits such as citrus, apple, strawberry and peach and in some cases, is known as post-harvest pathogens [5]. In recent years, molecular markers have been used for practical studies of many organisms. In a way, that exploring the various types of molecular markers has made major progresses in genetic studies [6]. DNA is the essence of genetic differences between two specified organism and DNA fingerprinting is now one of the methods for identification of biological organisms. DNA polymorphism is the base of many genetic studies [1]. Genetic diversity is the base of evolution of species and populations [7]. The ability of a population to adapt to different environmental conditions is dependent on the level of genetic diversity [8]. Populations with higher genetic diversity are stable against changes in environmental conditions. Therefore, the first step for managing a plant pathogen is to investigate the diversity. One of the reasons for lack of success in the management of plant diseases is due to lack of information about the structure of the pathogen populations; therefore enhancing knowledge in this area will certainly be helpful in adopting effective methods of managing a pathogen. Therefore being aware of the genetic diversity of populations and the structure of populations within a species, not only checks the evolutionary processes and mechanism of that, but also provides useful information about biological conservations [9].

There are several methods for evaluating genetic diversity. The choice of molecular markers depends on their high reproducibility, simplicity of method, low cost and high reliability. In 1994, a new type of molecular markers briefly known as ISSR was introduced by Zitkovich et al. and quickly was used in different fields. These markers have high similarity to RAPD markers and are widely dispersed throughout the genome [10]. ISSR is a multilocus marker that in addition to these features does not have the restrictions of other markers such as low reproducibility in RAPD, high costs and complexity in AFLP [11]. ISSR being able to create polymorphism patterns among near-organisms and their reproducibility caused to be known as informative markers with a wide range of applications including the study of genetic diversity [12]. Recent studies show the extremely diverse nature of these markers and their potential to study the different levels of population [2]. Although for some *Alternaria* species *Lewia* is known as sexual stage but the sexual stage of *A. alternata* is unknown [3,13]. These species are likely haploid which reproduce asexually in vegetative phase and is expected to have a low genetic diversity. Guo et al. [14] used 20 ISSR primers to study genetic diversity of 112 isolates of endophytic *Alternaria alternata* isolated from pine in China. They found out tha of these, only two primers showed high levels genetic diversity of this species in pine. Their results showed no correlation between the fungal genotype and host age and endophytes of *A. alternata* had great potential for development and maintenance of genetic diversity. Zhong-hui et al. [15] also studied genetic variation in eight isolates of *A. alternata* at five region of China on Tobacco using ISSR markers. Kale et al. [16] studied genetic diversity of 20 isolates of *A. alternata* on linseed at 14 districts of India using nine ISSR primers. Their research results showed that there was a high genetic diversity among these 20 isolates and no correlation between genetic diversity, geographic region, and intensity of pathogenicity.

Hamedan province in Iran, through having good and adapted

---

*Corresponding author: Doustmorad Zafari, Department of Plant Protection, College of Agriculture, University of Bu Ali Sina, Hamedan, Iran
E-mail: Zafari_d@yahoo.com

weather for the cultivation of potato has dedicated a large percentage of potato production in the country to itself, With cultivated area equal to 25,000 hectares and annual production of 900 thousand tons potato tubers has gained an special position in production of potato in the country.

Since *A. alternata* is one of the most important pathogens causing potato brown leaf spot in this region, the aim of this study was determined to investigate the genetic diversity of isolates of this species on potato using ISSR markers in Hamedan province, Iran.

Materials and Methods

Sampling, isolation and identification

Sampling was done during the spring, summer and autumn of 2012 and 2013 from potato fields, in different locations of Hamedan province, Iran. The collected samples were placed separately in paper bags and plant names, place and date of sample collection was recorded. In order to conduct further studies plant tissues were transported to the laboratory and stored in a refrigerator at 4°C. Plant organs with suspected symptoms of contamination were washed under running water for 5 minutes. Sections possess symptoms were cut from healthy sections and segmented into parts of 0.5 to 1 cm. These segments were surface sterilized with 10 percent sodium hypochlorite for 2 minutes and immediately were washed twice with sterile distilled water. The pieces on the dewatering filter paper were transferred to the petri dishes containing potato dextrose agar (PDA) medium. Petri dishes were kept in an incubator at 23-25 °C. Five to seven days after incubation, isolates were purified using single spore method and then were transferred to the tubes containing potato carrot agar (PCA) medium for storage and subsequent studies. Microscopic identification at the genus level was achieved using Lica microscope according to the imperfect fungi key [17].

Evaluation of morphological features

To study the morphological characteristics of purified isolates at the species levels, subcultures were transferred to the petri dishes containing PCA medium. These petri dishes were kept at 23 to 25°C under fluorescent light with light cycle of 8 hours light and 16 hours dark and were studied after five to seven days. For white light, two 40 W white fluorescent lamps were used at the distance of 40 cm from the surface of petri dishes. Microscopic identification was achieved using Lica microscope according to descripion of *Alternaria* species [18] After morphological identifications, *Alternaria* isolates with higher frequency were selected for further studies such as study of thier pathogenicity and genetic diversiy.

Pathogenicity test

To conduct the pathogenicity test, healthy potato tubers were cultured in sterilized pots containing pasteurized soil. Potato seedlings with 5-6 leaves, were sprayed separately with *A. alternata* spore suspension (10^6 spores per ml) of each isolate. 48 h before pathogen inoculation, plants were covered with plastic bags to keep the relative humidity at 100%. To avoid air trapped small holes was created at the top of the bags. After pathogen inoculation, the plants covered for 48 hours with plastic bags and were kept in greenhouse conditions with 14 hours light and 10 hours of darkness.

Molecular studies

Among 165 obtained isolates of *A. Alternata* from potato in different reagions of Hamedan province, screening was achived

according to the main reagiones of potato production in this province and kind of potato cultivatrs that isolates were isolated from those, therefore due to the great similarity among isolates from each reagion and considering to potato cultivare 11 isolates as representative were screened for study of gentics diversity.

DNA extraction

Mycelia of *A. alternata* isolates grown in potato dextrose broth (PDB) medium were collected, washed and were kept at -20°c. DNA was extracted using modified method of Sharma et al. [19]. Frozen mycelia were poured in a porcelain mortar which was pre-chilled in the freezer and were powdered in liquid nitrogen. 200 mg of powdered mycelia were transferred to each 1.5 ml tubes and 750 μl of extracted buffer stored in 60°c (2% PVP-40, pH 8.0 EDTA 20 mM, CTAB 5% (W/V) Tris-HCl pH 8.0 100 mM, NaCl 1.4 M, 2.0% mercaptoethanol) was added to each sample, mixed and for 35 minutes was kept in 60°C hot water bath and at this time the contents of the tubes were shaken gently several times. Equivalent to the volume of the tube, the mixture of chloroform – isoamyl alcohol (1:24) was added to each tube containing the sample and was mixed gently for one minute.

The mixture was centrifuged for 15 min at 7,000 rpm, then supernatant was taken and poured into a new sterile tube. Sodium acetate of 3 M, 5.2 PH and 0.6 volume of cold isopropanol solution was added to one to thirty of volume and the solution in the tubes was gently mixed several times, at this stage, the DNA strands were formed which were easily visible. The tubes containing DNA strands were centrifuged for 10 min at 7000 rpm at 4°C and the supernatant was emptied gently so that the DNA remained intact inside the tube. Then 500 ml 70% ethanol was added to the tubes containing DNA, and centrifuged at 13,000 rpm at 4°C for 5 min. The upper phase was discarded and tubes in the air upside down were placed on absorbent paper so that the deposition driedand finally 50 μl of sterile double- distilled water was added to each tube. The samples were stored overnight in the refrigerator until the mass of DNA distilled in water. To detect the extracted DNA, 1.2% Agarose gel in TBE buffer was used and then 5 μl of DNA with double amount of loading buffer was mixed and electrophorzed at a constant voltage of 80 volts for 1.5 hours. Quantity and quality of extracted DNA was determined by spectrophotometry and agarose gel electrophoresis.

Regulation of polymerase chain reaction (PCR)

To confirm the morphological identification, one of the isolates was selected as representative for molecular analysis and protein encoding gene of Glyceraldehyde 3-phosphate dehydrogenase (GAPDH) was amplyfied and sequenced. To assess the genetic diversity of isolates of this species, ISSR marker was selected and 15 UBC primers were used for conducting the test in polymerase chain reaction. The names and gpd_1, gpd_2 sequences of primers and specific UBC primer are listed in Table 1.

At the time of conducting tests, to prevent the time error, taking very small amounts of materials and to practice more easily and quickly, master mix of PCR reactions was prepared as shown in Table 2. Master mix includes all materials needed for the PCR reaction except DNA sample and primer.

Firstly, 2 μl of DNA of each sample was poured into PCR specific micro-tubes, then 2 μl primers (with concentration of 10 pmol/μl) was added and kept in the refrigerator. Immediately, 21 μl of the master mix was added to the each micro-tube containing DNA and primer. Thermocycler device of Techne model TC-512 was used

The name of primer	Sequence
UBC 807	5'-AGAGAGAGAGAGAGAGAGT-3'
UBC 808	5'-AGAGAGAGAGAGAGAGAGC-3'
UBC 809	5'-AGAGAGAGAGAGAGAGAGG-3'
UBC 818	5'-CACACACACACACACAG-3'
UBC 822	5'-TCTCTCTCTCTCTCTCA-3'
UBC 834	5'-AGAGAGAGAGAGAGAGAGCT-3'
UBC 835	5'-AGAGAGAGAGAGAGAGAGCC-3'
UBC 840	5'-GAGAGAGAGAGAGAGAGATT-3'
UBC 841	5'-GAGAGAGAGAGAGAGAGACC-3'
UBC 842	5'-GAGAGAGAGAGAGAGAGATG-3'
UBC 846	5'-CACACACACACACACAAT-3'
UBC 849	5'-GTGTGTGTGTGTGTGTCA-3'
UBC 850	5'-GTGTGTGTGTGTGTGTCA-3'
UBC 856	5'-ACACACACACACACACACCA-3'
gpd1	5'-CAACGGCTTCGGTCGCATTG-3'
gpd2	5'-GCCAAGCAGTTGGTTGTGC-3'

Table 1: Names and sequences of primers used in this study.

Per one reaction	Final concentration	Basal concentration	Material
15.2 µl	-	-	Sterile double distilled water
2.5 µl	1X	10X	PCR buffer
1.6 µl	3.2 Mm	50 Mm	$MgCl_2$
1.5 µl	0.6 Mm	10 Mm	dNTPS
0.2 µl	1 Unit	5 Unit/µl	Taq DNA polymerase enzyme
21 µl	-	-	*Total volume of master mix

Table 2: Ingredients and amounts for the preparation of master mix used in the PCR reaction. *Final volume of PCR reaction: (2 µl DNA+2 µl Primer+21 µl master mix)=25 µl.

Number of cycles	Conducted Steps	Time	Temperature (°C)
1 Cycle	Initial denaturation	5 min	94
35 Cycles	Denaturation	1 min	94
	Annealing	75 second	Depending on the type of primer used
	Extension	2 min	72
1 cycle	Final extension	10 min	72

Table 3: Time and temperature required for conducting different stages of PCR for primers gpd_1, gpd_2 and specific UBC primers.

for amplification. According to the program given to the machine, polymerase chain reaction (PCR) was conducted using primers based on the conditions shown in Table 3.

Electrophoresis of PCR products

For electrophoresis of each sample, 5 µl of the initial PCR product was removed and after mixing with 2 µl of loading buffer, electrophorzed on the 1.2% agarose gel in TBE buffer with constant voltage of 80 kV for 2.5 hours. The samples were run on a solution of ethidium bromide for 30 minutes (5.0 mg per µl). After washing the gel with distilled water, the gel was subjected to image analysis with a gel document device (DIGI.DOC H101 model).

Sequenced regions of gpd

To determine the DNA sequence of a representative isolate, regions of gpd were amplified and sequenced. Obtained sequences were blasted with the sequences of these regions related to the isolates of this species in the GenBank.

Analysis of data obtained from UBC primers

The resulting band patterns of each isolate were scored for the presence or absence of amplified products. Cluster analysis of the data was done using UPGMA, by the use of Jaccard's similarity coefficient in the NTSYS-PC software (version 2) and a dendrogram was constructed for each primers and total primers. Decomposition to main coordinates on similarity matrix was obtained through Jaccard's similarity coefficient. To determine the optimal number of clusters in which the highest distinction between the groups is achieved, analysis of molecular variance (AMOVA) was used.

Results and Discussion

In this study, 300 isolates of *Alternaria* were obtained from different hosts and regions in Hamedan province and were identified at species level, based on morphological characteristics. Among identified isolates, 165 isolates were belonged to the *A. alternata* species isolated from potato crop (Table 4). Since isolates of *A. alternata* from potato were the most frequent spcies, genetic diversity of these isolates was studied by using ISSR marker. Among collected isolates from potato crops in this study, 11 isolates were selected from nine different regions and four different potato cultivars in Hamedan province (Table 5).

Item	Name of isolates	Host	Number of isolates
1	Alternaria alternata	Solanum tuberosum	165
2	Alternaria alternata	Medicago sativa	2
3	Alternaria alternata	Prunus persica	2
4	Alternaria alternata	Cucumis sativus	3
5	Alternaria alternata	Salix sp.	2
6	Alternaria alternata	Rosa sp.	2
7	Alternaria alternata	Rumex alpinus	3
8	Alternaria alternata	Solanum lycopersicum	3
9	Alternaria alternata	Lactuca sativa	2
10	Alternaria alternata	Amarantus albus	1
11	Alternaria alternata	Acroptilon repens	2
12	Alternaria alternata	Prunus domestica	3
13	Alternaria alternata	Juglans regia	4
14	Alternaria alternata	Althaea officinalis	2
15	Alternaria alternata	Lepidium draba	2
16	Alternaria alternata	Fraxinus excelsior	3
17	Alternaria alternata	Triticum aestivum	3
18	Alternaria alternata	Diospyros sp.	3
19	Alternaria arborescens	Solanum tuberosum	8
20	Alternaria arborescens	Juglans regia	3
21	Alternaria arborescens	Lactuca sativa	3
22	Alternaria arborescens	Solanum lycopersicum	6
23	Alternaria arborescens	Lepidium draba	2
24	Alternaria arborescens	Triticum aestivum	2
25	Alternaria arborescens	Carex sp.	2
26	Alternaria arborescens	Althaea officinalis	2
27	Alternaria dumosa	Solanum tuberosum	8
28	Alternaria dumosa	Althaea officinalis	2
29	Alternaria infectoria	Solanum tuberosum	3
30	Alternaria infectoria	Althaea officinalis	1
31	Alternaria rosae	Alisma plantago	2
32	Alternaria solani	Solanum tuberosum	15
33	Alternaria tenuissima	Malus domestica	2
34	Alternaria tenuissima	Medicago sativa	3
35	Alternaria tenuissima	Fragaria ananassa	24
36	Alternaria tenuissima	Solanum tuberosum	5

Table 4: List of hosts of *Alternaria* spp. isolates.

Item	Number of isolates	Name of fungus	Host	Cultivar	Region
1	16	*Alternaria alternata*	Potato	Esprit	Asadabad
2	218	*A. alternata*	Potato	Sante	Asadabad
3	15	*A. alternata*	Potato	Marfona	Hamedan
4	226	*A. alternata*	Potato	Sante	Hamedan
5	223	*A. alternata*	Potato	Agria	Nahavand
6	215	*A. alternata*	Potato	Agria	Qorveh
7	214	*A. alternata*	Potato	Agria	Gahavand
8	120	*A. alternata*	Potato	Agria	Bahar
9	91	*A. alternata*	Potato	Agria	Gheidar
10	269	*A. alternata*	Potato	Agria	Kabodrahng
11	286	*A. alternata*	Potato	Agria	Razan

Table 5: List of the used isolates to determine the genetic diversity of along with the host, cultivars and sample locations.

Pathogenicity test

After three weeks the symptoms of the disease caused by the 214, 269 and 286 isolates were observed on tested plants. Samples were taken from diseased plant and cultured on PDA and again after ten days of incubation, the pathogen resembles to those of the original isolates, *A. alternata* was identified. However, no symptoms were observed in other tested isolates.

Results of molecular studies

Blast search from the *gpd* gene sequences showed 99% similarity between representative and *A. alternata* in the gene bank (Accesion number: KP057228) that confirmed the morphological identification results. The results of molecular studies to determine the genetic diversity of *A. alternata* isolates using ISSR markers showed that theses markers are excellent choices for fingerprinting genomes of these fungi. Among 15 primers used in this study that they have been selected accroding thier good resultes in previous studies by reseachers, 10 primers did not have favorable results for genotyping in this study, therefore remaining 5 primers were used to study the genetic relationships of this species. Based on the results obtained from these tests, the binding temperature of the primer in most cases, were close to their melting temperature (Table 6).

In this study, ISSR primers, fragments with sizes of 100 to 3000 bp were amplified. Totally, ISSR primers showed 540 bands that 408 bands were polymorphic. The results showed that UBC807 primer with 157 bands had the highest number of bands and UBC809 primer with 77 bands had the lowest bands. According to the (Figures 1 and 2), UBC809 primer had the highest polymorphism (92%) and UBC807 primer had the lowest levels of polymorphism (75%). Finally, 78 bands loci were amplified by ISSR primers that 63 loci of them were polymorphic. In this study we have examined the same ISSR primers which Kale et al. [16] were used in their conducted research, in order to study the genetic diversity of *A. alternata* isolates in linseed. In the study of Kale et al. [16] mean percentage of polymorphic was 98%, whereas in this study, ISSR markers showed 83.3% polymorphic. The details of used primers in this study are listed in Tables 7 and 8.

Cluster analysis based on ISSR markers

Comparative analysis of molecular variance showed that the highest difference between the two groups at the cutting point of two groups was obtained at similarity level of 55%, indicating the genetic diversity among *A. alternata* isolates. Isolates regarding to their collected regions were placed in two groups. Isolates collected from Asadabad (Esprit *cv*), Hamadan (Marfona *cv*), Asadabad (Sante *cv*), Hamadan (Sante *cv*),

Qorveh (Agria *cv*), Gheidar (Agria *cv*) and Nahavand (Agria *cv*) in a separate group and isolates collected from Ghahavand (Agria), Razan (Agria) and Kabodrahng (Agria) were placed in another group.

Primer name	Primer Binding Temperature (Ta)	Primer Melting Temperature (Tm)
UBC807	45	43
UBC809	44	44
UBC834	48	47
UBC835	48	48
UBC842	44	44
gpd1, gpd2	55	58

Table 6: The name of primer, melting temperature and binding temperature of effective specific primers.

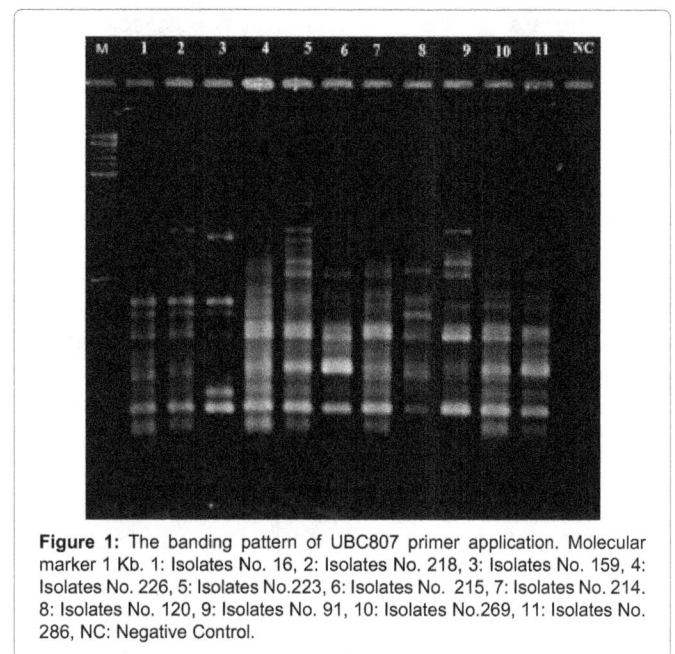

Figure 1: The banding pattern of UBC807 primer application. Molecular marker 1 Kb. 1: Isolates No. 16, 2: Isolates No. 218, 3: Isolates No. 159, 4: Isolates No. 226, 5: Isolates No.223, 6: Isolates No. 215, 7: Isolates No. 214. 8: Isolates No. 120, 9: Isolates No. 91, 10: Isolates No.269, 11: Isolates No. 286, NC: Negative Control.

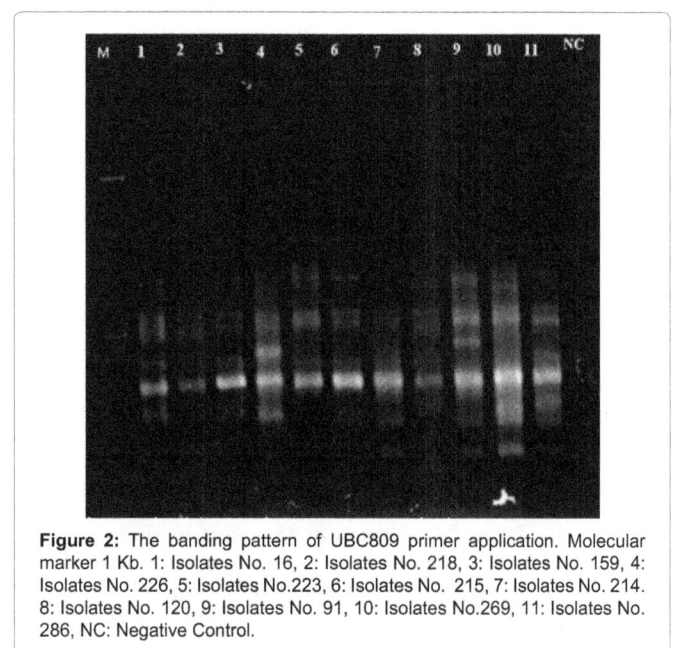

Figure 2: The banding pattern of UBC809 primer application. Molecular marker 1 Kb. 1: Isolates No. 16, 2: Isolates No. 218, 3: Isolates No. 159, 4: Isolates No. 226, 5: Isolates No.223, 6: Isolates No. 215, 7: Isolates No. 214. 8: Isolates No. 120, 9: Isolates No. 91, 10: Isolates No.269, 11: Isolates No. 286, NC: Negative Control.

The results also showed that isolates of *A. alternata* on potato had interaspecific variation which indicates that this marker could demonstrate the genetic diversity of this species. The results of cluster analysis showed that in the first group isolates with different cultivars and geographically dispersed areas and non-pathogenic were placed and in second group same cultivar and geographic region close to each other and pathogenic were placed, thus it can be said that there is some correlation between the genetic diversity of this species, geographical distribution, pathogenicity of potato variation. So this might be concluded that isolates from the same locations showed a tendency to group together compared to geographically farther ones. As can be seen in the dendrogram (Figure 3) isolates obtained from Razan, Kabodrahng and Ghahavand with Agria cultivar and pathogenicity in the north part of Hamedan Province placed in one group and the

Primer name	Number of band loci	Number of polymorphic loci	Total number of loci	Number of polymorphic bands
UBC807	20	15	157	102
UBC809	13	12	77	66
UBC834	17	14	120	87
UBC835	15	13	104	82
UBC842	13	11	82	60

Table 7: Data on locations of bands, polymorphic loci, total bands and polymorphic bands per each primer.

Studied Parameters	Results
Total number of bands	540
amplified polymorphic bands	408
number of amplified loci	78
The number of amplified loci that were polymorphic	65
The number of amplified loci that were not polymorphic	13
Percentage of polymorphic	%83/3
The highest percentage of polymorphic	%92 (UBC809)
The lowest percentage of polymorphic	%75 (UBC807)
The highest number of bands	UBC807))157
The lowest number of bands	UBC809))77
The range of band	100-3000 bp

Table 8: The summery of the results obtained for primer used in this study.

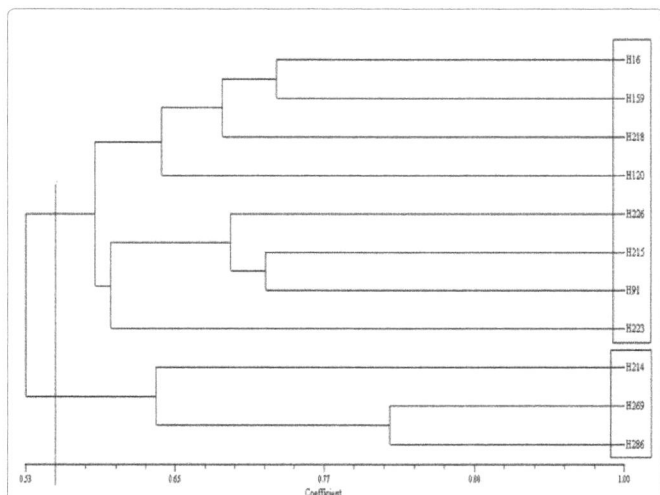

Figure 3: Dendrogram of total ISSR primers used in this study based on Jaccard's similarity coefficient and UPGMA cluster analysis.

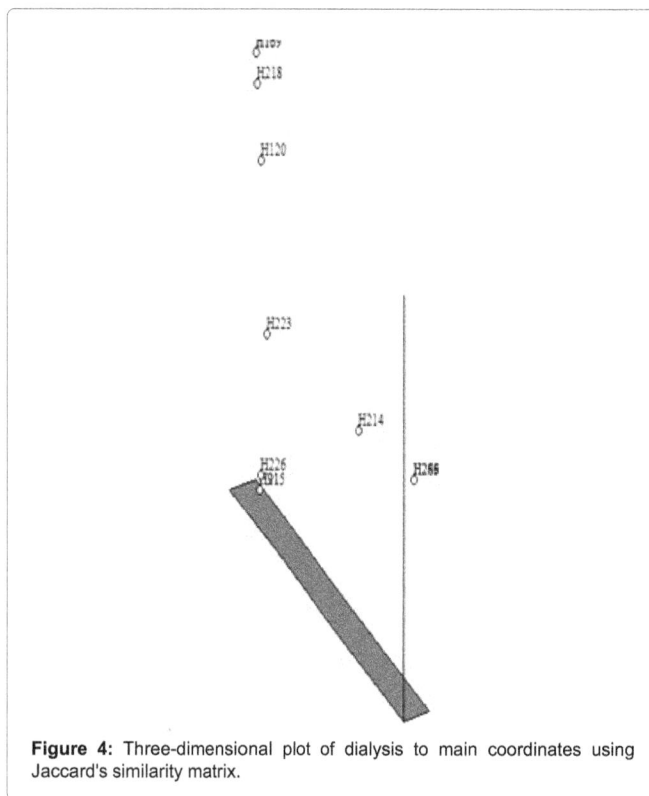

Figure 4: Three-dimensional plot of dialysis to main coordinates using Jaccard's similarity matrix.

remaining isolates on different cultivars and non-pathogenic obtained from other areas of the Hamadan province placed in another group.

Dialysis to main components partially confirmed the cluster analysis results. The first two components justified 70% of the changes that indicates favorable sampling of these remarks from the whole genome. Thus, each of the markers used in different parts of the genome had less correlation. This method also segregated some isolates from each other in different geographical areas (Figure 4).

This study demonstrated the existence of a narrow range of genetic diversity among potato isolates of *A. alternata* from Hamedan province, Iran. The isolates had mixed response for cultivars and pathogenicity characteri.

References

1. Krawczac M, Schmidtke J (1994) DNA fingerprinting. Bioss Scientific Publisher, First ed.

2. Chen JM, Gituru WR, Wang YH, Wang QF (2006) The extent of clonality and genetic diversity in the rare Caldesia grandis (Alismataceae): comparative results for RAPD and ISSR markers. Aguatic Botany 84: 301-307.

3. Simmons EG (1978) Alternaria - an exercise in diversity. Taxonomy of fungi (Subramanian CV). University of Madras, Madras, India 1: 130-135.

4. Tokumasu S, Aoiki T (2002) A new approach to studying microfungal succession on decaying pine needles in an oceanic subtropical region in Japan. Fungal Diversity 10: 167-183.

5. Thomma BP (2003) Alternaria spp.: from general saprophyte to specific parasite. Mol Plant Pathol 4: 225-236.

6. Cetano Analles G, Gresshof PM (1997) DNA Markers, Protocoles, Application and owerviews. New York: John Wiley & Sons.

7. Müller-Starck G, Baradat P, Bergmann F (1992) Genetic variation within European tree species. New Forests 6: 23-47.

8. Ayla FJ, Kiger JA (1984) Modern genetics. (2nded) California USA: Benjamin/ Cummings, Menlo Park.

9. Beiki AH, Abbaspour N, Mozafari J (2013) Genetic Diversity of Cultivated and Wild Crocus genus in Iran with ISSR Markers. Iranian Journal of Biology 26: 164-173.

10. Gupta PK, Varshney RK (2000) The development and use of microsatellite markers for genetic analysis and plant breeding with emphasis on bread wheat. Euphytica 113: 163-185.

11. Han YC, Teng CZ, Zhong S (2007) Genetic variation and clonal diversity in population of Nelumbo nucifera (Neloumbonaceae) in central China detected by ISSR markers. Aguatic Botany 86: 67-75.

12. Zietkiewicz E, Rafalski A, Labuda D (1994) Genome fingerprinting by simple sequence repeat (SSR)-anchored polymerase chain reaction amplification. Genomics 20: 176-183.

13. Simmons EG (2002) Alternaria themes and variations (305-309) Lewia/Alternaria revisited. Mycotaxon 83: 127-145.

14. Guo LD, Xu L, Zheng WH, Hyde KD (2004) Genetic variation of Alternaria alternata, an endophytic fungus isolated from Pinus tabulaeformis as determined by random amplified microsatellites (RAMS). Fungal Diversity16: 53-65.

15. Zhong-hui H, Li-hua Z, Lian-chun W, Bao-hua K, Jing-hue F, Hai-ru C, et al. (2011) Optimization of ISSR-PCR Reaction System for Alternaria alternate in Tobacco. Journal of Yunnan Agricultural Unversity 5: 602-606.

16. Kale SM, Pardesh VC, Gurjar GS, Gupta VS, Gohokar RT, et al. (2012) Inter-simple sequence repeat markers reveal high genetic diversity among Alternaria alternata isolates of Indian origin. Journal of Mycology and Plant Pathology 42: 194-200.

17. Barnett HL, Hunter BB (1998) Illustrated genera of imperfect fungi. 4th ed. USA: St Paul.

18. Simmons EG (1796-2007) Alternaria an identification manual fully illustrated and with catalogue raisonne CBS Biodiversity Series No. 6. Published by Utrecht, The Netherlands. CBS Fungal Biodiversity Centre.

19. Sharma AD, Prabhjot KG, Prabhjot S (2002) DNA isolation from dry and fresh samples of polysaccharide - rich plants. Plant Molecular Biology Reporter 20: 415a-415f.

13

Coffee Thread Blight (*Corticium koleroga*): A Coming Threat for Ethiopian Coffee Production

Kifle Belachew*, Demelash Teferi and Legese Hagos

Ethiopian Institute of Agricultural Research, Jimma Agricultural Research Centre, Plant Pathology Research Section, P.O. Box 192, Jimma, Ethiopia.

Abstract

Besides its importance coffee production constraints with number biotic factors of which diseases are major. Coffee is prone to a number of diseases that attack fruits, leaves, stems and roots, and reduce yield and marketability. Major coffee diseases in Ethiopia are Coffee berry diseases (*Colletotrichum kahawae*), Coffee wilt disease (*Gibberella xylarioides*) and coffee leaf rust (*Himalia vestatrix*) however, the rest diseases considered minor. Thread blight of coffee caused by *Corticium koleroga* is an important disease of Coffee in India. Thread blight diseases in Ethiopian coffee for first time recorded at Gera and Metu agricultural research sub-stations in 1978. However it sporadically occurs between June and September, but becoming important at high land coffee growing areas of southwestern, Ethiopia. Investigations including diagnostic surveys for assessing the disease occurrence, prevalence, incidence and severity was conducted and the sample was brought to Plant Pathology Laboratory of Jimma Agricultural Research Center. The results of study showed that the disease syndrome on detached coffee plants were similar with thread blight of coffee recorded so far and observed at the field. The disease invariably attacks coffee leaves, branches, twigs and berries with characteristic blight symptoms. White fungal threads were seen on the young stems and succulents tender tissues of coffee trees. These threads eventually become dark brown in color grow and spread to cover underside of leaves while coffee berries on infected braches are also completely destroyed leading to total crop failure. The isolation and identification of the causal pathogen from samples of leaves, berries, branches and shoots consistently produced fungal species which may be *Corticium koleroga* that further proved by pathogenicity tests. The disease mean incidence and severity during first outbreak (2008) at Limmu coffee plantation farm of "*Gummer*" was 49.02 and 9.8%, respectively. The second reported outbreak of the diseases was from Bebeka coffee estate of "*Disadis*" farm (2012). Current area wide outbreak at major coffee growing areas of southwest, west and south Ethiopia was in 2014 with mean incidence and severity of 58.44 and 32.59%, respectively, resulting in considerable damages. Among others, climatic factors prolonged rainfall with long period of wet favored the thread blight disease outbreaks implicating the present climate change scenarios are favoring challenging diseases on Arabica coffee production in Ethiopia.

Keywords: Black rot of coffee; *Ceratobasidium noxium*; *Corticium koleroga*; *Pellicularia koleroga*; Thread blight

Introduction

Coffee is not only one of the highly preferred international beverages, but also one of the most important trade commodities in the world next to petroleum [1]. It is currently grown over eighty countries in the tropical regions with an estimated 125 million people in Latin America, Africa and Asia depend on it for their livelihoods [2,3]. Global annual production in 2011 accounted to about 7.86 million tonnes of green beans [1]. Despite the largest percentage reduction in coffee production among 10 leading producers, Ethiopia remains the fifth leading coffee producer in the world and first from Africa. Ethiopia's coffee consumption also is 9th in the world ahead of UK and Inidia [4].

Coffee is prone to a number of diseases that attack fruits, leaves, stems and roots, which in turn reduce yield and marketability [5,6]. High rainfall and relative humidity are common in major coffee growing areas, which favors disease development and survival of inoculums on crop or alternate hosts over seasons. These conditions generally result in disease epidemics that reduce coffee yield. Very more than 13 types of diseases registered to affect coffee plant in Ethiopia. While major coffee diseases are Coffee Berry Diseases (CBD) caused by *Colletotrichum kahawae*, Coffee Wilt Disease (CWD) of *Gibberella xylarioides* and coffee leaf rust caused by *Hemileia vestatrix*, however the rest of diseases considered to be minor [5,6].

Thread blight of coffee caused by the phytopathogenic fugi (*Corticium koleroga*) is an important disease of Coffee in India,

Trindad and Tobego. The disease is fungal which attacks coffee berries and all plant parts except root. The disease associated with fungal pathogens from the *Ceratobasidium* species complex is considered an emerging one and has a very wide hosts ranging from annual herbaceous monocots to perennial woody fruit trees with many kinds of symptoms, which is extremely difficult to control and result in significant economic losses [7].

The first description of the white-thread blight and black rot pathogen was reported on coffee by Cooke in 1876 in India [8,9]. Cooke considered the pathogen a hyphomycete and named it *Pellicularia koleroga*. In 1910, Von Hoehnel re-described the pathogen naming the fungus as *Corticium koleroga* (Cooke) von Höhnel 1910 [8]. In addition to *Corticium koleroga*, synonymy of *P. koleroga* has included *Botryobasidium koleroga* (Cooke) Venkatarayan, *Hyphocnus koleroga*

---

*Corresponding author: Kifle Belachew, Ethiopian Institute of Agricultural Research, Jimma Agricultural Research Centre, Plant Pathology Research Section, P.O. Box 192, Jimma, Ethiopia, E-mail: kiflekef@gmail.com

Stevens & Hall, *Koleroga noxia* Donk and *Ceratobasidium noxium* (Donk) P. Roberts [10,11].

Thread blight diseases on Ethiopian coffee was observed for first time at Gera and Metu agricultural research sub-stations in 1978 and it might have been existed before [5,12]. However, this disease sporadically occurs between June and September, but increasingly becoming important at Gera, Metu and Limmu coffee plantation "*Gumer*". Currently, climate change not only favors the proliferation of certain diseases and pests, but also results in the spreading to regions where they were not existed. Thread blight of coffee outbreak was observed in different coffee plantations like at Limmu in 2008, at Bebeka in 2012 and at Limu horizon in 2014. Currently thread blight disease become significantly important disease in coffee growing areas of Ethiopia. Therefore, investigation including diagnostic survey on disease severity and incidence accompanied by infected coffee specimen's collections were conducted along with isolation and identification of the causal pathogen.

Materials and Methods

Survey areas

Survey was conducted during 2013/2014 cropping season at major plantation coffee farms of southwest and south Ethiopia to determine the distribution, incidence and severity of coffee thread blight. Surveyed farms are Agricaft of "*Duwina*" coffee plantation in Sheka zone of Southern Nations and Nationalities Peoples Region (SNNPR) and Limmu coffee plantation of Horizen PLC which is found in Jimma zone of Oromia Regional State. Additionally, coffee farms at Gera agricultural research sub center in southwestern Ethiopia and Awada agricultural research sub center in southern Ethiopia were observed. From each plantation five representative blocks and from each center different coffee trail plots were used for diseases assessment.

Disease assessment

Coffee diseases assessment was done to know the current occurrence and distribution of outbreak of thread blight diseases in the above mentioned areas (section 2.1). A clustered sampling procedure was used, where five coffee blocks were sampled at each farm. At existing coffee blocks, the assessment was readily undertaken following each row per sample plot in an ordinary manner for disease incidence and severity across the block. At the spot number of diseased and healthy coffee trees in each sample field was counted and recorded. Besides secondary information on coffee varieties, planting years (estimated age), agronomic practices (slashing, hoeing, pruning, shed and or open etc), field history (related to disease outbreak) and other relevant data were gathered during the survey.

In order to identify possible climatic and metrological data's that influence the occurrence and distribution of diseases outbreak, additional biophysical data, i.e., mean rainfall, temperature and relative humidity were collected from nearest metrological stations in each sample area.

Sampled trees were physically numbered from 1-30. Three pairs of branches each pair from upper, middle and lower canopy layers of the coffee plant, was selected and marked with label to assess diseases incidence and severity at each area.

Incidence

For each assessed farm blocks, disease incidence was determined by counting the number of diseased trees over total observed trees.

Disease incidence was obtained by the following formula.

$$\text{Disease incidence (I)} = \frac{\text{number of infected trees or leaves or berries}}{\text{Total number of assessed trees or leaves or berries}} \times 100$$

Severity

Disease severity on trees, leaves or berries was estimate based on percent area covered by lesions of the disease over total area observed.

$$\text{Disease severity (s)} = \frac{\text{plant part covered by lesions or rot}}{\text{Total plant part assessed}} \times 100$$

Isolation and Identification of diseases causing pathogen

Diseased twigs, leaves and berries were sampled from each farm. Which shows clear thread blight symptom was identified and semi infected leaves, branches and berries were collected and packed in perforated plastic bag. All samples were properly labeled with location (region, zone, district, localities, farm and altitude) cultural practices conducted and collection date, and then isolation was done at Jimma Agricultural Research Center plant pathology laboratory following standard procedures.

Standard procedures for isolation of pathogens were employed as, potato dextros agar (39 gm of potato dextrose agare with 1000 ml of distilled water) adjusted to pH 6.5-7.0 and amended with 20 ml (per liter medium) of 5% streptomycin sulphate (sigma) solution. The solutions were autoclaved at 120 °C and 121 Pascal for 25 minutes .The autoclaved media were then poured in sterile Petri dishes and solidified. Twigs, leaves and berries from infected plant parts specimen was carefully removed and cut in small sections (0.5 × 0.5 cm) excised from the intervening regions of discolored and healthy parts using a sterile scalpel. The sections (4-6 per specimen) was transferred in to plastic petridish having 5 ml of 10% of sodium hypochlorite (merck), uniformly agitated for one minute and then immediately rinsed in three changes (each change for 1 minutes) of sterile water. After surface disinfecting and blotting, 4 sections per specimen was aseptically plated using sterile forceps in to petridishes (9 cm) containing potato dextrose agar. The cultures were kept in to incubator adjusted with 21°C for 7 days. Growing cultures of each isolate were visually observed with the intention of detecting distinct colony growth, colony color, and growth habit. This was followed by microscopic observation to determine the type of the fungus based on the morphology of its mycelium, fruiting structures, and spores under microscope. Comparisons of morphological characteristics were made using appropriate information sources like web sites and books.

Pathogenecity test

The conidia used for inoculation was obtained by thoroughly rinsing of the petridish with good colony growth with sterile water in a sterile beaker and stirred up with magnetic stirrer and filtered through double layer cloth. The detached coffee leaves, berries and twigs of ten released coffee varieties were inoculated with a viable conidial suspension using pipettes. All coffee parts were misted with sterile water before inoculation, then after covered with transparent plastic sheet and kept in air conditioned growth room with high relative humidity (>90%) and optimum temperature (23-25°C) for 72 hr to favor infection. The experiments were laid out in completely randomized design (CRD) with three replications. Disease infection and symptom appearance were observed throughout the study period and finally re-isolation was conducted in the laboratory.

Data of diseases reaction on treated coffee plant parts were recorded using scale every five days for six times.

Results and Discussion

Symptoms of the disease on infected coffee trees in the field

The first symptoms of the disease were noticed mainly on the succulent twigs, berries, stems and leaves of the coffee trees. White fungal threads were seen on the young stems and succulents tender tissues of coffee tree. These threads eventually become dark brown in colour. The fungal threads grow and spread to cover underside of leaves (Figure 1).

The disease usually begins on the middle or lower canopies of coffee trees with apparently succulent and actively growing vegetative parts of the plant that spread down the vertical heads with primary or secondary branches leading to complete death. When the rain continues or weather became wet for some time (a week or more), the disease progress at alarming rate and fresh symptoms seen all over the farm, which frustrated growers. However, when the rain stops or became erratic for some time (a week or more), the disease progress also cease and fresh symptoms may not be seen. In such a way, multiple infections are possible per season that finally deteriorated or debilitated the coffee trees although it did not entirely kill the plant unlike vascular wilt or root rot diseases and also it did not affect only coffee berries unlike coffee berry diseases. Such symptoms are similar to that of the disease known as thread blight caused by *Corticium koleroga* documented so far [5,12-14] officially reported on Arabica coffee as minor and locally important coffee diseases in Ethiopia (Figure 1).

The infected leaves gradually turn dark brown, twigs, and berries followed blackening of nodes that progressed to internodes and darkening of petioles and basal parts of the leaves lamina then the dead leaves were hanging at the same node of infected branches or twigs from strong fungal mycelial threads that can be seen at back side of leaves, berry drop, and dieback of primary branches were seen. Further infection caused by the disease resulted in bare dead stems, dried coffee berries were turning to black brown and die. During the rainy and wet season, most of the infected parts show black rot and white mycelia growth. As the disease develops, these symptoms eventually cover the whole leaves in infected branches. In some cases, as the infected branches and twigs dieback, the axial leaves or tips turn yellow and finally become brownish black. The immature berries at pinhead and expanding stages, mature berries are also very susceptible and become shriveled, blanked and die.

Occurrence and distribution of thread blight of coffee

The disease syndrome was observed since 1978 at Metu and Gera agricultural research sub-centers, southwestern Ethiopia [5,12,13]. However, this disease sporadically occurs between June and September, but increasingly becoming important at high land coffee growing areas around Gera, Metu and "Gumer" Limmu. Thread blight of coffee outbreak was seen for first time in 2008 at Limmu coffee plantation farm of "Gumer" with mean diseases incidence and severity of 49.2 and 9.8, respectively. During the first disease outbreak assessment on four commercial varieties of coffee (741, 74110, 75227 and 744) diseases, severity percent was found to be 14.42, 21.08, 2.48 and 1.17, respectively (Figure 2). The second reported outbreak of the diseases was from Bebeka coffee estate of "*Disadis*" farm on which 34 hectares of coffee farm was devastated in 2012 (Jimma agricultural research center back to office report, 2012, unpublished). Currently in 2014 number of coffee farms such as *Duwina* coffee farm of AgriCeft PLC, Limmu coffee Plantation of Horizen PLC and coffee research sub centers such as *Gera, Haru, Mugi* and *Awada* reported similar coffee disease symptoms in the same season. At AgriCeft coffee plantation of *Duwina* farm disease incidence ranged from 5.12 to 92.0 percent per sample plot with average incidence of 50.4 percent and average disease severity of 30.92 percent (Figure 3). In the same year, in addition to its prevalence and increased intensity in already affected areas, the disease outbreak was further noted in another coffee growing farm called *Limmu Sintu and Gumer* with mean incidence and severity of 66.48 and 32.25%, respectively (Figure 4). Besides the same symptom was reported from *Metu, Gera, Haru, Mugi* and *Awada* Research sub-centers and experimental sites in South west, West and southern coffee growing areas of the country. According to Girma et. al. [14] thread

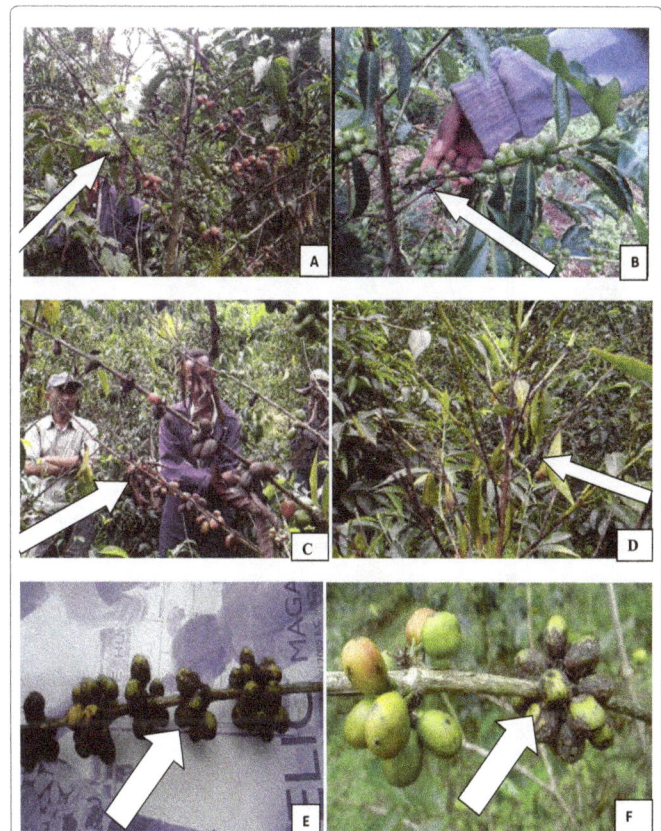

Figure 1: Infected coffee tree branches and stem(A), Infected coffee berries (B), Infected dead leaves were hanging on branches (C); dieback of primary branches (D), black decay and rot of mature berries (E &F).

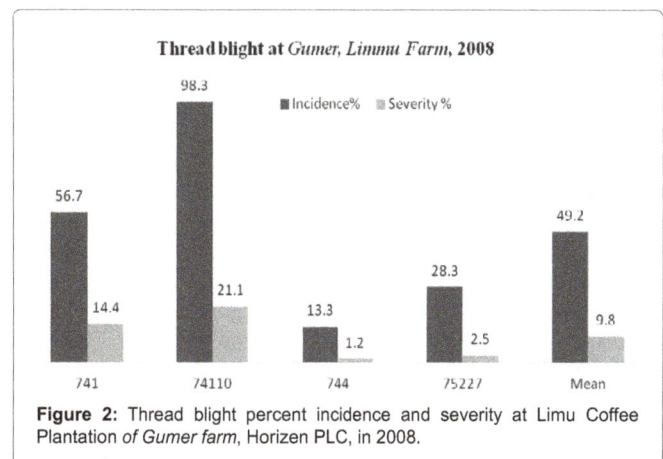

Figure 2: Thread blight percent incidence and severity at Limu Coffee Plantation of *Gumer farm*, Horizen PLC, in 2008.

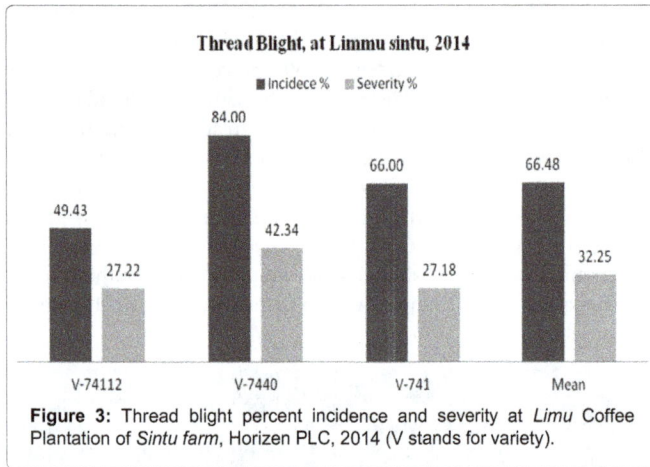

Figure 3: Thread blight percent incidence and severity at *Limu* Coffee Plantation of *Sintu farm*, Horizen PLC, 2014 (V stands for variety).

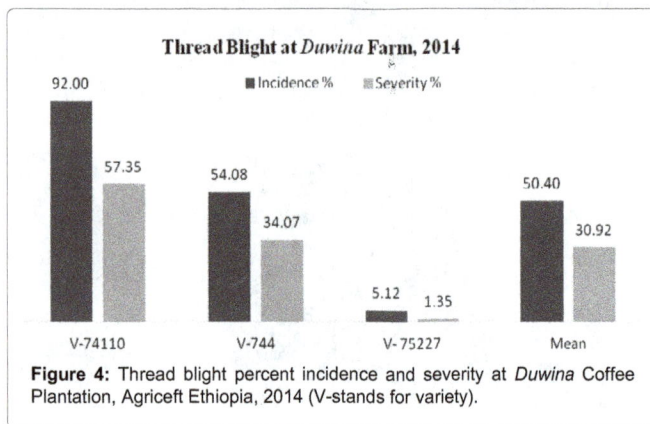

Figure 4: Thread blight percent incidence and severity at *Duwina* Coffee Plantation, Agriceft Ethiopia, 2014 (V-stands for variety).

blight diseases of coffee was one of locally important coffee diseases in Ethiopia. However, current scenario indicates wider distribution of the disease across geographical areas of coffee growing regions of Ethiopia.

The disease attacked the coffee trees at all stages although there were variations among the trees that perhaps attributed to genetic differences among the coffee types under field conditions where open coffee types affected less than compact ones. It was rather more severe and caused extensive damage on coffee varieties like 741 and 74110 and resulting in complete crop failure. Moreover, the severity was so serious in certain areas with heavy shade and long wet season. This observation indicates that fungal diseases are more sever when atmospheric relative humidity become very high in compact coffee varieties and under heavy shade coffee farms [16,17].

Predisposing factors for the occurrence and outbreaks of thread blight in Ethiopia

During this study, interviews with most farmers elucidated that they had not experienced such type of disease symptom that rot their coffee and they claimed severe and prolonged wet season for the disease occurrence. Some coffee farm experts also agreed that the occurrence and outbreak of the disease followed the unusual long rainfall with heavy wet season. These observations suggested that such weather calamities might have predisposed the coffee plants to thread blight infections. Matthew (1953) reported similar observations that heavy and continuous rainfall in coffee growing regions of Coorg, Chikmagalur and Hassan districts of Karnataka region of India has triggered berry dropping and fungal disease outbreak of black rot.

The same author reported that outbreaks of thread blight have been reported from Karnataka in previous years with extreme humidity and the name "koleroga" means "rotting disease" and is derived from the Indian Kannada language.

Heavy, long and continuous rainfalls as well as higher relative humidity from the month of june to September has triggered thread blight disease outbreak in 2014 at most coffee growing areas of Ethiopia (Figure 5). Besides other predisposing factors for the occurrence and outbreaks of thread blight are coffee genotypes, heavy shad and build up of diseases causing pathogens made favorable condition. Similarly, development of thread blight was favored by continuous and heavy rainfall, high atmospheric humidity (greater than 80 percent), shade, and overhanging branches. Moreover, the spread of the disease occur by wind, water, insects and other material, as well as mechanical means (Mathew, 1953; Girma et. al. 2009).

Colony morphology of thread blight isolates on culture plates

The sample infected coffee plant parts were collected from *Limmu Sintu*, *Duina* and *Gumer* coffee farms and from three different locations similar result was obtained. The result of colony morphology of thread blight isolates on culture plates which were collected from different locations was similar in color and growth pattern at early stage. All three days cultures were white in color and profuse fast growth character at obverse and reveres sides. However, pure isolates cultures color slightly changed over time (Figure 6).

Pathogenic reactions of thread blight isolates on detached coffee parts (Leaves, Twigs and Berries) of commercial coffee varieties

In the pathogenicity test, the thread blight isolates inoculated after sporulation on detached leaves, twigs and berries of different coffee varieties showed symptoms of thread blight disease on most coffee varieties (Figure 7). The symptoms were characteristically related to those observed on infected mature coffee trees where leaves gradually turn dark brown or black decay and rot, twigs and berries are rotted following blackening. There was significant difference among commercial coffee varieties tested for thread blight at Jimma agricultural research center pathology laboratory. The diseases development and severity varies among different coffee varieties and plant parts. Detached leaves, twigs and berry test coffee varieties 74110, 741, 7487, 7440, 754 and 74112, exhibited greater diseases severity which indicates highly susceptible reactions to thread blight under laboratory conditions (Table 1). However, F-59 and F-35 coffee varieties showed lowest diseases severity. From this result it is

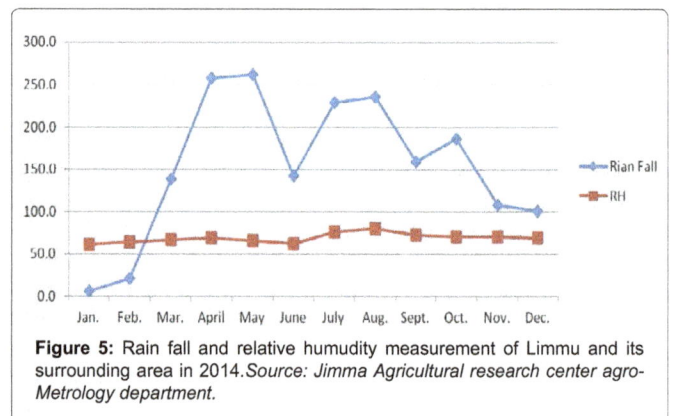

Figure 5: Rain fall and relative humudity measurement of Limmu and its surrounding area in 2014. *Source: Jimma Agricultural research center agro-Metrology department.*

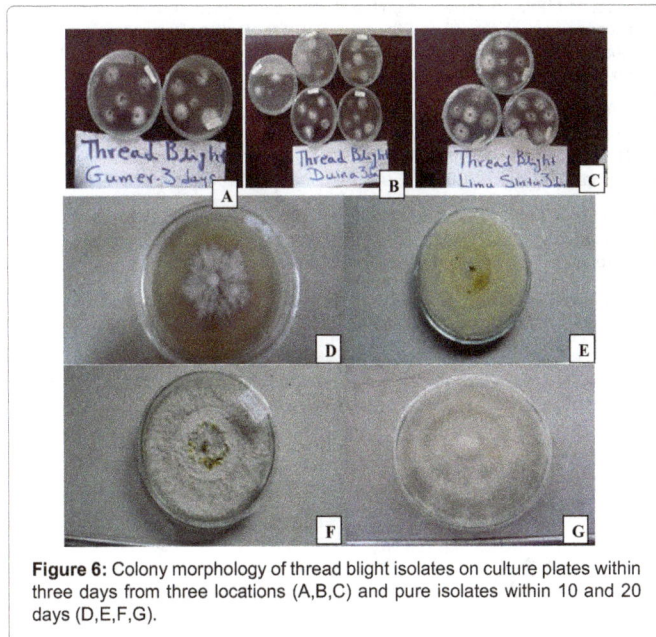

Figure 6: Colony morphology of thread blight isolates on culture plates within three days from three locations (A,B,C) and pure isolates within 10 and 20 days (D,E,F,G).

Figure 7: Pathogenic reactions of thread blight isolates on detached coffee parts (Leaves, Twigs and Berries) of different coffee varieties.

Treatment	Leaves (Disease index[a])	Twigs (Disease index[a])	Berries (Disease index[a])
V- 7487	+++	++	++
V-74110	+++	+++	+++
V-7440	+++	++	+++
V-F-35	–	++	+++
V-74112	+++	++	++
V-F-59	–	–	–
V-74140	++	++	+
V-741	+++	+++	+++
V-754	+++	+++	++
V-744	++	+++	+++

Table 1: Pathogenic reactions symptom of thread blight isolates on different coffee varieties on detached coffee parts. [a]scored as - =0% infection,+ =1-25% infection,++ =26-50% infection and +++ =greater than 50% infection.

possible to conclude that most of commercial coffee varieties showed susceptible reaction for *Corticium koleroga* on leaves, twigs and berries when favorable condition for diseases development happened.

Coffee varieties F-59 and F-35 showed tolerant reaction for coffee thread blight (*Corticium koleroga*). The origin of coffee variety F-59 is southwestern coffee forest at Kaffa zone. It is one of parent coffee verity for hybride coffee variety of "*Gawe*" which is one of three hybrid coffee varieties developed by Jimma agricultural research center for commercial production. F-35 or "*Meoftu*" coffee variety is released in 2002 for commercial utilization in Ethiopia. These two coffee varieties showed good hope for farther detailed genetical study against coffee thread blight (*Corticium koleroga*).

Conclusion

Thread blight diseases on Ethiopian coffee was known for long time and considered as minor coffee disease. However, it sporadically occurs every five to six years between June and September, but increasingly becoming important and observed in wide coffee growing regions as an epidemic in 2014. The disease epidemics is found to be favored by prolonged rainfall and high relative humidity and prevalence of wet and humid conditions, that perhaps reflects one of the climate change scenarios. The disease has been causing severe damage to Arabica coffee since 2008 in small extent but area wide outbreak was seen in 2014 from August to September thus it may be potential threat to coffee production in future. Thus, further in-depth research on the disease epidemiology and control practices are required along with exploring and developing resistant varieties against the emerging thread blight. While strong efforts have been implemented to contain its large-scale damage to major coffee growing areas in humid west and southwest Ethiopia. Optimization of standard protocol for proper pathogenecity test needs some effort for future screening activity. Moreover, detailed characterization of the thread blight causing organisms or pathogen of Arabica coffee is essential to clear out the present controversies in the pathogen population structure in relation to the disease symptoms.

Acknowledgements

The Authors acknowledge all staffs of plant pathology research laboratory, Jimma Agricultural Research Center, Ethiopia for their unlimited contribution of the work. We thank Mr. Sisay Tesfaye, Mr. Mamo Abye and Mr. Wondimu Bekele for their technical support throughout the experiment. Special gratitude goes to Ethiopian Institute of Agricultural Research and Jimma Agricultural Research Center for its facilitation for the experiment.

References

1. ICO (International Coffee Organization) (2012) ICO trade statistics, Exports by exporting countries to all destinations.

2. Osorio N (2002) The Global Coffee Crisis: A Threat to Sustainable Development. International Coffee Organization, London.

3. Musoli PC, Girma A, Hakiza GJ, Kangire A, Pinard F, et al. (2009) Breeding for resistance against Coffee Wilt Disease. In Flood J. (ed.). Coffee Wilt Disease. CABI international, UK.

4. ICO (International Coffee Organization) (2013) ICO trade statistics, Exports by exporting countries to all destinations.

5. Derso E, Gebrezigi T, Adugna G (2000) Significance of minor diseases of *Coffea arabica* L. in Ethiopia: A review. *In*: Proceedings of the workshop on control of coffee berry disease (CBD) in Ethiopia, Addis Ababa, Ethiopia, pp. 58-65.

6. Kifle B, Demelash T, Gabisa G (2015) Screening of some *Coffee arabica* Genotypes against Coffee Wilt Diseases (*Gibberella xylarioides* Heim and Saccus) at Jimma, Southwest Ethiopia. International Journal of Sustainable Agricultural Research 2: 66-76

7. Cavalcante M, Sales F (2001) Ocorrência da queima-do-fio (*Pellicularia*

koleroga) em cafezais em Rio Branco. Empresa Brasileira de Pesquisa Agropecuaria-Embrapa Acre, Rio Branco.

8.   Tims EC, Mills PJ, Exner B (1954) Thread-blight (*Pellicularia koleroga*) in Louisiana. Journal of Plant Disease 38: 634-637.

9.   Ceresini E, Costa S, Marcello Z, Edson F, Nilton L (2012) Evidence that the Ceratobasidium-like white-thread blight and black rot fungal pathogens from persimmon and tea crops in the Brazilian Atlantic Forest agroecosystem are two distinct phylospecies. Genetics and Molecular Biology 35: 480-497

10. Venkatarayan S (1949) The validity of the name *Pellicularia koleroga* Cooke. Indian Phytopath 2: 186-189.

11. CAB-International (2004) *Corticium koleroga* (Cooke) Höhn. CAB International, Wallingford.

12. Teferi D, Adunga G, Jefuka C, Tesfaye S, Zeru A, et al. (2008) Dynamics of sporadic diseases of coffee in Ethiopia: A review. In: Diversty and knowldage; Proceeding of National Work shop four decads of coffee research and development in Ethiopia, Addis Ababa, Ethiopia, pp. 14-17.

13. Ejeta M (1986) A review of coffee diseases and their control in Ethiopia. In: Proceedings of the First Ethiopian Crop Protection Symposium (Tsedeke Abate, ed.) Ethiopian Institute of Agricultural Research, Addis Ababa, Ethiopia, pp. 187-195.

14. Adugna G, Jefuka C, Zeru A, Tesfaye A (2009) Advances in Coffee Diseases Research in Ethiopia. Abraham Tadesse (ed.) Increasing Crop Production Through Improved Plant Protection-Volume II. Plant Protection Society of Ethiopia (PPSE). PPSE and EIAR, Addis Ababa, Ethiopia, pp. 542.

15. Furtado E (1997) Doenças do chá. In: Kimati H, Amorim L, Bergamim Filho A, Camargo L and Rezende J (eds) Manual de Fitopatologia. 3rd edition. Agronômica Ceres, São Paulo, pp. 257-260.

16. Mathew KT (1953) Studies on Black Rot of Coffee. Doctoral Dissertation. Universty of Madras Botany laboratory, India, Madras.

17. JARC (2014) Jimma Agricultural Research Center Research Progress Report for the Period of 2013/14. Getachew Weldemicheal, Yonas Belete, Worknesh Beker, Kifle Belachew (eds), EIAR/JARC, Jimma, pp: 261.

14

Biochemical Approach for Virulence Factors' Identification in *Xanthomonas Oryzae* Pv. *Oryzae*

Sylvestre Gerbert Dossa C[1,2]*, Petr Karlovsky[1] and Kerstin Wydra[3]

[1]*Molecular Phytopathology and Mycotoxin Research, University of Goettingen, Grisebachstr.6 37077, Germany*
[2]*International Rice Research Institute, Division of Plant Breeding, Genetics and Biochemistry, DAPO Box 7777, Metro Manila, Philippines*
[3]*Erfurt University of Applied Sciences, Altonaer Str. 25, 99085 Erfurt, Germany*

Abstract

Bacterial blight caused by *Xanthomonas oryzae* pv. *oryzae* (*Xoo*) leads to a substantial yield reduction of up to 50% in most rice-growing regions. Host plant resistance is an effective control method, and more than 30 resistance genes have been identified in rice genotypes. To understand the interaction of the pathogen in a susceptible reaction of the host plant, *Xoo* culture filtrate and treated culture filtrates were used to treat two rice genotypes using four strains Mai1, PXO88, Dak1 and Dak16. The study revealed that *Xoo* culture filtrate, heated culture filtrate and proteinase K treated culture filtrate induced typical bacterial blight symptoms on rice genotypes IRBB4 and FKR14 with a maximum lesion length of about 23.1 cm for culture filtrate. Heated culture filtrate phytotoxicity effects on both rice genotypes was with highest lesion length of about 6.9 cm, while 13.4 cm was the maximum length induced by a proteinase K treated fraction. After ethyl acetate treatment of the culture filtrate, a considerable reduction of the phytotoxicity was observed. Therefore we suggest that a low molecular-weight toxin may be present in the ethyl acetate extract should not play a major role in *Xoo* virulence and speculate that EPS, Xylanase, polygalacturonase, proteinaceouse contribute to *Xoo* virulence.

Keywords: *Xanthomonas oryzae* pv. *Oryzae*; Culture filtrate; Phytotoxicity; Rice; Virulence

Introduction

Xanthomonas oryzae pv. *oryzae* (*Xoo*) causes an important rice disease called bacterial blight. Rice bacterial blight was first reported in 1884 in Fukuoka, Prefecture of Japan and is today one of the most important rice diseases in Asia and Africa, implying huge economic consequences. Bacterial blight is the economically most important rice disease in the tropics Mew, Mew et al. [1,2] yield losses associated with the disease are up to 50% of the total yield [3]. During infection *Xoo* produces virulence factors such extracellular polysaccharides (EPS), extracellular enzymes, iron chelating side rophores and effectors of type III secretion [4,5]. These virulence factors were identified using molecular approach. The virulence factors play an important role in successful establishment of *Xoo* in the host plant. EPS such as xanthan and lipopolysaccharides (LPS) produced by *Xanthomonas* genus are involved in disease development [6]. It is also known that a diffusible signal factor (DSF) is required for virulence in *Xoo* [4,7-9].

In pathogenic fungi, a secondary metabolite production is well studied and their role in plant infection and their toxicity to humans are well described. Among plant pathogenic bacteria such as *Xanthomonas* spp. toxin production is reported in *X. albilineans* which produces albicidin, known as virulence factor [10]. A non-ribosomal peptide synthetase-polykitide synthase (NRPS-PKS) enzyme related to syringomycin (SyrE) responsible for syringomycin phytotoxin in *Pseudomonas syringae* is found in *X. axonopodis* pv. *citri*, but its toxins production is not known [11,12].

Erdman et al. [13] reported that Rhizobia strains induce chlorosis on young soybean leaves. The toxic compound was later purified from *Bradyrhizobium elkanii*, and called rhizobitoxine, its phtytotoxicity was demonstrated on new soybeans leaves [14]. Rhizobitoxine has been identified as enol-ether amino acid (2-amino-4-[2-amino-3-hydroxypropoxy]-trans-3-butenoic acid), with a molecular weight of 190. Rhizobitoxine has been reported to be an important virulence

factor in many human and animal pathogenic bacteria and is found in many plant pathogenic bacteria such as *X. fastidiosa*, *Rhizobium leguminosarum* and *Erwinia carotovora* [15,16]. In the *Xoo* genome, two apparent RTX toxins, rtxA and rtxC, were identified after genome sequencing of the Korean *Xoo* strain KACC10331[17], but to date the virulence role of the RTX toxins is not yet proved.

Many studies had identified *Xoo* virulence factors using the molecular approach while little is known of their biochemical properties. To better understand the *Xoo* interaction with its host plant, a biochemical approach is needed. Earlier studies revealed that *Xoo* may produce several toxins such as phenylacetic acid (PAA), trans-3-methylthio-acrylic acid (MTAA) and 3-methylthio-propionic acid, which can cause wilting and chlorosis on its host [18]. *Xoo* culture filtrate inhibited rice seed germination [19], but the phytotoxic compounds from *Xoo* and their role in pathogen virulence in the *rice-Xoo*-pathosystem is still not documented. Our study was conducted to identify virulence factors in *Xoo* through phytoxicity effects of four different treatments of culture filtrate.

Materials and Methods

To determine the phytotoxivity effects and identify virulence factors, experiments with liquid culture of the bacteria were conducted. Four *Xoo* strains (Mai1, PXO88, Dak1, and Dak16) were selected

*Corresponding author: Gerbert Sylvestre Codjo Dossa, Molecular Phytopathology and Mycotoxin Research, University of Goettingen, Grisebachstr.6 37077, Germany, E-mail: c.dossa@irri.org according to their virulence and location of origin, Mai1 from West Africa (African race 3), PXO88 from Philippines (race 9) and Dak1, Dak16 from Tanzania (East Africa). Two rice genotypes (FKR 14 and IRBB4) were used. The rice plants were grown for 3 weeks in the greenhouse and transferred into growth chamber for one week before inoculation under 28°C temperature and 78% relative humidity on the fourth week after transplanting.

The strains were grown on solid modified Wakimoto's medium (0.05 g of Fe_2SO_4, 0.82 g of NA_2PO_4, 0.5 g of Ca $(NO_3)_2$ $4H_2O$, 5 g of peptone, 15 g of agar and 20 g of sucrose dissolved in 1 liter of distilled water) for 48 h and a single colony was used to inoculate 300 ml modified Wakimoto's liquid medium and incubated on a shaker for three days at 28°C and 180 rpm. After incubation, the liquid culture was centrifuged at 4754 $x g$ for 10 min and the supernatant filtrated by passing it through a membrane filter with 0.45 µm pore size. For the following experiments, the culture filtrate was divided into 4 fractions: culture filtrate, boiled culture filtrate, culture filtrate with proteinase K and ethyl acetate extraction.

Culture filtrate experiment

Four week-old rice plants of rice genotypes IRBB4 and FKR14 were inoculated, by cutting the leaf tips with a pair of scissors priory dipped in the culture filtrate as described by [19,20].

Boiled culture filtrate

One fraction of the culture filtrate was heated at 80°C for 10 min in water bath to kill the bacteria if still present in the filtrated culture and to inactivate enzymes. The heated culture filtrate was cooled down before use to inoculate rice plants by the leaf clipping method as described above.

Culture filtrate with addition of proteinase

The third fraction of the filtrated culture was heated at 40°C for 20 min followed by addition of proteinase K (5 µg/ml) and heated at 70°C for 10 min. By addition of proteinase K, protein is degraded in the culture filtrate; the heated and treated mixture was centrifuged at 4754 x g for 10 min and the supernatant transferred into a new tube and used *to* inoculate rice plants by leaf clipping.

Ethyl acetate extraction

A culture filtrate of 100 ml volume was extracted with an equal volume of ethyl acetate. The aqueous phase (ethyl acetate phase) was concentrated using a rotary evaporator until 75% of ethyl acetate were evaporated. The 25% left of the aqueous phase were transferred into a 50 ml falcon tube and evaporated using speed vacuum. The product obtained after evaporation was re suspended with 2 ml of 100 mM NaCl and used to treat the rice plant by leaf clipping as described above. The control was performed with non-inoculated modified Wakimoto's medium. The bacteria suspension with 10^6 CFU/ml was used as positive control.

Evaluation and analyses of phytotoxicity and symptoms on rice genotypes IRBB4 and FKR14

For each experiment, phytotoxicity or disease development were assessed by measuring the phytotoxicity effect on leaves as symptom length and the leaf length. Plants were checked daily and the data from 14 and 21 days post treatment were used to build a graph with mean lesion length induced by each strain on rice genotypes IRBB4 and FKR14. For ethyl acetate extract, the lesion length induced by the

control (medium) was subtracted from the lesion length induced by each strain to receive the effect of the ethyl acetate extract. The whole experiment was replicated three times and the mean values were used for statistical analysis.

Statistical analyses were performed with Statistica software version 7. General Linear Models (GLM) were run for pair wise comparison of culture filtrate treatment to others and for all treatments.

Results

Virulence of *Xanthomonas oryzae* pv. *oryzae* strains from different origin on rice genotypes IRBB4 and FKR 14

The lesion length induced on leaves varied from 3.6 cm to 26.7 cm for IRBB4 and from 14.7 cm to 26.7 cm for FKR14, with the maximum lesion length of 26.7 cm at 14 and 21 dpi induced by the Asian strain PXO88. The West African strain Mai1 induced a lesion length of 14.7 cm on genotype FKR14, while the East African strains Dak1 and Dak16 reached medium values with 20 and 21.5 cm of maximum lesion length, respectively, on rice genotype FKR14 (Figure 1).

X. oryzae pv. *oryzae* strains Mai1, PXO88, Dak1 and Dak16, selected for the biochemical analyses of virulence factors were virulent on rice genotypes IRBB4 and FKR14. The highest virulence level was exhibited by the Asian strain PXO88. A difference was found on the virulence levels of PXO88 between genotypes IRBB4 and FKR14 at 14 dpi, and at 21 dpi, the virulence level was nearly the same on both genotypes with 26.7 cm of lesion length. The West African strain Mai1 did not show a high virulence level at 14 dpi and revealed to be the least virulent strain. All the strains showed a virulence progression on the rice genotypes except strain PXO88 which was highly virulent already after 14 dpi, infecting the whole leaf length of FKR14 after 14 dpi. The East African strains Dak1 and Dak16 were more virulent than the West African strain (Mai1), (Figure 1).

Phytotoxicity of culture filtrate

X. oryzae pv. *oryzae* culture filtrate of strains Mai1, PXO88, Dak1

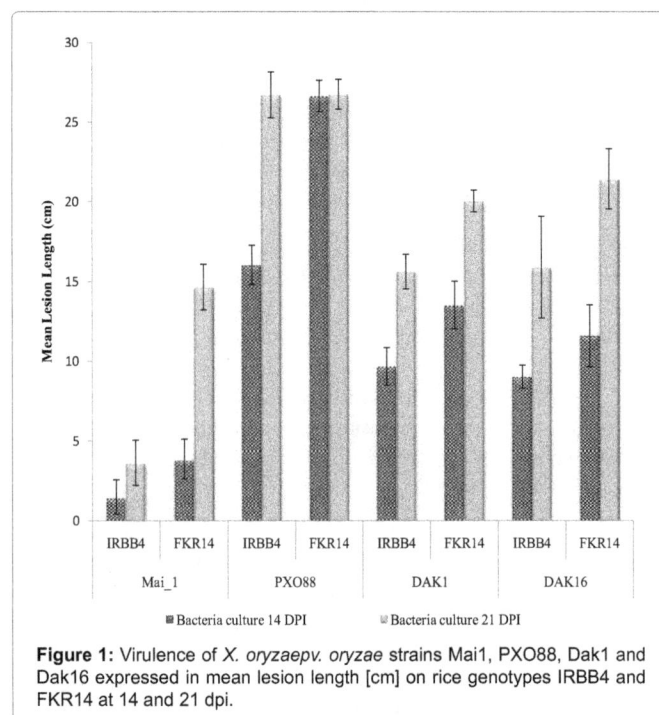

Figure 1: Virulence of *X. oryzae* pv. *oryzae* strains Mai1, PXO88, Dak1 and Dak16 expressed in mean lesion length [cm] on rice genotypes IRBB4 and FKR14 at 14 and 21 dpi.

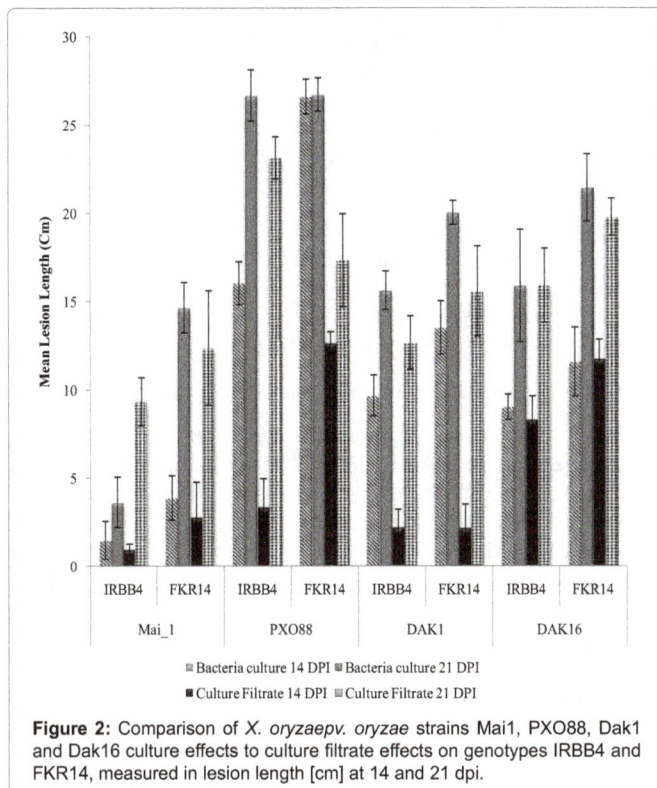

Figure 2: Comparison of *X. oryzaepv. oryzae* strains Mai1, PXO88, Dak1 and Dak16 culture effects to culture filtrate effects on genotypes IRBB4 and FKR14, measured in lesion length [cm] at 14 and 21 dpi.

and Dak16 induced phytotoxicity effects 14 and 21 days after leaf clipping treatment on both rice genotypes IRBB4 and FKR14. The phytotoxicity effects of *Xoo* culture filtrates on both genotypes in comparison to bacteria culture effects are presented as lesion length on leaves (Figure 2). Generally, lesion lengths at 14 dpi were lower in treatment with culture filtrate than in treatments with bacteria culture, with exception of strain Dak16, where lesion lengths from culture filtrate were similar to those from bacteria culture already at this time point. At 14 dpi the phytotoxicity effect of culture filtrate varied from 0.95 cm to 12.7 cm lesion length with the highest lesion length induced by strain PXO88 on FKR14. The phytotoxicity effects scored at 21 dpi showed a variation from 9.3 cm to 23.1 cm induced by strains Mai1 and PXO88, respectively, on rice genotype IRBB4. At 21 dpi lesions induced by culture filtrate had reached higher values and were comparable to the lesion induced by bacterial cultures in treatments with strain Dak16 on both rice genotypes, while the other strains values from culture filtrate treatment remained lower than the value from bacterial culture, except on the interaction between strain Mai1 and rice genotype IRBB4.

The statistical analysis of data derived from the 21 dpi measurement revealed significant differences between bacterial culture and culture filtrate treatments, strains and rice genotypes: bacteria culture induced higher lesion lengths than culture filtrate, strain PXO88 demonstrated higher effect on rice genotypes than the other strains, while rice genotype FKR14 was more susceptible than genotype IRBB4 (Table 1). Significant differences were also observed within treatments and strains, with culture filtrate of strain PXO88 inducing lower lesion length than its bacteria culture, and between treatments and rice genotype, with generally higher lesion lengths induced by culture filtrate on rice genotype FKR14 than on genotype IRBB4. The interaction of treatments, strains and rice genotypes was not significant.

Phytotoxicity of heated culture filtrate

The phytotoxicity effect of heated culture filtrates of *Xoo* strains Mai1, PXO88, Dak1 and Dak16 inoculated to rice genotypes IRBB4 and FKR14 was first observed one week after treatment. Lesions induced by heated culture filtrate were generally lower at 14 dpi and in all treatments lower at 21 dpi than values from non-heated culture filtrate (Figure 6). Lesion length from heated culture filtrate generally increased from 14 to 21 dpi (Figure 3). The lesion length induced by heated culture filtrate varied from 1.8 cm to 5.1 cm with rice genotype IRBB4 and from 3.3 cm to 6.9 cm on FKR14 at 21 dpi (Figure 6).

Figure 6 and statistical results (Table 2) show significant differences between culture filtrate and heated culture filtrate treatments, with higher phytotoxicity effect of culture filtrate compared to heated culture filtrate on both rice genotypes. Significant differences were also observed between strains and rice genotypes, with strains PXO88 and Dak1 inducing the highest phytotoxicity effect on both rice genotypes, while strain Mai1 showed a lower lesion length on rice genotype IRBB4. Significant interactions with p-value 0.00 were also recorded between treatments and strains, with strain PXO88 culture filtrate showing a higher divergence on lesion length compared to heated culture filtrate on rice genotype IRBB4 than on genotype FKR14.

Phytotoxicity of culture filtrate treated with proteinase K

Inoculation of culture filtrate of strains Mai1, PXO88, Dak1 and Dak16 treated with proteinase K to rice genotypes IRBB4 and FKR14 lead to a significant reduction in phytotoxicity lesion lengths compared to inoculation with untreated culture filtrate (Figure 6). Differences were higher at 21 dpi than at 14 dpi. Only in the interaction strain

	SS	DF	MS	F	P
Treatments	64.98	1	64.98	18.134	0.000169***
Strains	1122.08	3	374.03	104.372	0.000000***
Rice Genotypes	116.53	1	116.53	32.518	0.000003***
Treatments x Strains	112.22	3	37.41	10.438	0.000061***
Treatments x Rice Genotypes	53.57	1	53.57	14.949	0.000509***
Strains x Rice Genotypes	161.94	3	53.98	15.063	0.000003***
Treatments x Strains x Rice Genotypes	22.55	3	7.52	2.097	0.120142
Error	114.68	32	3.58		

Note:***p-value highly significant at 0.01; SC: Sum of Square; DF: Degree of freedom, MS: Mean square; F: F of Fisher; p: probability

Table 1: Statistics of comparison of *Xanthmonas oryzae* pv. *oryzae* culture effects to culture filtrate effects at 21 dpi.

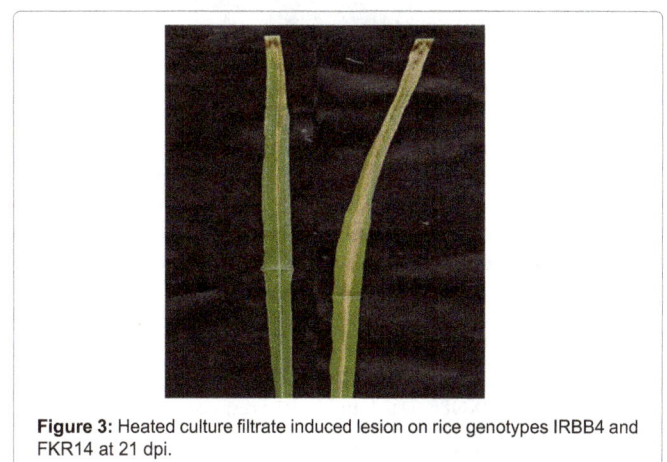

Figure 3: Heated culture filtrate induced lesion on rice genotypes IRBB4 and FKR14 at 21 dpi.

	SS	DF	MS	F	P
Treatments	1467.330	1	1467.330	306.030	0.000000***
Strains	170.705	3	56.902	11.868	0.000022***
Rice Genotypes	36.453	1	36.453	7.603	0.009548***
Treatments x Strains	164.669	3	54.890	11.448	0.000029***
Treatments x Rice Genotypes	6.564	1	6.564	1.369	0.250631
Strains x Rice Genotypes	46.760	3	15.587	3.251	0.034465**
Treatments x Strains x Rice Genotypes	63.189	3	21.063	4.393	0.010683**
Error	153.431	32	4.795		

Note:***p-value highly significant at 0.01; **p-value significant at 0.05; SC: Sum of Square; DF: Degree of freedom, MS: Mean square; F: F of Fisher; p: probability

Table 2: Statistics of comparison of *Xanthmonas oryzae* pv. *oryzae* culture filtrate effects to heated culture filtrate effects at 21 dpi.

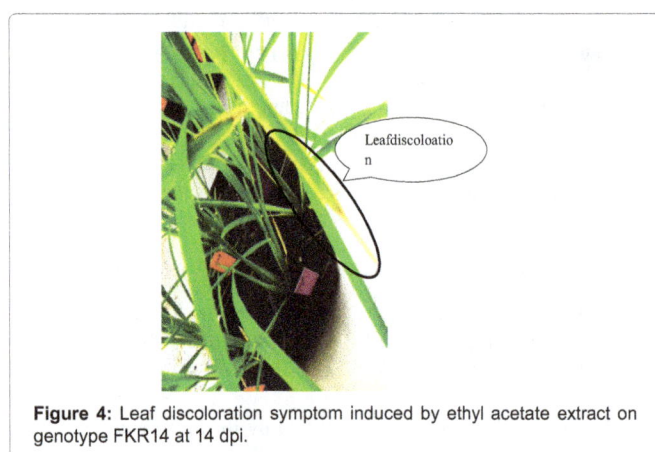

Figure 4: Leaf discoloration symptom induced by ethyl acetate extract on genotype FKR14 at 14 dpi.

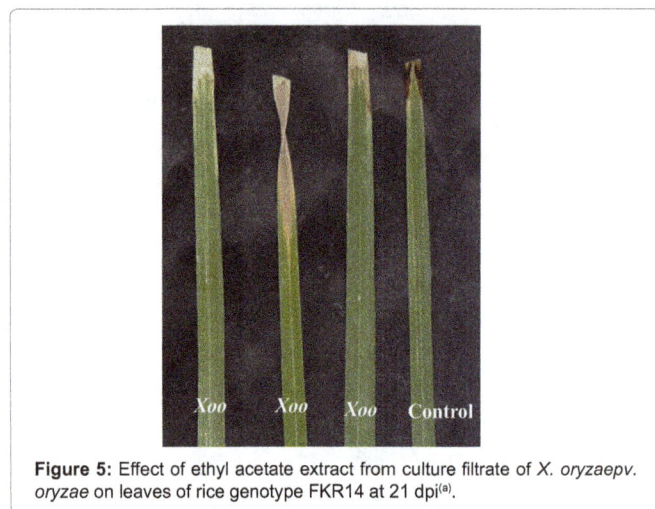

Figure 5: Effect of ethyl acetate extract from culture filtrate of *X. oryzaepv. oryzae* on leaves of rice genotype FKR14 at 21 dpi[a].

Dak1 on rice genotype FKR14, values from untreated and proteinase K treated culture filtrate were similar. At 14 dpi, inoculation with culture filtrate treated with proteinase K resulted in minimum and maximum lesion lengths of 0.6 cm and 1.7 cm induced by strain Dak1 on rice genotypes IRBB4 and FKR14, respectively. The phytotoxicity effect of proteinase K treated culture filtrate increased at 21 dpi, varying on rice genotype IRBB4 from 1.5 cm to 5.4 cm and on rice genotype FKR14 from 4.1 cm to 13.4 cm. at 21 dpi, with maximum lesion lengths of 5.4 cm, 5.9 cm, 13.4 cm and 4.2 cm, induced by treated culture filtrates of *Xoo* strains Mai1, PXO88, Dak1 and Dak16 respectively.

The comparison between culture filtrate effects to culture filtrate

treated with proteinase K effects (Figure 5) and the statistical analysis at 21 dpi (Table 3) revealed significant differences (p=0.00) between culture filtrate and proteinase K treated culture filtrate, with a lower phytotoxicity effect induced by proteinase K treated culture filtrate than with untreated culture filtrate, while the strain Dak1 induced similar lesion lengths with both treatments on rice genotype FKR14 at 21 dpi.

Significant differences were observed between treatments and rice genotypes (p=0.03), where both treatments induced higher phytotoxicity effects on rice genotype FKR14 than genotype IRBB4 at 21 dpi. Only strain PXO88 culture filtrate induced higher lesion lengths on genotype IRBB4 than on genotype FKR14.

Phytotoxicity of ethyl acetate extract from culture filtrate

X. oryzae pv. *oryzae* ethyl acetate fraction fromculture filtrate was used to investigate a possible role of small molecules as virulence factors of *Xoo*. Inoculation of culture filtrate extracted with ethyl acetate by leaf clipping resulted in a highly to totally reduced phytotoxicity induction (Figure 6). Nevertheless the phytotoxicity effects of the ethyl acetate fraction after leaf clipping treatment varied from 0.6 cm to 3.4 cm and 1.9 cm to 3.99 cm on rice genotype IRBB4 and on FKR14, respectively, at 21 dpi. The phytotoxicity effects were first observed 3 days after treatment with leaf discoloration symptoms (Figures 4 and 5). At 14 dpi the maximum lesion length of 2.03 cm was recorded, induced by strain Dak1 on rice genotype FKR14. At 21 dpi the maximum mean lesion lengths 3.4 cm and 3.9 cm were induced by Dak1 on genotypes IRBB4 and FKR14, respectively. All the strains demonstrated increases of phytotoxicity effects on both rice genotypes comparing 14 and 21 dpi.

The comparison of phytotoxicity of the ethyl acetate fraction from culture filtrate to non-extracted culture filtrate at 21 dpi revealed

	SS	DF	MS	F	P
Treatments	1298.960	1	1298.960	420.467	0.000000***
Strains	157.310	3	52.437	16.974	0.000001***
Rice Genotypes	52.585	1	52.585	17.021	0.000246***
Treatments x Strains	251.885	3	83.962	27.178	0.000000***
Treatments x Rice Genotypes	14.257	1	14.257	4.615	0.039366**
Strains x Rice Genotypes	121.773	3	40.591	13.139	0.000009***
Treatments x Strains x Rice Genotypes	74.693	3	24.898	8.059	0.000386***
Error	98.859	32	3.089		

Note:***p-value highly significant at 0.01; **p-value significant at 0.05; SC: Sum of Square; DF: Degree of freedom, MS: Mean square; F: F of Fisher; p: probability

Table 3: Statistics of comparison of *Xanthmonas oryzae* pv. *oryzae* culture filtrate effects to culture filtrate treated with proteinase K at 21 dpi.

	SS	DF	MS	F	P
Treatment	1724.74	1	1724.74	555.366	0.000000***
Strains	139.209	3	46.403	14.942	0.000011***
Rice Genotypes	9.825	1	9.825	3.164	0.087962
Treatment x Strains	125.881	3	41.96	13.511	0.000023***
Treatment x Rice Genotypes	0.001	1	0.001	0	0.988431
Strains x Rice Genotypes	51.367	3	17.122	5.513	0.005024***
Treatment x Strains x Rice Genotypes	26.518	3	8.839	2.846	0.058835
Error	74.534	24	3.106		

Note: *** p-value highly significant at 0.01; SC: Sum of Square; DF: Degree of freedom, MS: Mean square; F: F of Fisher; p: probability

Table 4: Comparison of effects of *Xanthmonas oryzae* pv. *oryzae* culture filtrate to culture filtrate extracted with ethyl acetate at 21 dpi.

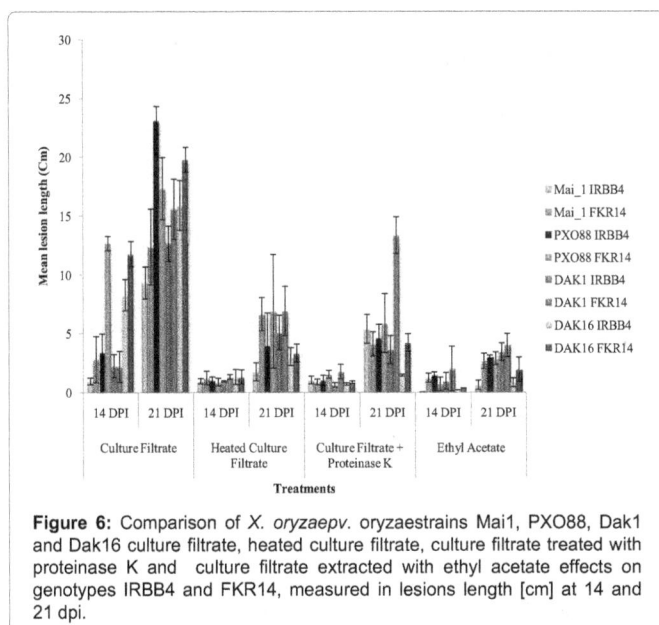

Figure 6: Comparison of *X. oryzaepv. oryzae*strains Mai1, PXO88, Dak1 and Dak16 culture filtrate, heated culture filtrate, culture filtrate treated with proteinase K and culture filtrate extracted with ethyl acetate effects on genotypes IRBB4 and FKR14, measured in lesions length [cm] at 14 and 21 dpi.

significantly lower lesion lengths induced by ethyl acetate extract compared to culture filtrate effects (Table 4). Significant differences were found between strains, and also the interaction between strains and rice genotypes was significant (p=0.005). Treatments and rice genotypes interaction revealed not significant at 21 dpi.

Discussion

Strain virulence

The virulence analysis result showed that Asian strain PXO88 is highly virulent compared to African strains used in this study. Also, within African strains, phenotypic differences were found. It has been reported that African *Xoo* strains are genetically distant from Asian strains [21]. Further studies with a higher number of strains are recommended. Two groups have been reported within the West African *Xoo* population [22]. The variability in the *Xoo* population is a consequence of the development of resistance durability of the bacterial races in time and space [23-25].

Near isogenic lines (NILs) with specific resistance genes have been identified and among them, lines with a single resistance gene are widely used to characterize virulence and differentiate pathogenic races of *Xoo*. Classification of the pathogenic race of *Xoo* is based on the interaction between its avr gene and rice *Xa* gene [23,26-28]. In this study, we used IRBB4 line carrying Xa4 gene and the susceptible genotype FKR14. The phenotypic reactions of the bacteria strains and rice line IRBB4 were similar to the reaction observed on FKR14 except with strain Mai1. IRBB4 line shows susceptibility to *Xoo* strains from Iran [28], while in China; it is susceptible to 50% of Chinese *Xoo* strains [29]. In earlier 1970, the *Xoo* populations were dominated by those strains that could not infect rice lines with Xa4 gene [30], while the newly developed rice lines with Xa4 gene are nowadays susceptible to several strains [31]. Strain Mai1 revealed non-virulent on IRBB4, confirming previous study on African *Xoo* strains, which reported that African *Xoo* race 3 is not virulent on IRBB4 [21].

Phytotoxicity effects from *Xoo* culture filtrate and treated culture filtrate on rice

X. oryzae pv. *oryzae* strains (Mai1, PXO88, Dak1 and Dak16) culture

filtrate, heated culture filtrate, culture filtrate treated with proteinase K and ethyl acetate extract were used to study their phytotoxicity effect on rice genotypes IRBB4 and FKR14. Culture filtrate had the highest phytotoxicity effect against rice genotypes compared to heated culture filtrate, culture filtrate treated with proteinase K and ethyl acetate extract. The least square mean (LSM) lesion length evoked by bacteria culture was 18.1 cm, while with culture filtrate, heated culture filtrate, culture filtrate treated with proteinase K and ethyl acetate extract the lesion lengths LSM were 15.8, 4.7, 5.4 and 2.4 cm, respectively (data not shown). This result revealed, that the phytotoxicity of culture filtrate is similar to the phytotoxicity produced by bacteria, where strain Dak16 demonstrated no significant difference between bacteria culture lesion (15.9 cm and 21.4 cm) and its culture filtrate phytotoxicity lesion (15.9 cm and 19.8 cm) on rice genotypes IRBB4 and FKR14, respectively. The bacteria cultures showed variations in their reactions on rice genotypes, and the same variations were also found by inoculation with culture filtrate, revealing that phytoxicity on rice is possibly produced by *Xoo* virulence factors. Significant differences were found between bacteria culture and culture filtrate effects, with higher effects of bacteria culture than culture filtrate with strains Dak16, strains PXO88 and Dak140, with phytotoxicity of culture filtrate (9.3 cm) being higher than bacteria effect (3.6 cm) of strain Mai1 on rice genotype IRBB4.

While the bacteria culture filtrate induced a phytotoxicity effect with a maximum lesion length of up to 23 cm, the heated culture filtrate induced 6.9 cm of mean length. Thus, the heating treatment caused a reduction of the phytotoxicity of the culture filtrate. This results, possibly indicating that the culture filtrate effect was related to lipopolysaccharides, xylanase, polygalacturonase that *Xoo* produced and are reported as typical bacteria virulence factors [6,32] - and which were inactivated or killed by heating. Generally, enzymes are inactivated at higher temperature. The optimum temperature for the activity of *Bacillus subtilis* xylanase is 55°C, and 0-45°C for *Aspergillus flavus* polygalacturonase, and while at 80°C 25% activities remained [33,34]. Therefore, the low phytotoxicity observed with heated culture filtrate in this study with 80°C of heating temperature should have affected the enzymes' activity in the culture filtrate.

The culture filtrate treated with proteinase K revealed generally lower phytotoxicity effect compared to other treatments, though the lesions were slightly larger than with heated culture filtrate. This finding supported the hypothesis that *Xoo* culture filtrate phytotoxicity is due to enzymes' activities or to proteins. Higher effect of culture filtrate treated with proteinase k could be due to presence of proteinaceous in the culture fitrate which perhaps inhibits the activity of related enzymes. Proteins including adhesion proteins are secreted via the Two Partner Secretion System (TPS) and contribute to virulence by bacteria attachment to the host surface [7,35]. *Xoo* produces proteins that are required for virulence, and the inactivation of proteins in heat-treated *Xoo* culture filtrate would be related to the low observed phytotoxicity effect on the rice genotypes. Chatterjee and Sonti [36] reported that *Xoo* protein phytase A (*PhytA*) is required for virulence. *Xoo* strain Dak1 showed 13.3 cm lesion length with culture filtrate treated with proteinase K on rice genotype FKR14, while other strains showed less effect. This result is in contrast to the fact that *Xoo* secreted proteins are required for virulence. Plant-pathogen interactions are related to the plant defense response from the initial recognition event to defense or susceptibility reactions. Generally, plant elicitors in response to a stress can be race specific. This could explain the fact, that *Xoo* strain Dak1 culture filtrate treated with proteinase K overcame the rice genotype FKR14 defense barriers in absence of proteins and induced a similar phytotoxicity effect compared to culture filtrate.

The ethyl acetate extract showed the lowest effect on rice genotypes compared to other treatment revealing that the ethyl acetate extract did not contain an active virulence factor inducing phytotoxicity. Therefore, we suggest, that *Xoo* does not produce low molecular-weight secondary metabolites (toxins) significantly involved in the bacterial virulence against the rice plant. The only phytotoxicity effect observed on rice leaves after treatment was leaf discoloration. Previous study (Shao and Wang, 1997) on *Xoo* culture filtrate revealed that *Xoo* culture filtrate contains some elicitors, which induced a toxicity effect on rice, and induced rapid necrosis on tobacco leaves. Our result also corroborates the data from Kong et al. [18] who found after extracting from bacteria culture with ethyl acetate, that *Xoo* toxins induced leaf discoloration and cell death on rice. A necrotic lesion as observed in our study has been reported from many bacteria species producing phytotoxic compounds. A phytotoxic compound of *Pseudomonas syringae* pv. *sesame* induced necrotic symptoms [37], which were similar to the phytotoxicity of mangotoxin from *P. syringae* pv. *syringae* strain UMAF0158 on mango tree [38]. Several toxins similar to lipodepsipeptide produced by strains of *P. syringae* pv. *syringae* were reported [39-41] such as syringomycins and syringopeptins which induced necrotic symptoms on plants [42,43]. Ethyl acetate extract treatments induce also chlorosis and necrosis on both rice genotypes IRBB4 and FKR14. The phytotoxin coronatine and antimetabolite toxins induced chlorosis and an increase in disease symptoms, respectively, and are considered as virulence factors in *P. syringae* [44-47], while *Xoo* ethyl acetate extract, did not increase disease symptoms, and therefore it is not, that the ethyl acetate extract contains virulence factor with phytotoxic activity.

Differences were seen between treatments, strains and rice genotypes. The culture filtrate effect was highest, and a considerable decrease of phytotoxicity was observed with the other treatments. We do not rule out that *Xoo* might produce a compound to induce phytotoxicity on rice leaves, but this compound may only to a minor extent contribute to the bacterial virulence. On the contrary, Noda et al. [48] identified seven toxic substances in ethyl acetate soluble acidic material: 3-methylthioproprionic acid, *trans*-3-methylthioacrylic acid, phenylacetic acid, isovaleric acid, tiglic acid, succinic acid and fumaric acid, isolated from *Xoo* culture suspension. His culture suspension induced necrosis and chlorosis on rice leaves at higher concentration (2000 µg/ml). We suppose that the contradictory results are due to the fact that Noda et al. [48] extracted the toxic substances from culture suspensions which contain bacteria, but we extracted from culture filtrate where bacteria were removed. Thus, when we also treated bacteria culture suspension with heat and proteinase K instead of treating the culture filtrate as in our general trials, we also found a higher phytotoxicity effect than with culture filtrate treated in the same way. Therefore, we believe that the *Xoo* virulence factors remain in association with the bacteria. Furthermore, secreted proteins including adhesins have been reported to be involved in the bacterial virulence, as well as the diffusible factor signal (DFS) synthase, which belongs to the enoyl-co-A-hydratase family and plays an important role in attachment, biofilm formation and virulence of *Xoo* [49].

The difference between strains could be related to the strains' virulence and/or to the genetic or geographic distance between strains. Gonzalez et al. [21] found that African strains are genetically distant from Asian strains, while Onasanya et al. [22] found two groups within West African *Xoo*. The differences observed between rice genotypes which showed highest effect against rice genotype FKR14 than IRBB4 corroborate the study of Shao and Wang [19]. The biochemical

approach for virulence factors identification in *Xoo* revealed no small metabolites playing a major role in *Xoo* virulence.

Conclusion

Biochemical assays for virulence factor identification in cultures, culture filtrate, heated culture filtrate, culture filtrate treated with proteinase K and in the ethyl acetate extracted fraction of *Xoo* strains Mai1, PXO88, Dak1 and Dak16 were carried out on rice genotypes IRBB4 and FKR14. Culture filtrate induced phytotoxicity effects on both rice genotypes with lesion lengths comparable to lesion lengths induced by the bacteria culture. A considerable reduction in phytotoxicity effects was observed with proteinase K treated culture filtrate, after heating culture filtrate and with ethyl acetate extract, in reducing order. First symptoms were visible after inoculation of *Xoo* cultures and ethyl acetate fraction at 3 dpi, with culture filtrate at one week and with heated culture filtrate and culture filtrate treated with proteinase K at 10 dpi. Statistical analyses confirmed that the phytotoxicity effects were significantly influenced by strains and by rice genotypes, and also the interactions treatments strains, strains rice genotypes and treatments strains rice genotypes were significant. We suggest, additionally to LPS, polygalacturonase and Xylanase, *Xoo* produces a proteinaceous virulence factor. Further factors possibly attached to the bacterial cells were not the target of the present study.

Acknowledgment

We thank the German Federal Ministry for Economic Development Cooperation (BMZ) for financing these studies in the framework of the mitigating the impact of climate change on rice diseases resistance in East Africa (MICCORDEA) project of Africa Rice Centre. We also thank Dr. Negussie Zenna for his help in samples collection in Tanzania.

References

1. Mew TW (1987) Current status and future prospects of research on bacterial blight of rice. Annu. Rev. Phytopathol. 25: 359-382.

2. Mew TW, Reyes RC, Vera Cruz CM (1989) Screening for bacterial blight resistance in rice. In Methods in Phytobacteriology 338-341 (Eds.): Z. Klement, K. Rudolph and D.C. Sands. Akademiai, Kiado, Budapest.

3. Robert KW, Pamela SG (1992) Foliar diseases: bacterial blight. Compendium of Rice Disease 1992; University of California, Davis. pp. 10-11.

4. Ray SK, Rajeshwari R, Sonti RV (2000) Mutants of Xanthomonas oryzae pv. oryzae deficient in general secretory pathway are virulence deficient and unable to secrete xylanase. Mol Plant Microbe Interact 13: 394-401.

5. Jha G, Rajeshwari R, Sonti RV (2007) Functional interplay between two Xanthomonas oryzae pv,. oryzae secretion systems in modulating virulence on rice. Mol Plant Microbe Interact 20: 31-40.

6. Wang CL, Xu A, Gao Y, Fan YL, Liang YT, et al. (2008) Generation and characterization of Tn5-tagged Xanthomonas oryzae pv. oryzae mutants that overcome Xa23 -mediated resistance to bacterial blight of rice. Eur J Plant Pathol. 123: 343-351.

7. Ray SK, Rajeshwari R, Sharma Y, Sonti RV (2002) A high-molecular-weight outer membrane protein of Xanthomonas oryzae pv. oryzae exhibits similarity to non-fimbrial adhesions of animal pathogenic bacteria and is required for optimal virulence. Mol Microbiol 46: 637-647.

8. Shen Y, Ronald P (2002) Molecular determinants of disease and resistance in interactions of Xanthomonas oryzae pv. oryzae and rice. Microbes Infect 4: 1361-1367.

9. Das A, Rangaraj N, Sonti RV (2009) Multiple adhesin-like functions of Xanthomonas oryzae pv. oryzae are involved in promoting leaf attachment, entry, and virulence on rice. Mol Plant Microbe Interact 22: 73-85.

10. Pieretti I, Royer M, Barbe V, Carrere S, Koebnik R, et al. (2009) The complete genome sequence of Xanthomonas albilineans provides new insights into the reductive genome evolution of the xylem-limited Xanthomonadaceae. BMC Genomics 10: 616.

11. Van Sluys MA, Monteiro-Vitorello CB, Camargo LE, Menck CF, Da Silva AC,

et al. (2002) Comparative genomic analysis of plant-associated bacteria. Annu Rev Phytopathol 40: 169-189.

12. Ryan RP, Vorhölter FJ, Potnis N, Jones JB, Van Sluys MA, et al. (2011) Pathogenomics of Xanthomonas: understanding bacterium-plant interactions. Nat Rev Microbiol 9: 344-355.

13. Erdman JG, Marlett EM, Hanson WE (1956) Survival of Amino Acids in Marine Sediments. Science 124: 1026.

14. Sugawara M, Haramaki R, Nonaka S, Ezura H, Okazaki S, et al. (2007) Rhizobitoxine production in Agrobacterium tumefaciens C58 by Bradyrhizobium elkanii rtxACDEFG genes. FEMS Microbiol Lett 269: 29-35.

15. Oresnik IJ, Twelker S, Hynes MF (1999) Cloning and characterization of a Rhizobium leguminosarum gene encoding a bacteriocin with similarities to RTX toxins. Appl Environ Microbiol 65: 2833-2840.

16. Simpson AJ, Reinach FC, Arruda P, Abreu FA, Acencio M, et al. (2000) The genome sequence of the plant pathogen Xylella fastidiosa. The Xylella fastidiosa Consortium of the Organization for Nucleotide Sequencing and Analysis. Nature 406: 151-159.

17. Lee BM, Park YJ, Park DS, Kang HW, Kim JG, et al. (2005) The genome sequence of Xanthomonas oryzae pathovar oryzae KACC10331, the bacterial blight pathogen of rice. Nucleic Acids Res 33: 577-586.

18. Kong F, Xu Z, Chunhong M, Wei J (1998) On the specific toxicity of toxin produced by Xanthomonas oryzae pv. oryzae to cytoplasmicmale sterile rice zhenshan 97a. Acta Phytopathol Sinica 28: 113-116.

19. Shao M, Wang JS (1997) Extraction, bioassay and composition analysis of the toxin from Xanthomonas oryzae pv. oryzae. Acta Phytopathol Sinica 27: 315-320.

20. Kauffman HE, Reddy APK, Hsieh SPY, Merca SD (1973) An improved technique for evaluating resistance of rice varieties to Xanthomonas oryzae. Plant Dis Rep 57: 537-541.

21. Gonzalez C, Szurek B, Manceau C, Mathieu T, Séré Y, et al. (2007) Molecular and pathotypic characterization of new Xanthomonas oryzae strains from West Africa. Mol Plant Microbe Interact 20: 534-546.

22. Onasanya A, Ekperigin MM, Nwilene FE, Sere Y, Onasanya RO (2009) Two pathotypes of Xanthomonas oryzae pv. oryzae virulence identified in West Africa. Cur Res Bact 2: 22-35.

23. Adhikari TB, Basnyat RC, Mew TW (1999) Virulence of Xanthomonas oryzae pv. oryzae on rice lines containing single resistance genes and gene combinations. Plant Dis 83: 46-50.

24. Nodal T, Li C, Li J, Ochiai H, Ise K, Kaku H (2001) Pathogenic diversity of Xanthomonas oryzae pv. oryzae strains from Yunnan province, China. Jpn Agric Res Q 35: 97-103.

25. Jagjeet SL, Yogesh V, Mandeep SH, Ravinder KG, Tajinder SB et al. (2011) Genotypic and pathotypic diversity of Xanthomonas oryzae pv. oryzae, the cause of Bacterial blight of rice in Punjab State of India. J of Phytopathol 159: 479-487.

26. Mew TW, Alvarez AM, Leach JE, Swings J (1993) Focus on bacterial blight of rice. Plant Disease 77: 5-12.

27. Jeung JU, Heu SG, Shin MS, Vera Cruz CM, Jena KK (2006) Dynamics of Xanthomonas oryzae pv. oryzae Populations in Korea and Their Relationship to Known Bacterial Blight Resistance Genes. Phytopathology 96: 867-875.

28. Khoshkdaman M, Ebadi AA, Kahrizi D (2012) Evaluation of pathogencity and race classification of Xanthomonas oryzae pv. oryzae in Guilan province-Iran. Agr Sci 3: 557-561.

29. Gang L, Cong-Feng S, Xiao-Mia P, Yue Y, Jin-Sheng W (2009) Analysis of pathotypic and genotypic diversity of Xanthomonas oryzae pv. oryzae in China. J of Phytopathology 157: 208-218.

30. Mew TW, Vera Cruz CM, Medalla ES (1992) Changes in race frequencies of Xanthomonas oryzae pv. oryzae in response to the planting of rice cultivars. Plant Dis 76: 1029-32.

31. Vera Cruz CM, Bai J, Ona I, Leung H, Nelson RJ, et al. (2000) Predicting durability of a disease resistance gene based on an assessment of the fitness loss and epidemiological consequences of avirulence gene mutation. Proc Natl Acad Sci U S A 97: 13500-13505.

32. Rajeshwari R, Jha G, Sonti RV (2005) Role of an in planta-expressed xylanase of Xanthomonas oryzae pv. oryzae in promoting virulence on rice. Mol Plant Microbe Interact 18: 830-837.

33. Sanghi A, Garg N, Gupta VK, Mittal A, Kuhad RC (2010) One-step purification and characterization of cellulase-free xylanase produced by alkalophilic Bacillus subtilis ash. Braz J Microbiol 41: 467-476.

34. Gewali MB, Maharjan J, Thapa S, Shrestha JK (2007) Studies on polygalacturonase from Aspergllus flavus. Sci World 5: 19-22.

35. Tseng TT, Tyler BM, Setubal JC (2009) Protein secretion systems in bacterial-host associations, and their description in the Gene Ontology. BMC Microbiol 9 Suppl 1: S2.

36. Chatterjee S, Sonti RV (2002) rpfF mutants of Xanthomonas oryzae pv. oryzae are deficient for virulence and growth under low iron conditions. Mol Plant Microbe Interact 15: 463-471.

37. Firdous SS, Asghar R, Haque MI, Afzal SN, Murtaza G et al. (2010) Isolation, purification, characterization and identification of virulence factors of Pseudomonas syringae pv. sesame and Xanthomonas campestris pv. sesami. Pak J Bot 42: 4191-4198.

38. Arrebola E, Cazorla FM, Duran VE, Rivera E, et al. (2003) Mangotoxin: a novel antimetabolie toxin produced by Pseudomonas syringae inhibiting ornithine/arginine biosynthesis. Physiol Mol Plant Pathol 63: 117-127.

39. Gross DC, DeVay JE (1977) Production and purification of syringomycin, a phytotoxin produced by Pseudomonas syringae. Physiol Plant Pathol 11: 13-28.

40. Gross DC, Hutchinson ML, Scholz BK, Zhang JH (1997) Genetic analysis of the role of toxin production by Pseudomonas syringae pv. syringaein plant pathogenesis. In: Pseudomonas syringae pathovars and Related Pathogens pp.163-169.

41. Ballio A, Barra D, Bossa F, Collina A, Grgurina I, et al. (1991) Syringopeptins, new phytotoxic lipodepsipeptides of Pseudomonas syringae pv. syringae. FEBS Lett 291: 109-112.

42. Sinden SL, DeVay JE, Backman PA (1971) Properties of syringomycin, a wide spectrum antibiotic and phytotoxin produced by Pseudomonas syringae and its role in the bacterial canker disease of peach trees. Physiol Plant Pathol 1: 199-213.

43. Iacobellis NS, Lavermicocca P, Grgurina I, Simmaco M, Ballio A (1992) Phytotoxic properties of Pseudomonas syringae pv. Syringae toxins. Physiol Mol Plant Pathol 40: 107-116.

44. Bender CL, Young SA, Mitchell RE (1991) Conservation of Plasmid DNA Sequences in Coronatine-Producing Pathovars of Pseudomonas syringae. Appl Environ Microbiol 57: 993-999.

45. Bender CL, Alarcón-Chaidez F, Gross DC (1999) Pseudomonas syringae phytotoxins: mode of action, regulation, and biosynthesis by peptide and polyketide synthetases. Microbiol Mol Biol Rev 63: 266-292.

46. Arrebola E, Cazorla FM, Codina JC, Gutiérrez-Barranquero JA, Pérez-García A, et al. (2009) Contribution of mangotoxin to the virulence and epiphytic fitness of Pseudomonas syringae pv. syringae. Int Microbiol 12: 87-95.

47. Carrión VJ, Arrebola E, Cazorla FM, Murillo J, de Vicente A (2012) The mbo operon is specific and essential for biosynthesis of mangotoxin in Pseudomonas syringae. PLoS One 7: e36709.

48. Noda T, Sato Z, Kobayashi H, Iwasaki S, Okuda S (1980) Isolation and structural elucidation of phytotoxic substances produced by Xanthomonas campestris pv. oryzae (Ishiyama) Dye. Ann Phytpath Soc Japan 46: 663-666.

49. Rai R, Ranjan M, Pradhan BB, Chatterjee S (2012) Atypical regulation of virulence-associated functions by a diffusible signal factor in Xanthomonas oryzae pv. oryzae. Mol Plant Microbe Interact 25: 789-801.

15

Efficacy and Economics of Fungicides and their Application Schedule for Early Blight (*Alternaria solani*) Management and Yield of Tomato at South Tigray, Ethiopia

Mehari Desta[1]* and Mohammed Yesuf[2]

[1]*Woldia University, Woldia, P.O.Box 400, Ethiopia*
[2]*Melkassa Agricultural Research Center, Ethiopia Institute of Agricultural Research (EIAR) P.O. Box 436, Nazareth, Ethiopia*

Abstract

Early blight, caused by Alternaria solani, is the most pressing problem of tomato Production and productivity in Tigray region. However, only limited attempts have been made to tackle this problem in the study area. Therefore, this study was conducted to (1) investigate the efficacy and spray frequencies of fungicides (2) determine yield loss incurred due to early blight and (3) assess cost benefit of the fungicides. Three fungicides (ridomil gold, Agrolaxyl and Mancozeb) each with three spray frequencies (every 7, 14 and 21 days) were evaluated using moderately susceptible variety, Melkashola in a Randomized Complete Block Design (RCBD) with three replications. Significant differences were observed among the treatments in-terms of disease incidence (DI), disease severity (DS), and AUDPC and disease progress rate (DPR) and; yield and yield components. Mancozeb among the fungicides and weekly spray among the spray frequencies were found the most effective in controlling the disease and improving the yield of tomato. Application of Mancozeb every week minimizes the disease by (47.75%) and consequently improves the yield by (112.48%). Weekly application of Mancozeb was found the most effective in controlling the disease with minimum values of DS (10.45%), AUDPC (266.0%-days), and DPR (0.09) and; higher marketable yield of (355.68 q/ha) and most economical with maximum marginal rate of return (MRR) (2,671.3%). Bi-weekly spray of Mancozeb also gave next higher MRR (1,724.3%). Maximum yield loss (52.94%) as compared to the most protected plot was incurred on untreated plots. Therefore, from the findings it can be conclude that application of mancozeb at weekly interval can be considered as the best management strategy to reduce disease epidemics and improve tomato yield.

Keywords: Alternaria solani; AUDPC; Disease incidence; Disease severity; Fungicide

Introduction

Tomato is the second most important vegetable crop next to potato worldwide with almost 4.5 million hectares of cultivated land [1-5] and with a yield potential of up to 42.1 tons/ha [6]. The total production of tomato in Ethiopia has shown a marked increase [7], since it became the most profitable crop providing a higher income to small scale farmers compared to other vegetable crops. According to Maheswari et al. [8] productivity of 9 tons/ha under farmers' practice and about 24 and 40 tons/ha under demonstration and research plots, respectively was realized. However, the national average yield of tomato in the country is very low which is around 7 tons/ha [9] and less than 50% of the current world average yield of about 27 tons/ha. Tomato is a key food and cash crop for many low income farmers in the tropics [10]. Tomato fruit is rich in vitamins A and C and contains an antioxidant, lycopene [11]. Despite of its importance, in the world as well as in Ethiopia the production and productivity of the crop is very low which mostly attributes to disease.

Among the major diseases of tomato, early blight caused by Alternaria solani is the worst damaging one [1,3] and cause reduction in quantity and quality of the crop. It is an economically important disease throughout the Southeastern United States and much of the world wherever tomato crops are grown under hot and humid conditions [12]. According to Batista et al. and Markham et al. [13,14] early blight epidemics are particularly severe in tropical countries during warm and wet seasons. Nevertheless, the disease is becoming more severe in all regions partly due to warmer temperatures experienced worldwide

[15]. During severe cases early blight can lead to complete defoliation and is most damaging on tomato [16] in regions with heavy rainfall, high humidity and fairly high temperature (24-29°C). Alternaria solani (Ellis and Martin) is a soil inhabiting air-born pathogen responsible for leaf blight, collar and fruit rot of tomato disseminated by fungal spores [17]. The pathogen causes infection on leaves, stem, petiole, twig and fruits as well as leads to the defoliation, drying of twigs and premature fruit drop which ultimately reduce the yield [18].

Like in other parts of the world in Ethiopia early blight which is caused by *A. solani* is the most destructive disease of tomato next to late blight [19]. Even though, the yield loss was not quantified, early blight is the bottle neck of tomato production in many tomatoes growing areas of Tigray region. It significantly reduces the production and market value of the crop.

Currently, sanitation, long crop rotation to reduce the spore concentration on decaying plant material and routine application of

*Corresponding author: Mehari Desta, Woldia University, P.O.Box 400 Woldia, Ethiopia, E-mail: mehari.des12@yahoo.com fungicides are the most common early blight management options in tomato production [20]. Management of early blight in tomato through fungicides such as ridomil gold, sulphur, copper oxychloride, carbendzim and mancozeb has been also reported [21,22]. In the tropics, management of early blight relies mostly on the intensive use of fungicides [13]. Apart from the application of fungicides, emphasis should be given to their time of application and/or frequency. In this regard Lemma et al. [23] have studied different fungicide application schedules such as weekly basis spray, SC-IPM, and TOM CAST and reported that TOM-CAST schedule reduced the number of spray application to 6 compared with 10 at weekly interval and the AUDPC of early blight was lowest with TOM-CAST schedule while highest with no fungicide. According to the same author weekly fungicide applications delayed early blight development by 8 days compared with non sprayed and SC-IPM treatments. AUDPC and EB severity at the end of the season were lower in the weekly and TOM-CAST treatments. Under Ethiopian condition there are also different fungicides recommended for the management of early blight. However, fungicide efficacy and their appropriate spray frequency are not properly studied in the study area. Therefore, the aim of this study was to (1) evaluate the efficacy of registered fungicides and their spray frequency on disease epidemics; (2) determine the yield loss incurred due to early blight of tomato and (3) determine cost effectiveness of the fungicides against early blight of tomato.

Materials and Methods

Description of the study area

Field experiment on management of early blight using fungicides was conducted during 2010 cropping season with supplement irrigation at alamata, south Tigray which is located 12° 15′ N latitude and 39° 35′ E longitude. It lies at an altitude of 1600 m.a.s.l. The mean annual rainfall is 663 mm.

Experimental design and treatments

Planting material was obtained from Melkassa agricultural research center and seedlings of improved tomato variety (Melkashola) were raised on a standard seed bed size of 1 m × 5 m. Apparently healthy seedlings were then transplanted into the experimental field with a plot size of 4.8 m × 3.3 m. Spacing between plants and rows were maintained as 30 cm and 80 cm, respectively. Each plot and block was separated by a buffer zone of 1.5 m and 2 m, respectively to prevent fungicide drift or cross contamination [24]. The experiment was consisted a total of six rows and four harvestable middle rows. The treatments (spray frequencies of each fungicide) were arranged in a Randomized Complete Block Design (RCBD) with three replications.

Three fungicides, ridomil gold, agrolaxyl and mancozeb were evaluated against tomato early blight at different frequencies with the rates of 2.5, 3 and 2 kg/ha, respectively. Spray frequencies were scheduled as every 7, 14, 21 days for all fungicides and unsprayed plot was included as a control. Spray was started soon after the initial appearance of symptoms. A total of four, three and two times application was practiced for every week, two week and three week spray frequencies of each fungicide, respectively.

Disease assessment

Disease incidence and severity were recorded every week starting from the first appearance of disease symptoms (i.e. at flowering and early fruiting stage) and a total of six assessments were conducted during the crop season. Incidence of early blight was assessed by counting the number of infected plants on the middle four rows and expressed as percentage of total plants assessed as:

$$DI(\%) = \frac{\text{Number of diseased plants}}{\text{Total number of plants inspected}} \times 100$$

Five plants were selected randomly from each replication per treatment, and then five leaves of each plant were used to determine the disease severity [2]. Severity of early blight was then recorded on the basis of 1-6 rating scales by modifying the scales adopted by [25] where 1=trace to 20% leaf infection, 2=21-41% infection, 3=41-60% infection, 4=61-80% infection, 5=81-99% infection and 6=100% leaf infection or the entire plant defoliation. The per cent severity index was then calculated as:

$$PSI = \sum \frac{\text{Individual numerical ratings} \times 100}{\text{Total number of leaves assessed} \times 6}$$

Where 6 is the highest numerical rating on the scale

Assessment of yield and yield components

Fruits were considered ready for picking when 50% of fruits turned yellow or red. Harvested fruits were categorized as clean marketable fruits (smooth, glossy surface and firm skin) or unmarketable if they had symptoms of damage by insects, disease infection or other physiological disorder. Fruits from each plot were sorted in to infected and healthy groups.

Data analysis

Data on early blight incidence and severity from each assessment; yield and yield components, PSI, AUDPC and disease progress rate were subjected to analysis of variance using Statistical Analysis System (SAS) version 9 Software [26]. Fisher's protected Least Significant Difference (LSD) values were used to separate differences among treatment means (P<0.05). AUDPC values were calculated from per cent severity index for each plot according to [27] and subjected to analysis of variance.

$$AUDPC = \sum_{i=1}^{n} \left[0.5 \left(x_{i+1} + x_i \right) \left[t_{i+1} - t_i \right] \right]$$

Where, x_i is the cumulative disease severity expressed as a proportion at the ith observation, t_i is the time (days after planting) at the ith observation and n is total number of observations.

Logistic, ln [(Y/1-Y)], [28] and Gompertz, -ln [-ln(Y)] [29] models were compared for estimation of disease progression parameters from each treatment. The transformed data of disease severity were regressed over time (days after planting) to determine the model. The goodness of fit of the models was tested based on the magnitude of the coefficient of determination (R^2). The appropriate model was then used to determine the apparent rate of disease increase (r) and the intercept of the curve.

Yield loss estimation

The relative losses in yield of each treatment were determined as percentage of that of the protected plots of the experiment according to [30] as:

$$RL(\%) = \frac{\left(YP - YT \right)}{YP} \times 100$$

Where, RL: Relative Loss (reduction of the yield parameter), YP: Mean Yield of the Protected Plots (plots with maximum protection - from mancozeb sprayed at weekly interval) and YT: Mean Yield in

Unprotected Plots (i.e. unsprayed plots or sprayed plots with varying level of disease).

Cost and benefit analysis

Price of tomato fruits (Birr/kg) was assessed from the local market and total price of the commodity obtained was computed on hectare basis. Input costs like fungicides and labor costs/ha were recorded. The price of fungicides was calculated based on their frequencies used on plot basis and converted in to hectare. The price of ridomil gold, agrolaxyl and mancozeb was 500, 160 and 115 Birr/kg, respectively. The total amount of these fungicides used for the experiment was computed and their price was converted into hectare basis. Cost of labor for spraying these fungicides from the first spray up to the final was 30 Birr per man-day and this was converted on hectare basis.

Before doing the economic analysis (partial budget) statistical analysis was done on the collected data to compare the average yields between treatments. Since there was difference between treatment means, the obtained economic data were subjected to analysis using the partial budget analysis method [31] Marginal rate of return was calculated as:

$$MRR(\%) = \frac{DNI}{DIC} \times 100$$

Where, MRR is marginal rate of returns, DNI, difference in net income compared with control, DIC, difference in input cost compared to control.

Results

Disease incidence and severity

Disease data before spray of fungicides indicated uniform spread of the disease in all experimental plots. However, in all the treatments there was an increase in disease incidence starting from the second assessment (104th DAP) to the last assessment (132th DAP). The rate of increase in the per cent disease index was slow in case of fungicide treated plots as compared to unsprayed plot. In the subsequent sprays, all the fungicide treated plots had recorded significantly less disease index over the control. Treatments were significantly different (P<0.05) on the per cent final disease incidence of early blight. All the treatments significantly reduced the final per cent disease incidence of tomato early blight recorded on the 132th DAP as compared to the unsprayed

control. However, the highest disease incidence reduction was observed on the most frequently sprayed fungicides (Table 1). The lowest disease incidence was obtained from mancozeb and ridomil gold each sprayed every 7 days and mancozeb sprayed every 14 days interval, respectively. The highest disease incidence, however, was recorded from unsprayed plot and agrolaxyl sprayed every 14 and 21 days; and ridomil gold sprayed every 21 days, respectively (Table 1). Frequently applied fungicides by far reduced disease severity as compared to the less frequently sprayed fungicides and unsprayed plots. All the fungicides reduced the severity significantly over unsprayed control. Four times applications at weekly interval of the fungicides mancozeb followed by ridomil gold at the same application interval significantly reduced early blight disease severity compared to other fungicide treatments and untreated plots (Table 1). The minimum disease severity was recorded on plots treated with mancozeb, ridomil gold and agrolaxyl sprayed at weekly interval, respectively followed by every two week application of same fungicides. On the contrary, the highest disease severity was recorded on unsprayed plot and the low frequently applied fungicides (i.e. at every 21 days interval). The overall fungicide treatments reduced the severity of early blight being 5.55 to 47.75% as compared to the unsprayed control.

Area under disease progress curve (AUDPC)

Early blight symptom appeared to start in both sprayed and unsprayed plots at about the same time, but subsequent disease progress was rapid in the non-sprayed plots and on the less frequently applied fungicides as indicated by their higher mean AUDPC values (Table 1). The AUDPC value of early blight on tomato exhibited highly significant difference (P<0.05) among the treatments. Minimum AUDPC was recorded on plots treated with different fungicides having different spray frequencies as compared to the untreated plot. Likewise, every week spray reduces the AUDPC values significantly as compared to all other spray interval as well unsprayed control. The lowest AUDPC values were obtained from mancozeb and ridomil gold treated plots at every 7 days interval and; mancozeb sprayed every 14 days interval, respectively. However, the maximum AUDPC value was recorded on the untreated control (Table 1).

Generally, when compared the three fungicides among each other and their spray frequencies mancozeb was found very effective than the other in reducing the AUDPC and; weekly and two week spray frequencies of all the fungicides were also best in minimizing the AUDPC value of the disease.

Fungicides	Spray frequency(Days)	DI (%)	DS (%)	Reduction of DS compared to control (%)	AUDPC (%-days)	Disease progress rate (r)
Ridomil Gold	7	37.07g	14.45f	27.75	357.00f	0.11
	14	50.93e	16.45d	17.75	431.67d	0.12
	21	68.38c	18.45b	7.75	472.89c	0.13
Agrolaxyl	7	54.09e	14.89fe	25.55	394.33e	0.11
	14	71.70c	17.55c	12.25	475.22c	0.12
	21	80.55b	18.89b	5.55	514.11b	0.12
Mancozeb	7	29.58h	10.45g	47.75	266.0g	0.09
	14	41.37f	15.33e	23.35	381.89fe	0.12
	21	61.04d	17.11dc	14.45	439.44d	0.12
Unsprayed	-	89.24a	20.00a	-	567.00a	0.28
Mean	-	58.40	16.36	-	429.96	
CV (%)	-	3.99	3.08	-	3.38	
LSD (5%)	-	4.01	0.86	-	24.93	

Table 1: Effect of different fungicide and their frequency on disease incidence (DI), disease severity, AUDPC and disease progress rate of early blight of tomato under field condition at south Tigray, Ethiopia. CV: Coefficient of Variation; LSD: Least Significant Difference; DI: Disease Incidence; ns: Not Significant; DS: Disease Severity, Means in a column followed by the same letter(s) are not significantly different.

Efficacy and Economics of Fungicides and their Application Schedule for Early Blight...

91

Disease progress rate (r)

Based on Gompertz model, the regression equation used to describe the rate of early blight progress was significant for all treatments as compared to the control (Table 1). On unsprayed plots early blight was increased at a rate of 0.28 units per day. However, all the fungicides sprayed at weekly interval was reduced the progress rate significantly. Disease progress rate on mancozeb treated plots at weekly interval was retarded by about 0.19 units per day which was more than half as compared to unsprayed control. Similarly, the disease progress rate was reduced by about 0.17 units per day on plots treated with ridomil gold and agrolaxyl sprayed at weekly interval (Table 1).

Generally, variation in early blight disease progress rate due to different fungicide application at different intervals was clearly observed. Early blight was progressed more rapidly on unsprayed plots and on the less frequently sprayed plots than those plots sprayed most frequently.

Fruit yield

There was a significant increase in fruit yields in the fungicide treated plots in contrast to untreated plots. Highly significant difference (P<0.05) was observed among treatments in-terms of harvested marketable fruit yields expressed per hectare basis. The efficacy of fungicides was reflected on the produced fruit yield. In this regard, plots sprayed with fungicides produced the highest fruit yield, being 355.68, 305.12, 291.24, 260.35 and 195.44 q/ha obtained from mancozeb, ridomil gold and agrolaxyl sprayed every week and; mancozeb sprayed every two and three week interval, respectively. However, the lowest yield was recorded from unsprayed plot (Table 2). Yield advantage which ranged from 6.79 to 112.48% was observed among treatments as compared to the non-sprayed plots. The maximum yield increase (112.48%) was recorded from every 7 days mancozeb sprayed plots followed by ridomil gold sprayed at the same interval. While least yield increase (6.79%) was observed on plots sprayed with agrolaxyl at every 21 days (Table 2).

Yield loss estimation

The variation in fruit yield losses was observed among the different frequencies of the fungicides in comparison to the most protected plot (Table 2). In unsprayed plots, fruit yield losses were notably higher than the protected plots. Fruit yield losses were significantly reduced by all fungicides at each spray frequencies as compared to the unsprayed

control of the variety Melkashola. Maximum relative fruit yield loss (52.94%) as compared to the most protected plot (mancozeb treated plots at weekly interval) was recorded on unsprayed plots and relatively minimum fruit yield losses was recorded on ridomil gold and agrolaxyl treated plots each sprayed every 7 days interval and; mancozeb sprayed every 14 days interval, respectively.

Cost benefit

Partial budget analysis indicated that every week and every two week spray interval of the fungicide ridomil gold had the highest total variable costs (Table 3). The highest gross field benefits were obtained in every week spray of mancozeb, ridomil gold and agrolaxyl, respectively. The net benefit obtained from sales of the produce from each spray frequencies ranged from 6233.55 to 12,460.20 US dollar. The highest net benefit in comparison with unsprayed plot and other treatments was obtained from weekly treated plots with the fungicide mancozeb. Ridomil gold and agrolaxyl each sprayed at weekly interval were ranked second and third. However, the least net benefit was obtained from plots treated with agrolaxyl at tri-weekly interval (Table 3).

Marginal analysis indicated that the highest marginal rate of return in comparison with unsprayed plots was obtained where mancozeb at weekly interval was used. Next highest marginal rate of return was attained from application of mancozeb and agrolaxyl at bi-weekly and weekly interval, respectively. However, the least marginal rate of return was recorded from plots treated with ridomil gold and agrolaxyl each at tri-weekly interval, respectively (Table 3).

Discussion

In this study symptom of early blight was appeared at early fruiting stage and this showed tomato plants are more susceptible at fruiting stage of the plant than early at the vegetative stage. This observation is in line with [18,32] who states that plants are more susceptible to infection by the disease during fruiting stage. Infected leaves were begun to defoliate starting two weeks after the appearance of symptom on those plots severely attacked. Jones [33] also reported that infected leaves eventually wither, die, and fall from the plant. In this study more defoliated leaves were observed on the unprotected plots than those plots treated with different fungicides having different frequency levels. Maximum fruit rot also observed on unprotected plots than the protected ones. This resulted in fruit yield losses up to 52.94% on untreated plots as compared to the most effective fungicide (mancozeb)

Fungicides	Spray frequency (days)	Marketable fruit yield (q/ha)	Unmarketable fruit yield (q/ha)	Yield advantage over the control (%)	Yield loss (%)
Ridomil gold	7	305.12[b]	22.70[i]	82.28	14.22
	14	234.50[e]	43.27[f]	40.09	34.07
	21	182.10[h]	65.80[c]	8.79	48.80
Agrolaxyl	7	291.24[c]	31.71[h]	73.98	18.12
	14	220.47[f]	53.02[e]	31.71	38.01
	21	178.77[h]	71.35[b]	6.79	49.74
Mancozeb	7	355.68[a]	16.59[j]	112.48	-
	14	260.35[d]	35.58[g]	55.53	26.80
	21	195.44[g]	61.49[d]	16.76	45.05
Unsprayed check	-	167.39[i]	81.18[a]	-	52.94
Mean		239.11	48.27	-	
CV (%)	-	2.01	4.64	-	
LSD (5%)	-	8.26	3.84		

Table 2: Effect of fungicides and their spray frequency on yield and yield components of tomato. CV: coefficient of variation, LSD: least significant difference, Means in a column followed by the same letter(s) are not significantly different.

Fungicides	Spray frequency(days)	Gross field benefit	Total variable Cost(TVC)	Net benefit ($ ha^{-1})	MRR (%)
Ridomil Gold	7	10897.14	485.51	10411.63	913.12
	14	8375	364.13	8010.87	558.22
	21	6503.57	242.75	6260.82	116.4
Agrolaxyl	7	10401.43	302.18	10099.25	1363.8
	14	7873.93	226.63	7647.30	736.47
	21	6384.64	151.09	6233.55	169.00
Mancozeb	7	12702.86	242.65	12460.20	2671.3
	14	9298.21	181.99	9116.23	1724.3
	21	6980	121.33	6858.67	725.70
Control		5978.21	-	5978.21	-

Table 3: Partial budget analysis for three fungicides and spray frequencies for the control of tomato early blight at south Tigrat, Ethiopia. MRR: marginal rate of return.

of the tested fungicides sprayed at weekly interval. This is because plants on the less protected plots fail to set fruits due to defoliation and drop their fruits due to fruit rot. This finding is in confirmation with Gwary and Nahunnaro [34] as he reported yield losses of 30-50% of the harvest due to fall of infected fruits. This observation was also agreed with Deahl et al. [35] who reported that yield reduction is observed when plants losses their leaves; because the plants fail to set fruits.

The overall yield losses of tomato due to early blight in this study were relatively lower compared to previous reports. Significant yield reduction (35 to 78%) in USA, Australia, Israel, UK and India has been reported by Deahl [35]. Yield losses up to 79% due to early blight damage were also reported from Canada, India, USA, and Nigeria [36-38]. This yield loss variation is likely to be occurred because various interrelated factors are attributed. These factors might be environmental condition, under which the experiment is conducted, the season when the study is carried out, the genotypes used and disease epidemics under the area.

The tested fungicides control early blight effectively and increase yield and yield components as compared to the control. Four times application of mancozeb at weekly interval minimizes the severity of early blight by (47.75%) and increases fruit yield by (112.48%) as compared to unprotected control. This is agreed with the findings of Mantecon [39] who reported that among the tested commercial fungicides mancozeb followed by Kavach were found to be very effective in controlling early blight with more than 50% disease control compared to the untreated control. Mantecon [39] also reported the most effective control of A. solani was achieved by copper oxychlorode (64.7%) followed by mancozeb (61.7%). According to Prior et al. [40] three spray of mancozeb reduces the disease intensity significantly compared to other chemicals and botanicals and gave the highest economic benefit. FAOSTAT [41] also reported that the highest usable yields of tomato with greater financial benefits obtained in chlorothalonil or mancozeb at 7 and 10 days interval was primarily due to suppression of Alternaria and other fruit rot.

The highest per cent disease incidence reduction (66.85%) was observed on plots treated four times with mancozeb sprayed at weekly interval. Three times application of mancozeb at two weeks interval was also resulted in 53.64% disease incidence reduction. This result is in line with Niederhauser [42] who noticed that best control of leaf blight disease of tomato caused by A. solani was achieved by three foliar sprays of mancozeb at 15 day interval. The incidence of blight was significantly lower in the said treatment [43,44]. However, Praveen Kumar Chourasiya et al. [45] reported that among the four fungicides sprayed 4 times at 15 days interval after the first appearance of tomato early blight and thereafter at 10 days interval for control of A. solani mancozeb gave effective control of the disease. SAS Institute Inc., [46]

also reported that among the non-systemic and systemic fungicides in controlling early blight of tomato mancozeb treatment gave the highest cost-benefit ratio of 1:11.4 in addition to reducing the disease incidence.

In addition to the appearances of more aggressive isolates, and isolates that are no longer inhibited by chemical protectants, the burden on the environment due to application of fungicide is high. Subsequently, plant pathogens are responsible for large amounts of chemical fungicides applied annually exacerbating control strategies [47,48]. Besides environmental problems unplanned and wide use of fungicides affects the health of users and consumers. To cope with these problems and due to the increase of public concern about adverse effects of agrochemicals on food safety and environment, there is need to stimulate the search for control strategies that are more durable and preferably based on natural products. So that, biological control agents which include effective microorganisms and microbial products, and organic fertilizers, have been attracting attention as alternatives to chemical agents [49]. Zhang et al. [49-51] reported that based on the whole plant tests, foliar spray with *Paenibacillus macerans*-GC subgroup A, *Serra-tia plymuthica*, *Bacillus coagulans*, *Serratia marcescens*-GC subgroup A, *Bacillus pumilis* -GC subgroup B and *Pantoea agglomerans* bacterial isolates reduced the disease severity of early blight significantly when compared with control. Such bio agents as *T. harzianum*, *T. viride*, *B. subtilis*, *P. fluorescens and S. cerivisae* have also been reported in reducing early and late blight of tomato significantly [6]. However, under Ethiopian condition, management of early blight through biological agents and botanicals has not been reported. So as to minimize the problems related to application of fungicides, biological control and botanicals should be considered in the future perspectives under Ethiopian condition as it has been practiced in other parts of the world.

Acknowledgements

The authors would like to thank Rural Capacity Building Project (RCBP) Ethiopia and Tigray Agricultural.

References

1. Abada KA, Mostafa SH, Hillal, Mervat R (2008) Effect of some chemical salts on suppressing the infection by early blight disease of tomato. Egypt J Appl Sci 23: 47-58.

2. Abd-El-Khair H, Karima HE, Haggag Nadia G, El-Gamal (2004) Biological control of wilt disease caused by Fusurium oxysporum in fennel under organic farming system. J Adv Agric Res (Fac Ag Saba Rasha) 9: 527-538.

3. Abdel-Sayed MHF (2006) Pathological, physiological and molecular variations among isolates of Alternaria solani the causal of tomato early blight disease: 181.

4. Pandey KK, Pandey PK, Kalloo G, Banerjee MK (2003) Resistance to early blight of tomato with respect to various parameters of disease epidemics. J Gen

Plant Pathol 69: 364-371.

5. Fravel DR (2005) Commercialization and implementation of biocontrol. Annu Rev Phytopathol 43: 337-359.

6. Mohammad SE (1988) Control of tomato early blight under plastic house conditions in Nineveh province, Iraq. Mesopotamia journal of agriculture 20: 359-366.

7. Lemma D (2002) Tomatoes; research experiences and production prospects. EARO, Report No. 43. Addis Ababa, Ethiopia.

8. Maheswari SK, Gupta PC, Gandhi SK (1991) Evaluation of different fungi toxicants against early blight of tomato. Agricultural Science Digest (Karnal) 11: 201-202.

9. CSA (Central Statistics Authority) (2009) Agricultural sample survey 2008/2007. Report on area and production of crops (Private peasant holdings, main season). Stat. Bull., Addis Ababa, Ethiopia, 01-446.

10. Robert GD, James HT (1991) A Biometrical Approach. Principles of Statistics. New York, USA.

11. Kapsa JS (2008) Important threats in potato production and pathogen/ pest management. Potato Res 51: 385-401.

12. Tesfaye T, Habtu A (1985) Review of research activities on vegetable crop diseases in Ethiopia. A review of crop protection research in Ethiopia (ed: Tsedeke Abate). IAR, Addis Ababa, Ethiopia.

13. Batista DC, Lima MA, Haddad F, Maffia LA, Mizubuti ESG (2006) Validation of decision support systems for tomato early blight and potato late blight, under Brazilian conditions. Crop Protection. 25: 664-670.

14. Markham L, Julie (1999) Biological activity of tea tree oil. In: Tea tree the genus Melaleuca (medicinal and aromatic plant: industerial profiles). Harwood Academic Publisher 2006.

15. Keinath AP, DuBose-VB, Rathwell PJ (1996) Efficiency and economics of three fungicidal application schedules for early blight control and yield of fresh market tomato. Plant Disease 80: 1277-1282.

16. Prasad Y, Naik MK (2003) Evaluation of genotypes, fungicides and plant extracts against early blight of tomato caused by Alternaria solani. Ind J Pl Protec, 31: 49-53.

17. Datar VV, Mayee CD (1981) Assessment of loss in tomato yield due to early blight. India Phytopathology 34: 191-195.

18. Naveenkumar S, Saxena RP, Pathak SP, Chauhan SKS (2001) Management of alternaria leaf disease of tomato. Indian phytopathology 54: 508.

19. Van der Plank JE (1963) Plant Diseases: Epidemics and Control. Academic Press, New York, London. pp. 344.

20. Mate GD, Deshmukh VV, Jiotode DJ, Chore NS, Dikkar M (2005) Efficacy of plant products and fungicides on tomato early blight caused by Alternaria solani. Research on crops. 6: 349-351.

21. Abdel-Kader MM, El-Mougy NS, Aly MDE, Lashin SM, Abdel-Kareem F (2012) Greenhouse Biological Approach for Controlling Foliar Diseases of Some Vegetables. Advances in Life Sciences 2012, 2: 98-103.

22. Lemma D, Yayeh Z, Herath E (1992) Agronomic Studies in Tomato and Capsicum. In: Herath and Lemma (eds) Horticulture Research and Development in Ethiopia: Proceedings of the Second National Horticultural Workshops of Ethiopia. 1-3 December. Addis Ababa, Ethiopia. pp. 153-163.

23. Dillard HR, Jhonston SA, Cobb AC, Hamilton GH (1997) An assessment of fungicide benefits for the control of fungal diseases of processing tomatoes in New York and New Jersey. Plant Disease, 81: 677-681.

24. Jones JB, Jones JP, Stall RE, Zitter TA (1991) Diseases caused by fungi. In: Compendium of Tomato Diseases. The American Psychopathological Society, St. Paul, MN. pp. 9-25.

25. Sherf AF, MacNab AA (1986) Vegetable diseases and their control. John Wiley and Sons, New York, pp. 634-640.

26. Campbell CL, Madden LV (1990) Introduction to plant disease epidemiology. J Wiley, New York

27. Yamaguchi M (1983) World vegetables: Principals, production and nutritive values. AVI Publishing Co., Westport, CT.

28. Berger RD (1981) Comparison of the Gompertz and Logistic equation to describe plant disease progress. Phytopathology 71: 716-719.

29. SAS Institute Inc., (2002) SAS/ Stat Guide for Personal Computers, SAS Institute Inc., Carry, NC.

30. CIMMYT (1988) Farm agronomic data to farmer's recommendations: a training manual. International maize and wheat center, Mexico.

31. Cerkauskas R (2005) Early blight. AVRDC, the world vegetable centre, www. avrdc.org.

32. Jones JB, Jones JP, Stall RE, Zitter TA (1997) Compendium of Tomato Diseases. APS press. pp 28-29.

33. Jones JB (1999) Tomato Plant Culture In the Field, Greenhouse and Home Garden. CRC Press LLC.

34. Gwary DM, Nahunnaro H (1998) Epiphytotics of early blight of tomatoes in Northeastern Nigeria. Crop Protection 17: 619-624.

35. Deahl K, Inglis D, DeMuth S (1993) Testing for resistance to metalaxyl in Phytophthora infestans isolates from northwestern Washington. American Journal of Potato Research 70: 779-795.

36. Basu PK (1974) Measuring early blight, its progress and influence on fruit losses in nine tomato cultivars. Can Plant Disease Survey 54: 45-51

37. Simmons EG, Roberts RG (1993) Alternaria Themes and Variations. Mycotaxon 48: 109-140.

38. Abhinandan D, Randhawa HS, Sharma RC (2004) Incidence of Alternaria leaf blight in tomato and efficacy of commercial fungicides for its control. Annual Biological 20: 211-218.

39. Mantecon JD (2007) Potato yield increases due to fungicide treatment in Argentinian early blight (Alternaria solani) and late blight (Phytophtora infestans) field trials during the 1996–2005 seasons. Plant Health Prog.

40. Prior P, Grimault V, Schmith J (1994) Resistance to Bacterial Wilt (Pseudomonas solanacearum) in tomato: Present status and prospects. In: Hayward AC., 57 Hartman, G. L. Bacterial Wilt. The Diseases and its Causative Agent, Pseudomonas solanacearum. Cab International.

41. Food and Agriculture Organization, (2004) FAOSTAT.

42. Niederhauser JS (1993) International cooperation in potato research and development. Annu Rev Phytopathol 31: 1-25.

43. Glasscock HH, Ware WM (1944) Alternaria blight of tomatoes. Agriculture 51: 417-420.

44. Momel TM, Pemezny KL (2006) Florida plant disease management guide: Tomato. Florida Cooperation Extensive Service, Institute of Food and Agriculture Sciences, Gaine ville, 32611 (http\\edis.infas.ufl.edu).

45. Chourasiya PK, Abhilasha A. Lal, Sobita Simon (2013) Effect of certain fungicides and botanicals against early blight of tomato caused by Alternaria solani (Ellis and Martin) under Allahabad Uttar Pradesh, India conditions. IJASR.

46. Dhingra OD, Sinclair JB (1995) Basic Plant Pathology Methods. CRC Press, Inc. USA.

47. Nuez F, Diez MJ, Pico B, Fernández P (1996) Catálogo de semillas detomate. Banco de germoplasma de la Universidad Politécnica de Valencia. Instituto Nacional de Investigación y Tecnología Agraria y Alimentación. Colección Monografías INIA No. 95. Ministerio de Agricultura, Pesca y Alimentación, Madrid.

48. Furgo PA, Mandokhot AM (2002) Management of early blight of tomato. Pestology 26: 38-40.

49. Zhang LP, Lin GY, Nino LD, Foolad MR, (2003) Mapping QTLs conferring early blight resistance in a Lycopersicon esculentum x L. hirsutum cross by selective genotyping. Mol Breed 12: 3-19.

50. Peralta IE, Knapp S, Spooner DM (2005) New species of wild tomatoes (Solanum section Lycopersicon: Solanaceae) from Northern Peru. Sys Bot 30: 424-434.

51. Yazici S, Yanar Y, Karaman I (2011) Evaluation of bacteria for biological control of early blight disease of tomato. AJB, 10: 1573-1577.

16

Bio Control Potential of *Pseudomonas fluorescens* against Coleus Root Rot Disease

S Vanitha[1] and R Ramjegathesh[2]*

[1]*Department of Sericulture, Centre for Plant Protection Studies, Tamil Nadu Agricultural University, Coimbatore-641 003, Tamil Nadu, India*
[2]*Department of Plant Pathology, Centre for Plant Protection Studies, Tamil Nadu Agricultural University, Coimbatore-641 003, Tamil Nadu, India*

Abstract

Ten different strains of *Pseudomonas fluorescens* were isolated from coleus rhizosphere except the pf1 strain and identified by biochemical tests. These strains were screened against *Macrophomina phaseolina* (Tassi) Goid, the causal organism of coleus root rot. The results revealed that Pf1 strains recorded maximum inhibition of mycelial growth against control. The mechanism of *Pseudomonas* strains namely the iron-chelating agent (siderophore), volatiles (HCN) and antibiotic (Fluorescein and pyocyanin) production tests were studied and reacted for siderophore, antibiotic and HCN production. The talc-based formulation of Pf1 and CPF1 was prepared and the bio-efficacy was tested under greenhouse conditions. The stem cutting and soil application of the talc-based formulation of Pf1 significantly reduced the root rot incidence and increased shoot and tuber length also.

Keywords: Antibiotics; Coleus; HCN; *Pseudomonas fluorescens*; Root rot; Siderophore

Introduction

Coleus (*Coleus forskohlii* Briq.) is cultivated mainly for their medicinal values in India. In India it is cultivated on about 2,500 hectares with an annual production of 1500 tonnes, especially in parts of Rajasthan, Maharashtra, Karnataka and Tamil Nadu. In Tamil Nadu alone, it is cultivated on more than 1000 hectares across. The crop is subjected to attack by many fungal diseases, namely leaf spot (*Botryodiplodia theobromae*), stem blight (*Phytophthora nicotianae* var. *Nicotianae*), collar rot (*Sclerotium rolfsii*, and *Rhizoctonia bataticola*), root rot (*Fusarium chlamydosporum, Rhizoctonia solani* and *Macrophomina phaseolina*) downy mildew (*Peronospora* sp) [1,2].

Root rot of coleus caused by *M. Phaseolina* is widely distributed in many countries and it is a devastating pathogen right from the establishment of the crop. Many effective fungicides have been tested against soil borne pathogens, but are not considered as long term solutions because of concerns about exposure risks, health and environmental hazards, high cost, residue persistence, the development of resistance to pesticides and the elimination of natural enemies. Biological control is a potential non-chemical means for plant disease management by reducing the harmful effects of a parasite or pathogen through the use of other living entities. The utilization of a plant's own defense mechanism is a fascinating arena of research which can be systemically activated upon exposure of plants to PGPR strains or infection by the plant pathogen. This phenomenon is called induced systemic resistance (ISR). This mechanism is facilitated by PGPR organism and activates through various defense compounds at the site of pathogen attack. Among the PGPR, fluorescent pseudomonads are the most exploited bacteria for biological control of soil-borne and foliar plant pathogens. In the past three decades, numerous strains of fluorescent pseudomonads have been isolated from the rhizosphere soil and plant roots by several workers and their biocontrol activity against soil-borne and foliar pathogens were reported [3]. Fluorescent pseudomonads are non-pathogenic rhizobacteria which suppress the soil-borne pathogens through rhizosphere colonization, antibiosis, iron chelation by siderophore production and ISR. In my knowledge, there is no report available in the control of root rot disease by Pseudomonas sps in coleus, but the use of antagonistic microorganisms such as

Pseudomonas fluorescens against *Macrophomina phseolina* have been reported many workers [4,5]. In the present investigation, attempts were made to test the antagonistic activity of *Pseudomonas fluorescens* and its mechanisms for coleus root rot management.

Materials and Methods

Isolation of pathogen and *Pseudomonas strains*

The root rot pathogen *M. Phaseolina* was isolated from coleus plants showing typical root rot symptoms and pure cultures of the pathogen were obtained by the single hyphal tip method [6]. The biocontrol agent *P. fluorescens* strain Pf1 was obtained from the culture collection section, Department of Plant Pathology, Centre for Plant Protection Studies, Tamil Nadu Agricultural University, India.

Other native coleus rhizobacterial *P. fluorescens* (CPF1 to CPF10) strains were isolated from soil samples obtained from different parts of Tamil Nadu state. One gram of rhizosphere soil adhering to root surface was collected and transferred to a 250 ml conical flask containing 100 ml of sterile water. After thorough shaking for 15 minutes in a shaker, different dilutions were prepared. One ml of each 10^{-5} and 10^{-6} dilution was pipetted out and poured into the sterile petridishes. Later King's medium B (KB) [7] was poured, rotated and incubated at room temperature (28 $\pm$ 2°C) for 24 hours. After 24 hours of incubation, the bacterial growth was purified by the dilution plate technique [8]. The bacterial culture was maintained in King's B broth (KB) in 30 percent (v/v) glycerol at -80°C.

Characterization of the different cultures of antagonistic bacteria

***Corresponding author:** R Ramjegathesh, Department of Plant Pathology, Centre for Plant Protection Studies, Tamil Nadu Agricultural University, Coimbatore - 641 003, Tamil Nadu, India, E-mail: ramjegathesh@gmail.com was done according to the methods recommended in the laboratory guide for identification of plant pathogenic bacteria published by the American Phytopathological Society [9]. In each test, 24-48 hours-old cultures were used.

Siderophores production

Production of siderophores by *P. fluorescens* was assayed by the plate assay method as described by Schwyn and Neilands [10]. The tertiary complex chromeazurol S (CAS) served as an indicator. To prepare one litre of the blue agar, 60.5 mg CAS was dissolved in 50 ml water and mixed with 10 ml Fe^{3+} solution (l mM $FeCl_2$ $6H_2O$ in 10 mM HCI and HDTMA dissolved in 40 ml water was added by constantly stirring). A forty-eight hour-old culture of fluorescent pseudomonads was streaked onto the succinate medium (Succinic acid, 4.00 g; K_2HPO_4, 3.00 g; $(NH_4)_2SO_4.7H_2O$, 7H_2O$, 0.2 g: distilled water, 1000 ml, pH 7.0) amended with the indicator and incubated for three days.

Production of hydrogen cyanide

The production of HCN was determined using a modification of the [11] procedure. Bacteria were grown on Tryptic-soy-agar (TSA) (animal peptone-5.0 g, soy peptone-5.0 g, sodium chloride-5.0 g, glycine-4.4 g, distilled water l000 ml). Filter paper discs soaked in a picric acid solution (2.5 g of picric acid, 12.5 g of sodium carbonate, and l000 ml of distilled water) were placed in the lid of each Petri-plate. Dishes were sealed with parafilm and incubated at 28°C for 48 hours. A change from yellow to light brown, brown or reddish brown of the discs was recorded as an indication of weak, moderate or strong production of HCN.

Detection of fluorescein and pyocyanin

Pseudomonas agar F (Casein enzymic hydrolysate, 10 g; Protease peptone, 10 g; K_2HPO_4, 1.5g; $MgSO_4$, 1.5 g; distilled water, 1l) favours the formation of fluorescein whereas Pseudomonas agar P (Peptone, 20 g; $MgCl_2$, 1.4 g; K_2SO_4, 10 g; Agar, 15 g; Distilled water, 1l) stimulates pyocyanin production and reduces fluorescein formation [7].

Effects of volatile metabolites

The effect of volatile metabolites from fluorescent pseudomonads on the growth of

M. Phaseolina was studied by a paired Petri dish technique by Gagne et al. [12].

Screening of antagonistic bacteria under *in vitro* condition

The antifungal efficacy of *Pseudomonas fluorescens* strains was tested by dual culture technique [13] using PDA medium. A mycelial disc (9 mm dia) of the pathogen namely *M. Phaseolina* was placed at one end of the plate and the bacterial antagonists were streaked at the periphery of the Petri-dish just opposite to the mycelial disc of the pathogen. The plates were incubated at 28 ± 2°C. The mycelial growth of the pathogen and inhibition zone was measured after 72 h of incubation.

Preparation of talc-based formulation of bio control agents

A loopful of *P. fluorescens* was inoculated into the King's B broth and incubated in a rotary shaker at 150 rpm for 72 hours at room temperature (28 ± 2°C). After 72 hours of incubation, the broth containing 9×10^8 cfu/ml was used for the preparation of talc-based formulation. To 400 ml of bacterial suspension, one kg of the talc powder (sodium ammonium silicate), calcium carbonate 15 g (to adjust the pH to neutral) and carboxy methyl cellulose (CMC) 10 g (as an adhesive) were mixed under sterile conditions following the method [14]. The product was shade dried to reduce the moisture content to 20 per cent and then packed in polypropylene bags and sealed. At the time of application the population of bacteria in talc formulation was checked in 2.5 to 3×10^8 cfu/g.

Greenhouse studies

Coleus cuttings were treated with talc-based formulations of Pf1 and CPF1 at 0.2 per cent for each stem cutting and planted in pots. Twenty-five grams of the formulated product (2.5 kg talc-based formulation mixed with 50 kg of farmyard manure) was given as a soil application per pot at 30 days after planting (DAP). Premixture fungicide (Carbendazim+Mancozeb) at 0.1 per cent was used as a standard check fungicide. It was applied as stem cutting treatment @ 0.2 per cent and also as a soil application @ 0.05 per cent at 30 DAP. A pure culture of *M. Phaseolina* was introduced into a sand-maize (19:1) medium and incubated for 15 days at room temperature for multiplication [15]. The potting soil (red soil: sand: cow dung manure, 1:1:1 w:w:w) was incorporated with the fungus, cuttings of coleus were surface-sterilized with 0.1% mercuric chloride for 30 s, rinsed three times with sterile distilled water and sown at two cuttings per pot. Each treatment was maintained for three replications. All treatments were replicated three times in factorial completely randomized design (CRD).

Results and Discussion

The development of biological techniques using PGPR amended with suitable bioformulations is an emerging trend in plant protection to reduce the plant diseases caused by plant pathogens. Production of antibiotics viz., Siderophore, HCN, pyrrolnitrin, phenazine and 2,4-diacetyl phloroglucinol and lytic enzymes by *P. fluorescens* against fungal pathogens were reported by many workers [3-5,16]. The above facts suggest that the inhibition of root rot pathogen, *M. Phaseolina* by *P. fluorescens* Pf1 may be due to the production of antibiotics, siderophore mediated competition and lytic enzymes, viz., chitinase, β-1,3-glucanase which degraded the fungal cell wall and restricted the growth of fungus under *in vitro* conditions. In the present study, among all the *P. fluorescens* strains, the Pf1 strain had maximum inhibition of mycelial growth and produced more amount of HCN, siderophore, pyocyanin and fluorescein in comparing to all the strains, so only the Pf1 strain was selected for this study. The biocontrol potential and production of volatile metabolites in Pf1 strain has a higher inhibitory effect followed by other *Pseudomonas* strains (Figure 1).

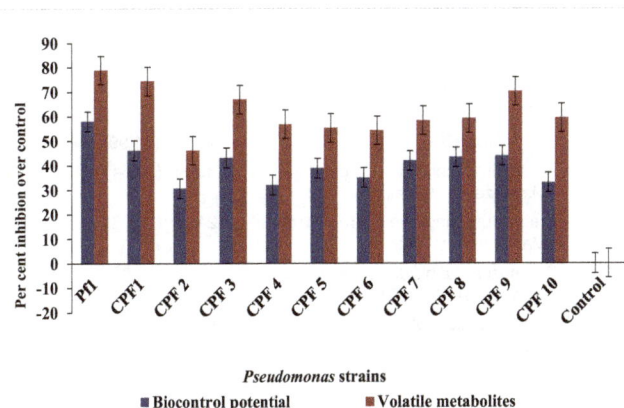

Figure 1: Bio control potential and volatile production of *P. fluorescens* against *M. phaseolina* under *in vitro* condition.

P. chlororaphis strain PA23 demonstrated excellent biocontrol in the canola phyllosphere through production of the non-volatile antibiotics and volatile antibiotics nonanal, benzothiazole and 2-ethyl-1-hexanol [17]. Secondary metabolites and volatile compounds contributed to the toxicity of the bacteria, with hydrogen cyanide efficiently repelling the nematodes and both hydrogen cyanide and 2, 4-DAPG functions as nematicides. Under certain conditions, antibiotics improve the ecological fitness of these bacteria in the rhizosphere which can further influence long-term biocontrol efficacy [18].

All the bacterial isolates produced siderophore by the CAS plate assay method. But all isolates except CPF6 and CPF10 produced a yellow color in the blue colored medium. Siderophore produced by *Pseudomonas* spp. and other rhizobacterial organisms (*Bacillus, Enterobacter*) have been used in the biological control of damping-off of cotton caused by *Pythium ultimum* [19]. Siderophore-mediated competition for iron by *Pseudomonas* sp. as well as induced resistance are primary mechanisms shown to be responsible for suppression of Fusarium wilt and this *Pseudomonas* culture and purified siderophore showed good antifungal activity against the plant deleterious fungi namely *Aspergillus niger, A. flavus, A. oryzae, F. oxysporum* and Sclerotium rolfsii [20]. Among the different *Pseudomonas fluorescens* tested, the intensity of HCN production was strong in *P. fluorescens* strains Pf1 and CPF1 followed by *P. fluorescens* strains CPF6 and CPF8. Normally the HCN production is directly related to inhibition of pathogen, the *P. fluorescens* strains Pf1 and CPF1 produced same

amount, but Pf1 strain produced some sideophores, it's also inhibit the most pathogens. HCN from a *P. fluorescens* strain CHA0 not repressed by fusaric acid played a significant role in the disease suppression of *F. oxysporum* f.sp. *radicis-lycopersici* in tomato [21]. HCN production by several strains of *P. fluorescens* and their efficacy in controlling root rot of groundnut caused by *M. Phaseolina* [22]. *Pseudomonas aeruginosa* LES4, an isolate of tomato rhizosphere was found to be positive for HCN against wilt disease [23]. All the *P. fluorescens* strains produced fluorescein and pyocyanin antibiotics in the Pseudomonas agar F and p plate assay method. But all isolates except CPF3 and CPF7 were surrounded with a yellow to greenish-yellow zone on Pseudomonas agar F and surrounded by a blue to green zone or with a red to dark brown zone due to pyocyanin production (Table 1). A marine isolate of fluorescent *Pseudomonas* sp. which has the ability to produce the pyoverdine has been found to inhibit the growth of *A. niger* under *in vitro* conditions [20]. Similar results were [24] reported the specific production of fluorescein and pyocyanin by the *Pseudomonas* bacterium against *Pythium*.

The standardization of the mode of application of bacterial antagonists indicated that dipping of the stem cuttings together with soil application were highly effective in inhibiting root rot incidence rather than the individual application of antagonists either through stem cutting dipping or soil application. Among the bacterial antagonists tested, delivery of Pf1 as stem cutting, dipping and soil application recorded the lowest disease incidence and highest per cent

S.No	PGPR strains	Fluorescein	Pyocyanin	Siderophore production	HCN
1	Pf1	+	+	+	+++
2	CPF1	+	+	+	+++
3	CPF 2	-	-	+	+
4	CPF 3	-	-	+	-
5	CPF 4	+	+	+	+
6	CPF 5	+	+	+	+
7	CPF 6	+	+	-	++
8	CPF 7	-	-	+	-
9	CPF 8	+	+	+	++
10	CPF 9	+	+	+	+
11	CPF 10	-	-	-	+

+: Produced;   - : Not produced   (+++: Strong,   ++: Medium,   +: Low production)

Table 1: Antibiotics, siderophore and HCN production of P. fluorescens strains.

S. No	Treatments	Per cent disease incidence	Per cent reduction over control	Shoot length (cm)	Tuber length (cm)	No. of tubers/plant
1	Pf1 (SCD alone)	37.26ᵈ (37.62)	41.74	46.65ᶠ	26.54ᵈ	7.62ᵈᵉ
2	Pf1 (SA alone)	26.53ᵇ (31.00)	58.74	58.54ᵇ	29.96ᵇ	8.52ᶜ
3	Pf1 (SCD+SA)	24.12ᵇ (29.42)	62.50	63.56ᵃ	32.86ᵃ	10.65ᵃ
4	CPF1 (SCD alone)	39.42ᵉ (38.89)	21.66	40.26ʰ	25.92ᶜᵈ	6.94ᵉ
5	CPF1 (SA alone)	33.23ᵈ (35.20)	23.44	50.93ᵉ	27.56ᶜ	7.02ᵈᵉ
6	CPF1 (SCD+SA))	32.56ᶜᵈ (34.99)	29.16	55.26ᶜ	30.92ᵃᵇ	8.82ᶜ
7	Pre mixture fungicide (Carbendazim +Mancozeb) (SCD alone)	30.53ᶜ (33.54)	52.52	42.26ᵍ	27.65ᶜ	7.54ᵈ
8	Pre mixture fungicide (Carbendazim +Mancozeb) (SA alone)	24.36ᵇ (29.58)	62.12	53.65ᵈ	27.12ᶜ	8.59ᶜ
9	Pre mixture fungicide (Carbendazim +Mancozeb) (SCD+SA)	18.56ᵃ (22.96)	71.14	58.72ᵇ	29.87ᵇ	9.56ᵇ
10	Inoculated Control	64.31ᶠ (53.32)	-	35.89ⁱ	20.14ᵉ	5.54ᶠ
	SEd	0.79		0.84	0.94	0.28
	CD (0.05)	1.64		1.76	1.97	0.57

SCD-Stem cuttings dip; SA-Soil application All Values are mean of three replications; Data followed by the same letter in a column are not significantly different from each other according to Duncan's multiple range test at P=0.05

Table 2: Effects of PGPR on the root rot incidence, growth characters and yield under green house condition.

inhibition over the control (24.12 and 62.50%). It was followed by the application of CPF1 both as stem cutting, dipping and soil application. The chemical premixture fungicide (Carbendazim+Mancozeb) (0.1%) recorded 71.14 per cent reduction over the control. The highest disease incidence was recorded in the inoculated control (64.31%) (Table 2). The positive colonization ability of Pseudomonas GRC2 lies in it being the successful colonizer of the spermosphere, increasing seedling emergence, and its establishment in the rhizosphere of peanuts giving protection against *M. Phaseolina* resulting in enhanced yield [25]. It was further reported by Thilagavathi et al. [26] that the combined application of *P. fluorescens* (Pf1) in seed and soil applications was effective in reducing the root rot disease in green gram under greenhouse and field conditions.

The bio control agents not only controlled dry root rot, but also promoted plant growth and this gives them an advantage over the use of chemical fungicides against root rot in disease management. The studies showed that the PGPR is capable of controlling the coleus root rot. Amongst the different *P. fluorescens* strains tried, stem cutting, dipping and soil application of Pf1 was found to be suitable for the management of coleus root rot under pathogen (*M. Phaseolina*) inoculated soil in greenhouse condition. However, field evaluation is necessary to determine its efficacy under natural ecosystem.

Acknowledgement

I thank to Dr T. Raguchander, Professor, Dept. of Plant Pathology, TNAU for providing biocontrol agents for this work.

References

1. Kamalakannan A, Mohan L, Amutha G, Chitra K, Parthiban VK, et al. (2003) Effect of volatile and diffusible compounds of biocontrol agents against Coleus forskohlii root rot pathogens. In: Symposium on Recent Development in the Diagnosis and Management of Plant Diseases.18-20 Dec. Dharwad, Karnataka: 92.

2. Daughtrey M, Eshenaur B, Holcomb G (2006) Source: Greenhouse Product News 6: 6

3. Vivekananthan R, Ravi M, Ramanathan A, Samiyappan R (2004) Lytic enzymes induced by *Pseudomonas fluorescens* and other biocontrol organisms mediate defence against the anthracnose pathogen in mango. World J Micro Biotech 20: 235-244.

4. Saravanakumar D, Harish S, Loganathan M, Vivekananthan R, Rajendran L, et al. (2007) Rhizobacterial bioformulation for the effective management of Macrophomina root rot in mungbean. Archives of Phytopathology and Plant Protection 40: 323-337

5. Sendhilvel V, Buvaneswari D, Kanimozhi S, Mathiyazhagan S, Kavitha K, et al. (2005) Management of cowpea root-rot caused by *Macrophomina phaseolina* (Tassi) Goid. using plant growth promoting rhizobacteria. J Biol Control 19: 41-46.

6. Rangaswami G (1972) Diseases of crop plants in India. Prentice Hall of India Pvt. Ltd., New Delhi, p520.

7. King EO, Ward MK, Raney DE (1954) Two simple media for the demonstration of pyocyanin and fluorescin. J Lab Clin Med 44: 301-307.

8. Waksman SA, Connick Jr WJ (1952) A tentative outline of the plate method for determining the number of microorganisms in the soil. Soil Science 14: 27-28.

9. Schaad NW (1992) Laboratory guide for identification of plant pathogenic bacteria. In: NW Schad (Ed.), The American Phytopathological Society. International Book Distributing Co., Lucknow. India.

10. Schwyn B, Neilands JB (1987) Universal chemical assay for the detection and determination of siderophores. Anal Biochem 160: 47-56.

11. Miller RL, Higgins VJ (1970) Association of cyanide with infection of birds' foot trefoil by Stemphylium rot. Phytopathology 60: 104-110.

12. Gagne SD, Quere L, Aliphat S, Lemay R, Fournier N (1991) Inhibition of plant pathogenic fungi by volatile compounds produced by some PGPR isolates (Abstr.). Can J Plant Pathol 13: 277.

13. Dennis C, Webster J (1971) Antagonistic properties of species groups of *Trichoderma* L. Production of non-volatile antibiotics. Transaction of British Mycological Society 57: 25-39.

14. Vidhyasekaran P, Muthamilan M (1995) Development of formulation of *Pseudomonas fluorescens* for control of chickpea wilt. Pl Dis 79: 780-782.

15. Riker AJ, Riker RS (1936) Introduction to Research on Plant Diseases. John Swift, New York, NY, USA.

16. Bharathi R, Vivekananthan R, Harish S, Ramanathan A, Samiyappan R (2004) Rhizobacteria based bioformulations for the management of fruit rot infection in chillies. Crop Protect 23: 835-843.

17. Sarangi NP, Athukorala WG, Dilantha F, Khalid Y, Rashid Teresa de K (2010) The role of volatile and non volatile antibiotics produced by *Pseudomonas chlorophis* strain PA23 in its root colonization and control of *Sclerotinia sclerotiorum*. Biocon Sci Techn 20: 875-890.

18. Neidig N, Paul RJ, Scheu S, Jousset A (2011) Secondary metabolites of *Pseudomonas fluorescens* CHA0 drive complex non-trophic interactions with bacterivorous nematodes. Microb Ecol 61: 853-859.

19. Laha G, Singh RP, Venna JP (1992) Biocontrol of Rhizoctonia solani in cotton by fluorescent pseudomonads. Indian Phytopath 45: 412-415.

20. Manwar AV, Khandelwal SR, Chaudhari BL, Meyer JM, Chincholkar SB (2004) Siderophore production by a marine *Pseudomonas aeruginosa* and its antagonistic action against phytopathogenic fungi. Appl Biochem Biotechnol 118: 243-251.

21. Duffy B, Schouten A, Raaijmakers JM (2003) Pathogen self-defense: mechanisms to counteract microbial antagonism,. Annu Rev Phytopathol 41: 501-538.

22. Meena B, Marimuthu T, Vidhyasekaran P, Velazhahan R (2001) Biological control of root rot of ground nut with antagonistic *Pseudomonas fluorescens* strains. J Pl Dis Prot 108: 369-381.

23. Sandeep Kumar M, Pandey P, Maheshwari DK (2009) Reduction in dose of chemical fertilizers and growth enhancement of sesame (*Sesamum indicum* L.) with application of rhizospheric competent *Pseudomonas aeruginosa* LES4. Euro J Soil Biol 45: 334-340.

24. Rachid D, Ahmed (2005) Effect of iron and growth inhibitors on siderophore production by *Pseudomonas fluorescens*. African J Biotech 4: 697-702.

25. Gupta CP, Dubey RC, Maheshwari DK (2002) Plant growth enhancement and suppression of *Macrophomina phaseolina* causing charcoal rot of peanut by fluorescent pseudomonads. Biol Fertil Soils 35: 399-405.

26. Thilagavathi R, Saravanakumar D, Ragupathi N, Samiyappan R (2007) A combination of biocontrol agents improves the management of dry root rot (*Macrophomina phaseolina*) in greengram. Phytopathol Mediterranea 46: 157-167.

17

Induction of Systemic Resistance in Sugar-Beet against Root-Knot Nematode with Commercial Products

Mostafa Fatma AM*[1], Khalil AE[2], Nour El Deen AH[1], Ibrahim[1] and Dina S[2]

[1]Agricultural Zoology Department, Faculty of Agriculture, Mansoura University, Mansoura, Egypt
[2]Nematology Division, Plant Pathology Res. Institution, Giza, Egypt

Abstract

The potentials of Bio-arc (a commercial formulation of the *Bacillus megaterium*) at the rate of 5, 10, 15 and 20 ml and Nemastrol (a commercial formulation of active ingredients) at the rate of 0.25 ml, for induction of systemic resistance to sugar-beet var. Negma infected with *M. incognita* were conducted in two soil types. Results revealed that all treatments with tested rates were found to have nematicidal activity against nematode infection and improved plant growth parameters of sugar-beet with various levels of success. The dual application of Bio-arc+Nemastrol at the rate of 20 ml +0.25 ml proved to be the best and showed significant improvement in plant growth parameters in terms of shoot length (92.6,127.5%) and total plant fresh weight (91.7, 370.4) of sugar-beet grown either in clayey or sandy soil, respectively. Among all treatments Nemastrol ranked next to oxamyl and performed the best and significantly suppressed total nematode population (Rf=1.9, 2.2), root galling (RGI=3.0, 3.0), number of egg masses (EI=3.0, 3.0) and number of eggs / egg mass (Red. %=76.5, 74.5) in clayey and sandy soil, respectively. However, concomitant treatment showed better results than did Bio-arc alone at four tested rates. The greatest suppression in total nematode population was recorded with clayey and sandy soil receiving the dual application of Nemastrol (0.25 ml) and Bio-arc (20 ml) with reproduction factor 2.2, 2.6 and reduction percentages reached 92.8, 92.6% respectively. Leaves of sugar-beet were assayed for their biochemical profiles with respect to NPK, total cholorophyll, total carbohydrates, proteins, and phenols. Moreover, remarkable induction in such chemical constituents except phenol content was recorded with the application of Bio- arc+Nemastrol (20 ml+0.25 ml). On the other hand, activities of related enzymes i.e. Peroxidase (PO) and Polyphenol Oxidase (PPO) were evaluated in roots of sugar-beet infected with *M. incognita*. The enzymes accumulation was much greatest in Bio-arc+Nemastrol (20+0.25 ml) treated plants compared to control as they reached their peak at day 9th from nematode inoculation.

Keywords: Induced resistance; Enzyme activity; *Meloidogyne incognita*; Sugar-beet; Biochemical activities; Soil type

Introduction

Sugar-beet (*Beta vulgaris* L.) is considered an important root crop, which is ranked second to sugar-cane for supporting the expansion of Egyptian sugar industry. Root-knot nematodes (RKNs) *Meloidogyne* spp. are among the most deleterious plant pathogens since these organisms play a detectable role in limiting the productivity of such economic agriculture crop. The root –knot nematode *Meloidogyne incognita* (Kofoid & White) Chitwood is among the most important nematode infesting sugar-beet. Many efforts to protect such crop from root-knot nematodes infestation are crucial. Because of the lack of plant resistance to most species of root-knot nematode as well as the environmental restrictions on nematicidal use for controlling plant parasitic nematodes, biological control and other eco-friendly disease control measures have gained recently increasing interest. The activation of the plant's own defense system through biotic and abiotic agents, called elicitors, has been considered as a focus of research only in recent years for the control of plant pathogens. The resulting elevated resistance due to an inducing agent upon infection by a pathogen is called induced systemic resistance (ISR) or systemic acquired resistance (SAR) [1].

However, induced resistance to plant parasitic nematodes has not been as extensively studied as that of fungi and bacteria. Ibrahim, et al. [2] recorded the capability of humic acid as well as thiamine at the concentration of 2000 ppm to induce resistance in sugar-beet against *M.incognita* and increase the activity of polyphenol oxidase (PPO) and peroxide oxidase (PO) enzymes compared with non-infected plants. On the other hand, plant growth promoting rhizobacterium (PGPR) belonging to *Bacillus* spp. are being exploited commercially

for plant protection to induce systemic resistance against various pests and pathogens. PGPR mediated rhizobacteria is often associated with the onset of defense mechanisms by expression of various defense related enzymes such a glucanase, chitinase, phenylalanine ammonia lyase (PAL), peroxidase (PO), and polyphenol oxidase (PPO) and accumulation of phenols [3] .In this point of view, the present work was carried out in order to study the impact of promoting growth rhizobacterium (PGR), *Bacillus megaterium*, Nemastrol active ingredients extract as resistance inducers to sugar-beet plant infected with *M. incognita* under greenhouse conditions.

Materials and Methods

Two greenhouse experiments were conducted at Nematological Research Unit (NERU), using sandy and clayey soil in order to evaluate the nematicidal properties of the commercial formulation of rhizobacterium, *Bacillus megaterium* (Bio-arc), the commercial biocide, Nemastrol against the root-knot nematode, *M. incognita*

---

*Corresponding author: Mostafa Fatma AM, Agricultural Zoology Department, Faculty of Agriculture, Mansoura University, Msansoura, Egypt
E-mail: mohsenfatma@hotmail.com

and the resulting effect on plant growth parameters of sugar-beet var. Negma. Induced resistance (IR) of such bio-agents was assayed through chemical composition and enzyme activities.

Tested bio-agents

Bio-arc: A native commercial formulation of phosphorus soluble bacterium, *Bacillus megaterium* (25×10^6 cfu/g) @ 2.5 g/L of distilled water, was obtained from Agricultural Research Institute, Giza, Egypt and enrolled by the Ministry of Egyptian Agriculture under No. 1087.

Nemastrol: A native commercial formulation of active ingredients containing glycosynolates (12%) , chitinase (12×10^5 IU) , cytokinins (200 ppm), flavonoids (5%) and β 1-3, Glucanase (2×10^5 IU) @ the rate of 5 L/ feddan, was obtained from Royal Company, Egypt.

Tested bio-agents rates: The tested bio- agent, Nemastrol was applied @ the rate of 0.25 ml/pot. However, Bio-arc was added using four different rates of 5, 10, 15 and 20 ml/pot in single application.

The chemical nematicide: Oxamyl 10%G; S-methyl -1(dimethylcarbamoyl)-N-[(methylcarbamoyl) oxyl] thioforminidate, was applied as a standard nematicide @ the rate of 0.3 g /pot

Experimental design: For each soil type, forty eight plastic pots (15-cm-d) containing 800 g steam- sterilized soil were planted with 3-5 seeds/pot of sugar-beet var. Negma, irrigated with water as needed then thinned to one seedling/pot after one month from germination. Fifteen days later, plant seedlings were inoculated with 2000 viable eggs of root-knot nematode, *M. incognita*. One week later, plants were treated with the selective materials as soil drench at the previous mentioned rates. For each soil type, four pots were treated with oxamyl @ the rate of 0.3 g /pot. However, four pots were left free of nematode infection and any treatment to serve as control (Ck1). Another four pots were received nematode alone and served as control (Ck2). Pots were then arranged in a randomized complete block design in a greenhouse @ 27 ± 3ºC, with four replicates and received water as needed. Therefore, treatments for each soil type were as follows:- 1-Bio-arc @ 5 ml/pot; 2- Bio-arc @ 10 ml/pot;3- Bio-arc @ 15 ml/pot;4- Bio-arc @ 20 ml/pot;5- Nemastrol @ 0.25 ml/ pot;6- Bio-arc @ 5 ml/pot +Nemastrol @ 0.25 ml/ pot ;7- Bio-arc @ 10 ml/pot+Nemastrol @ 0.25 ml/ pot;8- Bio-arc @ 15 ml/pot +Nemastrol @ 0.25 ml/ pot;9- Bio-arc @ 20 ml/pot+Nemastrol @ 0.25 ml/ pot;10- Oxamyl (O),11- Untreated Uninoculated plants (Ck1) and 12- Nematode alone (Ck2). Plants were harvested 45 days after nematode inoculation and roots were washed free from adhering soil. Data dealing with fresh shoot and root weight, dry shoot weight, shoot and root length, were recorded. Nematodes were extracted from soil using sieving and modified Baermann technique [4]. Roots were stained in 0.01 acid fuchsin [5] and examined for the developmental stages, females, galls and egg masses under stereomicroscope. Root galling or egg masses were rated on a scale of 0-5 where 0=no galls or egg masses, 1=1-2 galls or egg masses, 2=3-10 galls or egg masses, 3=11-30 galls or egg masses, 4=31-100 galls or egg masses, 5=more than 100 galls or egg masses per root system [6].

For each treatment, dry weight of shoot (1 g) was subjected to chemical analysis in order to evaluate total nitrogen, crude protein, total carbohydrate and total phenol. Samples of dried leaves were ground, wet digested and nitrogen (N), phosphorus (P), potassium (K) contents were determined according to kjeldahl methods [7] A.O.A.C. (1980) described by number of researchers (Pregl, Jackson, John) [8-10].

Determination of enzymatic activities: Sugar-beet plants treated with Bio-arc (20 ml/pot) and Nemastrol (0.25 ml/pot) singly and concomitantly were inoculated with 2000 second juveniles of *M.*

incognita and tested for enzymatic activities. The same protocol as outlined before was repeated. Roots were collected at different intervals (0, 3, 9 and 15 days after treatment and nematode inoculation) and assayed for activities of Peroxidase (PO) and Polyphenol Oxidase (PPO).

Preparation of enzyme extract: Enzyme extracts were prepared following the method described by [3] Maxwell and Bateman (1967(. Dry root tissues (0.5 g) of each treatment were ground in 3 ml Na-phosphate buffer at pH 6.8 in a mortar and then centrifuged at 1.500 g/20 min at 6°C. The resultant supernatant fluids were processed for enzyme assays.

Peroxidase activity (PO): Peroxidase was assayed using photochemical method as described by [11] Amako et al. The reaction mixture was added as the following sequences, 1500 ml phosphate buffer., 480 ml hydrogen peroxidase., 1000 ml pyrogallol, 20 ml sample extract. The increasing in the absorbance at 430 nm was recorded against blank with phosphate buffer instead of enzyme extract. One unit of enzyme activity was defined as the amount of the enzyme, which changing the optical density at 430 nm per min. at 25°C under standard assay conditions. Specific activity was expressed in units by dividing it to mg protein.

Polyphenol oxidase (PPO): Polyphenol oxidase was assayed using photochemical method as described by Coseteng and Lee [12]. The reaction mixture was added as the following sequences: 2.7 ml potassium phosphate buffer 90.05 M, pH 6.2, 0.25 ml of 0.25 M catechol, 0.05 ml of enzyme extract. The increasing in absorbance at 420 nm was measured. One unit of enzyme activity is defined as the amount of the enzyme that causes an increase of 0.001 absorbance unit per minute at 25°C.

Data analysis: Statistically, the obtained data were subjected to analysis of variance (ANOVA) [13] (Gomez and Gomez) followed by Duncan's multiple range tests to compare means [14].

Results

The influence of the two bio-control agents namely Bio-arc (a commercial formulation of *B. megaterium*) at four tested rates (5, 10, 15, and 20 ml) and Nemastrol @ 0.25 ml singly and concomitantly on plant growth response of sugar-beet plant var. Negma infected with *M. incognita* and grown in two soil types i.e. clayey and sandy is summarized Table 1. (Results revealed that *M. incognita* infection caused a significant reduction in plant growth parameters (shoot and root length, shoot weight) with reduction percentage in total plant fresh weight reached 35.0 and 64.0% in clayey and sandy soil respectively. Irrespective to soil type and tested rates, all treatments showed remarkable increase in plant growth parameters in terms of shoot length, shoot and root weight with various degrees. In single application, it was evident that the effectiveness of bio-arc to enhance plant growth parameters increased with the increase of addition in the two soil types. Plant growth response of sugar-beet infected with *M. incognita* was pronounced in sandy soil more than clayey soil. In sandy soil, a significant improvement in shoot length (85.8%), plant fresh weight (174.1%) and shoot dry weight (350.0%) was recorded with Bio-arc @ the rate of 20 ml/plant. Similar trend was noticed with sugar-beet grown in clayey soil with percentage of increase in shoot length, total plant fresh weight and dry shoot weight reached 70.4, 41.7 and 180.0%, respectively. However, Nemastrol at the rate of 0.25 ml resulted a pronounced improvement in plant growth parameters in terms of shoot length (100.0, 10.8%), total plant fresh weight (70.5, 44.4%) and shoot dry weight (200.0, 150.0%) of sugar-beet grown in clayey and sandy

Treatments	Rates/ml	*Plant growth response						Shoot Dry wt.(g)	Inc.%
		Shoot Length (cm)	Inc.%	Shoot fresh wt. (g.)	Root fresh wt. (g.)	Total Plant fresh wt.(g)	Inc. %		
Clayey Soil									
Bio-arc	5	18.3e	35.6	10.9de	5.8d	16.7	26.5	1.7ef	70.0
	10	21.0d	55.6	11.3c-e	5.8d	17.1	29.5	1.7ef	70.0
	15	22.0cd	63.0	11.7c-e	6.2d	17.9	35.6	2.2de	120.0
	20	23.0c	70.4	12.0cd	6.7cd	18.7	41.7	2.8cd	180.0
Nemastrol	0.25	27.0a	100.0	13.7a-c	8.8a	22.5	70.5	3.0b-d	200.0
Bio-arc+Nemastrol	5+0.25	25.0b	85.2	12.4b-d	8.5ab	20.9	58.3	3.2a-c	220.0
	10+0.25	25.3ab	87.4	13.6a-c	8.7a	22.3	68.9	3.5a-c	250.0
	15+0.25	25.5ab	88.9	14.9ab	8.5ab	23.4	77.3	3.7ab	270.0
	20+0.25	26.0ab	92.6	15.9a	9.4a	25.3	91.7	3.9a	290.0
Oxamyl		18.8e	39.3	11.9cd	7.6bc	19.5	47.7	1.3f	30.0
Plant free of N		13.8f	2.2	11.1de	9.2a	20.3	53.8	1.2f	20.0
N alone		13.5f	0.0	9.3e	3.9e	13.2	0.0	1.0f	0.0
Sandy Soil									
Bio-arc	5	12.9ef	7.5	2.1f	2.5ef	4.6	70.4	0.6g-i	50.0
	10	14.5e	20.8	2.1ef	2.9d-f	5.0	85.2	0.9f-h	125.0
	15	17.3d	44.2	2.9de	2.5e-g	5.4	100.0	1.5d	275.0
	20	22.3c	85.8	4.4c	3.0de	7.4	174.1	1.8cd	350.0
Nemastrol	0.25	13.3ef	10.8	2.1f	1.8gh	3.9	44.4	1.0fg	150.0
Bio-arc+Nemastrol	5+0.25	18.0d	50.0	3.1d	2.2f-h	5.3	96.3	1.0ef	150.0
	10+0.25	22.5c	87.5	5.4b	4.5c	9.9	266.7	2.2ab	450.0
	15+0.25	25.5ab	112.5	5.7b	5.3b	11.0	307.4	2.1bc	425.0
	20+0.25	27.3a	127.5	6.7a	6.0a	12.7	370.4	2.5a	525.0
Oxamyl		18.5d	54.2	1.3g	1.6h	2.9	7.4	0.6hi	50.0
Plant free of N		24.0bc	100.0	4.2c	3.3d	7.5	177.8	1.5de	275.0
N alone		12.0f	0.0	1.0g	1.7h	2.7	0.0	0.4i	0.0

Means in each column followed by the same letter(s) did not differ at P≤0.05 according to Duncan`s multiple range test'Each value presented the mean of four replicates N = *M. incognita* (2000 eggs/ plant)

Table 1 : Impact of Bio-arc (*Bacillusmegaterium*) and Nemastrol (a mixture of active ingredients) on plant growth response of sugar-beet var. Negma grown in two soil types infected with *M. incognita* under greenhouse conditions (27 ± 3°C).

Treatments	Rates/ ml	No. of galls*	RGI**	No. of eggmasses	EI*	No. of eggs/ eggmass	Nematode population in Soil	Females population in Root	Final population (Pf)	Rf***
Clayey soil										
Bio-arc	5	32.0b	4.0	22.3b	3.0	1009.95b	584.5b	34.5b	23141.4	11.6
	10	29.8b	3.0	20.8b	3.0	874.95c	439.5c	33.5b	18672.5	9.3
	15	20.5c	3.0	17.5c	3.0	689.95d	399.5d	22.5c	12496.4	6.2
	20	20.0cd	3.0	14.5cd	3.0	587.45e	388.5e	21.3cd	8927.8	4.5
Nemastrol	0.25	13.8e	3.0	11.3e	3.0	312.45j	299.5k	15.8e	3846.0	1.9
Bio-arc +Nemastrol	5+ 0.25	19.3cd	3.0	14.3cd	3.0	489.95f	372.5f	20.5cd	7399.3	3.7
	10+ 0.25	18.0c-e	3.0	13.0de	3.0	417.45g	361.2g	19.5c-e	5807.6	2.9
	15+ 0.25	16.0de	3.0	11.8de	3.0	392.45h	359.5h	18.3c-e	5008.7	2.5
	20+ 0.25	15.8de	3.0	11.3e	3.0	349.95i	337.5i	17.5de	4309.4	2.2
Oxamyl		7.3f	2.0	10.0f	2.0	249.95k	300.5j	10.3f	2810.3	1.4
N alone		53.5a	4.0	43.8a	4.0	1339.95a	1399.5a	54.8a	60145.6	30.0
Sandy soil										
Bio-arc	5	41.5b	4.0	33.3b	4.0	1074.95b	1443.8ab	45.5b	37285.9	18.6
	10	41.3b	4.0	30.5c	3.0	964.95c	1000.5a-c	44.0b	30476.0	15.2
	15	17.3d	4.0	30.3c	3.0	862.45d	810.0bc	37.0c	26979.7	13.5
	20	24.8d	3.0	22.5d	3.0	736.25e	507.6bc	26.3d	17099.5	8.5
Nemastrol	0.25	14.3fg	3.0	11.5fg	3.0	351.25j	279.5c	15.0f	4333.9	2.2
Bio-arc +Nemastrol	5 + 0.25	23.8d	3.0	22.3d	3.0	574.95f	400.5c	24.8de	13246.7	6.6
	10+ 0.25	18.5e	3.0	14.5e	3.0	509.95g	384.5c	21.8de	7800.6	3.9
	15+ 0.25	18.3ef	3.0	12.8ef	3.0	449.95h	362.5c	21.3e	6143.2	3.0
	20+ 0.25	16.3ef	3.0	11.8ef	3.0	417.45i	347.5c	21.0e	5294.4	2.6
Oxamyl		11.5g	3.0	11.0g	3.0	117.45k	360.5j	12.5f	1665.0	0.8
N alone		58.5a	4.0	50.5a	4.0	1379.95a	2069.5a	61.0a	71819.8	35.9

*Each value presented the mean of four replicates N = *M. incognita* (2000 eggs/ plant) . Means in each column followed by the same letter(s) did not differ at P ≤ 0.05 according to Duncan`s   multiple range test. * * Root gall index (RGI) or egg masses index (EI) was determined according to the scale given by Taylor and Sasser as follows : 0= no galls or egg masses, 1= 1-2   galls or egg masses , 2= 3-10 galls of egg masses, 3= 11-30 galls or egg masses, 4= 31-100 galls or egg masses and 5= more than 100 galls or egg masses ** Reproudction factor (Rf) = No. of eggs per root + Nematode population in soil + No. of developmental stages + No. of females/ No. of eggs inocula

Table 2: Impact of Bio-arc and Nemastrol (a mixture of active ingredients) singly and concomitantly on the development and reproduction of *M. incognita* infecting sugar-beet var. Negma grown in two soil types under greenhouse conditions   at 27 ± 3°C.

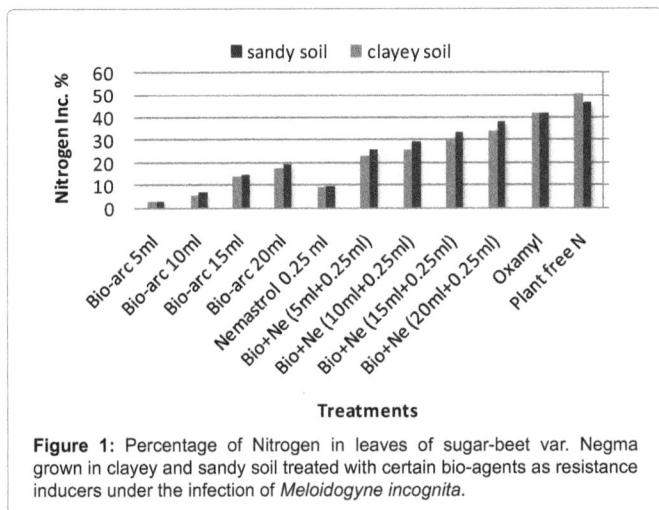

Figure 1: Percentage of Nitrogen in leaves of sugar-beet var. Negma grown in clayey and sandy soil treated with certain bio-agents as resistance inducers under the infection of *Meloidogyne incognita*.

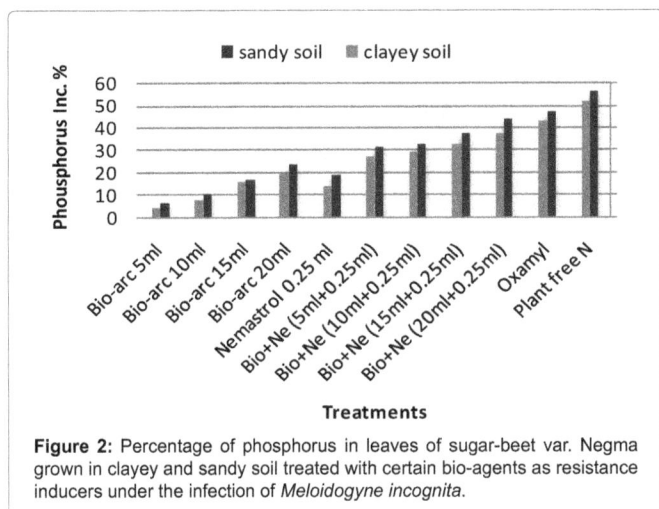

Figure 2: Percentage of phosphorus in leaves of sugar-beet var. Negma grown in clayey and sandy soil treated with certain bio-agents as resistance inducers under the infection of *Meloidogyne incognita*.

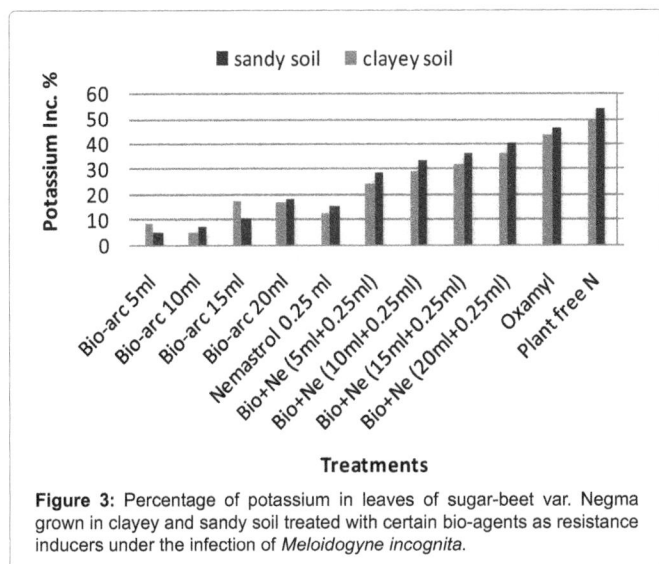

Figure 3: Percentage of potassium in leaves of sugar-beet var. Negma grown in clayey and sandy soil treated with certain bio-agents as resistance inducers under the infection of *Meloidogyne incognita*.

soil respectively. In concomitant treatment, Bio- arc (20 ml)+Nemastrol (0.25 ml) was the best and showed significant improvement in plant growth parameters in terms of shoot length (92.6, 127.5%) and total

plant fresh weight (91.7, 370.4) of sugar-beet grown either in clayey or sandy soil, respectively. Oxamyl as a standard nematicide showed moderate improvement in pervious criteria of sugar-beet grown in clayey soil with increase percentages 39.3, and 47.4 respectively. Similar trend was noticed with shoot length (54.2%) and dry shoot weight (50%) of sugar-beet grown in sandy soil treated with oxamyl.

Regarding the impact of Bio-arc and Nemastrol singly and concomitantly on the development and reproduction of the root-knot nematode, *M. incognita* infecting sugar-beet grown in clayey and sandy soil is documented (Table 2). Irrespective to soil type and tested rates results revealed that total nematode population was significantly suppressed with all tested treatments with reproduction factor ranged from 1.4 to 11.6 in clayey soil and from 0.8 to 18.6 in sandy soil compared to inoculated plants (Rf=30.0, 35.9) respectively. Among tested treatments, Nemastrol significantly suppressed total nematode population (RF=1.9, 2.2), root galling (RGI=3.0, 3.0), number of egg masses (EI=3.0, 3.0) and number of eggs /10 egg masses (Red. %=76.5, 74.6) in clayey and sandy soil, respectively. However, concomitant treatment showed better results than did bio-arc alone at four tested rates. Among the concomitant treatment the greatest reduction in total nematode population was recorded in clayey and sandy soil which received the dual application of Bio-arc (20 ml) and Nemastrol (0.25 ml) with reproduction factor 2.2, 2.6 and reduction percentages reached 92.8, 92.6%. Meanwhile, total nematode population was significantly suppressed with oxamyl introduced to clayey soil (Rf=1.4) and sandy soil (Rf=0.8) relative to control plants where Rf=30.0 and 35.9, respectively. Nevertheless, number of eggs/ egg mass were significantly suppressed with oxamyl application with percentage of reduction amounted to 76.7 and 74.5% in clayey and sandy soil respectively.

Biochemicals activities

Nitrogen, phosphorus and potassium contents: NPK contents were significantly suppressed due to nematode infection with reduction percentages 34.1, 35.5 and 39.2% in clayey soil and 32.2, 37.5 and 39.1% in sandy soil. However, a remarkable induction in NPK content was recorded with the application of Bio- arc + Nemastrol (20 ml+0.25 ml) with % of increase amounted to 34.94, 41.40, 48.43 and 39.55, 45.0, 48.99 in clayey and sandy soil respectively (Figures 1-3).

Total chlorophyll content: Chlorophyll *a* and *b* were moderately affected due to nematode infection with reduction % in total chlorophyll reached 32.5 and 32.3% in clayey and sandy soil respectively. Application of such treatments revealed a considerable induction with the dual application of Bio + Nemastrol (20 ml+ 0.25 ml) and oxamyl as well with % of increased reached 37.0, 42.6 and 38.6, 40.9% in clayey and sandy soil consecutively (Figure 4).

Total carbohydrates: Total carbohydrates were significantly suppressed due to nematode infection with reduction percentage 20.38 and 16.9% in clayey and sandy soil respectively. The highest increase was recorded in leaves of sugar-beet treated with Bio-arc+Nemastrol (20 ml+0.25 ml) with values 13.4 and 15.3% in clayey and sandy soil respectively (Figure 5).

Crude proteins: Untreated sugar-beet infected with *M. incognita* exhibited significant reduction in total proteins as compared with untreated uninoculated plants with percentage of reduction reached 34.1 and 32.1% in clayey and sandy soil respectively. Among all tested compounds, the highest increase percentage in crude proteins was obtained with oxamyl (43.4 and 42.2%) in clayey and sandy soil, respectively (Figure 6) followed by Bio-arc+Nemastrol (20 ml+0.25 ml) with values averaged 35.0 and 39.2% in clayey and sandy soil,

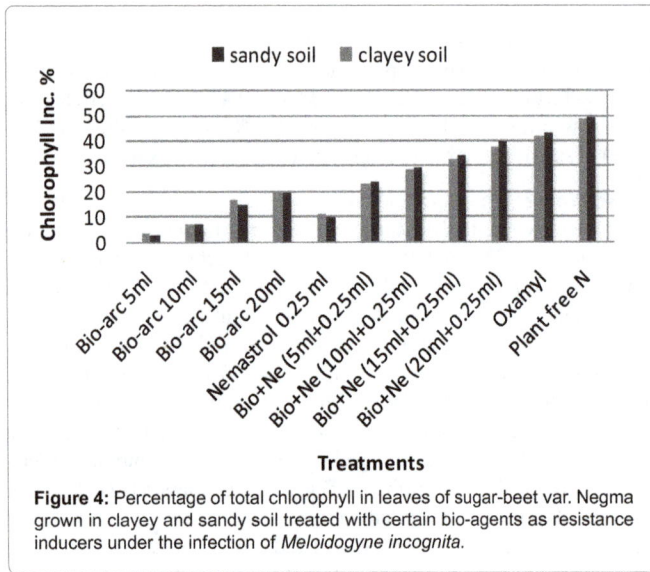

Figure 4: Percentage of total chlorophyll in leaves of sugar-beet var. Negma grown in clayey and sandy soil treated with certain bio-agents as resistance inducers under the infection of *Meloidogyne incognita*.

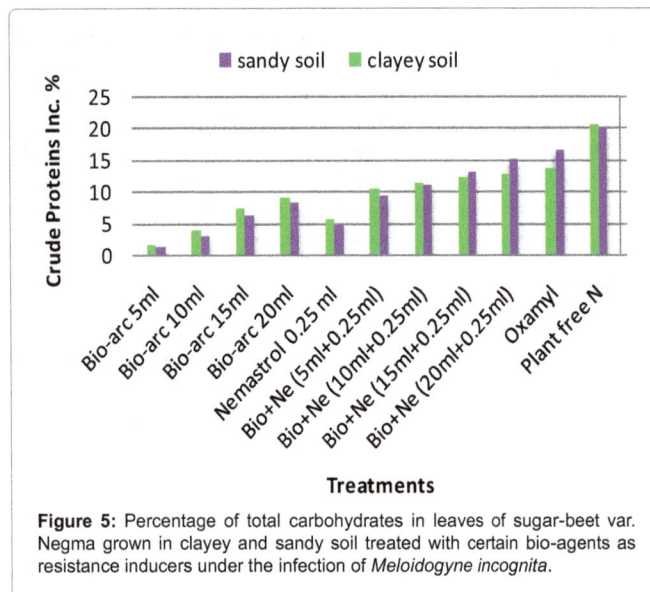

Figure 5: Percentage of total carbohydrates in leaves of sugar-beet var. Negma grown in clayey and sandy soil treated with certain bio-agents as resistance inducers under the infection of *Meloidogyne incognita*.

respectively.

Phenol content: The total phenol evaluated in leaves of sugar-beet infected with *M. incognita* revealed a moderate enhancement compared to control plants. However, phenol content showed different degrees of reduction in all treatments compared to untreated uninoculated plants grown in clayey and sandy soil. (Figure 7)

Defense related proteins: The tested materials viz. Bio-arc (20 ml), Nemastrol (0.25 ml), Bio-arc+Nemastrol (20 ml+0.25 ml) and oxamyl as well differed in their ability to stimulate Peroxidase (PO) and Polyphenol Oxidase (PPO) activities in sugar-beet plant inoculated with *M.incognita* (Figure 8). In untreated uninoculated plants, the activities of PO and PPO remained higher and attained their peak at the 9th day and thereafter a decline was noticed at 15th day. On the other hand, the least induction of PO and PPO was recorded with plants untreated and inoculated with nematodes and showed slight decline 3 days after nematode inoculation then increased and reached their peak at 9th day. However, increased PO and PPO activities were more pronounced in Bio-arc+Nemastrol (20 ml+0.25 ml) followed by oxamyl

then Nemastrol compared to untreated inoculated plants. In such treatments, the dual application Bio-arc + Nemastrol performed the best since PO& PPO activities were increased and reached their peak at the 3rd day after nematode inoculation then declined at 9th day followed by slightly increment at 15th day . Meanwhile, the increased activity of PO &PPO remained higher in plants treated with oxamyl and reached their peak at 9th day after nematode inoculation.

Discussion

Induction of systemic resistance (ISR) of plants against pathogens is a widespread phenomenon that has been intensively investigated in fungi and bacteria with respect to its potential use in plant protection. However, little attention has been given to nematode pests. Acquired or induced resistance can be achieved by inoculating a plant with incompatible or weak pathogens or by applying biotic or abiotic inducers [15]. The root-knot nematode, *M. incognita* caused a significant reduction in plant growth parameters (shoot and root length, shoot weight) with reduction percentage in total plant fresh weight reached 35.0 and 64.0% in clayey and sandy soil respectively. Application of

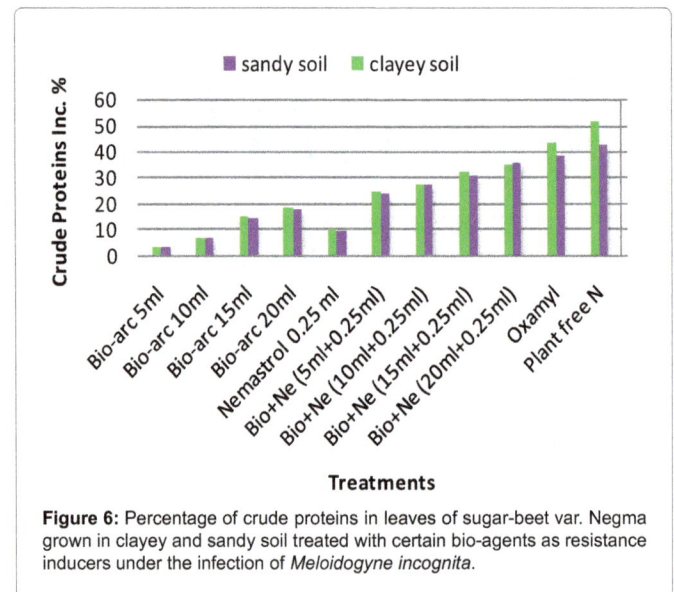

Figure 6: Percentage of crude proteins in leaves of sugar-beet var. Negma grown in clayey and sandy soil treated with certain bio-agents as resistance inducers under the infection of *Meloidogyne incognita*.

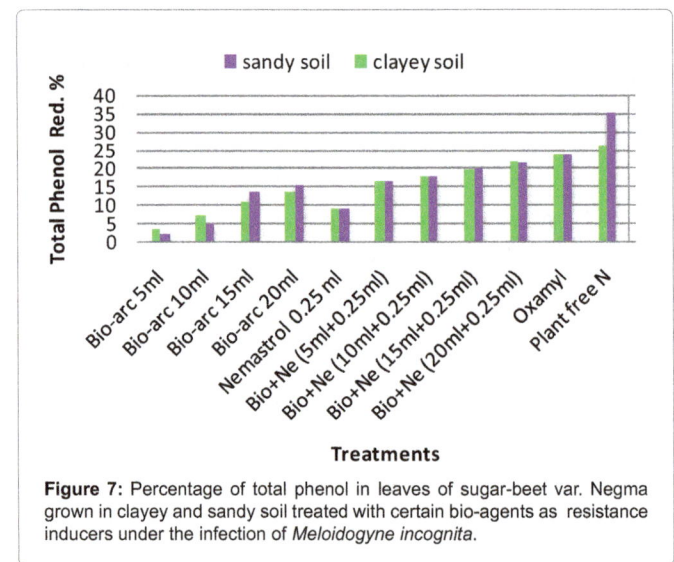

Figure 7: Percentage of total phenol in leaves of sugar-beet var. Negma grown in clayey and sandy soil treated with certain bio-agents as resistance inducers under the infection of *Meloidogyne incognita*.

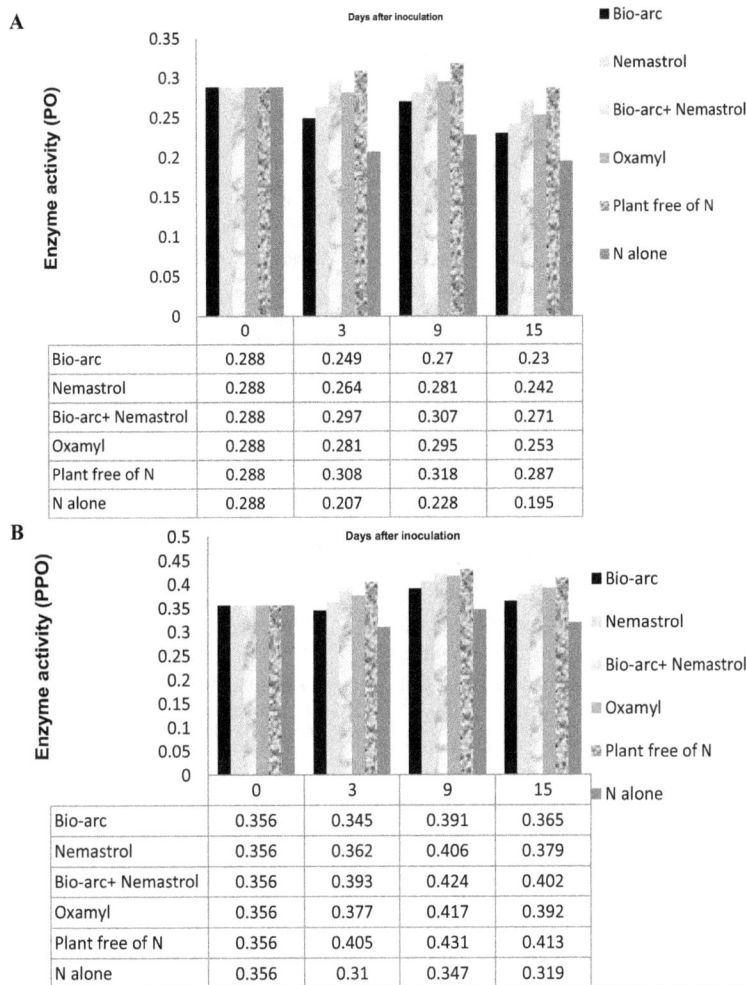

Figure 8: Impact of Bio-arc and Nemastrol as biotic resistance inducers on peroxidase (PO) and polyphenol oxidase (PPO) activities in roots of sugar-beet var. Negma after 0, 3, 9 and 15 days of *Meloidogyne incognita* inoculation. A= peroxidase activity, B= polyphenol oxidase activity.

phosphorus solubilizing bacterium (PSB), *B. megaterium* singly or concomitantly with Nemastrol has potential as a promising biocontrol candidate against root-knot nematode, *M. incognita* infecting sugar-beet var. Negma. As for single application, the effectiveness of Bio-arc to enhance plant growth parameters increased with rates increase in the two soil types.

Plant growth response of sugar-beet infected with *M. incognita* was more pronounced in sandy soil than clayey soil. In sandy soil, a significant improvement in shoot length, plant fresh weight and shoot dry weight was recorded with Bio-arc @ the rate of 20 ml/plant. This result support the findings of El-Deriny and Ibrahim [16,17]. However, in concomitant treatment, Bio-arc+Nemastrol (20 ml+0.25 ml) performed the best and showed significant improvement in plant growth parameters in terms of shoot length (92.6; 127.5%) and total plant fresh weight (91.7; 370.4%) of sugar-beet grown either in clayey or sandy soil. The presence of cytokinins in Nemastrol suggests a dynamic role for lateral root development. Irrespective to soil type and rates of application, total nematode population, root galling, number of egg masses and number of eggs/ egg mass were significantly suppressed with all treatments of Bio-arc and/or Nemastrol. The phosphate solubilizing bacterium (PSB) *B. megaterium* is considered a microorganism capable of dissolving the

unavailable phosphorus compounds in soil rendering them available for growing crops [18]. Increased phosphorus concentration may lead to reduction in root-knot nematodes. *B. megaterium* has been evaluated for their effects on a variety of root-knot nematodes [16,17,19,20] reported that *B. megaterium* greatly reduced numbers of galls, females and egg masses of *M. incognita* in the roots of sugar-beet followed by *B.subtilis, Paecilomyces lilacinus, P. fumosoreus* and *Trichoderma album* respectively. Furthermore, *B. megaterium* can extensively colonize the rhizosphere and reduce the sugar-beet cyst nematode infection under greenhouse trials [21]. *B. megaterium* produce antibiotic compounds [22] although no compounds from *B. megaterium* have been reported with activity against nematodes.

Nevertheless, Nemastrol performed the best and significantly suppressed total nematode population; root galling, number of egg masses and number of eggs /egg mass in clayey and sandy soil. The suppressive effect of such product could be attributed to the presence of mixture of enzymes i.e. chitinase and glucanase that dissolve chitin of nematode egg shell. However, concomitant treatment using Nemastrol+ Bio-arc showed better results than did Bio-arc alone at four tested rates. The greatest reduction in total nematode population was recorded with clayey and sandy soil receiving the dual application of Bio-arc (20 ml) and Nemastrol (0.25 ml) with reproduction factor and reduction percentage reached 2.2; 2.6 and 92.8; 92.6% respectively.

The impact of screened treatments on chemical components viz. NPK, chlorophyll, total carbohydrates, crude protein and total phenol in sugar-beet leaves infected with *M. incognita* revealed a remarkable induction in chemical constituents with the application of Bio-arc+Nemastrol (20 ml+0.25 ml). Conversely, the highest increase in total phenol percentage was recorded with untreated inoculated plants as a hypersensitive reaction (HR) to nematode infection. Plants are endowed with defense genes which are quiescent in healthy plants. When these genes are activated with various factors they induce systemic resistance against disease. Rhizobaceria induce systemic resistance by activation of various defense–related enzymes viz. PO, PPO and PAL. Recently, research work has demonstrated that the bio-agent *Pseudomonas fluorescens* might stimulate the production of biochemical compounds associated with the host defense [23]. Of these, early induction of peroxidase is more important as it is the first enzyme in the phenylpropanoid pathway, which leads to production of phytoalexin and phenolic substances leading the formation of lignin [24]. Conspicuously, the current investigation recorded the higher activity of Peroxidase in plants treated with Bio-arc+Nemastrol (20 ml+0.25 ml) and reached its peak at 9 days generating the speculation of induced defense responses in sugar-beet infected with *M. incognita*. Peroxidase activity in roots is important in the reinforcement of cell walls at the border of infection in resistant plants and that are considered as important components of active defense response of nematode invaded tissue [25]. The trend of increasing PPO activity was similar to that of PO in all treatments. Increased activity of Peroxidase (PO) or Polyphenol Oxidase (PPO) has been elicited by biocontrol agent strains in different plants [26,27]. Finally it can be concluded that use of such inducers viz. Bio-arc and Nemastrol singly and concomitantly represent a promising new approach for the control of the target root-knot nematode, *M. incognita* infecting sugar- beet within an environmental friendly integrated pest management via enhancing the resistance of plant to nematode.

References

1. Hammerschmidt R, Kuc S (1995) Induced resistance to disease in plants. Kluwer Academic Publishers, Dordrecht L., The Netherlands, p.182.

2. Dina SS, Ibrahim AH, Nour El-Deen, Khalil AE, Fatma AM Mostafa (2013). Induction of Systemic Resistance in sugar- beet infected with Meloidogyne incognita by humic acid, hydrogen peroxide, thiamine and two amino acids. Egyptian Journal of Agronematology, 1: 22-41.

3. Maxwell DP, Bateman DF (1967) Changes in the activity of oxidases in extracts of Rhizoctonia infected bean hypocotyls in relation to lesion maturation. Phytopathol, 57: 123-136.

4. Goodey JB (1957) Laboratory methods for work with plant and soil nematodes. Tech. Bull. No.2 Min. Agric. Fish Ed. London pp.47.

5. Bybd DW, Kirkpatrick T, Barker KR (1983) An improved technique for clearing and staining plant tissues for detection of nematodes. J Nematol 15: 142-143.

6. Taylor AL, Sasser JN (1978) Biology, identification and control of root-knot Nematodes (Meloidogyne spp.) Coop. pub. Dept. plant pathol. North Carolina State Univ. and U.S. Agency Int. Dev. Raleigh, N.C. 111 pp.

7. AOAC (1980) "Official methods of analysis" Twelfth Ed.Published by the Association of Official Analytical chemists, Benjamin, France line station, Washington. Dc.

8. Pregl E (1945) Quantative organic micro-analysis J. Chundril. London.

9. Jackson ML (1967) Soil Chemical Analysis. Printic Hall of India, New Delhi 144-197.

10. John MK (1970) Colorimetric determination of phosphorus in soil and plant material with ascorbic acid. Soil Sci. 109: 214-220.

11. Amako A, Ghen GX, Asala K (1994) Separate assays specific for the ascorpate peroxides and guaiacol peroxidase and for the chloroplastic and cytosolic isozyme of ascorbate peroxidase in plants. Plant cell Physiol., 53: 497-507.

12. Coseteng MY, Lee CY (1987) Change in apple polyphenoloxidase and polyphenol concentrations in relation to degree of browning. J. Food Sci. 52: 985-989.

13. Gomez KA, Gomez AA (1984) Statistical Procedures for Agriculture Research. June Wiley & Sons.Inc. New York.

14. Duncan DB (1955) Multiple range and multiple, F-test. Biometrics 11: 1-42.

15. Oka Y, Cohen Y, Spiegel Y (1999) Local and Systemic Induced Resistance to the Root-Knot Nematode in Tomato by DL-beta-Amino-n-Butyric Acid. Phytopathology 89: 1138-1143.

16. El-Deriny, Marwa M (2009) Studies on certain nematode pests parasitizing some ornamental plants. Ms. Thesis, Fac. Agric., Mansoura Univ., 135pp.

17. Ibrahim, Dina SS (2010) Studied on nematodes associated with sugar-beet plant roots in Dakahlia governorate. M. Sc. Thesis, Fac. Agric. Mansoura Univ., 108 pp.

18. Radwan SM (1983) Effect of inoculation with phosphate dissolving bacteria on some nutrients uptake from newly cultivated soil. M.Sc. Thesis, Fac. Agric. Ain Shams Univ. 184 pp.

19. El-Hadad ME, Mustafa MI, Selim ShM, El-Tayeb TS, Mahgoob AE, et al. (2011) The nematicidal effect of some bacterial biofertilizers on Meloidogyne incognita in sandy soil. Braz J Microbiol 42: 105-113.

20. El-Nagdi WMA, Haggag KHE, Abd-El-Fattah AI, Abd-El-Khair H (2011) Biological control of Meloidogyne incognita and Fusarium solani in sugar beet. Nematol. Medit. 39: 59-71.

21. Neipp PW, Becker JO (1999) Evaluation of Biocontrol Activity of Rhizobacteria from Beta vulgaris against Heterodera schachtii. J Nematol 31: 54-61.

22. Vary P (1992) Development of genetic engineering in Bacillus megaterium. Pp.251-310 in R.H.Doi and M.Mc Gloughlin, eds. Biology of bacilli: Applications to industry. Boston: Butterworth-Heinemann.

23. Kavino M, Harish S, Kumar N, Saravanakumar D, Domodaran T (2007) Rhizophere and endophytic bacteria for induction of systemic resistance of banana plantlets against bunchy top virus. Soil Biol. and Biochmis. 39: 1087-1098.

24. Bruce RJ, West CA (1989) Elicitation of lignin biosynthesis and is operoxidase activity by pectic fragments in suspension cultures of castor bean. Plant Physiol 91: 889-897.

25. Zacheo G, Bleve-Zacheo T, Pacoda D, Orlando C, Durbin RD (1995) The association between heat-induced susceptibility of tomato to Meloidogyne incognita and peroxidase activity. Physiological and Molecular Plant Pathol, 46: 491-507.

26. Hassan HAE, Buchenauer H (2007) Induction of resistance to fire blight in apple by acibenzolar-S- methyl and DL-3 aminobutyric acid. J. Plant Dis. Protec. 114: 151-158.

27. Govindappa M, Lokesh S, Ravishankar VR, Rudranaik V, Raju SC (2010) Induction of systemic resistance and management of safflower Macrophomina phaseolina root rot disease by biocontrol agents. Arch. Phytopathol. Plant Prot. 43: 26-40.

18

The Effect of Planting Time on Phytophthora Blight Disease Incidence and Severity on Cucumber (*Cucumis sativus*) in Nsukka, Derived Savanna, Agro Ecology, South Eastern Nigeria

Patience Ukamaka Ishieze*, Ugwuoke Kelvin I and Aba Simon C

Department of Crop science, University of Nigeria, Nsukka

Abstract

Field experiments were conducted to study the disease incidence and severity of six cucumber lines at different time of planting (April, 12th; May, 12th; June, 12th, July, 12th and September, 12th), Six varieties of cucumber (Marketter, Poinsett 76 Netherlands, Poinsett 76 Holland, Poinsett France, Poinsett Holland and super marketter) during 2013 farming season. It was observed from the results that the planting time of April had the lowest disease incidence (3.48%) and severity on leaves (1.20) and fruits (1.06), followed by September with lower disease incidence (3.60%), disease severity on leaves (1.22) and fruits (1.17) while July had the highest disease incidence (6.27%) and disease severity on leaves (1.56) and fruits (1.74). The yields during the April and September plantings were significantly ($p<0.05$) higher than other months. April and September had 481.60 and 483.60 tonnes per hectare respectively while July had 19.10 tonnes per hectare. The cucumber lines also showed.

significant ($p<0.05$) response to disease incidence and severity with the Supermarketer having the lowest disease incidence (4.15%) and severity on leaves (1.39) and fruits (1.17) and Poinsette Holland had the highest disease incidence (5.15%) and Poinsette 76 Holland had the highest severity both on leaves and fruits (1.45 and 1.35). The supermarketer gave the highest yield of 221.10 tonnes per hectare. From the observations farmers in derived savanna agro ecology can plant during the month of April and September for maximum yield and less usage of pesticides. Super marketer variety remains the best variety in this study.

Keywords: Disease incidence; Disease severity; Time of planting; Cucumber

Introduction

Cucumber (*Cucumis sativus* L.) belongs to the Cucurbitaceae family. Cucumber is a major vegetable crop worldwide and develops rapidly, with a shorter time from planting to harvest than for most crops [1]. It is a monoecious annual climber or creeper [2] that has been cultivated for over 3,000 years and is still widely cultivated today. It is soft, succulent with high water content. The plant has large leaves that form canopy over the fruit. The vines grow on stakes or on trellises. The fruit is roughly cylindrical, elongated with tapered ends and may be as long in diameter. The fruits are used in unripe matured state, usually eaten raw in salads or pickled, and are also stewed in tropical regions. Although cucumber is less nutritious than most fruits, it is still a very good source of vitamins A, C, K, B6, potassium and it also provides dietary fibres, pantothenic acid, magnesium, phosphorus, copper and manganese. It contains ascorbic acid and caffeic acid, both of which help to smoothen the skin irritation and reduces swelling. Its skin contains chlorophyll and silica, two beneficial elements that are lost when the vegetables are peeled. Its juice is often recommended as a source of silicon to improve the complexion and health of the skin. However, the production of the fruit in Nigeria is very low due to some constraints such as incidences of disease, which reduces pod yield. They are produced mainly in the northern states of Nigeria [2]. It is necessary to increase the production in order to supplement the high intake of carbohydrate in Nigeria, especially the southern parts of Nigeria where there is sparse and over dependence of its supply of salad vegetables and fruits on major suppliers from the north, resulting to relative higher price because of transportation cost and spoilage in the transportation of the fruit. The fruit constitute the dietary system of the populace in the cities as valuable ingredient in vegetable salads and fruit.

In Nigeria, adequate research has not been conducted to find out the best time to plant, fungicides and spray regimes to control the disease outbreak, which is common in the south due to warm weather. Fruit is usually under foliage, shading one another being allowed to creep on the ground thereby having a direct contact with some soil inhabiting disease-causing organisms. Because of its creeping nature, it pre disposes the plant for pests and disease infestation. Torrential rainfall sometimes splash sand on edible pod yield, leading to outright degradation of the market quality of pods.

Materials and Methods

The experiments was conducted at the Department of Crop Science Teaching and Research Farm, Faculty of Agriculture, University of Nigeria, Nsukka (07°29 N, 06°51E and 400 m.a.s.l.). Nsukka is characterized by lowland humid tropical conditions with bimodal rainfall pattern which starts in late March and ends in late July while the short rainy season extends from September to late October after a dry spell in August. It ranges between 1155 mm to 1955 mm with a mean annual temperature of 29°C to 31°C and a relative humidity of 69% to 79%.

*Corresponding author: Patience Ukamaka Ishieze, Department of Crop science, University of Nigeria, Nsukka
E-mail: makcynthia2003@yahoo.com

Field experiment

This experiment evaluated the effect of planting time on disease incidence, severity and pod yield on the six cucumber lines. The planting were carried out for a period of five months starting from April 12th through May 12th, to July 12th and September 12th.

Six cucumber lines including Super marketer, Poinsett France, Poinsett Holland, Poinsett 76 Holland, Poinsett 76 Netherlands and Marketer were used in the study. At each planting, the experiment was laid out in randomized complete block design (RCBD) of three replications.

A portion of land measuring 320 m² was cleared, ploughed and harrowed, before seed beds measuring 2 m × 3 m was prepared. The seeds were sown insitu at two seeds per hill at a spacing of 10 cm within rows and 20 cm between rows with planting depth of 2.5 cm on each bed. Prior to the planting, battery cage poultry manure was applied at 3tonnes per hectare on the readymade bed, one week before planting.

Data collection

In data collection the disease incidence and severity at 2, 4, 6, 8 and 10 weeks after planting and interaction of the various spray regimes used to check the disease incidence measured as percentage of plants showing visible symptom of phytophtora blight, disease severity and pod yield were collected. Disease severity (the proportion of area or amount of plant tissue that is diseased) on both leaves and fruits was also calculated in percentage and was later rated see Table 2. The following parameters were measured to determine the level of disease incidence and severity on these cucumber accessions keeping in mind some developmental qualities:

- Disease incidence at 2 weeks, 4weeks, 6 weeks, 8weeks and 10 weeks after planting.

- Disease severity on leaves at 2, 4, 6, 8, and 10 weeks after planting

- Disease severity on fruits at 6, 8, and 10 weeks after planting

- Variety response to disease incidence

- Variety response to disease severity on leaves

- Variety response to disease severity on fruits

- Yield per plant(kg)

Laboratory analysis

Apart from the visible signs of phytophthora blight, the diseased plant tissues were taken to the laboratory where they were cultured to know the fungal organism present. A strip of transparent sticky tape (10 cm long) is held in between the thumb and the forefinger. The sticky side of the tape is firmly pressed onto the surface of the sporulating colony cultivated on an appropriate medium in the Petri dish. After gently removing the cellophane tape, the sticky surface carrying fungal spores is carefully placed over drops of lacto phenol cotton blue stain kept at the centre of a clean glass slide. The tape is gently pressed and extended end of the tape was held over the ends of the slide. A light microscope 40 m magnification was used to examine the characteristics of spores.

Data analysis

Data collected were subjected to analysis of variance (ANOVA) using GenStat 10th edition release statistical software. Mean separation

was done using Fishers least significance test at 5% probability level as described by Obi [3].

Results and Discussion

Meteorological data of 2013 planting season

The meteorological data from the experimental site shows that there is a marked variation on the climatic elements being considered (Table 1). The distribution of rainfall shows the total rainfall (1537.28 mm) received during the period of the experiment (January to December 2013) with the highest rainfall amount and rain days being in July (283.96 mm and 19days) respectively. Moreover, the mean minimum and maximum temperatures were 21.14°C and 29.54°C for the year. The data also reveal that the mean monthly relative humidity at 10 a.m. and 4 p.m. were 74.67 % and 72.64% respectively for the year.

Aboratory results

The plant tissue analysed showed that there was growth of spores, which indicates the presence of fungal organism. When the characteristics of the spores were examined, the colour as observed showed that the organism present is *Phythophtora capsici*, the causal organism for phythophtora blight of cucumber.

The effect of planting time on incidence, severity on phytophtora blight disease leaves and fruits at 2, 4, 6, 8, and 10 weeks after planting on the six cucumber lines studied.

The results of the effect of planting time on the phytophtora blight disease incidence on six cucumber presented in Table 3 revealed that the month of July significantly (0.05) had the highest phytophtora disease (6.27%), followed closely by the month of May and June (5.57% and 5.31%) respectively. The disease incidence was lowest in April, followed by September (3.48% and 4.85% respectively). From the results, it shows that the best time to plant is April and September.

Month	Total rainfall (mm)	Rain day	Temperature(°C)		Relative humidity (%)	
			mini	max	10 a.m.	4 p.m.
January	21.84	2.00	20.55	31.23	75.00	75.00
February	0.00	0.00	22.18	32.86	75.00	75.00
March	38.10	5.00	22.58	32.81	72.74	62.94
April	183.81	10.00	22.30	30.67	74.00	68.90
May	198.63	11.00	21.61	29.52	74.77	69.87
June	168.60	11.00	21.17	28.67	75.67	72.70
July	283.96	19.00	20.71	27.35	74.90	73.61
August	219.18	12.00	20.26	26.61	76.13	76.16
September	197.60	16.00	20.50	27.43	77.00	77.00
October	167.90	11.00	20.74	28.55	77.00	77.00
November	41.91	2.00	21.70	30.37	77.00	77.00
December	15.75	2.00	19.39	29.35	66.77	66.03
Total	1537.28	101.00	253.69	354.42	895.98	871.66
Mean	128.10	8.42	21.14	29.54	74.67	72.64

Table 1: The total rainfall (mm), rain days, mean temperature (°C) and the relative humidity (%) of the site of the experiment in 2013.

Ratings	Percentage
0	0
1	1-25
2	26-50
3	51-75
4	76-100

Table 2: Disease severity Ratings.

	WEEKS AFTER PLANTING (WAP)										
MONTH	2WAP		4WAP		6WAP		8WAP		10WAP		MEAN	
APRIL	8.060	(2.419)	11.530	(3.139)	15.280	(3.667)	15.970	(3.825)	19.720	(4.360)	14.112	(3.482)
MAY	8.610	(2.733)	24.310	(4.860)	30.420	(5.475)	35.970	(5.988)	78.610	(8.810)	35.584	(5.573)
JUNE	11.670	(3.274)	17.220	(3.900)	18.750	(4.127)	40.000	(6.311)	80.560	(8.954)	33.640	(5.313)
JULY	11.670	(3.274)	17.220	(3.900)	18.750	(4.127)	100.000	(10.025)	100.000	(10.025)	49.528	(6.270)
SEPT	5.970	(2.099)	12.08	(3.245)	15.690	(3.762)	16.810	(4.015)	24.030	(4.898)	14.916	(3.604)
MEAN	9.196	(2.760)	16.472	(3.809)	19.778	(4.232)	41.750	(6.033)	60.584	(7.415)	29.556	(4.849)
F-LSD(0.05)		0.4025		0.3673		0.3347		0.2429		0.2413		

Values in parenthesis are square root transformed values to which LSD is applicable.

Table 3: The main effect of planting time on phytophtora blight disease incidence on the six cucumber lines at different weeks after planting.

	WEEKS AFTER PLANTING (WAP)										
MONTH	2WAP		4WAP		6WAP		8WAP		10WAP		MEAN	
APRIL	0.708	(1.056)	0.986	(1.188)	1.042	(1.219)	1.097	(1.248)	1.194	(1.294)	1.005	(1.201)
MAY	0.736	(1.088)	1.333	(1.343)	2.111	(1.591)	2.306	(1.640)	3.986	(2.118)	2.094	(1.556)
JUNE	0.875	(1.158)	0.944	(1.189)	1.028	(1.223)	2.306	(1.640)	4.000	(2.121)	1.831	(1.466)
JULY	0.875	(1.158)	0.944	(1.189)	1.028	(1.223)	4.000	(2.121)	4.000	(2.121)	2.169	(1.562)
SEPT	0.514	(0.973)	1.042	(1.213)	1.167	(1.284)	1.181	(1.289)	1.389	(1.363)	1.059	(1.224)
MEAN	0.742	(1.087)	1.050	(1.224)	1.275	(1.308)	2.178	(1.588)	2.914	(1.803)	1.632	(1.402)
F-LSD(0.05)		0.0653		0.0589		0.0607		0.0681		0.0305		

Values in parenthesis are square-root transformed values to which LSD is applicable.

Table 4: The main effect of planting time on Phytophtora blight disease severity on leaves of the six cucumber lines at different weeks after planting.

	WEEKS AFTER PLANTING (WAP)										
MONTH	2WAP		4WAP		6WAP		8WAP		10WAP		MEAN	
APRIL	-	-	-	-	0.556	(0.990)	0.722	(1.067)	0.806	(1.112)	0.695	(1.056)
MAY	-	-	-	-	0.958	(1.164)	1.722	(1.470)	1.431	(1.376)	1.370	(1.337)
JUNE	-	-	-	-	0.528	(0.978)	0.681	(1.057)	1.458	(1.387)	0.889	(1.141)
JULY	-	-	-	-	0.528	(0.978)	4.000	(2.121)	4.000	(2.121)	2.843	(1.740)
SEPT	-	-	-	-	0.694	(1.064)	0.986	(1.209)	1.069	(1.249)	0.916	(1.174)
MEAN		-		-	0.653	(1.035)	1.622	(1.385)	1.753	(1.449)	1.343	(1.290)
F-LSD(0.05)						0.0841		0.0635		0.0543		

Values in parenthesis are square-root transformed values to which LSD is applicable.

Table 5: The main effect of planting time on Phytophtora blight disease severity on Fruits of the six cucumber lines at different weeks after planting.

The results of the effect planting time on the phytophtora blight disease severity on cucumber presented in Table 4 revealed that the month of July significantly (0.05) had the most severe phytophtora disease (1.56%), followed closely by the month of May and June (1.56% and 1.47%) respectively. The severity of phytophtora blight of cucumber was lowest in April, followed by September (1.20% and 1.22% respectively). From the results, it shows that the best time to plant with lowest disease severity on leaves is April and September.

Table 5 shows the results of the effect of planting time on the phytophtora blight disease severity on cucumber fruits and it revealed that the month of July significantly (0.05) had the most severe phytophtora blight disease on the fruits (1.74%), followed closely by the month of May and September (1.34% and 1.17%) respectively. The severity of phytophtora blight of cucumber was lowest in April, followed by June (1.06% and 1.14% respectively). From the results, it shows that the best time to plant with lowest disease severity on Fruits is April and June.

Table 6 revealed that the yield during the April and September planting were significantly (0.05) higher than other months. April and

September recorded 481.60 and 483.60 tonnes per hectare respectively while May, June, July was 35.50, 32.30 and 19.10 tonnes per hectare respectively. The fruit girth, number of fruits per plant, fruit weight and fruit length were also statistically significant(0.05) and it also followed the above trend of April and September being the best and July having the lowest girth, length, number of fruits per plant, fruit weight and diseased fruits.

The effect of the six cucumber lines on the manifestation of phytophtora blight disease incidence, severity on leaves and fruits at 2, 4, 6, 8, and 10 weeks after planting for all the planting times considered.

In Table 7, the different lines of cucumber significantly affected the rate of disease incidence. The poinsett Holland PH, Poinsett 76 Holland P76H, Poinsett France PF, Poinsett 76 Netherland P76N, recorded higher incidences, (5.15, 5.15, 5.05 and 4.81) percentage respectively while marketer M and super marketer SM had a lower incidence at 4.79 and 4.15% respectively. The varieties when compared reveal that the super marketer is the best cucumber line that can resist phytophtora blight disease in this derived savannah.

MONTH	FG(cm)	FL(cm)	FW/Plt(kg)	NFPP	FY(t/ha)	DF(t/ha)
APRIL	5.162	20.010	2.408	5.505	481.6	2.408
MAY	2.383	12.359	0.178	2.271	35.5	0.178
JUNE	2.556	11.608	0.161	2.146	32.3	0.161
JULY	2.440	11.608	0.095	1.982	19.1	0.095
SEPT	5.162	20.010	2.443	5.577	483.6	2.419
MEAN	3.541	15.119	1.057	3.496	210.420	1.052
F-LSD(0.05)	0.1392	0.2097	0.0717	0.2023	12.92	0.0625

FG=Fruit girth, FL=Fruit length, FW=Fruit weight, NFPP=Number of fruit per plant, FY=Fruit yield, DF=diseased fruits

Table 6: The effect of planting time on some yield parameters of the six cucumber lines.

VARIETY	WEEKS AFTER PLANTING (WAP)											
	2WAP		4WAP		6WAP		8WAP		10WAP		MEAN	
M	9.000	(2.707)	17.000	(3.914)	19.170	(4.191)	39.330	(5.735)	61.170	(7.424)	29.134	(4.794)
P76H	10.670	(3.085)	20.170	(4.296)	22.330	(4.586)	42.500	(6.130)	64.000	(7.632)	31.934	(5.146)
P76N	10.500	(2.919)	14.330	(3.561)	18.500	(4.112)	42.000	(6.080)	60.000	(7.380)	29.066	(4.810)
PF	8.830	(2.820)	18.170	(4.061)	22.500	(4.591)	43.670	(6.219)	62.170	(7.571)	31.068	(5.052)
PH	9.830	(2.849)	19.170	(4.119)	23.830	(4.646)	46.000	(6.492)	63.500	(7.643)	32.466	(5.150)
SM	6.330	(2.177)	10.000	(2.903)	12.330	(3.265)	37.000	(5.541)	52.670	(6.844)	23.666	(4.146)
MEAN	9.193	2.760	16.473	3.809	19.777	4.232	41.750	6.033	60.585	7.416	29.556	(4.850)
F-LSD(0.05)		0.4409		0.4024		0.3667		0.2661		0.2643		

M=Marketer,P76H=Poinsett 76 Holland, P76N=Poinsett 76 Netherlands, PF=Poinsett France,PH=Poinsett Holland, S=Super marketer. Values in parenthesis are square-root transformed values to which LSD is applicable.

Table 7: The main effect of the six cucumber lines on phytophtora blight disease incidence at different weeks after planting for the five planting times considered.

VARIETY	WEEKS AFTER PLANTING (WAP)											
	2WAP		4WAP		6WAP		8WAP		10WAP		MEAN	
M	0.583	(1.006)	1.000	(1.211)	1.333	(1.325)	2.567	(1.694)	2.950	(1.816)	1.687	(1.410)
P76H	0.750	(1.090)	1.200	(1.293)	1.450	(1.377)	2.433	(1.673)	2.933	(1.810)	1.753	(1.449)
P76N	0.700	(1.069)	0.933	(1.180)	1.167	(1.272)	1.933	(1.515)	2.900	(1.798)	1.527	(1.367)
PF	0.917	(1.171)	1.233	(1.291)	1.333	(1.338)	2.050	(1.548)	2.933	(1.810)	1.693	(1.432)
PH	0.833	(1.136)	1.200	(1.284)	1.367	(1.336)	2.233	(1.608)	2.950	(1.816)	1.717	(1.436)
SM	0.667	(1.047))	0.733	(1.087)	1.000	(1.203)	1.850	(1.488)	2.817	(1.770)	1.413	(1.387)
MEAN	0.742	(1.094)	1.050	(1.224)	1.275	(1.309)	2.178	(1.588)	2.914	(1.803)	1.632	(1.404)
F-LSD		0.0715		0.0645		0.0665		0.0746		NS		

M=Marketer, P76H=Poinsett 76 Holland, P76N=Poinsett 76 Netherlands, PF=Poinsett France, PH=Poinsett Holland, S=Super marketer. Values in parenthesis are square-root transformed values to which LSD is applicable.

Table 8: The main effects of the six cucumber lines on disease severity on leaves at different weeks after planting for the five planting times considered.

VARIETY	WEEKS AFTER PLANTING (WAP)											
	2WAP		4WAP		6WAP		8WAP		10WAP		MEAN	
M	-	-	-	-	0.717	(1.073)	1.667	(1.392)	1.733	(1.446)	1.372	(1.304)
P76H	-	-	-	-	0.917	(1.167)	1.667	(1.421)	1.750	(1.454)	1.445	(1.347)
P76N	-	-	-	-	0.483	(0.952)	1.633	(1.388)	1.750	(1.451)	1.289	(1.264)
PF	-	-	-	-	0.733	(1.073)	1.650	(1.406)	1.783	(1.456)	1.389	(1.312)
PH	-	-	-	-	0.817	(1.114)	1.683	(1.404)	1.933	(1.514)	1.478	(1.344)
SM	-	-	-	-	0.250	(0.831)	1.433	(1.296)	1.567	(1.373)	1.083	(1.167)
MEAN	-	-	-	-	0.653	(1.035)	1.622	(1.385)	1.753	(1.449)	1.343	(1.290)
F-LSD(0.05)	-		-		0.0921		0.0696		0.0596		

M=Marketer,P76H=Poinsett 76 Holland, P76N=Poinsett 76 Netherlands, PF=Poinsett France,PH=Poinsett Holland, S=Super marketer. Values in parenthesis are square-root transformed values to which LSD is applicable.

Table 9: The main effect of the six cucumber lines on disease severity on fruit at different weeks after planting for the five planting times considered.

In Table 8, the response of the six lines of cucumber to disease severity on leaves was statistically significant at 2, 4, 6, 8 weeks after planting but not significant at 10 weeks after planting. The mean disease severity recorded on the leaves was 1.41, 1.45, 1.37, 1.43, 1.44, and 1.39 for M, P76H, P76N, PF, PH and SM respectively. From the table it indicates that super marketer can resist the phytophtora blight disease, that it is more severe on other varieties than the SM.

In Table 9, the response of the six lines of cucumber to disease severity on fruits was statistically significant at 6, 8 and 10 weeks after planting. The mean disease severity recorded on the fruits was 1.30,

1.35, 1.26, 1.31, 1.34, and 1.17 for M, P76H, P76N, PF, PH and SM respectively. From the table it indicates that super marketer can give out fruits that shows less symptoms of phytophtora blight disease, that other varieties shows more severity of the aforementioned disease on their fruits than the SM.

In Table 10, the six-cucumber lines response to yield and number of fruit per plant was not significant but the fruit girth, fruit length and fruit weight was statistically significant. Though the yield of the supermarketer remains the best but the yield of all the lines and the number of fruits per plant and the bad/diseased fruits were not significant statistically.

The effect of planting time and variety on the phytophtora blight disease incidence, and severity on leaves and fruits at 2, 4, 6, 8, and 10 weeks after planting.

In Table 11, the interaction between the variety and time of planting on phytophtora blight disease incidence was not significant at 2, 4, and 10 weeks after planting, but were highly significant at 6 and 8 weeks after planting which was the onset of fruiting and also the peak of fruiting respectively. Super Marketer recorded the lowest disease incidence (2.652% and 2.759%) at the month of April and September respectively, while the Poinsette France and Poinsette76 Holland recorded the highest disease incidence during the month of July (6.493 and 6.492 respectively). Showing from the arrays of mean displayed on the table, all varieties recorded lower disease incidence in the month of April and September, but the incidences were higher in the month of July. It is also high in May and slightly higher in June.

In Table 12, the interaction between the variety and time of planting on phytophtora blight disease severity on leaves was highly significant at 2, 4, 6 and 8 weeks after planting, but not significant at 10weeks after planting. The variety with highest disease severity (1.655) is the marketer during the month of May while the supermarketer recorded the lowest severity on the leaves (1.104) at the month of April. From the results obtained, the super marketer during the month of April had less severed diseased leaves thereby promoting photosynthetic ability and hence supplying the plant the needed growth nutrients for good yield and ability to resist the disease.

In Table 13, it was recorded that the interaction between the variety and time of planting on phytophtora blight disease severity on fruits was highly significant at 8 weeks after planting, but not significant at 6 and 10weeks after planting. The variety with highest severity (1.789) is the Poinsette 76 Holland during the month of July while the super marketer recorded the lowest severity on the fruits (0.880) during the month of April. From the results, it showed that the super marketer is the best variety and planting during the month of April is the best time to apply while July remains the worst time to plant.

In Table 14, the data collected showed that the interaction between the month and the variety was not significant for the yield and the number of fruit per plant. From the yields recorded, the super marketer planted during the month of April and September remains the best variety and the best time to plant. It recorded 508 and 506 tonnes per hectare for April and September respectively. Where super marketer is not available, Poinsette Holland planted at the month of April can make for the best alternative.

Discussion

Effects of planting time on phytophthora blight disease incidence, severity and the fruit yield of cucumber.

The variation in the phytophthora blight disease incidence and severity on cucumber planted at various months of April, May, June, July and September indicates that April is the best time to plant in Nsukka Agro- ecology. The high yield experienced during the April planting over the May, June, July and September plantings could be attributed to moderate rainfall at the flowering and fruiting stage of the crop, which began in the middle of May. High rainfall during flowering and fruiting can lead to bees' inactivity with subsequent flower abortion with resultant low yield, which may have resulted in the yields experienced during the July planting. This is in agreement with the finding of Papadopoulos [4], who stated that high moisture tend to discourage the activity of bees and resultant high relative humidity which influence water condensing on the plant leaves which may result in the development of pests and diseases.

The significant ($p < 0.05$) increase in disease incidence and severity during the month of July could be attributed to the high rainfall witnessed during that month (283.96 mm), low temperature and high relative humidity. These observations agrees with Babadoost, [5] that *Phytophthora capsici* causes pre and post-emergence damping-off in cucurbits under wet and warm (20-30°C) weather condition, therefore the lower the temperature, the faster the growth of the spores which will in turn increase the disease incidences and severity with low yield. Soil moisture conditions are important for disease development. Sporangia form when soil pores are drained, and they release zoospores when soil is saturated (soil pores are filled with water). The disease is usually associated with heavy rainfall, excessive-irrigation, or poorly drained soil. Frequent irrigation increases the incidence of the disease. Warm conditions are favourable for disease development.

The response of the six varieties of cucumber studied to phytophtora blight disease incidence, severity and yield

The results of this study showed that there were significant differences among the six varieties of cucumber under study. In all cases, Super marketer, Poinsette 76 Netherlands and Marketer varieties

VARIETY	FG(cm)	FL(cm)	FW/Plt(kg)	NFPP	FY(t/ha)	DF(t/ha)
M	3.398	14.580	1.034	3.317	206.7	1.034
P76H	3.368	14.711	1.077	3.473	215.4	1.077
P76N	3.462	15.151	1.001	3.477	200.1	1.001
PF	3.639	15.170	1.058	3.575	211.6	1.058
PH	3.515	14.857	1.037	3.519	207.5	1.037
SM	3.860	16.245	1.136	3.616	221.1	1.107
MEAN	3.540	15.119	1.057	3.496	210.400	1.052
F-LSD(0.05)	0.1525	0.2297	0.0786	0.2216	14.1500	0.0547

M=Marketer, P76H=Poinsett 76 Holland, P76N=Poinsett 76 Netherlands, PF=Poinsett France,PH=Poinsett Holland, S=Super marketer. FG=Fruit girth, FL=Fruit length, FW=Fruit weight, NFPP=Number of fruit per plant and FY=Fruit yield, DF=Diseased fruits

Table 10: The Impact of varieties on Some Yield Parameters Considered on the five planting times considered.

VARXMON	WEEKS AFTER PLANTING (WAP)											
	2WAP		4WAP)		6WAP		8WAP		10WAP		MEAN	
MAPRIL	5.830	(2.185)	7.500	(2.607)	8.330	(2.714)	8.330	(2.714)	16.670	(4.013)	9.332	(2.847)
MMAY	9.170	(2.925)	26.670	(5.141)	32.500	(5.675)	35.000	(5.922)	81.670	(9.029)	37.002	(5.738)
MJUNE	13.330	(3.437)	21.670	(4.608)	21.670	(4.608)	39.170	(6.236)	83.330	(9.122)	35.834	(5.602)
MJULY	13.330	(3.437)	21.670	(4.608)	21.670	(4.608)	100.000	(10.025)	100.000	(10.025)	51.334	(6.541)
MSEPT	3.330	(1.552)	7.500	(2.607)	11.670	(3.351)	14.170	(3.777)	24.170	(4.929)	12.168	(3.243)
P76HAPRIL	11.670	(3.082)	14.170	(3.611)	17.500	(4.120)	17.500	(4.120)	20.830	(4.476)	16.334	(3.882)
P76HMAY	10.000	(2.929)	31.670	(5.589)	34.170	(5.804)	38.330	(6.159)	87.500	(9.326)	40.334	(5.961)
P76HJUNE	12.500	(3.562)	20.830	(4.370)	21.670	(4.477)	40.000	(6.298)	85.830	(9.239)	36.166	(5.589)
P76HJULY	12.500	(3.562)	20.830	(4.370)	21.670	(4.477)	100.000	(10.025)	100.000	(10.025)	51.000	(6.492)
P76HSEPT	6.670	(2.292)	13.330	(3.541)	16.670	(4.050)	16.670	(4.050)	25.830	(5.095)	15.834	(3.806)
P76NAPRIL	10.000	(2.764)	10.830	(2.975)	15.830	(3.802)	16.670	(4.013)	19.170	(4.335)	14.500	(3.578)
P76NMAY	7.500	(2.503)	20.000	(4.417)	29.170	(5.353)	35.830	(5.991)	77.500	(8.794)	34.000	(5.412)
P76NJUNE	13.330	(3.358)	14.170	(3.611)	15.830	(3.802)	40.830	(6.360)	80.830	(8.984)	32.998	(5.223)
P76NJULY	13.330	(3.358)	14.170	(3.611)	15.830	(3.802)	100.000	(10.025)	100.000	(10.025)	48.666	(6.164)
P76NSEPT	8.330	(2.611)	12.500	(3.189)	15.830	(3.802)	16.670	(4.013)	22.500	(4.764)	15.166	(3.676)
PFAPRIL	5.000	(1.974)	14.170	(3.446)	19.170	(4.108)	20.000	(4.215)	24.170	(4.831)	16.502	(3.715)
PFMAY	10.830	(3.244)	22.500	(4.667)	29.170	(5.353)	35.000	(5.908)	78.330	(8.798)	35.166	(5.594)
PFJUNE	11.670	(3.455)	19.170	(4.265)	22.500	(4.694)	42.500	(6.521)	80.830	(8.964)	35.334	(5.580)
PFJULY	11.670	(3.455)	19.170	(4.265)	22.500	(4.694)	100.000	(10.025)	100.000	(10.025)	50.668	(6.493)
PFSEPT	5.000	(1.974)	15.830	(3.661)	19.170	(4.108)	20.830	(4.426)	27.500	(5.236)	17.666	(3.881)
PHAPRIL	12.500	(3.061)	15.000	(3.694)	23.330	(4.752)	23.330	(4.752)	24.170	(4.835)	19.666	(4.219)
PHMAY	8.330	(2.714)	28.330	(5.273)	34.170	(5.844)	40.830	(6.395)	80.830	(8.994)	38.498	(5.844)
PHJUNE	9.170	(2.822)	18.330	(3.830)	19.170	(3.937)	43.330	(6.597)	86.670	(9.308)	35.334	(5.299)
PHJULY	9.170	(2.822)	18.330	(3.830)	19.170	(3.937)	100.000	(10.025)	100.000	(10.025)	49.334	(6.128)
PHSEPT	10.000	(2.825)	15.830	(3.967)	23.330	(4.761)	22.500	(4.691)	25.830	(5.050)	19.498	(4.259)
SMAPRIL	3.330	(1.448)	7.500	(2.503)	7.500	(2.503)	10.000	(3.137)	13.330	(3.669)	8.332	(2.652)
SMMAY	5.830	(2.081)	16.670	(4.074)	23.330	(4.822)	30.830	(5.554)	65.830	(8.105)	28.498	(4.927)
SMJUNE	10.000	(3.008)	9.170	(2.718)	11.670	(3.247)	34.170	(5.852)	65.830	(8.105)	26.168	(4.586)
SMJULY	10.000	(3.008)	9.170	(2.718)	11.670	(3.247)	100.000	(10.025)	100.000	(10.025)	46.168	(5.805)
SMSEPT	2.500	(1.340)	7.500	(2.503)	7.500	(2.503)	10.000	(3.137)	18.330	(4.313)	9.166	(2.759)
MEAN	9.194	(2.760)	16.473	(3.809)	19.779	(4.232)	41.750	(6.033)	60.583	(7.415)	29.556	(4.850)
F-LSD(0.05)		NS		1.639		1.190		NS		NS		

M=Marketer,P76H=Poinsett 76 Holland, P76N=Poinsett 76 Netherlands, PF=Poinsett France,PH=Poinsett Holland, S=Super marketer. Values in parenthesis are square-root transformed values to which LSD is applicable.

Table 11: The effect of planting time and variety on phytophtora blight disease incidence at different weeks after planting.

VAR X MON	WEEKS AFTER PLANTING (WAP)											
	2WAP		4WAP		6WAP		8WAP		10WAP		MEAN	
MAPRIL	0.417	(0.909)	0.833	(1.125)	0.917	(1.155)	0.917	(1.155)	1.167	(1.284)	0.850	(1.126)
MMAY	0.833	(1.138)	1.333	(1.344)	2.583	(1.735)	3.333	(1.939)	4.000	(2.121)	2.416	(1.655)
MJUNE	0.833	(1.1380	1.000	(1.225)	1.000	(1.225)	3.333	(1.939)	4.000	(2.121)	2.033	(1.628)
MJULY	0.833	(1.138)	1.000	(1.225)	1.000	(1.225)	4.000	(2.121)	4.000	(2.121)	2.167	(1.566)
MSEPT	0.000	(0.707)	0.833	(1.139)	1.167	(1.284)	1.250	(1.314)	1.583	(1.433)	0.967	(1.175)
P76HAPRIL	0.750	(1.068)	1.083	(1.241)	1.167	(1.284)	1.167	(1.284)	1.167	(1.284)	1.067	(1.232)
P76HMAY	0.750	(1.095)	1.417	(1.373)	2.417	(1.690)	2.917	(1.837)	4.000	(2.121)	2.300	(1.623)
P76HJUNE	1.000	(1.225)	1.167	(1.284)	1.250	(1.314)	2.917	(1.837)	4.000	(2.121)	2.067	(1.556)
P76HJULY	1.000	(1.225)	1.167	(1.284)	1.250	(1.314)	4.000	(2.121)	4.000	(2.121)	2.283	(1.613)
P76HSEPT	0.250	(0.837)	1.167	(1.284)	1.167	(1.284)	1.167	(1.284)	1.500	(1.403)	1.050	(1.218)
P76NAPRIL	0.667	(1.052)	0.750	(1.095)	1.000	(1.211)	1.083	(1.254)	1.250	(1.314)	0.950	(1.185)
P76NMAY	0.667	(1.052)	1.167	(1.284)	1.750	(1.472)	1.750	(1.472)	4.000	(2.121)	1.867	(1.480)
P76N3XJUNE	0.750	(1.095)	1.000	(1.211)	1.000	(1.211)	1.750	(1.472)	4.000	(2.121)	1.700	(1.422)
P76NJULY	0.750	(1.095)	1.000	(1.211)	1.000	(1.211)	4.000	(2.121)	4.000	(2.121)	2.150	(1.552)
P76NSEPT	0.667	(1.0520	0.750	(1.095)	1.083	(1.254)	1.083	(1.254)	1.250	(1.314)	0.967	(1.229)
PFAPRIL	1.000	(1.171)	1.250	(1.273)	1.167	(1.271)	1.167	(1.271)	1.250	(1.314)	1.167	(1.260)
PFMAY	0.917	(1.182)	1.417	(1.373)	1.917	(1.535)	1.917	(1.517)	4.000	(2.121)	2.034	(1.546)
PFJUNE	1.000	(1.225)	1.000	(1.225)	1.167	(1.284)	1.917	(1.517)	4.000	(2.121)	1.817	(1.474)

PFJULY	1.000	(1.225)	1.000	(1.225)	1.167	(1.284)	4.000	(2.121)	4.000	(2.121)	2.233	(1.595)
PFSEPT	0.667	(1.052)	1.500	(1.362)	1.250	(1.314)	1.250	(1.314)	1.417	(1.373)	1.217	(1.283)
PHAPRIL	1.000	(1.211)	1.333	(1.344)	1.250	(1.314)	1.250	(1.314)	1.250	(1.314)	1.217	(1.299)
PHMAY	0.750	(1.095)	1.667	(1.457)	2.500	(1.714)	2.333	(1.646)	4.000	(2.121)	2.250	(1.607)
PHJUNE	0.750	(1.095)	0.833	(1.139)	0.917	(1.168)	2.333	(1.646)	4.000	(2.121)	1.767	(1.434)
PHJULY	0.750	(1.095)	0.833	(1.139)	0.917	(1.168)	4.000	(2.121)	4.000	(2.121)	2.100	(1.529)
PHSEPT	0.917	(1.182)	1.333	(1.344)	1.250	(1.314)	1.250	(1.314)	1.500	(1.403)	1.250	(1.311)
SMAPRIL	0.417	(0.923)	0.667	(1.052)	0.750	(1.082)	1.000	(1.211)	1.083	(1.254)	0.783	(1.104)
SMMAY	0.500	(0.966)	1.000	(1.225)	1.500	(1.403)	1.583	(1.427)	3.917	(2.100)	1.700	(1.424)
SMJUNE	0.917	(1.168)	0.667	(1.052)	0.833	(1.139)	1.583	(1.427)	4.000	(2.121)	1.600	(1.381)
SMJULY	0.917	(1.168)	0.667	(1.052)	0.833	(1.139)	4.000	(2.121)	4.000	(2.121)	2.083	(1.520)
SMSEPT	0.583	(1.009)	0.667	(1.052)	1.083	(1.254)	1.083	(1.254)	1.083	(1.254)	0.900	(1.165)
MEAN	0.742	(1.086)	1.050	(1.224)	1.275	(1.308)	2.178	(1.588)	2.914	(1.804)	1.632	(1.402)
F-LSD(0.05)		0.160		0.144		0.149		0.167		NS		

M=Marketer, P76H=Poinsett 76 Holland, P76N=Poinsett 76 Netherlands, PF=Poinsett France, PH=Poinsett Holland, S=Super marketer. Values in parenthesis are square-root transformed values to which LSD is applicable.

Table 12: The effect of planting time and variety on phytophtora blight disease severity on leaves at different weeks after planting.

WEEKS AFTER PLANTING (WAP)												
VARXMON	2WAP		4WAP		6WAP		8WAP		10WAP		MEAN	
MAPRIL	-	-	-	-	0.667	(1.052)	0.583	(1.009)	0.833	(1.139)	0.694	(1.067)
MMAY	-	-	-	-	0.917	(1.155)	2.250	(1.639)	1.417	(1.373)	1.528	(1.389)
MJUNE	-	-	-	-	0.583	(1.009)	0.667	(1.052)	1.417	(1.373)	0.889	(1.145)
MJULY	-	-	-	-	0.583	(1.009)	4.000	(2.121)	4.000	(2.121)	2.861	(1.750)
MSEPT	-	-	-	-	0.833	(1.138)	0.833	(1.138)	1.000	(1.225)	0.889	(1.167)
P76HAPRIL	-	-	-	-	0.917	(1.168)	0.833	(1.138)	0.917	(1.182)	0.889	(1.163)
P76HMAY	-	-	-	-	1.000	(1.192)	1.500	(1.397)	1.500	(1.397)	1.333	(1.329)
P76HJUNE	-	-	-	-	0.833	(1.125)	1.000	(1.225)	1.333	(1.344)	1.055	(1.231)
P76HJULY	-	-	-	-	0.833	(1.125)	4.000	(2.121)	4.000	(2.121)	2.944	(1.789)
P76HSEPT	-	-	-	-	1.000	(1.225)	1.000	(1.225)	1.000	(1.225)	1.000	(1.225)
P76NAPRIL	-	-	-	-	0.333	(0.866)	0.750	(1.082)	0.917	(1.168)	0.667	(1.039)
P76NMAY	-	-	-	-	0.833	(1.138)	1.750	(1.475)	1.500	(1.397)	1.361	(1.337)
P76N3XJUNE	-	-	-	-	0.333	(0.880	0.583	(1.009)	1.250	(1.314)	0.722	(1.162)
P76NJULY	-	-	-	-	0.333	(0.880)	4.000	(2.121)	4.000	(2.121)	2.778	(1.707)
P76NSEPT	-	-	-	-	0.583	(0.996)	1.083	(1.254)	1.083	(1.254)	0.916	(1.168)
PFAPRIL	-	-	-	-	0.667	(1.052)	1.000	(1.184)	0.667	(1.039)	0.778	(1.092)
PFMAY	-	-	-	-	1.000	(1.157)	1.333	(1.344)	1.417	(1.373)	1.250	(1.291)
PFJUNE	-	-	-	-	0.667	(1.052)	0.833	(1.138)	1.750	(1.492)	1.083	(1.227)
PFJULY	-	-	-	-	0.667	(1.052)	4.000	(2.121)	4.000	(2.121)	2.889	(1.765)
PFSEPT	-	-	-	-	0.667	(1.052)	1.083	(1.241)	1.083	(1.254)	0.944	(1.182)
PHAPRIL	-	-	-	-	0.583	(1.009)	0.750	(1.063)	1.083	(1.222)	0.805	(1.098)
PHMAY	-	-	-	-	1.333	(1.316)	1.833	(1.502)	1.583	(1.433)	1.583	(1.417)
PHJUNE	-	-	-	-	0.750	(1.095)	0.917	(1.168)	1.750	(1.487)	1.139	(1.250)
PHJULY	-	-	-	-	0.750	(1.095)	4.000	(2.121)	4.000	(2.121)	2.917	(1.779)
PHSEPT	-	-	-	-	0.667	(1.052)	0.917	(1.168)	1.250	(1.308)	0.945	(1.176)
SMAPRIL	-	-	-	-	0.167	(0.793)	0.417	(0.923)	0.417	(0.923)	0.334	(0.880)
SMMAY	-	-	-	-	0.667	(1.025)	1.667	(1.462)	1.167	(1.284)	1.167	(1.257)
SMJUNE	-	-	-	-	0.000	(0.707)	0.083	(0.750)	1.250	(1.314)	0.444	(0.924)
SMJULY	-	-	-	-	0.000	(0.707)	4.000	(2.121)	4.000	(2.121)	2.667	(1.650)
SMSEPT	-	-	-	-	0.417	(0.923)	1.000	(1.225)	1.000	(1.225)	0.806	(1.124)
MEAN	-	-	-	-	0.653	(1.035)	1.622	(1.385)	1.753	(1.449)	1.343	(1.290)
F-LSD(0.05)	-	-	-	-	NS		0.156		NS			

M=Marketer, P76H=Poinsett 76 Holland, P76N=Poinsett 76 Netherlands, PF=Poinsett France, PH=Poinsett Holland, S=Super marketer. Values in parenthesis are square-root transformed values to which LSD is applicable.

Table 13: The effect of planting time and variety on phytophtora blight disease severity on fruits at different weeks after planting.

were superior to Poinsette 76 Holland, Poinsette France and Poinsette Holland. Super marketer, Poinsette 76 Netherlands and Marketer varieties records were low in both the disease incidence, disease severity on both leaves and fruits and yield characters. The Super marketer variety had differential yield characters–fruit girth, weight of fruit plant and fruit lengths, which were significantly different from the other varieties. These differential growth and yield characters of cucumber have been reported by researchers in different parts of the world. The differences in vegetative and yield characters can be attributed to genetic composition of the varieties used; the Ashley variety may have

VAR X MON	FG(cm)	FL(cm)	FW/Plt(kg)	NFPP	FY(t/ha)	DF(t/ha)
MAPRIL	4.958	20.217	2.259	5.484	451.80	2.259
MMAY	2.392	10.781	0.209	1.950	41.90	0.209
MJUNE	2.425	10.843	0.185	1.692	37.00	0.185
MJULY	2.258	10.843	0.119	1.986	23.70	0.119
MSEPT	4.958	20.217	2.397	5.475	479.30	2.397
P76HAPRIL	4.833	19.590	2.371	5.614	474.20	2.371
P76HMAY	2.125	11.696	0.200	2.054	40.00	0.200
P76HJUNE	2.525	11.340	0.182	1.963	36.40	0.182
P76HJULY	2.525	11.340	0.117	1.946	23.40	0.117
P76HSEPT	4.833	19.590	2.514	5.788	502.70	2.514
P76NAPRIL	4.977	19.813	2.302	5.475	460.40	2.302
P76NMAY	2.425	12.536	0.162	2.314	32.40	0.162
P76N3XJUNE	2.467	11.798	0.151	2.328	30.20	0.151
P76NJULY	2.467	11.798	0.076	1.886	15.20	0.076
P76NSEPT	4.977	19.813	2.313	5.380	462.60	2.313
PFAPRIL	5.256	20.098	2.445	5.288	488.90	2.445
PFMAY	2.350	12.441	0.161	2.460	32.20	0.161
PFJUNE	2.800	11.606	0.157	2.379	31.50	0.157
PFJULY	2.533	11.606	0.079	2.134	15.70	0.079
PFSEPT	5.256	20.098	2.449	5.614	489.70	2.449
PHAPRIL	5.143	19.491	2.539	5.383	507.90	2.539
PHMAY	2.242	12.493	0.158	2.373	31.60	0.158
PHJUNE	2.550	11.406	0.122	2.314	24.50	0.122
PHJULY	2.500	11.406	0.075	2.042	15.10	0.075
PHSEPT	5.143	19.491	2.292	5.484	458.30	2.292
SMAPRIL	5.806	20.852	2.531	5.788	508.90	2.531
SMMAY	2.767	14.207	0.175	2.473	35.00	0.175
SMJUNE	2.567	12.658	0.171	2.200	34.10	0.171
SMJULY	2.354	12.658	0.107	1.898	21.50	0.107
SMSEPT	5.806	20.852	2.696	5.722	506.20	2.550
MEAN	3.541	15.119	1.057	3.496	210.41	0.075
F-LSD(0.05)	0.341	0.514	0.1757	NS	NS	NS

M=Marketer, P76H=Poinsett 76 Holland, P76N=Poinsett 76 Netherlands, PF=Poinsett France,PH=Poinsett Holland, S=Super marketer. FG=Fruit girth, FL=Fruit length, FW=Fruit weight, NFPP=Number of fruit per plant and FY=Fruit yield, DF=Diseased fruits

Table 14: The effect of planting time and variety on the yield of six varieties of cucumber.

been quicker in adapting to the environment than the other varieties or the vegetative characters of the Ashley variety may have been more active than the other varieties and therefore had a strong source to sink relationship which resulted in high yields experienced in the variety. The number of fruits per plant was higher than what was reported by Phu [6] and by Jizhe [7] in Thailand. The yields obtained in this varietal studies was higher than what Manyvong and Phu [8] reported in Thailand but the same as the yield obtained by Mingbao [9].

From the result of this study, it was found that cucumber can be planted at any time of the rainy season because of the consistent high yields recorded in the varietal trials which were carried out in different months of April, May, June, July and September in 2013. This is in consonance with the findings of Mas who stated that cucumber can be planted at any time of the year provided during the growing period, there is ample moisture and the soil is fertile. The results also agree with the findings of Thoa [10] who observed that cucumber can be planted at anytime of the year and that even in temperate regions, during the winter the crop can be grown under greenhouse conditions

using artificial lighting systems. The high yield which was consistently recorded by the Super Marketer variety throughout the five months of study could be attributed to the genetic composition and its ability to quickly adapt to this environment. Super marketer produced the highest yield per hectare, and the Poinsette Holland, the lowest.

The cumulative effect of variety and time of planting on phytophthora blight disease incidence, severity and yield of cucumber.

The observed significant (p<0.05) variation in the varieties of cucumber (marketer, super marketer, poinsette 76 Holland and Netherlands, poinsette France and Holland) and the time of planting (April, May, June, July and September) having Supermarketer planted in the month of April, is attributed to the genetic make up of this particular variety and environmental condition [5] during the time it was planted. Generally, the varieties planted during the month of April gave the best results as regards the disease incidence, severity on leaves and fruits and the total yield, closely followed by those planted during the month of September. The month of July gave the poorest results; this reflects the substantial contribution of environmental condition in controlling the introduction and subsequent spread of the disease if occurred in a field. The month of July had more rain days and length (19.00 and 283.96) respectively, and these played a significant role on the spread of the disease. Furthermore, the month of July had low temperature and high humidity which helps in spreading phytophthora blight very fast in a crop field [5]. This result also tallied with the assertion of Agrios [11] that the higher the moisture content of the soil and lower the temperature, the higher the spread of phytophthora blight disease of cucurbits. The yield (508.90 t/ha) recorded for the super marketer during the month of April remains the highest, therefore super marketer remains the best variety while April is the best time to grow cucumber in this derived savannah agro-ecology of Nsukka.

References

1. Wehner TC, Guner N (2004) Growth stage, flowering pattern, yield and harvest date prediction of four types of cucumber tested at 10 planting dates. Acta Hort 637.

2. Adetula O, Denton L (2003) Performance of vegetative and yield accessions of cucumber (Cucumis sativa L.) Horticultural Society of Nigeria (HORTSON) Proceedings of 21st annual conference.

3. Obi IU (2002) Statistical methods in detecting differences between treatment means and Research methodology issues in laboratory and field experiment. Second edition p. 41.

4. Papadopoulos AP (1994) Growing greenhouse seedless cucumbers in soil-less media Research centre, Harrow, Ontario Canada.

5. Babadoost M (2005) Phytophthora blight of cucurbits. The Plant Health Instructor.

6. Phu NT (1998) Nitrogen and potassium effect on cucumber yield. ARC/AVRDC Training Bangkok Thailand.

7. Jizhe C (1993) Cucumber evaluation trial .ARC/AVRDC Training Thailand.

8. Manyvong V (1997) Cucumber varietal trial ARC/AVRDC Training workshop, Thailand.

9. Mingbao L (1991) Cucumber varietal trial.ARC/AVRDC Training Bangkok Thailand.

10. Thoa DK (1998) Cucumber seed Multiplication and characterization A.R.C–AVRDC Research Report. Bangkok Thailand.

11. Agrios George N (2004) Plant pathology /5th ed. Elsevier Academic Press 30 Corporate Drive, Suite 400, Burlington, MA 01803, USA.

Maize (*Zea Mays* L.) Growth and Metabolic Dynamics with Plant Growth-Promoting Rhizobacteria under Salt Stress

Abd El-Ghany TM*, Masrahi YS, Mohamed A, Al Abboud, Alawlaqi MM and Nadeem I Elhussieny

Biology Department, Faculty of Science, Jazan University, 114, KSA

Abstract

Maize (*Zea mays* L.) biomass and its allied attributes were assessed under salinity stress and three plant growth-promoting rhizobacteria (*Pseudomonas fluorescens*, *Pseudomonas putida* and *Azotobacter vinelandii*) treatments. The three PGPRs inocula exhibited a different pattern of shoot growth under both normal and saline stress conditions. Plant biomass, carbohydrates, protein and chlorophyll content were reduced by saline stress, however application of PGPRs treatments improved them either in comparison to control samples or to untreated samples under saline stress. Lipids and antioxidant enzymes (catalase and peroxidase) increased as a response for saline stress as an indication of oxidative stress. Plant growth-promoting rhizobacteria treatment restored them to semi-normal levels. Sodium/ potassium balance was observed to be disturbed by saline stress through higher levels of Na^+ and lower levels of K^+, but treating samples balance was clearly restored close to normal conditions especially in the root system.

Keywords: *Zea mays* L.; Plant growth-promoting rhizobacteria; Salt stress

Introduction

Biofertilizers diminish the need for expensive chemical fertilizers in crop farming systems because of they are an inexpensive source of nitrogen that increase crop yields. Thus the extensive use of biofertilizers would provide economic benefits to farmers, improve the socio-economic condition of the people and preserve natural resources. Biofertilizers are ecofriendly inputs and are less damaging to the environment when compared to chemical fertilizers [1,2]. Beneficial rhizobacteria, often referred to as plant growth-promoting rhizobacteria (PGPR), affect plant growth either directly or indirectly through various mechanisms of action [3-6]. Although, the mechanisms by which PGPRs promote plant growth are not fully understood, some mechanisms include gibberellic acid and/or cytokinins production, nitrogen fixation, and solubilization of mineral phosphate and other nutrients [7].

Maize (*Zea mays* L.) is the third most important cereal after wheat and rice all over the world [8]. Subramaniyan *et al.* [9] stated, "the application of biofertilizers improved the total carbohydrate, protein, amino nitrogen and chlorophyll content of of *Zea mays*". Five growth-promoting strains (*Azotobacter* sp. Lx191, *Pseudomonas* sp. Jm92, *Bacillus* sp. LM4-3, *Bacillus* sp.LH12-3, and *Azospirillum* sp. LHS11) were previously isolated from rhizosphere of wheat, maize, oat in arid fields,. These strains were proofed to stimulate these plant growth under controlled conditions via *in vitro* and pot experiments [10,11].

Soil salinity decreases plant growth, reduces photosynthetic activity and results in nutrient imbalance in plants.It was reported that PGPR significantly increased shoot/root fresh weight, shoot/root dry weight, chlorophyll a, b and cartenoid contents of maize under salt stress. PGPR can induce plant tolerance to salinity by producing various hormones and enhancing the availability of nutrients from the soil matrix [12]. Hasnain and Sabri [13] reported that inoculation with *Pseudomonas* sp. stimulated plant growth by reduction of toxic ion uptake and production of stress-specific proteins in plant. PGPR strains can also produce exopolysaccharides (EPSs) to bind cations including sodium, thus help alleviating salt stress in plants grown under

saline environment [14]. The rhizosphere is the soil portion found around the root and under the influence of the root. It is the site with complex interaction between the root and associated microorganisms [15]. The rhizosphere harbors a multitude of microorganisms that are affected by both abiotic and biotic stresses. Among these are the dominant rhizobacteria that prefer living in close vicinity to the root or on its surface and play a crucial role in soil health and plant growth [16,17]. It has been noted by many workers that *Pseudomonas, Bacillus, Arthrobacter, Azospirillum, Klebsiella,* and *Enterobacter,* isolated from the rhizosphere of various crops, showed synergistic effects on plant growth [18]. Weller [19] reported that PGPR belong to several genera, e.g. *Agrobacterium, Alcaligenes, Arthrobacter, Actinoplanes,* Azotobacter, *Bacillus, Pseudomonas* sp., *Rhizobium, Bradyrhizobium, Erwinia, Enterobacter, Amorpho sporangium, Cellulomonas, Flavobacterium, Streptomyces* and *Xanthomonas* .These groups of bacteria are important as they are involved in various soil biochemical processes such as fixation of atmospheric nitrogen, solubilization of minerals such as phosphorus, production of siderophores that solubilize and sequester iron and/or production of plant growth regulators [20]. Nitrogen-fixing bacteria are widely distributed in nature where they reduce atmospheric nitrogen in soil or in association with plant [21]. They have been found in a wide variety of terrestrial and aquatic habitats in both temperate and tropical regions of the world [22]. Biofertilizer contains living microorganisms and promotes growth by increasing the availability of primary nutrients (nitrogen and phosphorus) to the host plant [23-26]. Among the free-living nitrogen-fixing bacteria those belonging to genus Azotobacter play a remarkable role, being broadly dispersed in

*Corresponding author: Abd El-Ghany TM, Biology Department, faculty of Science, Jazan University, KSA, E-mail: tabdelghany@yahoo.com different environments, such as soil, water and sediments [27]. Several authors have shown the beneficial effects of *Azotobacter chroococcum* on vegetative growth and yields of maize [27-29], as well as the positive effect of inoculation with this bacterium on wheat [30]. *Bacillus* group was the most dominant strain found in the three types of biofertilizer products. The other bacteria were *Azospirillum*, *Corynebacterium*, *Pseudomonas* and *Proteus mirabilis*. These bacteria have the potential to fix atmospheric nitrogen, able to produce IAA with the supplemented tryptophan, and showed some phosphorus solubilizing activity [15,31]. The aim of this work was to assess the effect of PGPR on the growth of *Z. mays*. Also it was carried out to elucidate the role of PGPR on growth and ion uptake of maize under salt stress condition.

Materials and Methods

Description of the study area

The present study focused on the area in Jazan. The study area is situated in Jazan, Kingdom of Saudi Arabia (Lat. 16°53′21″ N , Long. 42°33′03″ E and 19 m Elevation above sea level) with the significant features of evergreen forests and also it was a less explored ecosystem for the investigation of biofertilizers population.

Bacterial strains used

Plant growth-promoting rhizobacteria were isolated from maize rhizosphere growing in Jazan area. Selection of isolates was performed on the basis of the PGPR traits. Selected isolates identified according standard microbiological methods as described in Bergys Manual of Systematic Microbiology [32]. The physiological and biochemical characters, included: starch hydrolysis, gelatin liquefaction, indole production, nitrate reduction, urease activity, citrate utilization, production of oxidase, catalase, methyl red, voges proskauer, tryptophane deaminase, gelatinase, lysine decarboxylase, arginine dihydrolase, ß-galactosidase and fermentation oxidation of the following carbon sources (D-glucose, D-mannitol, inositol, D-sorbitol, rhamnose, D-sucrose, D-melibiose maltose, fructose, inulin and L-arabinose) were used for identification of bacterial isolates. Bacterial isolates were grown on yeast manitol agar (YMA) supplemented with different concentrations of NaCl for salt tolerance test. Identification of highly NaCl tolerance was done using biochemical analysis. Three isolates including *Pseudomonas fluorescens*, *Azotobacter vinelandii* and *Pseudomonas putida* were used in this study.

Inoculums preparation and *Zea mays* L. growth experimental design

Fresh cultures of selected isolates were inoculated in pikovisky media and shaken at 37°C for two days in an orbital shaker at 100 rpm. After two days bacterial cultures were centrifuged at 3000 rpm for 15 mins. Maize (*Z. mays* L.) grains were surface sterilised with 0.5% (v/v) NaOCl for 10 min and were subsequently washed with sterilized deionised water. The sterilized grains were soaked in distilled water in case of un-inoculated control. The rest of sterilized seeds were soaked in broth cultures of isolates form 4-5 hr prior to sowing at two different concentrations 10^4 (low concentration) and 10^8 (high concentration) bacterial cells. Grains were germinated in plastic pots (15 cm diameter) with 2 kg sterilized soil. After sowing, seedlings were reduced to three per pot. The pots were treated as the following: 1) Untreated control, 2) 10^4/ml bacterial cell suspension, 3)10^8/ml bacterial cell suspension, 4) 35 mM Na Cl solution, 5) 70 mM Na Cl solution, 6) 35 mM Na Cl +10^4/ml bacterial cell mixture, 7) 35 mM Na Cl +10^8/ml bacterial cell mixture, 8) 70 mM Na Cl +10^4/ml bacterial cell mixture and 9) 70 mM

Na Cl +10^8/ml bacterial cell.

Growth parameters including dry weight, height, leaves length and width of Maize plants were recorded.

Na+ and K+ content analysis

Oven - dried samples of *Zea mays* roots and shoots were powdered for estimation of Na+ and K+ by flame photometric method [33].

Quantitative determination of chlorophyll

Chlorophylls content was determined with using the following equations:

Chlorophyll a (mg)/tissue (g) = 11.63 (A 665) –2.39 (A 649)

Chlorophyll b (mg)/tissue (g) = 2.11 (A 649) –5.18 (A 665)

Where A denotes the reading of the optical density.

Antioxidant enzymes catalase and peroxidase of healthy and infected plant were determined according to Kar and Mishra [34].

Estimation of Total protein content

Total protein was estimated calorimetrically [35] by recording absorbance at 595 nm. Bovine serum albumin was used as standard. Protein content in plant samples was recorded as mg of protein per g of sample.

Estimation of Total carbohydrate content

Plant extract was taken in 25 ml test tubes and 6 ml anthrone reagent (150 mg of anthrone in 72 % H_2SO_4) was added, and then heated in boiling water bath for 10 min. The test tubes were ice cooled for 10 min and incubated for 20 min at 25°C. Optical density (OD) was read at 625 nm on a spectrophotometer. The carbohydrate content was calculated from the standard curve using glucose with the same method which mentioned above [36].

Estimation of total lipid content

Total lipid was estimated using Vanillin reagent (6.1 g of vanillin was dissolved in water and diluted to 1 liter). The OD was read on a spectrophotometer at 540 nm. The lipid content was calculated from the standard curve using standard solution of cholesterol with the same method which mentioned above.

Results

Plant biomass growth

Maize biomass in terms of plant height, stem diameter, leaf surface area, and plant dry weight was investigated in relation to saline stress (0, 35, and 70 mM) and PGPRs (*Azotobacter vinelandii*, *Pseudomonas fluorescens* and *Pseudomonas putida*) treatments. Plant height (Table 1) was reduced from 103.33 cm to 91 cm by saline stress in untreated samples. PGPRs treatments improved plant height in both stressed and normal conditions. *A. vinelandii* at both their concentrations significantly increased plant height from 103.33 to 144.33 cm under normal conditions. Under saline stress *A. vinelandii* increased plant height at 35 mM by 33.77% and 13.55% at 70 mM saline stress. *P. fluorescens* was inferior to *A. vinelandii* as it was able to increase plant height in untreated samples to 131 cm in normal conditions while it was 29.17% more than untreated sample at 35 mM saline stress and no significant increase at 70 mM. *P. putida* did not improved plant height significantly both in normal and stressed conditions.

PGPR(Cell ml⁻¹)		Saline stress (mM)		
		0	35	70
Control		103.33 ± 1.45cdef	101.67 ± 9.28def	91.00 ± 3.79f
A. vinelandii	10⁴	132.33 ± 7.84ab	114.33 ± 5.33bcde	102.67 ± 7.06cdef
	10⁸	144.33 ± 4.37a	136.00 ± 6.51ab	103.33 ± 11.67cdef
P. fluorescens	10⁴	125.00 ± 2.89abc	100.67 ± 8.35def	89.33 ± 7.31f
	10⁸	131.00 ± 4.58ab	131.33 ± 10.27ab	92.33 ± 8.88ef
P. putida	10⁴	119.33 ± 1.86bcd	105.67 ± 7.45cdef	85.33 ± 2.03f
	10⁸	119.33 ± 3.84bcd	115.67 ± 10.84bcd	98.67 ± 2.96def

Means followed by the same letter are not significantly different.

Table 1: Mean comparison and standard error of maize plant height (cm) under saline stress and PGPRs treatment.

PGPR(Cell ml⁻¹)		Saline stress (mM)		
		0	35	70
Control		2.93 ± 0.12efg	2.80 ± 0.06fg	2.17 ± 0.19g
A. vinelandii	10⁴	4.07 ± 0.22abcde	3.43 ± 0.23cdef	2.63 ± 0.19fg
	10⁸	4.67 ± 0.38ab	3.57 ± 0.23bcdef	3.23 ± 0.46defg
P. fluorescens	10⁴	4.27 ± 0.15abcd	3.60 ± 0.15bcdef	3.07 ± 0.35defg
	10⁸	4.80 ± 0.06a	4.03 ± 0.09abcde	3.43 ± 0.09cdef
P. putida	10⁴	4.60 ± 0.21abc	3.80 ± 0.44abcdef	2.77 ± 0.19fg
	10⁸	4.87 ± 0.09a	3.17 ± 0.67defg	3.37 ± 1.07def

Means followed by the same letter are not significantly different.

Table 2: Mean comparison and standard error of maize stem diameter (cm) under saline stress and PGPRs treatment.

PGPR(Cell ml⁻¹)		Saline stress (mM)		
		0	35	70
Control		231.25 ± 16.16bcdef	77.00 ± 3.82j	81.85 ± 17.82j
A. vinelandii	10⁴	258.68 ± 14.81abcde	157.70 ± 22.05fghij	128.78 ± 24.20hij
	10⁸	293.08 ± 25.93abc	179.88 ± 50.12efghi	134.30 ± 26.04hij
P. fluorescens	10⁴	243.88 ± 2.99abcde	190.30 ± 37.53defghi	125.05 ± 18.71ij
	10⁸	312.70 ± 1.76a	268.75 ± 33.11abcd	141.60 ± 17.33ghij
P. putida	10⁴	209.13 ± 35.58defgh	157.73 ± 21.21fghij	126.83 ± 22.48ij
	10⁸	298.35 ± 11.89ab	217.35 ± 27.07cdefg	152.38 ± 18.59fghij

Means followed by the same letter are not significantly different.

Table 3: Mean comparison and standard error of maize leaf surface area (cm²) under saline stress and PGPRs treatment.

Stem diameter was not significantly affected by saline stress (Table 2), however it was clearly influenced by PGPRs treatment. The three tested PGPRs (at 10⁴ and 10⁸ cell ml⁻¹) improve stem diameters by 38 – 66% under no saline stress. *P. fluorescens* was the best PGPR treatment that enhanced stem diameter from 2.8 to 4.03 cm at 35 mM saline stress and 10⁸ cell ml⁻¹; while stem diameter was 3.43 cm at 70 mM saline and 10⁸ cell ml⁻¹ against 2.17 cm at 70 mM saline untreated samples. *A. vinelandii* treatment improve plant stems diameter by 27.5% and 48.85 % at 35 and 70 mM saline stress respectively. *P. putida* both concentrations improve plant stems diameter to 3.8 at low saline stress and 3.37 at the higher one.

Leaf surface area of maize plant was significantly reduced by 66.7% due to saline stress (Table 3). All tested PGPRs treatment increased plant leaf surface area to 298.35 cm² under no saline stress. *A. vinelandii* and *P. putida* treatment limited leaf surface area reduction by saline stress from 66.7% to range between 31.8 and 6% at 35 mM stress and between 44.3 and 34.1% at 70 mM. *P. fluorescens* treatment was superior it was capable of increasing leaf surface area to 268.75 cm² that is 16% more than control leaf surface area

Plant dry weight (Table 4) was reduced from 6.4 g to 3.4 g by saline stress in untreated samples. PGPRs treatment improve plant dry weight in both stressed and normal conditions. *A. vinelandii* both concentrations significantly increased plant dry weight from 6.4 g to 11.2 g under normal conditions. Under saline stress *A. vinelandii*

increased plant dry weight at 35 mM to 12.8 g. *P. fluorescens* was able to increase plant dry weight under no saline stress to 13.1 g while it was 11.9 g at 35 mM saline stress and 9.9 g at 70 mM. *P. putida* improved plant height significantly both in normal and stressed conditions.

Biochemical contents

Maize plant carbohydrate content (Table 5) was investigated during the current study. However, the differences in carbohydrates content among treatments was limited; it showed high degree of significance. Carbohydrate content was 79.2 mg/g in the control plant. PGPRs treatments at no saline stress increased carbohydrates content of the plant up to 82.2 mg/g. Although saline stress reduced plant carbohydrate content in all treated and untreated samples as compared to control sample; it is obvious that except of *P. putida*, the other treatments improve plant carbohydrate content as compared to untreated samples (Table 6).

Maize proteins were 26.17 mg/g in control sample. PGPRs treatments significantly improve plant protein content up to 33.17 mg/g under no saline stress. Saline stress reduced maize protein content to 24.17 and 20.17 mg/g at 35 and 70 mM saline stress, respectively. *A. vinelandii* and *P. fluorescens* treatment improve palnt protein content by 7.5% as compared to untreated sample at 35 mM. PGPRs treatment showed no considerable differences in the plant protein content when compared to untreated samples at 70 mM. Unlike other investigated

PGPR(Cell ml⁻¹)		Saline stress (mM)		
		0	**35**	**70**
Control		6.40 ± 0.29f	3.40 ± 0.23h	4.00 ± 0.29h
A. vinelandii	**10⁴**	10.00 ± 0.23d	11.10 ± 0.29c	5.40 ± 0.29g
	10⁸	11.20 ± 0.29c	12.80 ± 0.29b	6.80 ± 0.35ef
P. fluorescens	**10⁴**	9.30 ± 0.17d	11.93 ± 0.20c	5.07 ± 0.26g
	10⁸	13.10 ± 0.58b	11.30 ± 0.23c	9.90 ± 0.06d
P. putida	**10⁴**	7.30 ± 0.17ef	14.00 ± 0.29a	6.20 ± 0.12f
	10⁸	13.00 ± 0.29b	14.20 ± 0.23a	6.80 ± 0.23ef

Means followed by the same letter are not significantly different.

Table 4: Mean comparison and standard error of maize dry weight (g) under saline stress and PGPRs treatment.

PGPR(Cell ml⁻¹)		Saline stress (mM)		
		0	**35**	**70**
Control		79.20 ± 0.13f	69.20 ± 0.13j	66.20 ± 0.13m
A. vinelandii	**10⁴**	81.00 ± 0.13c	69.95 ± 0.13i	67.18 ± 0.13l
	10⁸	81.69 ± 0.13b	70.98 ± 0.13h	67.89 ± 0.13k
P. fluorescens	**10⁴**	80.20 ± 0.13e	70.20 ± 0.13i	69.89 ± 0.13i
	10⁸	82.20 ± 0.13a	71.98 ± 0.13g	69.94 ± 0.13i
P. putida	**10⁴**	80.00 ± 0.13e	67.95 ± 0.13k	65.28 ± 0.12n
	10⁸	80.62 ± 0.12d	67.58 ± 0.13k	62.80 ± 0.13o

Means followed by the same letter are not significantly different.

Table 5: Mean comparison and standard error of maize carbohydrates (mg/g) under saline stress and PGPRs treatment.

PGPR(Cell ml⁻¹)		Saline stress (mM)		
		0	**35**	**70**
Control		26.17 ± 0.02e	24.17 ± 0.01h	20.17 ± 0.18l
A. vinelandii	**10⁴**	28.18 ± 0.01d	25.18 ± 0.06g	20.12 ± 0.18l
	10⁸	32.13 ± 0.06b	26.00 ± 0.02e	20.93 ± 0.01k
P. fluorescens	**10⁴**	30.17 ± 0.01c	25.17 ± 0.06g	21.10 ± 0.18k
	10⁸	33.17 ± 0.06a	25.69 ± 0.02f	21.99 ± 0.01j
P. putida	**10⁴**	26.19 ± 0.01e	22.12 ± 0.06j	18.94 ± 0.01m
	10⁸	28.11 ± 0.07d	23.33 ± 0.32i	20.12 ± 0.12l

Means followed by the same letter are not significantly different.

Table 6: Mean comparison and standard error of maize protein (mg/g) under saline stress and PGPRs treatment.

PGPR(Cell ml⁻¹)		Saline stress (mM)		
		0	**35**	**70**
Control		18.65 ± 0.13h	19.14 ± 0.14g	19.65 ± 0.13f
A. vinelandii	**10⁴**	18.02 ± 0.06i	20.41 ± 0.06de	23.05 ± 0.02b
	10⁸	20.56 ± 0.12d	21.15 ± 0.01c	23.95 ± 0.06a
P. fluorescens	**10⁴**	19.65 ± 0.12f	18.65 ± 0.05h	18.15 ± 0.05i
	10⁸	19.85 ± 0.14f	18.45 ± 0.12h	19.65 ± 0.12f
P. putida	**10⁴**	17.00 ± 0.06j	19.21 ± 0.07g	21.02 ± 0.07c
	10⁸	18.54 ± 0.12h	20.15 ± 0.02e	20.90 ± 0.12c

Means followed by the same letter are not significantly different.

Table 7: Mean comparison and standard error of maize lipid content (mg/g) under saline stress and PGPRs treatment.

primary metabolites, lipid content was significantly proportion to saline stress. Lipid content (Table 7) of maize was also improve by PGPRs treatments. *A. vinelandii* treatment raised lipid content in maiz plant from 18.65 to 20.56 mg/g at no saline stress, and to 23.95 mg/g at 70 mM saline stress. *P. fluorescens* treatment did not increase maize lipid content significantly as compared to untreated samples at no saline stress. *P. putida* improved lipid content in maize plant at 35 and 70 mM saline stress.

Chlorophyll content of maize plant was investigated in the current study. Under no saline stress *A. vinelandii* and *P. Fluorescens* significantly improve chlorophyll a content (Table 8) to 6.99 and 7.99 mg/g, respectively. Saline stress reduced plant chlorophyll a content in both PGPRs treated and untreated samples. *A. vinelandii* and *P. fluorescens* treatments significantly limited chlorophyll a content reduction from 46.86% to 31.59 and 27.99% at 35 mM saline stress, respectively. Chlorophyll b content (Table 9) was reduced due to saline stress from 3.21 to 2.1 mg/g. PGPRs treatments either did not influence or reduced chlorophyll b content under no saline stress, 35 mM and 70 mM saline stress.

Antioxidant enzymes as indicators for oxidative stress exerted on the plant were investigated. Saline stress increased catalase concentration (Figure 1) in untreated and PGPRs treated samples. PGPRs treatments reduced the amount of catalase in plants under saline stress when compared to untreated stressed plants. Salinity stress

PGPR(Cell ml⁻¹)		Saline stress (mM)		
		0	35	70
Control		6.68 ± 0.05c	3.55 ± 0.09i	3.50 ± 0.01i
A. vinelandii	10^4	6.77 ± 0.06c	4.31 ± 0.05f	3.62 ± 0.14hi
	10^8	6.99 ± 0.01b	4.57 ± 0.06e	3.65 ± 0.13hi
P. fluorescens	10^4	7.91 ± 0.03a	4.81 ± 0.03d	3.78 ± 0.07h
	10^8	7.99 ± 0.02a	4.98 ± 0.03d	3.99 ± 0.05g
P. putida	10^4	6.68 ± 0.06c	3.65 ± 0.02hi	3.50 ± 0.02i
	10^8	6.66 ± 0.04c	3.66 ± 0.06hi	3.62 ± 0.07hi

Means followed by the same letter are not significantly different.

Table 8: Mean comparison and standard error of maize chlorophyll a (mg/g) under saline stress and PGPRs treatment.

PGPR(Cell ml⁻¹)		Saline stress (mM)		
		0	35	70
Control		3.21 ± 0.01b	2.15 ± 0.12efg	2.10 ± 0.13fg
A. vinelandii	10^4	3.25 ± 0.12ab	2.14 ± 0.06efg	1.91 ± 0.06g
	10^8	3.32 ± 0.04ab	2.10 ± 0.12fg	1.91 ± 0.13fg
P. fluorescens	10^4	3.45 ± 0.12ab	2.45 ± 0.08de	2.12 ± 0.07fg
	10^8	3.52 ± 0.04a	2.52 ± 0.12d	2.01 ± 0.12fg
P. putida	10^4	2.19 ± 0.12efg	2.45 ± 0.08de	2.18 ± 0.07efg
	10^8	2.89 ± 0.04c	2.50 ± 0.12d	2.23 ± 0.12def

Means followed by the same letter are not significantly different.

Table 9: Mean comparison and standard error of maize chlorophyll b (mg/g) under saline stress and PGPRs treatment.

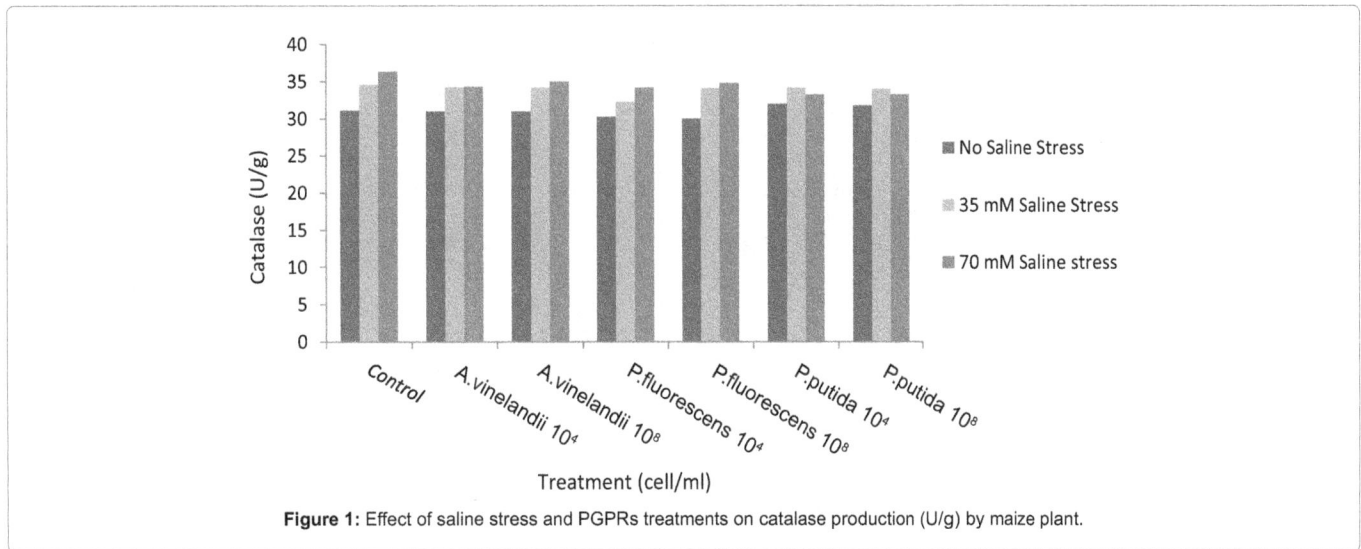

Figure 1: Effect of saline stress and PGPRs treatments on catalase production (U/g) by maize plant.

increased peroxidase concentration (Figure 2) in untreated samples. PGPRs treated samples showed irregular response to saline stress as peroxidase concentration increased at 35 mM saline stress over the untreated samples concentrations while it was reduced at 70 mM to minimum when compared to samples under no saline stress.

Sodium potassium flux

Root system sodium potassium balance (Table 10) was investigated in this study. It was found that Sodium potassium balance did not influence by PGPRs treatments under normal conditions. Salinity stress disturbed the Na⁺ K⁺ balance as Na⁺ jumped up to 8.14 at 35 mM and 9.25 mg/g at 70 mM while K⁺ withdrawn to be 4.43 and 4.22 at 35 and 70 nM saline stress, respectively. A. vinelandii treatment restored Na⁺ K⁺ balance at 35 and 70 mM saline stress close to normal balance. P. fluorescens treatment reduced Na content of plant roots when compared to untreated samples at both 35 and 70 mM saline stress; while the K⁺ content slightly increased. P. putida was disabled to restore the balance

as Na⁺ remained high and K⁺ low. Sodium and potassium contents in shoot system (Table 11) were influenced by saline stress as Na⁺ content increased up to 3.95 mg/g while the K⁺ content reduced to 23.85mg/g. PGPRs treatments slightly reduced Na content of the plant shoot system and increased its K⁺ content. Under 35 mM saline stress also, Na⁺ was slightly reduced by PGPRs treatments while the K⁺ content increased especially by A. vinelandii and P. fluorescens. The same pattern observed at 70 mM saline stress but both Na⁺ reductio and K⁺ improvement were limited.

Discussion

The effect of three PGPRs (Pseudomonas fluorescens, Pseudomonas putida and Azotobacter vinelandii) application on the Maize (Zea mays L.) growth and its allied attributes was assessed under saline stress in the current study. The results showed that the application of PGPRs significantly increased the shoot and root growth as compared to untreated plants. Parida and Das [37] reported that, the negative effects

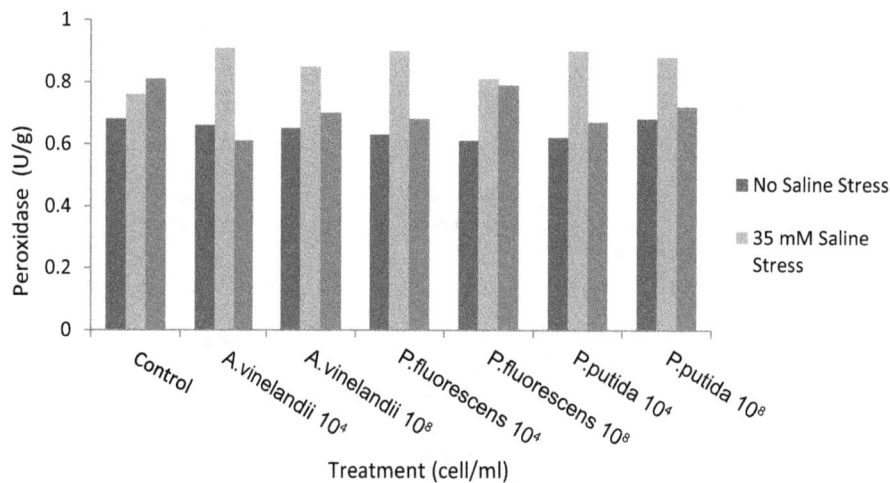

Figure 2: Effect of saline stress and PGPRs treatments on peroxidase production (U/g) by maize plant.

PGPR(cell ml⁻¹)		0		35		70	
		Na⁺	K⁺	Na⁺	K⁺	Na⁺	K⁺
Control		5.78 ± 0.032h	6.51 ± 0.01e	8.14 ± 0.027bc	4.43 ± 0.03l	9.25 ± 0.088a	4.22 ± 0.01m
A. vinelandii	10⁴	5.62 ± 0.024h	7.07 ± 0.04b	6.37 ± 0.032f	6.31 ± 0.05f	7.24 ± 0.038e	5.13 ± 0.01g
	10⁸	5.38 ± 0.029i	8.07 ± 0.10a	6.11 ± 0.009g	6.22 ± 0.05f	7.19 ± 0.003e	5.20 ± 0.03g
P. fluorescens	10⁴	5.63 ± 0.022h	6.59 ± 0.02e	6.32 ± 0.052f	4.75 ± 0.03i	7.21 ± 0.006e	4.54 ± 0.02l
	10⁸	5.39 ± 0.030i	6.96 ± 0.01c	6.38 ± 0.262f	4.91 ± 0.01h	7.21 ± 0.022e	4.44 ± 0.01l
P. putida	10⁴	5.72 ± 0.007h	6.53 ± 0.01e	8.04 ± 0.031c	4.60 ± 0.02jk	8.28 ± 0.044bc	4.43 ± 0.04l
	10⁸	5.67 ± 0.022h	6.75 ± 0.03d	7.77 ± 0.075d	4.66 ± 0.02ij	7.83 ± 0.092d	4.44 ± 0.03l

Na⁺ means followed by the same letter are not significantly different; K⁺ means followed by the same letter are not significantly different.

Table 10: Mean comparison and standard error of maize root system Na⁺ and K⁺ content (mg/g) under saline stress and PGPRs treatment.

PGPR(cell ml⁻¹)		0		35		70	
		Na⁺	K⁺	Na⁺	K⁺	Na⁺	K⁺
Control		2.95 ± 0.03gh	25.28 ± 0.04g	3.51 ± 0.03c	23.76 ± 0.03i	3.95 ± 0.08a	23.85 ± 0.08i
A. vinelandii	10⁴	2.29 ± 0.03j	27.54 ± 0.27d	3.15 ± 0.03ef	28.57 ± 0.01c	3.83 ± 0.03b	26.23 ± 0.12ef
	10⁸	2.15 ± 0.01k	31.91 ± 0.08a	3.12 ± 0.01f	29.25 ± 0.04b	3.32 ± 0.01d	26.54 ± 0.03e
P. fluorescens	10⁴	2.31 ± 0.03j	27.90 ± 0.03d	3.15 ± 0.03ef	28.47 ± 0.07c	3.81 ± 0.05b	26.16 ± 0.08ef
	10⁸	2.23 ± 0.09jk	32.17 ± 0.41a	3.11 ± 0.01f	29.31 ± 0.05b	3.33 ± 0.02d	26.42 ± 0.07ef
P. putida	10⁴	2.84 ± 0.03h	26.05 ± 0.07f	3.24 ± 0.04de	24.42 ± 0.05h	3.54 ± 0.06c	23.94 ± 0.07i
	10⁸	2.57 ± 0.01i	27.56 ± 0.06d	2.98 ± 0.05gh	26.28 ± 0.03ef	3.56 ± 0.02c	24.38 ± 0.04h

Na⁺ means followed by the same letter are not significantly different; K⁺ means followed by the same letter are not significantly different.

Table 11: Mean comparison and standard error of maize shoot system Na⁺ and K⁺ content (mg/g) under saline stress and PGPRs treatment.

of salinity stress on plant-growth include a reduction in growth rate and biomass, shorter stature, smaller leaves, osmotic effects, nutritional deficiency and mineral disorders. Therefore, according to Bacilio et al. [38] the use of PGPR to promote plant-growth in saline conditions is an important technology. The three PGPRs inocula exhibited a different pattern of shoot growth under both normal and salinity stressed conditions. Plant height, dry weight, stems diameter,and leaf surface area were clearly improved by PGPRs treatments in normal conditions. In fact, field trials have demonstrated that, inoculation with Azotobacter has beneficial effects on plant yields, due to the increase of fixed nitrogen content in agricultural soil [28,39-43], and to the

microbial secretion of stimulating hormones, like gibberellins, auxins and cytokinins [44,45]. Our results showed that P. fluorescens was the best inoculum for Zea mays growth enhancing. These results confirm previous findings where the enhancing effect of Zea mays inoculation with P. fluorescens on dry weight and yield of maize was reported [3,46], increase in plant height, root weight and total biomass were observed in response to inoculation.

Reduction in plant growth parameters under salt stress condition were recorded. Salinity is one of serious environmental problems that cause reduction in plant growth and yield productivity in irrigated areas of arid and semi-arid regions of the world [37]. The obtained adverse effects of salt stress on the *Zea mays* L. growth was alleviated by the PGPRs inoculation and decreased with concentration of inoculants. From the current study the effect of *P. fluorescens* was more pronounced than that of other two PGPRs. Our results confirm previous findings that inoculated plants grew better and had higher biomass compared to non-inoculated plants under salt stress conditions [38,47-49]. Jagnow [50] found that wheat and maize inoculated with Azotobacter increases both the plant biomass of the above ground (26- 50%) and the yield (19-30%). Recently, Zafar-ul-Hye *et al.* [51] found that maize productivity increased under salt stress inoculated with *P. syringae* and *P. fluorescens*.

Plants inoculated with PGPRs showed higher protein and carbohydrate content compared to control plants. Thus inoculation with *P. fluorescens* induced *Zea myes* L soluble protein and Carbohydrate yields (33.17 and 82.20 mg/gm respectively) compared with control (26.17 and 79.20 mg/gm respectively). On the other hand, under salt stress and without *P. fluorescens* treatment plant contents of protein and carbohydrate were decreased. The inoculation with *P. fluorescens* induce synthsis of protein and carbohydrate. Similar effects where also showed by the *A. vinelandii* and *P. putida*. Usually the increase of protein yield is related to higher nitrogen fixation activities, this knowledge was confirmed with many authors [52-54]. On the other hand, lipid content was observed to increase under salinity stress and reduced when PGPRs treatments were applied. The results indicate that *P. fluorescens* increased chlorophyll a and b (7.99 and 3.52 mg respectively) of *Zea mays* L. compared with the control (6.68 and 3.21 mg respectively) while *A. vinelandii* and *P. putida* were less effective on chlorophyll contents. Our results showed the co-inoculation of stress *Zea mays* L. markedly stimulated chlorophyll a and b content as compared to plants cultivated under salt stress without inoculation especially at low concentration 35 mM Na Cl. The effect of salinity on the synthesis of chlorophyll depended on the specific concentration of NaCl. Nevertheless, the inoculation with PGPRs of current study enhanced the content of chlorophyll revealing a positive effect on growth and plant development. A similar trend has also been observed in other researchers [12,55,56]. In the present investigation, the responses of *Zea mays*L. plant to high level of salinity were reflected by increased of catalase and peroxidase activities. Mittler [57] stated that antioxidant enzyme activities are usually affected by salinity and used as indicators of oxidative stress in plants. To protect against oxidative stress, plant cells produce both antioxidant enzymes such as peroxidase and catalase, and non-enzymatic antioxidants such as ascorbate, glutathione and tocopherol [58]. The results showed that the exogenous application of PGPRs decreased catalase and peroxidase activities of cultivated *Zea mays* under salt stress. A similar trend has also been observed in other researchers [59, 60].

Salinity causes an imbalance in the ion flux inside plants. The present results showed that during salinity stress, the plants had higher Na^+ and lower K^+ contents, compared with control plant in root and shoot system (Tables 10 and 11). This is also according to the results [61], salinity increases the uptake of Na^+ or decreases the uptake K^+ which lead to nutritional imbalances. Accumulation of excess Na^+ may cause metabolic disturbance in processes where low Na^+ and high K^+ are required for optimum plant function [62]. Increased K^+ concentration under salinity conditions may help to decrease Na^+ uptake and this can indirectly maintain the growth of the plant [63]. Based on the results obtained, applying PGPRs treatment significantly increased the K^+ content of maize under salt stress conditions. The higher K^+ uptake may demonstrate the role of K^+ in salt tolerance. This is also according to the results of [64] where PGPR strains from *Azotobacter sp.* increased

the maize plant growth and potassium and phosphorus intake under different levels of salinity stress. Recently, Sang-Mo et al. [65] found that the PGPR-applied plants had reduced sodium ion concentration; while the potassium was abundantly present as compared to control under stress of *Cucumis sativus* cultivation. K^+ play a key role in plant water stress tolerance and has been found to be the cationic solute responsible for stomata movements in response to changes in bulk leaf water status [66]. There are several reports of lower Na^+ concentrations in plants inoculated with PGPR under salinity conditions [63,67,68].

Conclusion

P. fluorescens, *A. vinelandii* and *P. putida* had significant impact on maize growth, suggesting that can be applied as biofertilizers for improved maize production under salinity stress. Further greenhouse studies should provide more definitive information about the movement and uptake of macro-elements (Na^+ and K^+) to plants with the impacts of PGPR–based inoculants.

Acknowledgement

Part of this study was conducted as part of the Project No. 30-G5-1434: "Modern integrated management of halothermophilic biofertilizers for enhancing agricultural soil fertility in Jazan", which is supported and funded by the Deanship of Scientific Research, Jazan University, KSA.

References

1. Kannaiyan S, Kumar K, Govindarajan K (2004) Biofertilizers technology for rice based cropping. Jodhpur, p: 450

2. Nuttall N (2006) Soil biodiversity key to environmentally friendly agriculture. Below ground biodiversity-from worms to bacteria-key to environmentally friendly agriculture. The 8th Conference of the Parties to the Convention on Biological Diversity, Curitiba, Brazil.

3. Shaharoona B, Arshad M, Zahir ZA, Khalid A (2006) Performance of Pseudomonas spp. containing ACC-ase for improving growth and yield of maize (Zea mays L.) in the presence of nitrogenous fertilizer. Soil Biology and Biochemistry 38: 2971-2975.

4. Awad EMM, Abd El-Hameed AM, Shall ZS (2007) Effect of glycine, lysine and nitrogen fertilizer rates on growth, yield and chemical composition of potato. J Agric Sci Mansoura Univ 32: 8541-8551.

5. Faten SA, Shaheen AM, Ahmed AA, Mahmoud AR (2010) Effect of foliar application of amino acids as antioxidants on growth, yield and characteristics of Squash. Research Journal of Agriculture and Biological Science 6: 583-588.

6. Maryam S, Mohammad RH, Mohammad TD (2014) Effects of amino acids and nitrogen fixing bacteria on quantitative yield and essential oil content of basil (Ocimum basilicum). Agriculture Science Developments 3: 265-268.

7. Benizri E, Baudoin E, Guckert A (2001) Root colonization by inoculated plant growth-promoting rhizobacteria. Biocontrol Sci Techn 11: 557-574.

8. Anonymous (2000) Agricultural Statistics of Pakistan, Government of Pakistan, Ministry of Food, Agriculture and Livestock, Economic Wing, Islamabad. p: 104.

9. Subramaniyan V, Krishna S, Malliga P (2012) Analysis of biochemical and yield parameters of Zea Mays (Corn) cultivated in the field supplemented with coir pith based cyanobacterial biofertilizers.

10. Yao T, Yasmin S, Hafeez FY (2008) Potential role of rhizobacteria isolated from Northwestern China for enhancing wheat and oat yield.The Journal of Agricultural Science 146: 49-56.

11. Rong LY, yao T, zhao GQ, (2011) Filtration of siderophores production PGPR bacteria and antagonism against to pathogens. Plant Protection 1: 59-64.

12. Nadeem SM, Zahir ZA, Naveed M, Arshad M, Shahzad SM (2006) Variation in growth and ion uptake of maize due to inoculation with plant growth promoting rhizobacteria under salt stress. Soil & Environ 25: 78-84.

13. Hasnain S, Sabri AN (1996) Growth stimulation of Triticum aestivum seedlings under Cr-stress by nonrhizospheric Pseudomonas strains. In: 7th International Symposium on nitrogen fixation with non-legumes. Faisalabad, Pakistan, p: 36.

14. Ashraf M, Berge SH, Mahmood OT (2004) Inoculating wheat seedling with

exopolysaccharide-producing bacteria restricts sodium uptake and stimulates plant growth under salt stress. Biology and Fertility of Soils 40: 157-162.

15. Sylvia DM, Fuhrmann JJ, Hartel PG (1998) Zuberer Principles and applications of soil microbiology. Prentice-Hall, Inc. Upper Saddle River, NJ, 1998:550.

16. Johri BN, Sharma A, Virdi JS (2003) Rhizobacterial diversity in India and its influence on soil and plant health. Adv Biochem Eng Biotechnol 84: 49-89.

17. Roesch LFW, Camargo FAO, Bento FM, Triplett, EW (2008) Biodiversity of diazotrophic bacteria within soil, root and stem of field grown maize. Plant Soil 302: 91-104.

18. Glick BR, Karaturovi DM, Newell PC (1995) A novel procedure for rapid isolation of plant growth-promoting pseudomonads. Can J Microbiol 41: 533-536.

19. Weller DM (1988) Biological control of soil borne plant pathogens in the rhizosphere with bacteria. Ann Rev Phytopathol 26: 379-407.

20. Kloepper JW (1997) Current status and future trends in Biocontrol Research and Development in the US. -In: International symposium on clean agriculture, Sapporo, DECD, pp: 49-52.

21. Skinner PS, Banfield JB (2005) The Role of Microorganisms in Nitrogen Fixation. Journal of Symbiotic Association 14: 600-700.

22. Yooshinan SU (2001) Biological Nitrogen Fixation. Journal of Biological Association 8: 50-55.

23. Ahmed ZI, Ansar M, Tariq M, Anjum MS (2008) Effect of different Rhizobium inoculation methods on performance of lentil in pothowar region. Int J Agric Biol 10: 85–88.

24. Vessey JK (2003) Plant growth promoting rhizobacteria as biofertilizers. Plant Soil 255: 571–586.

25. Das K, Dang R, Shivananda TN, Sekeroglu N (2007) Influence biofertilizers on the biomass yield and nutrient content in Stevia rebaudiana Bert. grown in Indian subtropics. J Med Plants Res 1: 5-8.

26. Verma NK (2011) Integrated nutrient management in winter maize (Zea mays L.) sown at different dates. Afr J Plant Breed Crop Sci 3: 161-167.

27. Mishra OR, Tomar US, Sharama RA, Rajput AM (1995) Response of maize to chemicals and biofertilizers. Crop Research 9: 233–237.

28. Palleroni NJ (1984) Gram negative aerobic rods and cocci. In: Krieg NR, (Ed), Bergey's Manual of Systematic Bacteriology, Williams and Wilkins, Baltimore, pp: 140–199.

29. Pandey A, Sharma E, Palni L (1998) Influence of bacterial inoculation on maize in upland farming systems of the Sikkim Himalaya. Soil Biology and Biochemistry 3: 379–384.

30. Radwan FI (1998) Response of some maize cultivars to VA-mycorrhizal inoculation, biofertilization and soil nitrogen application. Alexandria J Agricultural Research 43: 43–56.

31. Fares CN (1997) Growth and yield of wheat plant as affected by biofertilisation with associative, symbiontic N2-fixers and endomycorrhizae in the presence of the different P-fertilizers. Annals of Agriculture Science 42: 51–60.

32. Tan GH, Nordin MS, Kert TL, Napsiah AB, Jeffrey LSH (2009) Isolation of beneficial microbes from biofertilizer products. J Trop Agric and Fd Sc 37: 103–109.

33. Hensyl WR (1994) Bergey's 'Manual of Systematic Bacteriology (9thEdn). John G Holt and Stanley, T. Williams (Eds.) Williams and Wilkins, Baltimore, Philadeiphia, Hong kong, London, Munich, Sydney, Tokyo.

34. Ryan J, Estefan G, Rashid A (2001) Soil and analysis laboratory manual. ICARDA, p: 172.

35. Kar M, Mishra D (1976) Catalase, Peroxidase, and Polyphenoloxidase Activities during Rice Leaf Senescence. Plant Physiol 57: 315-319.

36. Bradford MM (1976) A rapid and sensitive method for the quantitation of microgram quantities of protein utilizing the principle of protein-dye binding. Anal Biochem 72: 248-254.

37. Hedge JE, Hofreiter BT (1962) In: Carbohydrate Chemistry, 17 (Eds. Whistler R.L. and Be Miller, J.N.), Academic Press, New York.

38. Parida AK, Das AB (2005) Salt tolerance and salinity effects on plants: a review. Ecotoxicol Environ Saf 60: 324-349.

39. Bacilio M, Rodriguez H, Moreno M, Hernandez JP, Bashan Y (2004) Mitigation of salt stress in wheat seedlings by a gfp-tagged Azospirillum lipoferum. Boil Fertil Soil 40: 188-193.

40. Sarig S, Blum A, Okon Y (1988) Improvement of the water status andyield of field-grown grain sorghum (Sorghum bicolor) by inoculation with Azospirillum brasilense. J Agric Sci 110: 271-277.

41. Zahir ZA, Arshad M, Hussain A, Sarfraz M (1996) Improving wheat yield by inoculation with Azotobacter under optimum fertilizer application. Pakistan Journal of Agricultural Science 11: 129–131.

42. Naz I, Asghari B, Bushra R, Simab P, Mazhar I, et al. (2012) Potential of Azotobacter vinelandii Khsr1 as bioinoculant. African J Biotechnol 1: 10368-10372.

43. Abd El Ghany TM, Alawlaqi MM, Al Abboud MA (2013) Role of biofertilizers in agriculture: a brief review. Mycopath 11: 95-101.

44. Nalawde A, Bhalerao SA (2015) Effect of Biofertilizer (Azotobacter sp.) on the Growth of Blackgram Vigna mungo (L.) Hepper Int J Curr Res Biosci Plant Biol 2: 20-23.

45. Salmeron V, Martinez Toledo MV, Gonzalez Lopez Jb (1990) Nitrogen fixation and production of auxins, gibberellins and cytokinin by Azotobacter chroococcum strain isolated from root of Zea mays in presence of insoluble phosphate. Chemosphere 20: 417–422.

46. Gonzales-Lopez J, Martinez Toledo MV, Reina S, Salmeron V (1991) Root exudates of maize on production of auxins, gibberellins, cytokinins, amino acids and vitamins by zotobacter chroococcum chemically defined media and dialysed soil media. Toxicology and Environmental Chemistry 33: 69–78.

47. Gupta S, Arora DK, Srivastava AK (1995) Growth promotion of tomato plants by rhizobacteria and imposition of energy stress on Rhizoctonia solani. Soil Biol Biochem 27: 1051-1058.

48. Diouf D, Duponnois R, Tidiane BA, Neyra M, Lesueur D (2005) Symbiosis of Acacia auriculiformis and Acacia mangium with Arbuscular mycorrhizal fungi and Bradyrhizobium spp. improves salt tolerance in greenhouse conditions. Funct Plant Biol 32: 1143–1152.

49. Ravikumar SK, Alikhan KL, Prakash S, Williams G, Gracelin NAA (2007) Growth of Avicennia marina and Ceriops decandra seedlings inoculated with halophilic azotobacters. Journal of Environmental Biology 28: 601-603.

50. Omid Y, Abolfaz B, Amin N (2013) The effects of Pseudomonas fluorescence and Rhizobium meliloti co-inoculation on nodulation and mineral nutrient contents in alfalfa (Medicago sativa) under salinity stress. Intl J Agri Crop Sci 5: 1500-1507.

51. Jagnow G (1987) Inoculation of cereal crops and forage grasses with nitrogen-fixing rhizosphere bacteria: possible causes of success and failure with regard to yield response. A review - Z. Pflanzenernahr bodenkd 150: 361-368.

52. Zafar-ul-Hye M, Farooq HM, Zahir ZA, Hussain M, Hussain A (2014) application of ACC-deaminase containing rhizobacteria with fertilizer improves maize production under drought and salinity stress. Int J Agric Biol 16: 591-596.

53. Zhang F, Dashti N, Hynes RK, Smith DI (1996) Plant Growth Promoting Rhizobacteria and Soybean [Glycine max (L.) Merr.] Nodulation and Nitrogen Fixation at Suboptimal Root Zone Temperatures. Ann Bot 77: 453-460.

54. Marius S, Neculai M, Vasile S, Marius M (2013) Effects of inoculation with plant growth promoting rhizobacteria on photosynthesis, antioxidant status and yield of runner bean. Romanian Biotechnological Letters 18: 8132-8143.

55. Prathibha KS, Siddalingeshwara KG (2013) Effect of plant growth promoting Bacillus subtilis and Pseudomonas fluorescence as Rhizobacteria on seed quality of sorghum. Int J Curr Microbiol App Sci 2: 11-18.

56. Swedrzynska D, Sawicka A (2000) Effect of Inoculation with Azospirillum brasilense on Development and Yielding of Maize (Zea mays ssp. Saccharata L.) under Different Cultivation Conditions. Polish Journal of Environmental Studies 9: 505-509.

57. Parida A, Das AB, Mittra B (2004) Effects of salt on growth, ion accumulation, photosynthesis and leaf anatomy of the mangrove, Bruguiera parviflora. Trees Struct Funct 18: 167-174.

58. Mittler R (2002) Oxidative stress, antioxidants and stress tolerance. Trends Plant Sci 7: 405-410.

59. del Río LA, Corpas FJ, Sandalio LM, Palma JM, Barroso JB (2003) Plant

peroxisomes, reactive oxygen metabolism and nitric oxide. IUBMB Life 55: 71-81.

60. Shaik ZA, Vardharajula S, Minakshi G, Venkateswar RL, Bandi V (2011) Effect of inoculation with a thermotolerant plant growth promoting Pseudomonas putida strain AKMP7 on growth of wheat (*Triticum* spp.) under heat stress, Journal of Plant Interactions 6: 239-246.

61. Yachana J, Subramanian RB (2013) Paddy plants inoculated with PGPR show better growth physiology and nutrient content under saline conditions. Chilean journal of agricultural research 73: 213-219.

62. Yildirim E, Taylor AG, Spittler TD (2006) Ameliorative effects of biological treatments on growth of squash plants under salt stress. Sci Hortic 111: 1–6.

63. Glenn EP, Brown JJ, Blumwald E (1999) Salt tolerance and crop potential of halophytes. Crit Rev Plant Sci 18: 227–255.

64. Giri B, Mukerji KG (2004) Mycorrhizal inoculant alleviates salt stress in Sesbania aegyptiaca and Sesbania grandiflora under field conditions, evidence

for reduced sodium and improved magnesium uptake. Mycorrhiza 14: 307-312.

65. Rojas-Tapias D, Moreno-Galván A, Pardo-Díaz S, Obando M, Rivera D, et al. (2012) Effect of inoculation with plant growth-promoting bacteria (PGPB) on amelioration of saline stress in maize (Zea mays). Appl Soil Ecol 61: 264–272.

66. Sang-Mo K, Abdul Latif K, Muhammad W, Young-Hyun Y, Jin-Ho K, et al. (2014) Plant growth-promoting rhizobacteria reduce adverse effects of salinity and osmotic stress by regulating phytohormones and antioxidants in *Cucumis sativus*. Journal of Plant Interactions 9: 673-682.

67. Caravaca F, Figueroa D, Barea JM, Azcón-Aguilar C, Roldán A (2004) Effect of mycorrhizal inoculation on the nutrient content, gas exchange and nitrate reductase activity of *Retama sphaerocarpa* and *Olea europaea* subsp. sylvestris under drought stress. Journal of Plant Nutrients 27: 57-74.

68. Sharifi M, Ghorbanli M, Ebrahimzadeh H (2007) Improved growth of salinity-stressed soybean after inoculation with salt pre-treated mycorrhizal fungi. J Plant Physiol 164: 1144-1151.

Occurrence and Importance of *Xanthomonas axonopodis* pv. *Phaseoli* in Common Bean (*Phaseolus vulgaris* L) Seed Produced under Different Seed Production System in Central Rift Valley of Ethiopia

Leta A[1]*, Lamessa F[2] and Ayana G[3]

[1]*Arsi University College of Agriculture and Environmental Sciences, Ethiopia*
[2]*Jimma University College of Agriculture and Veterinary Medicine, Ethiopia*
[3]*Ethiopian Institute of Agricultural Research, Malkassa Agricultural Research Center, Ethiopia*

Abstract

Common bacterial blight of bean caused by the seed-borne bacteria *Xanthomonas axonopodis* pv. phaseoli (*Xap*) (Smith) Vauterin and *X. axonopodis* pv. phaseoli var. *fuscans* (Burkholder) Starr and Burkholder is one of the most constraint of common bean production all over the world. The pathogen is seed-borne and survives as long as the seed remains viable. Use of pathogen-free seeds has been the main method used to control the disease in most bean production areas, and detection of this pathogen in seeds is essential for effective disease control. This study was carried out to detect and characterize *Xap* in seed lots collected from different seed dealers and local markets in Central Rift Valley of Ethiopia. A semi-selective medium *Xanthomonas campestris* pv. phaseoli (XCP1) and yeast extract-dextrose-calcium carbonate agar (YDCA) were used to recover the bacterium from whole bean seed extract and direct seed plating respectively. The pathogenicity test was done on Mexcan-142 bean cultivar to confirm pathogen identification. Colonies of the bacterium were yellow, mucoid and convex on XCP1 media and zone of hydrolysis formed around them. Further biochemical test results also confirm that the colonies were gram negative, rod shape and hydrolyze starch, casein and Tween80. The results confirmed the presence of seed borne *Xap* in all seed dealers and local market seed lots in the study area. The result reveled *Xap* was prevalent in 79.27% of the total seed samples collected. Lower prevalence (21.43%), seed infection percentage (1.643%) and bacterial population were resulted in seed lots from Melkassa Agricultural Research Center seed lots; while the higher prevalence, seed infection percentage and bacterial population were observed in cooperative union, local market, and seed producer's cooperative seed lots. From the result, it can be concluded that *Xap* was potentially recovered from naturally infected seeds using XCP1 media and the pathogen is highly distributed in seed lots in the study area with high prevalence in farmer's produced seed lots. Therefore, seed plating on semi selective medium XCP1 can be used as standard method for routine analysis of *Xap* from bean seeds. Seed dealers in the study area should follow strict disease free seed production programs and farmers in the study area should be encouraged not to use local market and/or their own saved seeds for planting purpose.

Keywords: Common bacterial blight; Seed detection; *Xanthomonas axonopodis pv. Phaseoli*

Introduction

Common bacterial blight of common bean (*Phaseolus vulgaris* L) caused by the seed-borne bacteria *X. axonopodis* pv. *phaseoli* (*Xap*) (Smith) Vauterin and *X. axonopodis* pv. *phaseoli* var. *fuscans* (Burkholder) Starr and Burkholder is one of the most constraint of common bean production all over the world [1,2]. The disease causes both quantitative and qualitative yield losses and the yield loss reach up to 40%, depending on bean cultivar susceptibility and environmental conditions [3]. The pathogen distributed in most regions where common bean is cultivated except in arid tropical areas. It is a major disease in African countries such as Malawi [4], Uganda, Kenya, Burundi [5] and Tanzania [6]. It is also present in other south-eastern and southern Africa countries [7]. In Ethiopia, it is ranked among the most important diseases of common bean [8,9], and predominantly severe in areas characterized by high temperature, relative humidity and amount and intensity of rain fall [10].

Common bacterial blight disease is a seed-borne [11], and the pathogen survives as long as the seed remains viable [12]. Seed transmission is the primary means by which the pathogen is disseminated [13-15]. Internally and externally infested seeds are important sources of primary inocula for *Xap* [14,16]. Sutton and Wallen [17] reported that approximately one diseased seed in 10000 seeds are capable of causing an outbreak of blight. Weller and Saettler [15] also report that 1000 to 10000 bacteria per seed is the minimum needed to produce infected plants under field conditions. Therefore, the

use of pathogen-free seeds has been the main method used to control the disease in most bean production areas [18-22] and detection of this pathogen in seeds is essential for effective disease control [23]. To limit this major inoculum source, specific seed production areas and seed certification were created in several countries [24]. These seed production areas should be located in areas where climate is considered to be non-conducive to diseases and/or where seed producers follow strict rules concerning the sanitary quality of stock seeds and cultural conditions like long rotations and isolated location of fields to limit the introduction and multiplication of inoculum.

However, in Ethiopia, most farmers retain bean seed for future planting and certified seed is seldom used even in the case of the new cultivars for which seed production has been organized with research centers, commercial seed production enterprise or farmer's seed production cooperatives. In most cases, certified seed is typically used in

---

*Corresponding author: Ararsa Leta, Arsi University College of Agriculture and Environmental Sciences, P.O. Box 193 Asella, Ethiopia E-mail: ararsaleta@gmail.com

Occurrence and Importance of Xanthomonas axonopodis pv. Phaseoli in Common Bean...

123

the first production year only. Then after, most farmers plant uncertified seed mainly saved from their own previous harvest, purchased from local markets or commercial seed dealers like farmers cooperative unions. Moreover, seeds from these commercial seed dealers were also uncertified for phytosanitary except cleaned by removing discolored and shriveled seeds during seed grading. However, symptomless, and slightly diseased seeds obtained from infected fields may rise to severely infected seedling. Therefore, there is a strong likelihood that such seed may act as sources of primary inoculum for seed borne diseases like common bacterial blight. This is particularly so for resource poor farmers who do not have access to certified seeds.

In line with the above-mentioned problems there was no strong work done and the pathogen status of bean seeds from different seed sources in the study area were not known. Therefore, this study was carried out to detect and characterize common bacterial blight pathogen from common bean seeds lots collected from different sources with specific objectives to evaluate their level of contamination with *Xap*.

Materials and Methods

Study area

This study was conducted in Central Rift Valley of Ethiopia during 2014 cropping season (Figure 1). The selection of Disticts/Woredas and Kebeles/Peasnt Association were made based on the potential of bean production and the presence of farmer's seed producer's cooperative with the intention to collect seed samples from different seed sources.

Seed sample collection

Common bean seeds were collected from various seed sources; which include seed producer's cooperative seed lots, farmer's cooperative unions, experimental site, and local markets. About 1 kg of seed samples were sampled from different parts of seed storage according to international seed health test (ISHT) sampling procedure. In seed producer's cooperatives and farmers cooperative unions seed lots, based on size of the seed lot one primary sample for each 500 kg were taken from each producer's seed lots. Markets in each locality where farmer's cooperative seed producers found were visited and about 1 kg of sample was collected randomly from 5 individuals in each locality. In case of experimental site about 1 kg of seeds from each cultivar was sampled. A total of 82 seed samples were collected from different seed sources in the study area. Each sample was collected separately in plastic bag and transported to Melkassa agricultural research center plant pathology laboratory and stored at +5°C in refrigerator till analysis. These samples were grouped in to four seed source as markets seeds, seed

producers' cooperatives seeds, farmers' cooperative unions seed and research trial seeds.

Seed assay

Seed soak method: A working sample of 500 bean seeds was drawn from each sample and used for assay in line with International Seed Testing Association (ISTA) standards [25]. Seeds were soaked in seed extraction solution (0.85% saline with Tween20) in the proportion of 1:2 (1 g of seed in 2 ml of solution) and kept overnight at 5°C. After incubation, each suspension was thoroughly agitated and 10-fold dilution series (to 10^{-5}) of the seed extract was prepared. Afterwards, 0.1 ml of each undiluted and diluted extracts were spread on three plates of semi-selective media, *Xanthomonas campestris* pv. *phaseoli* (XCP1) as describe by International Seed Testing Association [26]. The plates were incubated at 28°C for five days. Then, the plates were visually assessed and all colonies typical of *Xanthomonas* genus (yellow pigment, convex, mucoid colony with inter margin) were examined and counted for each sample to determine number of colony forming units (cfu) per ml of seed extract. Suspected colonies of the pathogen were purified by sub culturing single colony on yeast extract-dextrose-calcium carbonate agar (YDCA). One colony of the purified suspect pathogen from each sample was selected and maintained on NA and YDCA slant at 4°C for further test.

Direct plating method : Similar to the seed soaking method, 500 randomly selected seeds from each seed lot were used for this assay. Sub samples of five of 100 seeds used as a replicate and 10 seeds plated per plate. Seeds were first sterilized by dipping in a 1% sodium hypochlorite solution for 30 seconds and rinsed in three changes of sterile distilled water (SDW) for three minutes to remove traces of sodium hypochlorite. The seeds were then placed on sterile filter paper to dry. Then seed were plated hilum downwards, on YDCA plate [27] and incubated at 28°C ± 2°C. After five days, the plated seeds were visually assessed for the presence of *X. axonopodis* pv. *phaseoli (Xap)* colonies under a stereomicroscope based on their morphological characteristics typical to *Xap*. The seeds from which *Xap* recovered were recorded as positive (+) and those from which *Xap* not recovered was negative (-). Mean percent seed infection level was calculated for each replicate from infection proportion per plate.

Pathogen characterization: Suspected colonies obtained from seed assay were subjected to a number of tests which included the gram reaction, casein hydrolysis, tween80 hydrolysis, starch hydrolysis and pathogenicity tests on bean plant.

Gram reaction: Grease on the slide was removed by flaming it several times. A small drop of water was placed on the middle of the slide. A small amount of the yellow pigment colonies was removed from the culture using a sterile wire loop and placed in the water drop. After mixing the cells with the water on the slide, the smear was dried by holding the slide over the flame. After cooling crystal violet was pipetted onto the surface and left for one minute. The stain was poured off the slide, washed with 70% alcohol, and iodine solution added for one minute before being rinsed off with water. Then safranin was applied and left for three minutes and then rinsed off with water before drying over the flame. The mounted specimen was examined under a compound microscope with 100x lens using oil immersion with no cover slip.

Casein hydrolysis test: Casein hydrolysis was demonstrated by streaking yellow-pigmented colonies on Skim milk agar medium. A single line streak inoculation was made from each isolate culture and plates were incubated at 28°C. After 48 h all incubated plate were observed for any clearing around the line of growth.

Figure 1: Central rift valley of Ethiopia.

Starch hydrolysis test: The ability to degrade starch was performed by culturing the suspected isolate on starch agar media. A single streak inoculation of each isolate was made into the center of starch agar plates. The cultured plate was incubated at 28°C for 48 h in inverted position. After 48 h the surface of the plates were flooded with iodine solution and examined. A clear or yellow zone around a colony in otherwise blue media indicate a positive starch hydrolysis reaction.

Tween80 hydrolysis test: In Tween80 hydrolysis demonstration, a suspected colony of *Xap* isolate was streaked into the center of XCP1 plate. Plates were incubated at 28°C for 3-5 days. A milky zone around a colony growth indicates positive Tween80 hydrolysis.

Pathogenicity tests: For pathogenicity test four seeds of a susceptible bean variety; Mexcan-142 were planted in 20 cm diameter pots and after emergence two plants per pot maintained for inoculation. Selected isolates obtained from each sample in the seed assay experiments were cultured on nutrient agar NA, and then transferred onto nutrient broth NB and incubated on a shaker for 24 h at 25°C. Cells were suspended in distilled water and approximately adjusted to 10^8 CFU ml^{-1}. Plants were sprayed with water before inoculation to provide favorable conditions for infection. In addition, the floor of the greenhouse was covered with fiber sucks and kept wet to generate humidity in order to favor development of CBB. For inoculation, scissors contaminated with the bacterial suspension were used to cut the leaflets. Two leaflets of each plant were inoculated (always the middle leaflet). The plants were assessed for blight symptoms from seven days after inoculation.

Data analysis: Analysis of variance for colony populations/colony forming unit (cfu) and seed infection percentage were analyzed with SAS 9.2 computer software GLM procedure of nested design and mean separation test was performed by Duncan multiple range test.

Results

The seed assay Seed soak method: The seed extract plating result reveals that *Xap* was recovered from 65 samples of 82 total samples analyzed. This indicates that the pathogen is prevalent in 79.27% of the seed sample collected (Table 1). Low prevalence of Xap (21.43%) was recorded in seed samples collected from MARC trial sites seed lots while the highest prevalence was observed in seed samples collected from farmers' cooperative union seed lots. The result of *Xap* population confirmed that *Xap* colony populations were high for all positive samples ranging between 1.37×10^5 to 7.89×10^6 cfu/ml of seed extract. *Xap* colony population of MARC seed lots was significantly lower than the other seed sources (Table 2). There was also a significant difference in *Xap* population within markets and cooperative unions seed lots whereas there was no significant difference in *Xap* population within seed producer cooperatives and trial sites seed lots (Figure 2). Within the markets seed lots the lowest *Xap* population recovered from seed lots collected from Bofa local market, whereas the mean *Xap* populations of the other markets were statistically comparable.

Direct seed plating method: The result from direct seed plating

Seed source	No of sample collected	No of *Xap* positive sample	% *Xap* prevalence
Markets	35	30	85.71
Seed producers' cooperatives	24	23	95.83
Farmers' cooperative unions	9	9	100.00
MARC trials	14	3	21.43
Total	82	65	79.27

Table 1: Occurrence of *Xap* in bean seed samples from different seed sources.

Seed source	CFU		
Local markets	5.2375[a]		
Seed producer cooperatives	5.7378[a]		
Farmers' cooperative unions	6.5357[a]		
MARC trial sites	1.2375[b]		
CR	1.989	2.080	2.135
CV	1.0225		

Means with the same letter are not significantly different; The figures in the table are log transformed mean bacterial population

Table 2: Mean *Xap* population in bean seed sample from different seed sources.

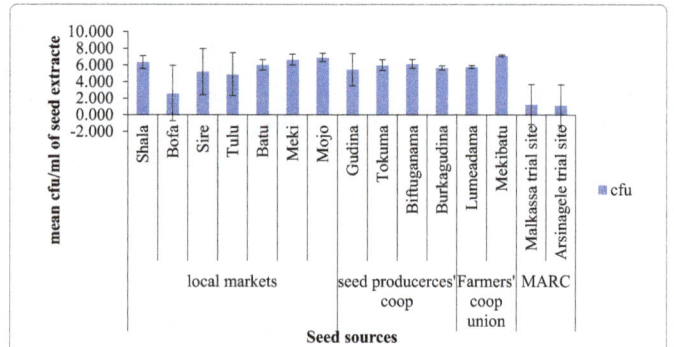

Figure 2: Mean log *Xap* population in seed lots within different seed sources.

Seed source	Infection%		
Local markets	7.006[a]		
Seed producer cooperatives	5.575[a]		
Farmers' cooperative union	9.156[a]		
MARC	1.643[b]		
CR	3.724	3.895	3.997
CV	18.27		

Means with the same letter were not significantly different

Table 3: *Xap* infection level of bean seed in different seed sources.

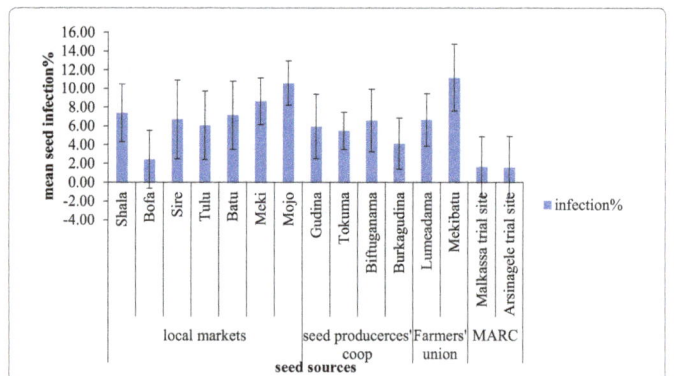

Figure 3: Mean bean seed infection in seed lots within seed sources.

assay reveals that *X. axonopodis* pv. *phaseoli (Xap)* was recovered from all sample positive in seed extract plating. The mean percentage of seed infection by Xap in seed samples significantly varied between seed sources (Table 3). The lowest seed infection percentage (1.643%) was recorded in MARC trial sites seed lots whereas the other seed sources had comparable seed infection level ranging from 5.575% to 9.156%. Within the seed sources, only markets seed lots were show significant difference in seed infection percentage. Low seed infection (2.44%) was observed in seed lots collected from Bofa local market, while the others market seed lots had comparable seed infection ranging from 6.08% to 10.60% (Figure 3).

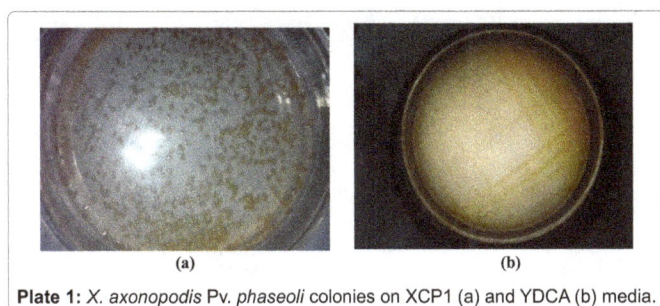

Plate 1: *X. axonopodis* Pv. *phaseoli* colonies on XCP1 (a) and YDCA (b) media.

Characteristics	*Xap* reaction
Gram reaction	-
Casein hydrolysis	+
Tween80 hydrolysis	+
Starch hydrolysis	+
Pathogenicity on bean	+
Key (+): *Positive* identification (-): Negative identification	

Table 4: Biochemical characteristics of *Xap*.

Pathogen characterization: The bacterial isolates recovered from seed samples were classified as Xanthomonas like, based on yellow pigment, convex mucoid morphology (Plate 1a and 1b).

The colonies of *Xap* on XCP1 medium appeared as bright yellow, mucoid, convex, smooth, round with entire margins and surrounded by zone of starch and tween80 hydrolysis. The biochemical test results also confirm that the colonies were gram negative, rod shape and hydrolyze starch, cuisine and Tween80. The observed characteristics of *Xap* were summarized in Table 4. Other unidentified bacteria with creamy white, flat circular colony were also recovered from the seed at the same time.

In the pathogenicity test on bean plant, all suspected isolate recovered from seed assay induced symptoms ten days after inoculation. The symptom was first appeared around growing tips as small water-soaked spots on the underside leaves. The lesions gradually enlarge and join to develop into large irregular shape lesions. Then the lesions become dry brown and surrounded by a narrow yellow border.

Discussion

Seed assays are the important steps in the seed system and the most reliable methods of determining whether or not seeds are infected with seed borne pathogens [28]. Although several techniques like serology [29,30] and polymerase chain reaction [31,32] have been developed for detection of CBB pathogen in seed, isolation on semis elective media remains the most widely used detection method [33-35]. In the present study, the morphological characteristics of the bacterium on the XCP1 medium, biochemical and pathogenicity test results confirmed that the pathogen recovered from seed was *X. axonopodis* pv. *phaseoli* (*Xap*) and this result was in agreement with the finding of [33]. In this study the bacterium, *Xap* was successfully isolated by plating the extract obtained from the whole seeds onto the semi-selective medium XCP1. This indicated the suitability of XCP1 for detection of *Xap* from seed lots. Remeeus and Sheppard [33] report that ISTA/ISHI evaluate several semi-selective media for use in the detection of *Xap* in bean seed including BBD, MT, YSSM, MXP and PTSA media and find that no significant difference was observed in the detection of *Xap* between media. However, characteristics of some media proved more suitable for routine testing of bean seed than others and XCP1 was consistently found to have good selection for both fuscans and non-fuscans types and was preferred by participants for ease of detection.

The current study also revealed that *Xap* was prevalent in all kinds of seed system in the study area including seed from research center seed multiplication site. The average prevalence was 79.27% and it varies from seed source to seed source and within the seed sources. This variation was probably because different seed source follow different seed production management including cultural practices during seed production and handling. This result was in agreement with the finding of Kedir et al. [36] where they reported that *Xap* is associated with 59.1% and 75.0% of bean seed sample produced respectively in intercropping and sole cropping system in Eastern Ethiopia.

Results of seed infection level and bacterial population of the seed lots in this study also showed that there was heavy infection of *Xap* in bean seed lots in the study area. The lowest seed infection level (1.643%) was found in seed lots from Melkassa Agricultural Research Center (MARC) while the other seed sources had higher seed infection level ranging from 5.575% to 9.156% with heavy bacterial population. Karavina et al. [37] report seed borne *X. phaseoli* is common in both retained and certified common bean seed lots in Zimbabwe with the former having significantly higher bacterial population levels. In Canada, 0.5% of seed infection level has led to disease epidemics [38]. Weller and Saettler [15] reported that the minimum population of *Xap* to initiate common bacterial blight infection is 10^3 to 10^4 which, shows bean seed from all seed sources in current study was heavily infected by *Xap* and capable to cause CBB epidemic when there is favorable environmental conditions. Moreover, the result shows that the seed samples from farmer's seed lots (seed collected from local market, seed producer cooperatives and cooperative unions) had higher levels of pathogen infection compared to seed from trial site. This might be because of seed production management system the seed dealers' follow including site selection, field sanitation and other disease management practices and seed grading problems. During the study, it was observed that all seed dealers in the study area have less seed grading practices and they only remove sherveled and seeds with visible disease symptom. However, contamination of seeds without symptom expression during the growing season represents a risk for eventual disease outbreaks [39].

Conclusion

The experiments confirmed the presence of *Xap* as a seed borne pathogen in all different seed sources in the study area. This wide spread distribution of *Xap* in all seed production system and the fact that CBB pathogen build up over time which may result in high risk of disease outbreak indicate that, strict seed phytosanitary certification and disease free seed production program as a primary *Xap* management in the study are in particular and as a country in general should be given apriority.

Farmer's seeds (seed collected from local market, seed producer cooperatives and cooperative unions) had higher levels of pathogen infection compared to seed from trial site; therefore, seed dealers should have to get expertise/extension services and farmers in the area should have to be encouraged not to use uncertified seeds for planting purpose. Moreover, as seed inoculum is the primary source of infection seed dealers in the study area should have to follow exclusion disease management strategies like producing seeds in off season and in dry cooler areas so that they able to supply disease free seeds. A specific standard method for detection of *Xap* from bean seeds has not reported in Ethiopia so far, and the result shows that the method used in this study was effective and suitable in isolating this bacterium from naturally infected bean seeds. Therefore, seed plating on semi selective medium XCP1 can be used as standard method for routine analysis of *Xap* from bean seeds.

Acknowledgement

First of all, I would like to give glorious thanks to Almighty God for his favor in all ups and downs I face during the research works. I am greatly thankful to Mr. Amare Fufa, Endrias G/Kirstos, Nakechew Alemu, and other staff members of Malkassa Agricultural Research Center crop protection case team for their undeserved cooperation and support. My thanks also go to Mr. Habtamu Haile and Ambo University applied biology department lab technician for their many help during lab work. My thanks also extended to Mr. Debale Abera for sketching the study site.

References

1. Mabagala RB, Saettler AW (1992) An improved semi-selective media for the recovery of Xanthomonas *campestris* pv. phaseoli. Plant Dis 76: 443-446.

2. Opio AF, Allen DJ, Teri JM (1996) Pathogenic variation in *Xanthomonas campestris* pv. phaseoli, the causal agent of common bacterial blight in Phaseolus Beans. Plant Pathol 45: 1126-1133.

3. Mutlu N, Miklas PN, Steadman JR, Vidaver AV, Lindgren D, et al. (2005) Registration of pinto bean germplasm line ABCP-8 with resistance to common bacterial blight. Crop Sci 45: 806-807.

4. Edje OT, Mughogho LK, Rao VP, Msuku WAB (1981) Bean production in Malawi. In: Amaya S, Motta FM, (Eds). Regional Workshop on Potential of Field Beans in Eastern Africa, Proceedings, Cali, Colombia, Centro Internacional de Agricultura Tropical, East Africa.

5. Opio AF, Teri JM, Allen DJ (1993) Studies on seed transmission of *Xanthomonas campestris* pv. phaseoli in common beans in Uganda. Afr Crop Sci J 1: 59-67.

6. Karel AK, Ndunguru BJ, Price M, Semuguruka SH, Singh BB (1981) Bean production in Tanzania. In: Amaya S, Motta FM (Eds). Regional Workshop on Potential for Field Beans in Eastern Africa, Proceedings, Cali, Colombia, Centro Internacional de Agricultura Tropical, East Africa.

7. EPPO (European and Mediterranean Plant Protection Organization) (1997) Data Sheets on Quarantine Pests: *Xanthomonas axonopodis* pv. phaseoli. Prepared by CABI and EPPO for U.

8. Habtu A, Sache I, Zadoks JC (1996) A survey of cropping practices and foliar diseases of common bean in Ethiopia. Crop Prot 15: 179-186.

9. Fininsa C (2003) Relationship between common bacterial blight severity and common bean yield loss in pure stand and maize common bean intercropping. Int J Pest Manag 49: 177-185.

10. Habtu A, Bosch F, Zadoks J C (1995) Focus expansion of common bean rust in cultivar mixture. Plant Pathol 44: 503-509.

11. Schaad NW (1982) Detection of seed borne bacterial plant pathogens. Plant Dis 66: 885-890.

12. Hirano SS, Upper CD (1983) Ecology and epidemiology of bacterial pathogens. Ann Revi Phytopathol 21: 243-269.

13. Cafati CR, Saettler AW (1980) Transmission of *Xanthomonas phaseoli* in seeds of resistant and susceptible Phaseolus genotypes. Phytopathol 70: 638-640.

14. Weller DM, Saettler AW (1980a) Colonization and distribution of *Xanthomonas phaseoli* in field-grown navy beans. Phytopathol 70: 500-506.

15. Weller DM, Seattler AW (1980b) Evaluation of seed borne *Xanthomonas phaseoli* and *Xanthomonas phaseoli* var fuscans as primary inoculum in bean blights. Phytopathol 70: 148-152.

16. Hall R (1994) Compendium of bean diseases. The American Phytopathological Society, New York, USA.

17. Sutton MD, Wallen VR (1970) Epidemiological and ecological relations of *Xanthomonas phaseoli* and *Xanthomonas phaseoli* var. fuscans on beans in south western Ontario, 1961-1968. Can J Bot 48: 1329-1334.

18. Yoshii K (1980) Common and fuscous blights. Pages 155-172 in: bean Production Problems. Schwartz, H.F. and Galvez GE (Eds.) Center of International Agriculture Tropic (CIAT). California. Colombia, 424 pp.

19. Saettler AW, Cafati R, Weller DM (1986) Non-over wintering of *Xanthomonas bean* blight bacteria in Michigan. Plant Dis 70: 285-287.

20. Abo-Elyousr KAM (2006) Induction of systemic acquired resistance against common blight of bean (Phaseolus vulgaris) caused by Xanthomonas campestris pv. phaseoli. Egypt J Phytopathol 34: 41-50.

21. Berova M, Stoeva N, Zlatev Z, Stoilova T, Chavdarov P (2007) Physiological change in bean (*Phaseolus vulgaris* L.) leaves infected by the most important bean disease. J Cent Eur Agric 8: 57-62.

22. Darrasse A, Bureau C, Samson R, Morris C, Jacques MA (2007) Contamination of bean seeds by *Xanthomonas axonopodis* pv. phaseoli associated with low bacterial densities in the phyllosphere under field and greenhouse conditions. Eur J Plant Pathol 119: 203-215.

23. Lahman LK, Schaad NW (1985) Evaluation of the "Dome Test" as a reliable assey for seedborne bacterial blight pathogens of bean. Plant Dis 69: 680-683.

24. Webster DM, Temple SR, Galvez GE (1983) Expression of resistance to *Xanthomonas campestris* pv. phaseoli in *Phaseolus vulgaris* under tropical conditions. Plant Dis 67: 394-396.

25. Draper SR (1995) International Rules for Seed Testing. Seed Science and Technology 3(ii) 331.

26. Grimault V, Olivier V, Rolland M, Darrasse A, Jacques MA (2014) Seed health testing methods. 7-021: Detection of Xanthomonas axonopodis pv. phaseoli on Phaseolus vulgaris. In: ISTA International rules for seed testing. Annexe to Chapter 7: Seed health methods 7-021. International Seed Testing Association, Bassersdorf, Switzerland, 1-20.

27. Schaad NW (1988) Laboratory Guide for Identification of plant pathogenic bacteria. (2ndedn), The American Phytopathological Society, Minnesota, USA.

28. Rideout MS, Roberts SJ (1997) Improving quality control procedures for seed borne pathogens by testing sub-samples of seeds. Seed Sci Technol 25: 195-202.

29. Malin EM, Roth DA, Belden EL (1983) Indirect immunofluorescent staining for detection and identification of *Xanthomonas campestris* pv. phaseoli in naturally infected bean seed. Plant Dis 67: 645-647.

30. Sheppard JW, Roth DA, Saettler AW (1989) Detection of *Xanthomonas campestris* pv. phaseoli in Bean. In: Saettler AW, Schaad NW, Roth DA (Eds.) Detection of bacteria in seed and other planting material. The American Phytopathological Society, Minnesota, USA. 17-29 p.

31. Audy P, Braat CE, Saindon G, Huang HC, Laroche A (1996) A rapid and sensitive PCR-based assay for concurrent detection of bacteria causing common and halo blights in bean seed. Phytopathol 86: 361-366.

32. Molouba F, Guimier C, Berthier C, Guenard M, Olivier V, et al. (2001) Detection of bean seed-borne pathogens by PCR. Acta Horticult 546: 603-607.

33. Schaad NW (1982) Detection of seedborne bacterial plant pathogens. Plant Dis 66: 885-890.

34. Sheppard JW, Kurowski C, Remeeus PM (2007) International Rules for Seed Testing, 7-021: Detection of *Xanthomonas axonopodis* pv. phaseoli and *Xanthomonas axonopodis* pv. phaseoli var. fuscans on Phaseolus vulgaris. Bassersdorf, Switzerland: International Seed Testing Association (ISTA).

35. Balaz J, Popovi T, Vasi M, Nikoli Z (2008) Elaboration method for the detection of Xanthomonas axonopodis pv. phaseoli in bean seeds. Pestic Fitomed 23: 89-98.

36. Kedir O, Setegn G, Kindie T (2014) Assessment of common bean (*Phaseolus vulgaris* l.) seed quality produced under different cropping systems by Smallholder farmers in eastern Ethiopia. Afr J Food Agric Nutr Dev 14: 8566.

37. Karavina C, Chihiya J, Tigere TA (2008) Detection and characterization of *Xanthomonas phaseoli* (e. f. sm) in common bean (*Phaseolus vulgaris* L) seeds collected in Zimbabwe. J Sustain Dev 10: 105-119.

38. Zaumeyer WJ, Thomas HE (1957) A monographic study of bean diseases and methods for their control. US Department of Agriculture Technical Bulletin, 868, 255 pp.

39. Grum M, Camloh M, Rudolph K, Ravnikar M (1998) Elimination of bean seed-borne bacteria by thermotherapy and meristem culture. Plant Cell Tissue Organ Cult 52: 79-82.

21

Shikimic and Salicylic Acids Induced Resistance in Faba Bean Plants against Chocolate Spot Disease

Heshmat Aldesuquy*, Zakaria Baka and Nahla Alazab

Botany Department, Faculty of Science, Mansoura University, Egypt

Abstract

Surveys for faba bean chocolate spot disease covering 6 districts in Delta of Egypt were conducted. Out of these surveys, six isolates of the pathogen were obtained and purified using single spore technique. These isolates were identified as *Botrytis fabae*. All isolates were subjected to pathogenicity tests to determine the most aggressive one. All isolates appear to have the potency to cause chocolate spot disease, but the isolate from Sherbin was considered to be the most aggressive one and was selected for further studies. *In vitro*, the effect of provided phenolic compounds on inhibition of mycelia growth and the growth rate of *B. fabae* was investigated and arranged as follows: salicylic acid > Shikimic acid > shikimic acid+salicylic acid, as compared to control values. Furthermore, significant reduction in the disease incidence (%) and severity (%) were recorded in faba bean plants treated with salicylic acid followed by shikimic acid then their interaction.

Botrytis fabae infection caused noticeable increase in the activity of defense enzymes (i.e., peroxidase, polyphenol oxidase and phenyl alanine ammonia lyase) in infected faba bean plants. In the majority of cases, the applied phenolic compounds induced additional increase in such enzymes than that sprayed with fungicide. This increment was concomitant with the increase in the endogenous total phenols, shikimic acid and salicylic acid content. In addition, the most effective treatment in enhancement faba bean resistance against *B fabae* infection was salicylic acid at 0.7 mM.

Keywords: Faba bean; Chocolate spot disease; Oxidative enzymes; Shikimic acid; Salicylic acid

Introduction

The faba bean (*Vicia faba* L.) is considered as one of the most profitable field crops in Egypt. The plant is grown mainly for its green pods and dried seeds, which are rich in a protein (18.5 to 37.8%) that can substitute for animal protein in humans, as well as other compounds [1]. However, the faba bean is susceptible to attack by many diseases, such as chocolate spot disease caused mainly by *Botrytis fabae* and diseases are responsible for considerable losses in seed yield [2]. Peroxidases oxidize phenols to quinines, which are toxic to pathogens. Peroxidases participate in a broad range of physiological processes, such as the formation of lignin and suberin, the cross-linking of cell wall components, and phytoalexin synthesis. Peroxidase also functions in the metabolism of Reactive Oxygen Species (ROS) and Reactive Nitrogen Species (RNS), thus activating the Hypersensitive Response (HR), a form of programmed cell death at the infection site that is associated with limiting pathogen development [3]. Polyphenol oxidases participate in the oxidation of aromatic substrates and dihydroxyphenolic compounds in the presence of oxygen in host tissues, producing quinones that are toxic to pathogens [4].

The effect of *B. fabae* infection in the two faba bean varieties caused biochemical changes in the content of phenolic, oxidative enzymes, phytoalexins and free amino acids. HPLC analysis showed that the resistant cultivar Giza 461 demonstrated significantly higher amounts of oxidative enzymes, free, conjugated and total phenols and phytoalexins compared with the highly susceptible cultivar Giza 429. The results of this study may improve our understanding of the biochemical basis of resistance to *B. fabae* in the faba bean [5]. The beneficial effects derived from phenolic compounds have been attributed to their antioxidant activity [6]. Phenolic compounds could be a major determinant of antioxidant potentials of foods [7] and could therefore be a natural source of antioxidants. The antioxidant activity of phenolic compounds is due to their ability to scavenge free radicals, donate hydrogen atoms or electron, or chelate metal cations [8].

The aim of this study was to evaluate shikimic and salicylic acids as chemical elicitors compared with Ridomil MZ as a fungicide on disease incidence (%), severity (%) and oxidative enzyme activity as well as phenolic compounds in response to infection of *Vicia faba* by *B. fabae*.

Materials and Methods

The diseased leaves of faba bean plants were collected from different sites covering 6 districts (Sherbin, Meet ghamer, Talkha, Mansoura, Dekernis and Bani Ebaid) belong to Dakahlia Governorate, Nile Delta Egypt. Infected leaves were thoroughly washed in running tap water followed by sterile water. The leaves were cut into small pieces (1-2 cm^2) using sterilized scalpel, and immersed in 0.01% $HgCl_2$ for 1-3 min. for surface sterilization. Surface sterilized pieces were then rinsed in sterilized water several times to remove the remaining disinfectant solution and dried on sterilized filter papers. Using sterilized forceps, the dried pieces were then transferred into Petri-dishes containing Potato-Dextrose Agar (PDA) medium [200 g potato; 20 g dextrose and 15 g Difco purified agar, made up to 1 L with distilled water, pH was adjusted to 6.1] supplemented with broad-spectrum antibacterial agent

*Corresponding author: Heshmat Aldesuquy, Botany Department, Faculty of Science, Mansoura University, Egypt
E-mail: heshmat-aldesuquy@hotmail.com

(streptomycin sulphate, 200 ppm). The dishes were then incubated at 20°C+2 for 7-10 days and checked for microbial growth two days after inoculation. Purification of the resulting isolates was done by using single spore technique to obtain them in pure cultures [9]. Pure cultures of the isolated fungus were identified according to the cultural characteristics, morphological and microscopic features (mycelia development and spore formation) as described by Ellis [10]. Stock cultures were stored on PDA medium under sterile mineral oil at 4°C.

Pathogenicity test

Pathogenicity test was carried out on healthy faba bean plants to determine the pathogenic potentialities of *Botrytis fabae* isolates, which were isolated from different sites. The most aggressive isolate was used throughout this investigation. Artficial inoculation with conidial concentration of 40×10^4 spore's ml^{-1} was used to spray plants.

Inoculum preparation

The isolated purified fungus was grown on faba bean-sucrose-sodium chloride- agar (FSSA) medium [100 ml faba bean extract, 20 g sucrose, 20 g sodium chloride and 20 g agar dissolved in one liter of distilled water] as proposed by Leach and Moore [11]. The medium was autoclaved and poured into sterilized petri-dishes. Plates were inoculated with 5 days old discs, 6 mm in diameter, of the isolated fungus grown on PDA medium. Plates were then incubated at 20°C+2 for 12 days [12] under light regime of UV and normal fluorescent (12 hr/12 hr) to enhance sporulation. Conidia were harvested using a camel-hair brush and 10 ml of sterilized water for each plate. Conidial concentration was estimated using a haemocytometer and adjusted to about 40×10^4 spore's ml^{-1}.

Artificial inoculation

Seeds of faba bean susceptible cultivar (Giza 429) were sown in plastic pots (35 cm in diameter) containing 12-13 kg clay: sand (2:1, v/v) soil. Five seeds were sown in each pot; plants were left to grow and irrigated with normal tap water when necessary. After 45 days, the intact plants were sprayed with the spore suspension of the pathogen and covered with polyethylene bags to maintain enough humidity around the plants. Control plants were sprayed with sterilized water. All pots were kept in a glass house under natural conditions. Disease assessment was recorded after 10 days of microbial spraying.

Disease assessment

The disease severity (DS) was recorded according to the disease index which based on the standard scale of Gondran [13] as follows:

Disease severity (%) was calculated using the equation adopted by Hanounik [14]:

$$\% \text{ disease severity} = \frac{(n \times v)}{gN} \times 100$$

Where: n=number of plants in every sequence (v)

N=total number of examined plants.

g=maximum disease grade.

Then efficacy percentage (E%) of each treatment in reducing disease, severity percentage of faba bean was assessed according to the equation adapted by Rewal and Jhooty [15] as follow:

$$E\% = \frac{\% \text{ disease severity in control} - \% \text{ disease severity in treatment}}{\% \text{ Disease severity in control}}$$

Effect of phenolic compounds on the linear growth of *Botrytis fabae*

Salicylic acid, shikimic acid or their interaction were used to study their effects on the linear growth of *B. fabae* on agar plates. These phenolic compounds were incorporated in PDA medium at concentration of 0.4 mM for shikimic acid and 0.7 mM for salicylic acid by adding the appropriate amount of each substance aseptically to the melted PDA medium just before solidification. Plates containing 20 ml of the medium for each treatment were prepared. Disks (1.2 cm in diameter) taken from the growing edge of 5- day-old colony of *B. fabae* were used to inoculate the prepared plates. For each treatment, five replicates (plates) were used. The plates were incubated at 20°C+2 for 6 days. The linear growth of *B. fabae* in each treatment was measured after 2, 4 and 6 days of inoculation. Linear growth was measured in cm by taking the average of two perpendicular diameters. The percentage of inhibition (I%) was calculated according to Tops and Wain equation [16] as follows:

$$I\% = (A-B)/A \times 100$$

Where: I%=Percentage of inhibition.

A=Mean diameter of growth in the control.

B=Mean diameter of growth in treatment.

Planting and growth conditions

Seeds of faba bean (*Vicia fabae* L.) susceptible [G_{429}] were surface sterilized in 0.01% mercuric chloride for 3 min., subsequently rinsed with sterilized water several times. Sterilized seeds were divided into 4 sets. Seeds of the 1st, 2nd, 3rd and 4th set were soaked in distilled water, 0.4 mM Shikimic Acid (SH), 0.7 mM Salicylic Acid (SA) and Shikimic Acid+Salicylic Acid (SH+SA), respectively, for 12 hrs. Seeds of each set were planted in plastic pots (5 seeds per pot) filled with 12-13 kg mixed soil (2 clay:1 sand, v/v). All plants were watered regularly to near field capacity with tap water. Plants were maintained under natural conditions (mean day temperature 22°C, night temperature 18°C and 16 hr. photo period). Forty five days later from planting, faba bean plants were inoculated with a spore suspension (40×10^4 spores ml^{-1}) of *B. fabae*. Untreated plants (control) were sprayed with sterilized water at the same time. Chemical fungicide Ridomil MZ at the rate of (250 g/100 L) was applied as spray treatments four times at 15-day intervals.

The experimental design can be summarized as follows:

1- Uninoculated treatment (control) (Cont)

2- *B. fabae* treatment (pathogen) (P)

3- Shikimic acid treatment (SH)

4- Shikimic acid+*B. fabae* treatment (SH+P)

5- Salicylic acid treatment (SA)

6- Salicylic acid+*B. fabae* treatment (SA+P)

7- Salicylic acid+Shikimic acid treatment (SA+SH)

8- Salicylic acid+Shikimic acid+*B. fabae* treatment (SA+SH+P)

9-Fungicide (Ridomil MZ) (F)

10-Fungicide+*B. fabae* treatment (F+P)

Samples from the 3rd compound leaf were harvested after 10 weeks from planting for measurements of enzymes activity and phenolic compounds.

Enzymes activity

Estimation of peroxidase activity (POD): 0.5 g of the leaf material was homogenized in a mortar with 30-40 ml of (0.02M) phosphate buffer (pH 7) then filtered through cheese cloth and centrifuged at 2000 rpm for 10 min then the extract was made up to 100 ml with the buffer. All operations were carried out at 4°C. Peroxidase activity was assayed according to the methods adopted by Devi [17].

Estimation of polyphenol oxidase activity (PPO): The extract was prepared by using the method suggested for the peroxidase. Polyphenol oxidase activity was assayed as the increase in absorbance at 420 nm due to the formation of purpurogallin [17].

Estimation of phenylalanine ammonia lyase activity (PAL): 0.5 g of the leaves was homogenized in a mortar with 5 ml cold (25 mM) borate buffer (pH 8.8) containing 5 mM mercaptoethanol (0.4 ml L^{-1}). The homogenate was centrifuged in a refrigerated centrifuge at 12,000 rpm for 20 min. The supernatant served as enzyme extract. All operations were carried out at 4°C. Phenylalanine ammonia lyase activity was determined according to Brueske [18].

Estimation of total phenols

Total phenols estimation was carried out with folin ciocalteau reagent according to the method described by Malik and Singh [19].

Estimation of salicylic acid and shikimic acid: The salicylic acid and shikimic acid as phenolic compounds were extracted according to Tomas-Barberan et al. [20] with some modifications. Five gram of leaves was homogenized in a Poltroon (2 min on ice) with 10 mL of extraction solution (water/methanol 2:8 containing 2 mM NaF to inactivate polyphenol oxidases and prevent phenolic degradation due to browning). Homogenates were kept in ice until centrifuged (11500 rpm, 15 min, 2-5°C); the supernatant was recovered carefully to prevent contamination with the pellet, and the volume was measured. Samples were run through a HPLC system (Agilent 1100 series) coupled with UV–Vis detector (G1315B) and G1322A DEGASSER. Sample injections of 20 μL were made from an Agilent 1100 Series auto-sampler; the chromatographic separations were performed on ZORBAX-EclipseXDB-C$_{18}$ column (4.6 × 250 mm, particle size 5 μm). Optimum efficiency of separation was obtained using acetonitrile (solvent A), and the flow rate of methanol (solvent B) with flow rate 1 ml/min. Ten min of equilibration is required before the next injection. All the stock solutions were prepared in Me OH: H$_2$O 80/20 and the standard dilutions with deionized water.

Statistical analysis

A test for significant differences between means at P ≤ 0.05 was performed using least significant difference (LSD) test [21] by SPSS program.

Results and Discussion

Because of hazards of pesticides in general, and fungicides in specific, on public health and environmental balance, moreover, faba bean cultural practice modifications using fungicides provided only partial crop protection without attention to subsidiary adverse effects of fungicides on the host plant. Thus, this study was planned to replace the undesirable and unsafe chemical control by a relatively recent direction for faba bean chocolate spot disease control, the so called "induced resistance" which is a promising modern approach with a broad spectrum in plant disease control [22,23]. The use of shikimic or salicylic acids and their combination to increase the resistance of infected faba bean plants against *B. fabae* were investigated.

Surveys for faba bean chocolate spot disease covering 6 districts (Sherbin, Meet ghamer, Talkha, Mansoura, Dekernis and Bani Ebaid) belong to Dakahlia Governorate, Nile Delta Egypt were conducted. Out of these surveys, six isolates of the pathogen were obtained and purified using single spore technique. These isolates were identified as *B. fabae*. All isolates were subjected to pathogenicity tests to determine the most aggressive one. All isolates appeared to have the potency to cause chocolate spot disease, but the isolate (B1) from Sherbin was considered to be the most aggressive one and was selected for further studies (Table 1). Shikimic acid or salicylic acid were found to be the most effective inhibitors for the linear growth of *B. fabae in vitro* than their combination (Table 2). These results were in good agreement with those obtained by many investigators on different pathogens [24,25]. Furthermore, Abd El-Hai et al. [26] showed that salicylic acid at 10 mM alone or in combination with citric acid completely inhibited the linear growth (*in vitro*) of *Rhizoctonia solani* and *Macrophomina phasolina* the causal pathogens of damping-off and charcoal rot diseases of sunflower plant, respectively.

It was also reported that salicylic acid as an allelochemical greatly inhibited *Fusarium oxysporum* f. sp. *niveum* (the causal pathogen of water melon Fusarium wilt) growth and conidia formation and germination, though stimulated mycotoxin production and activities of hydrolytic enzymes by *F. oxysporum* f. sp. *niveum* [27].

Phenolic are well-known antifungal, antibacterial and antiviral compounds occurring in plants [28]. Cowan [29] explained the mechanisms thought to be responsible for the phenolic toxicity to microorganisms on the basis of enzyme inhibition by the oxidized compounds, possibly through reaction with sulfhydryl groups or through more nonspecific interactions with the proteins. The site(s) and number of hydroxyl groups on the phenol compound are thought to be related to their relative toxicity to microorganisms, with evidence that increased hydroxylation results in increased toxicity. In addition, some investigators have found that the more highly oxidized phenols the more inhibitory effect to a pathogen [30].

The antimicrobial action of salicylic acid may be due to inhibition of functions of several enzymes by oxidized compounds, dissolved membrane lipids and interfere with membrane functions including transport of nutrients and interferes with proteins, RNA and DNA synthesis [31].

The role of salicylic acid is well-known at the beginning of defense responses against different abiotic and biotic stress [32]. Thus, the tested phenolic compounds (shikimic acid, salicylic acid or their combination) were highly effective in controlling chocolate spot disease incidence and severity when applied as seed treatment before planting more than fungicide which provided only partial crop protection (Table 3). These results are in a close parallelism with those of Hassan et al. [23] who reported that application of chemical inducers such as

Isolate	Location	Disease incidence (%)	Disease severity (%)
B1	Sherbin	75	5
B2	Meet Ghamer	65	4
B3	Talkha	67	2
B4	Mansoura	54	2
B5	Dekernis	51	2
B6	Bani Ebaid	55	3

Table 1: Pathogenicity test for the different isolates of *B. fabae* collected from different locations in Dakahlia governorate, Egypt.

Treatments	Days of incubation						Growth rate (cm/day)
	2days		4 days		6 days		
	Colony diameter (cm)	%Inhibition	Colony diameter (cm)	%Inhibition	Colony diameter (cm)	%Inhibition	
Control	3.35	0.00	6.55	0.00	8.88	0.00	1.38
Shikimic acid (SH)	2.70	19.40	3.40	48.14	5.5	38.06	0.70
Salicylic acid (SA)	2.40	28.35	3.30	49.67	5.2	41.44	0.70
SH+SA	3.10	7.46	3.55	45.85	6.8	23.42	0.93
LSD at P≤0.05	0.46	-	0.41	-	0.56	-	0.19

Each value represents the mean of 5 replicates.

Table 2: Effect of shikimic acid, salicylic acid and their combination on the linear growth, inhibition (%) and growth rate (cm/day) of *B. fabae* after 2, 4 and 6 days of incubation.

Treatments / Parameters	Disease incidence (%)	Disease severity (%)	Efficacy (%)*
Uninoculated treatment (control)	0.00	0.00	-
B. fabae	82.66	50.7	-
Fungicide (Ridomil MZ)	0.00	0.00	-
Fungicide +*B. fabae*	54.33	28.6	43.42
Shikimic acid	0.00	0.00	-
Shikimic acid + *B. fabae*	21.46	13.4	73.42
Salicylic acid	0.00	0.00	-
Salicylic acid + *B. fabae*	18.90	11.3	77.63
Salicylic acid +Shikimic acid	0.00	0.00	-
Salicylic +Shikimic + *B. fabae*	23.43	14.8	70.65
LSD at P ≤ 0.05	3.09	4.75	-

$$\text{*Efficacy\%=}\frac{\text{\% disease severity in control - \% disease severity in treatment}}{\text{\% Disease severity in control}}$$

Table 3: Effect of shikimic acid, salicylic acid and their combination on Disease incidence (%), disease severity and efficacy (%) of faba bean plants infected with chocolate spot disease.

salicylic acid at two different concentrations can decrease the disease severity of chocolate spot disease in faba bean plants. Furthermore, Abbas et al. [33] showed that application of salicylic acid as a foliar spray to faba bean plant led to reduction in the incidence of chocolate spot disease caused by *Botrytis fabae*.

El-Hendawy et al. [34] noticed that presoaking of faba bean seeds in chemical inducers such as salicylic acid give the highest reduction in disease severity of chocolate spot at different periods of inoculation. In addition, Ghazanfar et al. [35] found that 1.5 mM salicylic acid caused a 56.8% disease reduction of chickpea blight disease caused by *Ascochyta rabiei*. Hadi and Balali [36] recorded similar results in case of potato plants infected with *Rhizoctonia solani*.

The effect of salicylic acid to induce resistance in groundnut plants against *Alternaria alternata* was investigated by Chitra et al. [37]. Foliage spray of 2 mM salicylic acid significantly reduced (40.3%) leaf blight of onion caused by *Stemphylium vesicarium* under greenhouse conditions 15 d after inoculation [38]. The ability of the four chemicals salicylic acid, sodium salt of salicylic acid, isonicotinic acid, and DL-β-amino-*n* butyric acid, as well as the yeast antagonist *Cryptococcus flavescens* were evaluated against *Fusarium* head blight of wheat under greenhouse conditions showing that sodium salicylate(NaSA) and 2,6-dichloroisonicotinic acid (INA) at 10 mM significantly reduced *Fusarium* Head Blight (FHB) severity compared with the non-treated disease control 3 d prior to challenge with the pathogen [39]. The significant trait of the disease resistance inducer is that it persists longer in the host by increasing its biochemical constituents. Under infection

with *Phytophthora cryptogea*, applying a higher concentration of salicylic acid at 8 g L^{-1} to potato plants caused more vigorous growth than in the control [40]. Furthermore, a higher accumulation of salicylic acid at the infection site was found, which may be attributable to transcriptional activation of PR genes in the inoculated and uninoculated leaves [41].

The superiority effect of fungicide in reducing infection may attributed to its toxicity against the pathogen, while the positive effect of shikimic acid or salicylic acid or their combination may be due to their action as plant defense activators [42].

Salicylic acid has an affinity to bind with the enzymes involved in reactive oxygen species metabolism and in redox homeostasis. Alteration in this homeostasis leads to induction of a defense response in plants [43]. SA also affects the lipid peroxidation, which plays a key role in initiating defense response [44].

Salicylic acid-induced pathway is characterized by the production of a cascade of pathogenesis related proteins. Where, a number of biochemical and physiological changes has been associated with pathogen infection, these include:

1. The production of antifungal chitinases, glucanases and thaumatins, and oxidative enzymes such as PODs, PPOs and lipoxygenases.

2. Low molecular weight compounds with antimicrobial properties (phytoalexins) can also accumulate [45].

3. Cell death and deposition of callose and lignin [46].

4. Forming novel proteins [47].

Phenolic seem to inhibit disease development through different mechanisms involving the inhibition of extracellular fungal enzymes (cellulases, pectinases, lactase, xylanase), inhibition of fungal oxidative phosphorylation, nutrient deprivation (metal complexion, protein insolubilisation), and antioxidant activity in plant tissues [48]. Many phenolic compounds are known to be antimicrobial, function as precursors to structural polymers such as lignin, or serve as signal molecules.

Increase in the activity of the defense enzymes has been reported in plants treated with various biotic and abiotic inducers of resistance [49]. The data presented in Table 4 indicated that presoaking of faba bean seeds in phenolic compounds led to an increase in the activity of oxidative enzymes (POD, PPO) in plant leaves than untreated infected plants. These results are parallel to those obtained by El-Hendawy et al. [34] who stated that there was a significant increase in the activity of POD and PPO after treatment of wheat, anise, sugar beet and fava bean plants with salicylic acid. This might be due to the role of salicylic acid in generation of the oxidative burst by inducing a rapid transient generation of O_2^- which is responsible for regulation of POD activity [50].

Activities of oxidative enzymes in any infected plant tissues are known to contribute to disease resistance mechanisms through the oxidation of phenols [51]. Nawar and Kuti [52] mentioned that, POD activity in leaves of resistant cultivars of broad bean infected with *B. fabae* was 10 times higher than that of susceptible cultivars and eight times higher in uninfected leaves of the resistant than the susceptible cultivars. PODs are ionically bound to the cell walls and are involved in the polymerization of phenylpropanoid lignin precursors [53]. Oxidation of PODs makes the cell wall more mechanically rigid by cross-linking matrix polysaccharide and glycoprotein molecules, thus modifying the mechanical properties of the cell wall. In general, cross linking of matrix polysaccharides in cell walls are also likely to inhibit cell wall degrading enzymes of the pathogen.

POD causes the oxidation of a wide variety of substrates, using H_2O_2, such as phenols which play a considerable role in lignin synthesis [54]. Tarrad et al. [51] reported that the increase in POD activity enhanced lignification in response to chocolate spot disease, which may restrict fungal penetration. Another supportive suggestion was made by Nawar

and Kuti [53] who stated that, an increase in POD activity is considered to be a preliminary indicator for resistance of broad beans to chocolate spot disease. These findings indicate a positive relationship between resistance and POD activity.

PODs have roles in both the production and scavenging of reactive oxygen species [55]. Some PODs are responsible for the production of H_2O_2 [55]. H_2O_2 produced by these PODs might serve as the substrate for other PODs or act as an antimicrobial agent and signals to trigger self-defense responses like hypersensitive reaction [56,57].

The accumulation of these enzymes is dependent on the fungal growth; the newly synthesized polyphenol and their oxidized products may limit the fungal activity on the plant tissues. POD present in plant tissues participate in the biosynthesis of ethylene which acts as an inducer for the resistance by stimulating the production of phytoalexins in some plants [58].

Polyphenol oxidase generally catalyzes the oxidation of phenolic compounds to quinones (antimicrobial compounds) using molecular oxygen as an electron acceptor which are toxic to the invading pathogens and pests [59]. PPOs are suggested to be indirectly involved in IAA biosynthesis because the O-quinones produced can react with tryptophan to form indole-3-acetic acid [60]. POD and PPO are important in the defense mechanism against pathogens, through its role in the oxidation of phenolic compounds to quinones [61]; the quinones formed may act in several ways leading to the protection of plants through:

(1) Their high capacity for reacting with other cellular compounds, quinones can limit the development of diseases at the infected sites; accelerating the cellular death of cells close to the infection site; preventing the advance of infection and/or by generating a toxic environment which will inhibit the growth of the pathogen inside the cells.

(2) Their ability to alkylate proteins, mainly by becoming covalently linked to amino acids susceptible to alkylation, such as lysine, histidine, cysteine and methionine, thereby reducing the bioavailability of such proteins.

(3) Their ability to react with other phenolic compounds; increasing the formation of polymers, covalent linkages and condensation with more proteins, leading to additional barriers that can prevent the pathogens from entering the cell.

Lamb and Dixon [62] reported that salicylic acid causes an increase in the accumulation of H_2O_2 in plant tissues which plays a key role in initiating hypersensitive responses and providing SAR against pathogenic microbes. Salicylic acid is found to alter the activity of a mitochondrial enzyme, alternative oxidase, which mediates the oxidation of ubiquinol/ubiquinone pool and reduction of oxygen to water, without the synthesis of ATP in mitochondria and this altered activity of enzyme alternative oxidase affects the ROS levels in mitochondria and in turn induces an antiviral defense response in plants [63].

The data in Table 4 clearly show that shikimic acid accelerate the production of the defense enzymes (POD and PPO) which increase the oxidation of phenol to its corresponding quinones which have a significant role in the rigidity of cell wall by increasing the lignification. Furthermore, shikimic acid acts as a precursor of lignin biosynthesis [64].

Phenylalanine Ammonia-Lyase (PAL) is a crucial enzyme in the

Parameters / Treatments	POD (U min^{-1} g^{-1} fresh wt)	PPO (U min^{-1} g^{-1} fresh wt)	PAL (μmole t-cinnamic acid hr^{-1}g^{-1}fresh wt)
Uninoculated treatment (control)	17.56	1.46	397.41
B. fabae	25.28	2.51	588.37
Fungicide (Ridomil MZ)	17.19	1.38	391.73
Fungicide +*B. fabae*	23.90	2.46	584.49
Shikimic acid	19.86	1.97	433.33
Shikimic acid + *B. fabae*	31.41	2.93	632.29
Salicylic acid	22.74	2.14	442.89
Salicylic acid + *B. fabae*	32.82	3.08	666.15
Salicylic acid +Shikimic acid	17.40	1.50	396.12
Salicylic +Shikimic + *B. fabae*	27.33	2.81	628.68
LSD at P≤0.05	0.57	0.08	2.69

Table 4: Effect of shikimic acid, salicylic acid and their combination on peroxidase (POD), polyphenol (PPO) and phenylalanine ammonia lyase (PAL) enzymes activities in the shoot of faba bean plants infected with chocolate spot disease.

phenylpropanoid pathway, catalyzing the formation of *trans*-cinnamic acid via the L-deamination of phenylalanine. It is commonly considered the principal enzyme in the biosynthesis of phenolic compounds [65]. PAL activity in plant tissue may rapidly change under the influence of various factors, e.g. pathogen attack and treatment with elicitors [66]. The present results showed that, the activity of PAL was increased with infection by *B. fabae* (Table 4). Furthermore, seed presoaking in phenolic compounds caused additional increase in the PAL activity, which is in agreement with the finding of El-Khallal [67], who showed that PAL activity significantly increased in tomato plants treated with salicylic acid in response to *Fusarium oxysporum*.

The obvious increase in PAL activity as a result of shikimic acid application may probably due to shikimic acid acts as a precursor for major of phenolic compounds within plant tissues which consequently increase the production of PAL within the plant tissues especially the infected ones.

Phenols have been recorded to offer resistance to diseases and pests in plants, and grains containing high amount of polyphenols are resistant to several plant diseases [19]. There were many researchers established that higher level of phenolic content was positively proportional to the degree of plant resistance against various fungal diseases [38]. Total phenols, salicylic and shikimic acids contents were progressively increased with infection (Table 5). Moreover, shikimic acid, salicylic acid and their combination induced additional increase in the endogenous total phenols, salicylic and shikimic acids in healthy and infected plant tissues. These findings are in a good accordance with those of El-Hendawy et al [34] who found that faba bean plants treated with salicylic acid either foliar spray or seed soaking showed the maximum accumulation of total phenols, as compared with the untreated control ones. Furthermore, similar results were obtained by El-Khallal [67]; Abou-Elyousr et al. [38].They reported that there is a positive correlation between the application of phenolic compounds and the increase in PAL activity was found, which in turn led to the accumulation of total phenolic. In addition, Abd El-Hai et al. [26] reported that the total phenols content was increased significantly in sunflower plant infected with damping-off and charcoal rot disease under application of salicylic acid treatment.

The increased level of phenolic provides an adequate substrate to oxidative reactions catalyzed by PPO and/or POD that, consuming oxygen and producing fungi toxic quinones, make the medium unfavorable to the further development of pathogens. Antifungal

Parameters / Treatments	Total phenols (µg g⁻¹ fresh wt)	Salicylic acid (ng g⁻¹ fresh wt)	Shikimic acid (ng g⁻¹ fresh wt)
Uninoculated treatment (control)	11.41	46.08	117.46
B. fabae	13.09	49.41	129.66
Fungicide (Ridomil MZ)	12.05	47.93	121.89
Fungicide + *B. fabae*	13.76	50.71	138.34
Shikimic acid	15.97	63.57	156.88
Shikimic acid + *B. fabae*	18.53	67.64	170.56
Salicylic acid	17.67	71.18	171.24
Salicylic acid + *B. fabae*	19.90	75.82	180.54
Salicylic acid +Shikimic acid	14.20	52.19	131.22
Salicylic +Shikimic + *B. fabae*	15.83	57.61	164.42
LSD at P≤0.05	1.42	1.85	1.41

Table 5: Effect of shikimic acid, salicylic acid and their combination on total phenols, salicylic acid and shikimic acid contents in the shoot of faba bean plants infected with chocolate spot disease.

activity of oxidized phenolic may also be related to the necrotic reaction, e.g. the oxidative polymerization involving phenolic compounds, amino acids and proteins that yields brown melanin. This reaction results in the formation of an impermeable barrier to pathogenesis by plant parasites, and in a decrease of nutrients essential to the fungal development [68]. It is well known that synthesis of phenols occurs as an early response of plants to attempt infection by pathogens, as antimicrobial compounds, signal molecules, and cell wall strengthening components [69]. It was found in the present study that the activity of POD and PPO enzymes were higher in treated leaves, these enzymes act by oxidation of phenolic compounds to quinones that have toxicity 100 times more than the corresponding phenolic compounds [70]. Phenols are essential for the biosynthesis of lignin, which is considered an important structure component of plant cell walls [71]. Antibiotic phenols have the ability to bind to some proteins *in vitro*, forming soluble and insoluble complexes [72]. These phenolic-protein interactions are thought to be, in part, responsible for putative function of phenolic as plant defense compounds. However, several phenolic compounds are directly antifungal in their free state as well [73].

In this study, the dramatic increase in total phenols of faba bean plants (Table 5) as a result of infection by *B. fabae* are in conformity with the finding of Petkovsek et al. [74] who found that the level of total phenolic in the infected tissue was 1.3–2.4 times higher than in the healthy leaves and fruit of apple plant. The production and accumulation of phenolic occurs in healthy plant cells surrounding wounded or infected cells, and they are stimulated by alarm substances produced and released by the damaged cells and diffusing into the adjacent healthy cells. Therefore, the activity of many phenol-oxidizing enzymes is generally higher in the infected tissue than in the uninfected tissue of healthy plants. Changes at the phenolic level can play a role in the protection of the plant.

In the present study, endogenous salicylic acid increased in faba bean plants due to inoculation of *B. fabae* (Table 5). This increment of salicylic acid concentration in the leaf tissues might have contributed for enhanced resistance to the pathogen as was evidenced by significantly reduced chocolate spot appearance. These results agree with those of Mandal et al. [75] who noticed that the content of endogenous salicylic acid was increased significantly in tomato plants treated with salicylic acid as a foliar spray and through root treatment under *Fusarium oxysporum* infection. Salicylic acid accumulation due to the pathogen or phenolic compounds application was shown to result from the enhanced activity of PAL [76]. Chen et al. [77] proposed that high activity of PAL in salicylic acid -treated plants indicated that the development of acquired resistance by salicylic acid may be attributed at least partly, to the salicylic acid induced PAL gene expression and activation. Where, salicylic acid could activate PAL by enhancing the accumulation of PAL mRNA, the synthesis of new PAL protein and its activity and consequently enhances the accumulation of phenylpropanoids such as phenolic acids. In this respect, results suggested that phenyl propanoid compounds are more rapidly synthesized in leaves of infected faba bean plants especially treated with phenolic compounds.

The data showed that both the PAL activity and phenolic compounds accumulations were enhanced dramatically after the phenolic compounds treatments, indicating that PAL is the key enzyme in regulating the salicylic acid elicited phenolic acid accumulations in faba bean plants. These results are in good agreement with those obtained by Dong et al. [78]. Chitra et al. [37] revealed that foliar applications of salicylic acid induced in PAL, POD and PPO activities and caused an increase in phenolic content of groundnut plants upon challenge inoculation with *Alternaria alternata*.

Conclusion

The application of shikimic acid may act in the same manner of salicylic acid in enhancing the production of PAL and endogenous phenolic compounds (Total phenols, salicylic and shikimic acids) but, the lack of data about the effect of shikimic acid on different infected plant tissues as the authors are aware makes this postulation not decisive. Therefore, the prospect for the future is good for the application of phenolic compounds because of the lessened availability of fungicides to protect crop plants. Moreover, we recommended the application of such phenolic compounds separately because of the insignificance effect of their combination which may be attributed to it's over dose on the plant. In addition, the most effective treatment in enhancement faba bean resistance against *B fabae* infection was salicylic acid.

References

1. El-Sayed F, Nakoul H, Williams P (1982) Distribution of protein content in the collection of faba bean (vicia faba L.). FABIS 5: 37-41.

2. Mazen MMM (2004) Resistance induction against disease of faba bean crop. Ph.D. Thesis, Fac. Agric., Suez Canal Universityp.159.

3. Almagro L, Gómez Ros LV, Belchi-Navarro S, Bru R, Ros Barceló A, et al. (2009) Class III peroxidases in plant defence reactions. J Exp Bot 60: 377-390.

4. Mayer AM (2006) Polyphenol oxidases in plants and fungi: going places? A review. Phytochemistry 67: 2318-2331.

5. Mohamed AM, Saleh AA, Monira RA, Abeer R, Abd El-Aziz M (2012) Biochemical screening of chocolate spot disease on faba bean caused by Botrytis fabae. African J Microbiol Res 6: 6122-6129.

6. Heim KE, Tagliaferro AR, Bobilya DJ (2002) Flavonoid antioxidants: chemistry, metabolism and structure-activity relationships. J Nutr Biochem 13: 572-584.

7. Parr AJ, Bolwell GP (2000) Phenols in the plant and in man. The potential for possible nutritional enhancement of the diet by modifying the phenols content or profile. J Sci Food Agric 80: 985-1012.

8. Amarowicz R, Peg RB, Rahimi-Moghaddam P, Barl B, Weil JA (2004) Free-radical scavenging capacity and antioxidant activity of selected plant species from the Canadian prairies. Food Chem 84: 551-562.

9. Shabana YM (1987) Biological control of water weeds by using plant pathogens. M.Sc. Thesis, Fac. Agric Mansoura Univ Egypt. pp. 78.

10. Ellis MB (1971) Dematiaceous Hyphomycetes.Common wealth Mycological Institute.Kew. Surrey, England.

11. Leach R, Moore KG (1966) Sporulation of Botrytis fabae on agar culture. Trans Brit Mycol Soc 49: 593-601.

12. Last FT, Hamley RE (1956) A local lesion technique for measuring the infectivity of conidia of Botrytis fabae Sard Ann Appl Biol 44: 410-418.

13. Gondran J (1986) Resistance de la vase de narbonne et de la Feverole a Botrytis fabae V. eme Journee de phytiatrie de phytoarmacie circum. Mediterrancennes. 15-20 Mai, Rebat, Maroc. 1977; [C.F. Fabis Newsletter N 16: 46-52].

14. Hanounik SB (1986) Screening Techniques for Disease Resistance in faba bean. International Centre for Agricultural Research in the Dry Areas (ICARDA). Aleppo, Syria, pp.59.

15. Real HS, Hoot JS (1985) Differential response of wheat varieties to systemic fungicides applied to Stillage triticale (Pers.). Roster Indian J Agaric Sic 55: 548-549.

16. Topps J, Hand Win RL (1957) Investigation of fungicides. III. The fungi toxicity and 5-alkyl salicylic anal ide and para chloroanilines. Ann Appl Biol 45: 506-511.

17. Devi P (2002) Principles and methods in plant molecular biology, biochemistry and genetics. Agrobios Ind 41: 57-59.

18. Brueske CH (1980) Phenylalanine ammonia-lyase activity in tomato roots infected and resistant to the root-knot nematode, Meloidogyne incognita. Physiol Plant Pathol 16: 409-414.

19. Malik CP, Singh MB (1980) Estimation of Total Phenols in Plant Enzymology and Histo-enzymology. Kalyani Publishers, New Delhi.

20. Tomás-Barberán FA, Gil MI, Cremin P, Waterhouse AL, Hess-Pierce B, et al. (2001) HPLC-DAD-ESIMS analysis of phenolic compounds in nectarines, peaches, and plums. J Agric Food Chem 49: 4748-4760.

21. Snedecor GW, Cochran WG (1976) Statistical Methods. 6th ed. Oxoford IBH Publishing Co. New Delhi.

22. Reglinski T, Whitaker G, Cooney JM, Taylor JT, Pooles PR, et al. (2001) Systemic acquired resistance to Sclerotinia sclerotiorum in kiwi fruit vines. Physiol Mol Plant Pathol 58: 111-118.

23. Hassan MEM, Abd El-Rahman SS, El-Abbasi IH, Mikhail MS (2006) Inducing resistance against faba bean chocolate spot disease. Egypt J Phytopathol 34: 69-79.

24. Shahda WT (2000) The use of antioxidants for control of tomato damping off. Alex J Agric Res 45: 307-316.

25. Shabana YM, Abdel-Fattah GM, Ismail AE, Rashad YM (2008) Control of brown spot pathogen of rice (Bipolaris oryzae) using some phenolic antioxidants. Braz J Microbiol 39: 438-444.

26. AbdEl-Hai KM, El-Metwally MA, El-Baz SM, Zeid AM (2009) The use of antioxidants and microelements for controlling damping-off caused by Rhizoctonia solani and charcoal rot caused by Macrophomina phasoliana on sun flower. Plant Pathol J 8: 79-89.

27. Wu HS, Raza W, Fan JQ, Sun YG, Bao W, et al. (2008) Antibiotic effect of exogenously applied salicylic acid on in vitro soilborne pathogen, Fusarium oxysporum f.sp.niveum. Chemosphere 74: 45-50.

28. Hayat S, Ahmad A (2007) Salicylic Acid: A Plant Hormone. Springer, Dordrecht, Netherlands.

29. Cowan MM (1999) Plant products as antimicrobial agents. Clin Microbiol Rev 12: 564-582.

30. Scalbert A (1991) Antimicrobial properties of tannins. Phytochem 30: 3875-3883.

31. Nesci A, Rodriguez M, Etcheverry M (2003) Control of Aspergillus growth and aflatoxin production using antioxidants at different conditions of water activity and pH. J Appl Microbiol 95: 279-287.

32. Catinot J, Buchala A, Abou-Mansour E, Métraux JP (2008) Salicylic acid production in response to biotic and abiotic stress depends on isochorismate in Nicotiana benthamiana. FEBS Lett 582: 473-478.

33. Abbas EE, Ghoneem KM, Ali AA, El-baz SM (2006) Yield maximization and chocolate spot control of some faba bean cultivars by antioxidants. J Agric Sci Mansoura Univ 31: 7605-7615.

34. El-Hendawy S, Shaban W, Sakagami JI (2010) Does treating faba bean seeds with chemical inducers simultaneously increase chocolate spot disease resistance and yield under field conditions? Turk J Agric For 34: 475-485.

35. Ghazanfar MU, Wakil W, Sahi ST (2011) Induction of resistance in chickpea (Cicer arietinum L.) against Ascochyta rabiei by applying chemicals and plant extracts. Chil J Agr Res 71: 52-62.

36. Hadi MR, Balali GR (2010) The effect of salicylic acid on the reduction of Rhizoctonia solani damage in the tubers of marfona potato cultivar. Am-Euras J Agric Environ Sci 7: 492-496.

37. Chitra K, Ragupathi N, Dhanalakshmi K, Mareeshwari P, Indra N, et al. (2008) Salicylic acid induced systemic resistant on peanut against Alternaria alternata. Arch Phytopathol Plant Protect 41: 50-56.

38. Abo-Elyousr KAM, Hashem M, Ali EH (2009) Integrated control of cotton root rot disease by mixing fungal biocontrol agents and resistance inducers. Crop Prot 28: 295-301.

39. ZhangY, Tian Z, Xi R, Gao H Qu P (2002) Effect of salicylic acid on phenolics metabolization of Yali pear growing fruits. J Agric Univ Hebei 25: 33-36.

40. Quintanilla P, Brishammar S (1998) Systemic induced resistance to late blight in potato by treatment with salicylic acid and Phytophthora cryptogea. Potato Res 41: 135-142.

41. Durner J, Shah J, Kessig DF (1997) Salicylic acid and disease resistance in plants. Trends Plant Sci 2: 266-274.

42. Ata AA, El-Samman MG, Moursy MA, Mostafa MH (2008) Inducing resistance against rust disease of sugar beet by certain chemical compounds. Egypt J Phytopathol 36: 113-132.

43. Durrant WE, Dong X (2004) Systemic acquired resistance. Annu Rev Phytopathol 42: 185-209.

44. Anderson MD, Chen Z, Klessig DF (1998) Possible involvement of lipid peroxidation in salicylic acid-mediated induction of PR1 gene expression. Phytochem 47: 555-566.

45. Soylu S, Bennett MH (2002) Mansfiels JW. Induction of phytoalexin accumulation in broad bean (Vicia faba L.) cotyledons following treatments with biotic and abiotic elicitors. Turk J Agric 26: 343-348.

46. Green S, Bailey KL, Tewari JP (2001) The infection process of Alternaria cirsinoxia on Canada thistle (Cirsium arvense) and host structural defense responses. Mycol Res 105: 344-351.

47. Myers GA, Grinvalds R, Booth S, Hutton SI, Binks M (2000) Expression of two novel proteins in Chlamydia trachomatis during natural infection. Microb Pathog 29: 63-72.

48. Hammerschmidt R (2005) Phenols and plant-pathogen interactions: The saga continues. Physiol Mol Plant Pathol 66: 77-78.

49. Raghvendra VB, Lokesh S, Govindappa M, Vasanth KT (2007) Dravya as an organic agent for the management of seed borne fungi of sorghum and its role in the induction of defense enzymes. Pes Biochem Physiol 89: 190-197.

50. Rao MV, Paliyath G, Ormrod DP, Murr DP, Watkins CB (1997) Influence of salicylic acid on H2O2 production, oxidative stress, and H2O2-metabolizing enzymes. Salicylic acid-mediated oxidative damage requires H2O2. Plant Physiol 115: 137-149.

51. Tarrad AM, El-Hyatemy YY, Omar SA (1993) Wyerone derivatives and activities of peroxidase and polyphenol oxidase in faba bean leaves as induced by chocolate spot disease. Plant Sci 89: 161-165.

52. Nawar HF, Kuti, JO (2003) Wyerone acid phytoalexin synthesis and peroxidase activity as markers for resistance of broad beans to chocolate spot disease. Phytopathol J 151: 564-570.

53. Cvikrová M, Malá J, Hrubcová M, Eder J (2006) Soluble and cell wall-bound phenolics and lignin in Ascocalyx abietina infected Norway spruces. Plant Sci 170: 563-570.

54. Goldberg R, Liberman M, Mathieu C, Pierron CM, Catesson AM (1987) Development of epidermal cell wall peroxidase along the bean hypocotyls: Possible involvement in the cell wall stiffening process. J Exp Bot 38: 1378-1390.

55. Bolwell GP, Bindschedler LV, Blee KA, Butt VS, Davies DR, et al. (2002) The apoplastic oxidative burst in response to biotic stress in plants: a three-component system. J Exp Bot 53: 1367-1376.

56. Wojtaszek P (1997) Oxidative burst: an early plant response to pathogen infection. Biochem J 322 : 681-692.

57. Gozzo F (2004) Systemic acquired resistance in crop protection. Outlooks on Pest Manag. pp 20-23.

58. Ahmed SM (2010) Effects of salicylic acid, ascorbic acid and two fungicides in control of early blight disease and some physiological components of two varieties of potatoes. J Agric Res Kafer El-Shiekh Univ Egypt 36: 220-237.

59. Weir TL, Park SW, Vivanco JM (2004) Biochemical and physiological mechanisms mediated by allelochemicals. Curr Opin Plant Biol 7: 472-479.

60. Mayer AM, Harel E (1979) Polyphenoloxidases in plants. Phytochem. 18: 193-215.

61. Melo GA, Shimizu MM, Mazzafera P (2006) Polyphenoloxidase activity in coffee leaves and its role in resistance against the coffee leaf miner and coffee leaf rust. Phytochemistry 67: 277-285.

62. Lamb C, Dixon RA (1997) THE OXIDATIVE BURST IN PLANT DISEASE RESISTANCE. Annu Rev Plant Physiol Plant Mol Biol 48: 251-275.

63. Singh DP, Moore CA, Gilliland A, Carr JP (2004) Activation of multiple antiviral defence mechanisms by salicylic acid. Mol Plant Pathol 5: 57-63.

64. Aldesuquy HS, Makarios AT, Awad HA (1998) Effect of two antitranspirants on growth and productivity of salt treated wheat plants. Egypt J Sci Physiol 22: 189-211.

65. Ali MB, Hahn EJ, Paek KY (2007) Methyl jasmonate and salicylic acid induced oxidative stress and accumulation of phenolics in Panax ginseng bioreactor root suspension cultures. Molecules 12: 607-621.

66. Mandal S, Mitra A (2007) Reinforcement of cell wall in roots of Lycopersicon esculentum through induction of phenolic compounds and lignin by elicitors. Physiol Mol Plant Pathol 71: 201-209.

67. El-Khallal SM (2007) Induction and modulation of resistance in tomato plants against Fusarium wilt disease by bioagent fungi (Arbuscular Mycorrhiza) and/or hormonal elicitors (jasmonic Acid & salicylic Acid): 2- changes in the antioxidant enzymes, phenolic compounds and Pathogen Related- proteins. Aust J Basic Appl Sci 1: 717-732.

68. Beckman CH, Mueller VC, Mace ME (1974) The stabilization of artificial and natural cell wall membranes by phenolic infusion and its relation to wilt disease resistance. Phytopathol 64: 1214-1220.

69. Kruger WM, Carver TW, Zeyen RJ (2002) Phenolic inhibition of penetration resistance to Blumeria graminis f. sp. hordei in barley near isogenic lines containing seven independent resistance genes or alleles. Physiol Mol Plant Pathol 61: 41-51.

70. Carrasco AE, Serrano ML, Zapata JM, Sabater B, Martín M (2001) Oxidation of phenolic compounds from Aloe barbadensis by peroxidase activity: Possible involvement in defence reactions. Plant Physiol Biochem 39: 521-527.

71. Hahlborck K, Sheel D (1989) Physiology and molecular biology of phenylpropanoid metabolism. Plant Mol Biol 40: 347-369.

72. Hagerman AE, Robbins CT (1987) Implications of soluble tannin-protein complexes for tannin analysis and plant defense mechanisms. J Chem Ecol 13: 1243-1259.

73. Sarma BK, Singh UP (2003) Ferulic acid may prevent infection of Cicer arietinum by Sclerotium rolfsii. World J Microbiol Biotechnol 19: 123-127.

74. Petkovsek MM, Stampar F, Veberic R (2009) Accumulation of phenolic compounds in apple in response to infection by the scab pathogen, Venturia inaequalis. Physiol Mol Plant Pathol 74: 60-67.

75. Mandal S, Mallick N, Mitra A (2009) Salicylic acid-induced resistance to Fusarium oxysporum f. sp. lycopersici in tomato. Plant Physiol Biochem 47: 642-649.

76. Panina YS, Gerasimova NG, Chalenko GI, Vasyukova NI, Ozeretskovskaya OL (2005) Salicylic acid and phenylalanine ammonia-lyase in potato plants infected with the causal agent of late blight. Russ J Plant Physiol 52: 511-515.

77. Chen JY, Wen PF, Kong WF, Pan QH, Zhan JC (2006) Effect of salicylic acid on phenylpropanoids and phenylalanine ammonialyase in harvested grape berries. Posthar Biol Tech 40: 64-72.

78. Dong J, Wan G, Liang Z (2010) Accumulation of salicylic acid-induced phenolic compounds and raised activities of secondary metabolic and antioxidative enzymes in Salvia miltiorrhiza cell culture. J Biotechnol 148: 99-104.

22

In Vitro Evaluation of Botanicals, Bio-Agents and Fungicides against Leaf Blight of *Etlingera linguiformis* caused by *Curvularia lunata* Var. Aeria

Chijamo Kithan* and **Daiho L**

Department of Plant Pathology, School of Agricultural Sciences and Rural Development, Nagaland University, Medziphema Campus, Nagaland, India

*Corresponding author: Chijamo Kithan, Department of Plant Pathology, School of Agricultural Sciences and Rural Development, Nagaland University, Medziphema Campus, Nagaland-797106, India, E-mail: chijamokyong@ymail.com

Abstract

The present study was attempted with the objective of screening the potential antifungal activities of eight fungicides, nine plant extracts and five bio-agents *in vitro* against leaf blight of *E. linguiformis* caused by *C. lunata* var. aeria. Mancozeb at 0.3% (97.37% inhibition) was found to be significantly superior among the non-systemic fungicides evaluated at three concentrations (0.1, 0.2 and 0.3%). Among the systemic fungicides, metalaxyl showed 98.48% inhibition of mycelial growth at 0.3% concentration followed by carbendazim (95.25% inhibition at 0.3%). Among the plant extracts, *Millettia pachycarpa* root extracts (55.78) at 10 per cent was superior followed by *Acorus calamus* with 53.40% inhibition at 10 per cent. Among the bio-agents tested, *Trichoderma harzianum* showed maximum inhibition of 68.85 per cent inhibition. The study indicated that suitable integration of more efficient eco-friendly treatments like bio-agents and botanicals with lesser use of fungicides may provide effective management of the disease.

Keywords: *Etlingera linguiformis*; *Curvularia lunata* var. *aeria*; Leaf blight; Fungicides; Bio-agents; Plant extracts

Introduction

Etlingera linguiformis (Roxb.) R.M.Sm of Zingiberaceae family is an important indigenous medicinal and aromatic plant found in Nagaland. The plant species was identified from Botanical Survey of India, Eastern Regional Centre, Shillong. A voucher specimen (BSI/ERC/2014/Plant identification/883) has been deposited in the herbarium of the Botanical Survey of India, Shillong. It grows well in warm climates with loamy soil rich in humus. The entire plant is aromatic. The rhizome of the plant is used medicinally to treat sore throat, stomachache, rheumatism and respiratory complains etc. The aromatic rhizome of *E. linguiformis* contains an essential oil used in perfumery. Young leaves and shoots are used as leafy vegetables. Rhizomes are used as medicine to cure jaundice [1]. Recently, severe symptoms typically of Curvularia leaf blight were observed. The disease is mainly a foliar disease. Leaf blight of Curvularia species have been reported as casual agents of leaf spots and leaf blight [2]. Initial symptoms of the disease are small brown water soaked flecks appears on the upper leaf surface with diameter ranging from 0.5 to 3 cm which later coalesced to form dark brown lesions with a well-defined border. Lesions often merged to form large necrotic areas, covering more than 90% of the leaf surface, which contributed to plant death. The pure culture of the fungus was examined under microscope and identified as *Curvularia lunata* var.aeria. The morphological authenticity of the isolated fungus was confirmed from the Indian Type Culture Collection, Division of Plant Pathology, Indian Agriculture Research Institute, New Delhi (ITCC Accession No. 7895.10). Further molecular identification was done through DNA sequencing by using primers ITS4 and ITS5 from the Microbial Type Culture Collection & Gene Bank at the Institute of Microbial Technology (IMTECH), Chandigarh, India (Accession No. MTCC11875). The disease significantly reduces the number of functional leaves. Late in the disease progression, stems and rhizomes were also affected thereby reducing oil yield and quality. Leaf blight of

Java citronella (*Cymbopogon winterianus* Jowitt.), caused by *Curvularia* spp., and was found to cause a dramatic change in oil yield and its constituents [3]. It was also observed that Curvularia leaf blight infection increased with the age of the crop. Therefore, in view of the magnitude of damage caused by the fungus on this important plant, the present investigation was undertaken to screen out the most efficient fungicides, botanicals and bioagents against *C. lunata* var. aeria for field management of the disease.

Materials and Methods

The present investigation on *in vitro* bioassay of different fungicides, botanicals, and bioagents was carried out at the Department of Plant Pathology, School of Agricultural & Rural Development, Nagaland University during 2011-2012. Eight fungicides consisting of four systemic and four non-systemic were assayed for their efficacy against *C. lunata* var. aeria under *in vitro* condition. Each of the fungicides were tested at 0.1, 0.2, 0.3 per cent concentrations. The poisoned food technique was adopted for *in vitro* testing of fungicides [4]. The calculated quantities of fungicides were thoroughly mixed in the medium before pouring into Petriplates so as to get the desired concentration of active ingredient of each fungicide separately. Twenty ml of fungicide amended medium was poured in each of 90 mm sterilized Petriplates and allowed to solidify. The plates were inoculated centrally with 8 mm disc of 10 days old young sporulating culture of *C. lunata* var. aeria. Controls without fungicides were also maintained. The experiment was conducted in Completely Randomised Design (CRD) with three replications in each treatment. The inoculated petriplates were incubated at 25±2°C. The colony diameters were measured after 10 days when the control plates were full of fungal growth. Per cent inhibition of growth was calculated by using formula given by Vincent [5].

$$I = [(C - T)/(C)] \times 100$$

Where, I=Per cent inhibition; C=Colony diameter in control; T=Colony diameter in treatment

Plant tissues was grounded using pestle and mortar by adding equal amount of sterilized distilled water (1:1 w/v). The extract was filtered through muslin cloth. The supernatant was taken as standard plant extract solution (100%). Further, the extract was diluted by adding sterilised water to get 5, 7.5 and 10 per cent concentrations respectively. The plant extracts were subjected to heating to 50°C in water bath to avoid contamination and then incorporated into PDA medium by transferring 2 ml of each type of plant extracts into a Petridish containing 20 ml melted warm PDA medium and gently shaken for thorough mixing of the extract. The PDA plates containing the plant extracts were inoculated aseptically with *C. lunata* var. aeria by transferring 8mm diameter agar disc of 10 days old culture of the pathogen to the center of PDA medium in Petridish. Three replications were maintained for each treatment. The basal medium (PDA) without any plant extract was served as control. All the inoculated Petridishes were incubated at 25 ± 2°C. The radial growth of the test fungus in the treated plates was measured when the pathogen growth touched the periphery in the control Petridishes. The per cent inhibition of fungal growth was estimated by using the formula given by Vincent [5].

Five bioagents, *Trichoderma harzianum*, *Trichderma viride*, *Trichoderma koningii*, *Pseudomonas fluorescens* and *Bacillus subtilis* at 10^8 spores/ml were evaluated against *C. lunata* var. aeria by dual culture technique. The bio-agents as well as the test pathogen were inoculated equidistant on PDA medium aseptically and incubated at 28 ± 2°C. Three replications were maintained along with untreated control. After obtaining the maximum growth in the control, the zone of inhibition was recorded by measuring the clear distance between the margin of the test fungus and antagonistic organisms. The colony diameter of pathogen in control plate was also recorded. The per cent inhibition of the growth of the pathogen was calculated by using the formula suggested by Vincent (Figures 1-5) [5].

Statistical analysis

The statistical analysis of obtained results was carried out by using SPSS (Statistical Package for the Social Sciences) Version 16.0 software. Per cent data were transformed to arc sine values and analyzed statistically.

Figure 1: Showing the following- Plate 1: A-Healthy Plant, B-Leaf blight infection, C- *Curvularia lunata* var. aeria

Sl.No	Treatments	Per cent inhibition Concentration (%)			
I	Non systemic fungicides	0.001	0.002	0.003	Mean
1	Captan	29.86	44.96	76.56	50.46
		-33.1	-42.11	-61.06	
2	Chlorothanil	24.13	40.65	61.82	42.2
		-29.41	-39.61	-51.84	
3	Copper oxychloride	33.16	52.79	91.6	59.18
		-35.15	-46.6	-73.18	
4	Mancozeb	43.79	64.62	97.37	68.59
		-41.43	-53.5	-80.77	
II	Systemic fungicides				
1	Carbendazim	51.32	76.78	95.25	74.45
		-45.76	-61.21	-77.47	
2	Hexaconazole	41.49	53.94	90.34	61.92
		-40.09	-47.26	-71.99	
3	Metalaxyl	60.87	79.87	98.48	79.74
		-51.29	-63.36	-83.01	
4	Mycobutanil	25.16	41.08	51.06	39.1
		-30.1	-39.86	-45.6	

	CD (p=0.05)	3.57
	S.Em±	4.72
	CV%	3.65

Table 1: *In vitro* evaluation of fungicides against *C. lunata* var.aeria, *E. linguiformis* leaf blight pathogen. *Figure in parenthesis is arcsine transformations.

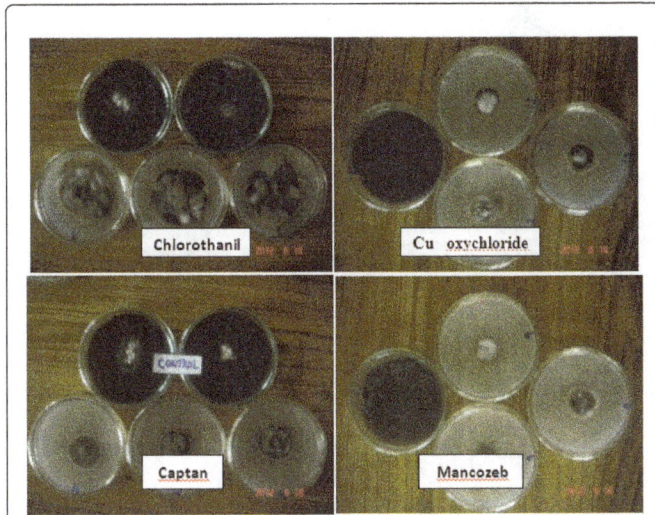

Figure 2: Non-Systemic fungicide

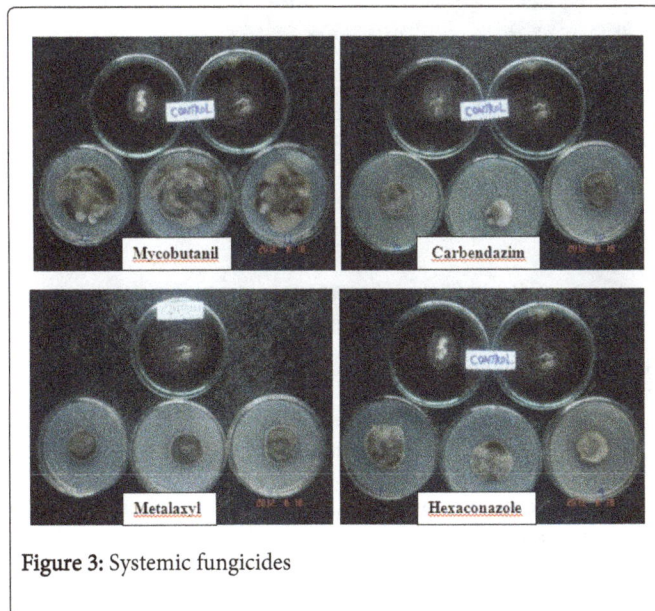

Figure 3: Systemic fungicides

Results and Discussion

In vitro evaluation of provides useful and preliminary information regarding efficacy of fungicides. Plant extracts and bio-agents against pathogen within a shortest period of time and therefore, serves as a guide for field testing. Chemical control measures have been tested and found effective in the control of diseases [6,7]. Certain protective fungicides although hazardous to environment are still used for the control of fungal diseases [8].

In the present investigation, four systemic fungicides and four non-systemic fungicides were tested at three concentrations (Tables 1-3). Among the non-systemic fungicides tested, mancozeb and copper oxychloride was found to be most effective in inhibiting the mycelia growth at all three concentrations (0.1, 0.2 and 0.3%) with inhibition of 97.37% and 91.60% respectively.

SI.No	Plant extracts	Per cent inhibition of mycelia growth Concentrations (%)			
		Parts used	5	7.5	10	Mean
1	*Acorus calamus*	Rhizome	29.61	45.21	53.4	42.74
			-32.95	-42.3	-47	
2	*Allium sativum*	Bulb	22.82	30.58	44.11	32.5
			-28.51	-33.6	-41.6	
3	*Azadirchta indica*	Leaf	17.7	27.82	42.19	28.23
			(22/43)	-31.8	-40.5	
4	*Curcuma caesia*	Rhizome	27.95	42.55	47.85	39.45
			-31.91	-40.7	-43.8	
5	*Curcuma longa*	Rhizome	13.15	29.86	37.33	26.78
			-21.2	-33.1	-37.7	
6	*Lawsonia inermis*	Leaf	19.41	31.11	38.32	29.61
			-26.12	-33.9	-38.2	
7	*Millettia pachycarpa*	Root	34.03	47.28	55.78	
			-35.58	-43.4	-48.3	45.69
8	*Oroxylum indicum*	Bark	17.47	23.68	33.55	
			-24.68	-29.1	-35.4	24.9
9	*Zingiber officinale*	Rhizome	13.13	30.34	37.42	
			-21.21	-33.4	-37.7	26.96
CD (p=0.05)	3.29				
S.Em±	5.75				
CV%	7.27				

Table 2: *In vitro* evaluation of different plant extracts against mycelia growth of *C. lunata* var.aeria. * Figure in parenthesis is arcsine transformations.

Figure 4: Plant extracts

Sl.No	Biocontrol agents	Per cent inhibition of mycelia growth
1	*Trichoderma harzianum*	68.85 (56.10)
2	*Trichoderma viride*	57.82 (49.50)
3	*Trichoderma koningii*	48.35 (44.05)
4	*Pseudomonas fluorescens*	51.36 (45.78)
5	*Bacillus subtilis*	30.32 (33.39)
	CD (p=0.05)	5.29
	S.Em±	8.47
	CV%	5.67

Table 3: *In vitro* evaluation of biocontrol agents against mycelia growth of *C. lunata* var. aeria, the causal agent of leaf blight of *E. linguiformis.* *Figure in parenthesis are arcsine transformations.

It was followed by captan (76.56%) at 0.3 per cent. Pawar et al. [9] also reported that eight fungicides were screened against *C. lunata* and *C. pallescens*. Out of them, mancozeb (0.2%), tricyclazole (0.1%) and campainion (mancozeb + carbendazim) (0.25%) were found effective against *C. lunata*. Jackson [10] reported that blight disease can be controlled by spraying with copper fungicides. Singh et al. [11] reported that mancozeb was effective against *C. lunata* var. aeria. The least effective among the non-systemic fungicides was chlorothanil (61.82% inhibition). Among the systemic fungicides, metalaxyl and carbendazim were found to be highly effective in inhibiting the growth of *C. lunata* at 0.3% concentrations (98.48% and 95.25% inhibition) followed by hexaconazole (90.34%) at 0.3 per cent. Adejumo [12] reported that the use of fungicides such as copper and copper metalaxyl-based compounds is the most reliable and popular with farmers because of the quick and effective action. Efficacy of these fungicides was previously reported by Sumangala et al. [13] and Jackson [10]. Mycobutanil was found to be the least effective (51.06% inhibition) among the systemic fungicides evaluated.

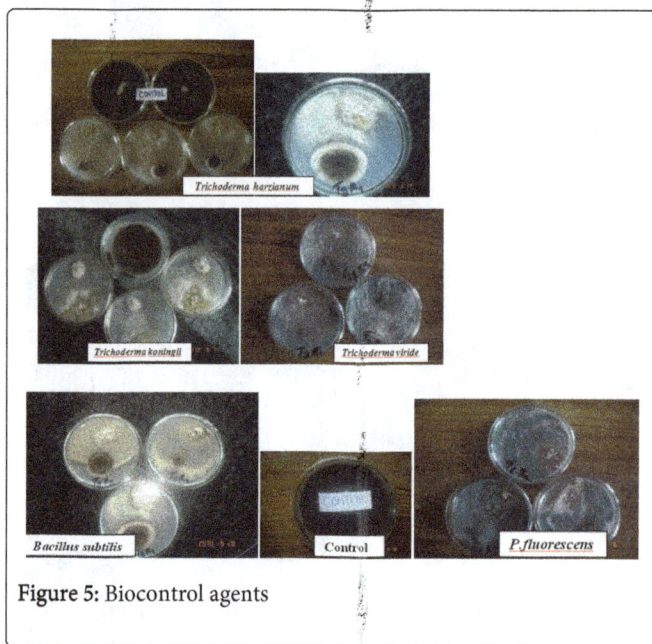

Figure 5: Biocontrol agents

Plant metabolites and plant based pesticides appear to be one of the better alternatives as they are known to have minimal environmental impact and danger to consumers in contrast to synthetic pesticides [14]. Extracts of many higher plants have been reported to exhibit antifungal properties under laboratory [15,16]. Plant extracts show antifungal activity against a wide range of fungi [17]. Plants in their natural state possess a relative stable biological balance with microbes on their surface. In the present investigation, though complete inhibition of the fungus was not observed in any of the nine botanicals used, considerable amount of inhibition of growth was noticed in some of the botanicals. Root extracts of *M. pachycarpa* was found to be most effective at all the three concentrations 5% (34.03%), 7.5% (47.28%), 10% (55.78%) followed by *A. calamus* rhizome extract (53.40% inhibition) at 10 per cent. The fungicidal effects of *Acorus calamus* (sweet flag) has already been investigated by Parinthawong

[18]. Bulb extracts of *Allium sativum, Azadirchta indica , Lawsonia inermis* and *Azadirchta indica* are also effective to some extent. Later, Choi et al. [19], Bandara et al. [20] reported *Acorus calamus, Zingiber zerumbet* and *Curcuma longa* to possessed several important biological activities including antifungal. Crude extracts of *Curcuma longa, Zingiber officinale, Allium sativum,* tested *in vitro* showed significant antifungal activity against *Curvularis* sp. [21]. Significant inhibition of *Dreschlera graminae, Curvularia lunata, Aspergillus fumigatus* and *Candida albicans* were observed with petroleum ether and benzene fractions of the leaf of the leaf of *Lawsonia inermis* [22]. Pawar et al. [9] reported that Garlic (*Allium sativum*) completely inhibited the mycelial growth of both pathogenic fungi followed by *Azadirachta indica.*

Biological control through the use of antagonistic microorganisms is a potential non-chemical means of controlling plant disease by reducing inoculum levels of the pathogens. Such a management would help in preventing the pollution and also health hazards. In the present investigation, the antagonistic effect of different bioagents assessed against *C. lunata* var.aeria by dual culture technique, *T. harzianum* has inhibited maximum growth of fungus with 68.85% followed by *T. viride* (57.82%), and *P. fluorescens* (51.36%). *Bacillus subtilis* was found to be the least effective (30.32%). The inhibitory effect of these bio-agents may be due to competition and antibiosis. Pawar et al. [9] reported that *T. harzianum* and *T. viride* significantly inhibited growth of *C. lunata in vitro.* The present investigations are in agreement with Sumangala et al. [13], who found effectiveness of *Trichoderma* sp. against *C. lunata.* This could be obviously due to several possibilities of existence of microbial interactions such as stimulation, inhibition, mutual intermingling of growth of antagonistic isolate over test pathogen etc. have been enumerated by many workers [23-25]. The antagonistic activities of *T. harzianum* against several pathogenic fungi have also been reported by many workers [26-28]. The antagonism of *Trichoderma* spp. against many fungi is mainly due to production of acetaldehyde compound [29]. Sumangala et al. [13] reported that *B. subtilis , T. viride* and *T. harzianum* were found effective in inhibiting the radial growth of fungus.

Conclusion

The present study revealed that Metalaxyl and mancozeb @ 0.1 per cent found most effective in inhibiting the growth of *C. lunata* var. aeria. Carbendazim and Hexaconazole both at 0.1 per cent were also found equally effective which can be used as an alternative to mancozeb and metalaxyl. Among the bioagents, *Trichoderma harzianum* was found to be most effective in inhibiting the mycelial growth. *Millettia pachycarpa* and *Acorus calamus* extracts at 5.0 per cent showed maximum growth inhibition among the plant extracts. Thus the study indicated that suitable integration of more efficient eco-friendly treatments like bio-agents and botanicals with lesser use of fungicides may provide a better management of the disease.

References

1. Ramana MV, Tagore JK, Bhattacharjee A (2012) Report of two medicinal and aromatic gingers from Andaman and Nicobar Islands. India J Threatened Taxa 4: 2582–2586.

2. Benoit MA, Mathur SB (1970) Identification of species of Curvularia on Rice seed. Contribution from the Danish Government Institute of seed pathology for developing countries, Copenhagen, Denmark.

3. Chutia M, Mahanta JJ, Sakia RC, Baruah AKS, Sarma TC (2006) Influence of leaf blight disease on yield and its constituents of Java

citronella and in vitro control of the pathogen using essential oils. World Journal of Agricultural Sciences 2: 319- 321.

4. Nene YL, Thapliyal PN (1993) Fungicides in plant diseases control. Oxford and IPH. Publishing Co. Pvt. Ltd., New Delhi PP.531.

5. VINCENT JM (1947) Distortion of fungal hyphae in the presence of certain inhibitors. Nature 159: 850.

6. Ogundana SK, Denis C (1981) Assessment of fungicides for prevention of storage rot of yams. Pesticide Science 11: 491-494.

7. Plumbley RA (1985) Benomyle tolerance in strain of Penicillium sclereteginum infecting yams and use of imazalid as a means of control. Tropical Agriculture Trinidad 61: 182-185.

8. Vaish DK, Sinha AP (2003) Determination of tolerance in Rhizoctonia solani, Trichoderma virens and Trichoderma sp. (isolate 20) to systemic fungicides. Indian journal of Plant Pathol 21: 48-50.

9. Pawar DM, Arekar JS, Borkar PG (2012) In vitro evaluation of bioagents and botanicals against Curvularia blight of gladiolus. Journal of Plant Disease Sciences 7: 70-72.

10. Jackson GVH (1996) Strategies for taro leaf blight research in the region. Taro Leaf Blight Seminar. Proceedings. Alafua pp. 95–100.

11. Singh AK, Sinha RKP, Sinha RP (2002) Curvularia lunata: a major pathogen for leaf blight of wheat in Bihar, India. Journal of Applied Biology 5: 83-84.

12. Adejumo TO (1997) Identification, incidence, severity and methods of control of the causal organism of false smut disease of cowpea (vignia unguiculata L.) Walp. PhD Thesis. University of Ibadan, Nigeria 201.

13. Sumangala K, Patil MB, Nargund VB, Ramegowda G (2008) Evaluation of fungicides, botanicals and bio-agents against Curvularia lunata, a causal agent of gain discoloration in rice. Journal of Plant Disease Sciences 3: 159-164.

14. Verma J, Dubey NK (1999) Prospectives of botanical and microbial products as pesticides of tomorrow. Current Science 76: 172-179.

15. Mohana DC, Raveesh KA, Lokanath R (2008) Herbal remedies for the management of seed-borne fungal pathogens. Phytopathology and Plant protection 41: 38-49.

16. Ashwani Tapwal, Nisha, Shipra Garg, Nandini Gauta, Rajesh Kumar (2011) In vitro antifungal potency of plant extracts against five phytopathogens. Braz arch biol technology, 54: 6.

17. Wilson CL, Solar JM, Ghaouth AE, Wisniewski ME (1997) Rapid evaluation of plant extracts and essential oils for antifungal activity against Botrytis cinerea. Plant Disease 81: 201-210.

18. Parinthawong N, Tansian P, Youngnit C (2010) Effects of three plant crude extracts on fungal spore germination and hyphal growthof Curvularia sp. Dept. of Plant Protection Technology, Bangkok, Thailand.

19. Choi GJ, Jang KS, Kim JS, Lee SW, Cho JY, et al. (2004) In vivo antifungal activities of 57 plant Extracts against six plant pathogenic fungi. Journal of Plant Pathology 3: 184-191.

20. Bandara JMRS, Wijayagunasekera NNP (1988) Antifungal activity of some rhizomatous plant extracts. Proceedings of the 5th International congress of plant pathology, Kyoto, Japan.

21. Sanit S (2012) Antifungal activity of twenty four medicinal crude extracts against Curvularia sp., The pathogen of dirty panicle disease in rice. Faculty of Science and Technology. 37th Congress on Science and Technology of Thailand.

22. Ankita S, Kanika S (2011) Assay of Antifungal Activity of Lawsonia inermis Linn. and Eucalyptus citriodora Hook. Journal of Pharmacy Research 4: 1313-1314.

23. Ghaffar A (1969) Biological control of white rot of onion. Interaction effect of soil microorganisms with Sclerotium cepivorum. Mycopathologica at Mycological Application 38: 101-111.

24. Naik MK, Sen B (1995) Biocontrol of plant disease caused by Fusarium spp. In : Recent Development in Biocontrol of Plant Diseases 37-51.

25. Basha SUK (2002) Antagonism of Bacillus species (strain BC121) toward Curvularia lunata. Current Science 82: 1457-1463.

26. Jat JG, Agalave HR (2013) Antagonistic properties of Trichoderma species against oil seed-borne fungi. Science Research Reporter 3: 171-174.

27. Kakde RB, Chavan AM (2011) Antagonistic properties of Trichoderma viride and Trichoderma harzianum against storage fungi. Elixir Appl. Botany 41: 5774-5778.

28. Patale SS, Mukadam DS (2011) Management of plant pathogenic fungi by using Trichoderma species. Bioscience Discovery 2: 36-37.

29. Dennis C, Webster J (1971) Antagonistic properties of species groups of Trichoderma II. Production of volatile antibiotics. Trans British Mycol Sac 57: 41-48.

The Effect of Tartaric Acid-Modified Enzyme-Resistant Dextrin from Potato Starch on Growth and Metabolism of Intestinal Bacteria

Slizewska Katarzyna[1]*, Barczynska Renata[2], Kapusniak Janusz[2] and Kapusniak Kamila[2]

[1]*Institute of Fermentation Technology and Microbiology, Faculty of Biotechnology and Food Sciences, Technical University of Lodz, 171/173 Wolczanska Street, 90-924 Lodz, Poland*
[2]*Institute of Chemistry, Environmental Protection and Biotechnology, Jan Dlugosz University, 13/15 Armii Krajowej Avenue, 42-200 Czestochowa, Poland*

Abstract

In present study, enzyme-resistant dextrin, prepared by heating of potato starch in the presence of hydrochloric (0.1% dsb) and tartaric (40% dsb) acid at 130°C for 2 h (TA-dextrin), was tested as the source of carbon for probiotic lactobacilli and bifidobacteria cultured with intestinal bacteria isolated from faeces of three heathy 70-year old volunteers. The dynamics of growth of bacterial monocultures in broth containing tartaric acid (TA)-modified dextrin was estimated. It was also investigated whether lactobacilli and bifidobacteria cultured with intestinal bacteria in the presence of resistant dextrin would be able to dominate the intestinal isolates. Prebiotic fermentation of resistant dextrin was analyzed using prebiotic index (PI). Fermentation products were determined by HPLC. It was shown that all of the tested bacteria were able to grow and utilize TA-modified dextrin as a source of carbon, albeit to varying degrees. In co-cultures of intestinal and probiotic bacteria, the environment was found to be dominated by the probiotic strains of *Bifidobacterium* and *Lactobacillus*, which is a beneficial effect.

Keywords: Resistant dextrin; Prebiotic; Intestinal bacteria

Introduction

Nowadays, one's lifestyle is indicative of one's future health. Many factors determine the risk of disease, or conversely, the possibility of remaining healthy. Physical activity and an appropriate diet are examples of daily routines that may influence one's health. A lack of physical activity, particularly if associated with overconsumption, increases the risk of development of nutrition-related chronic diseases, such as obesity, hypertension, cardiovascular diseases, osteoporosis, type II diabetes, and several types of cancer. Over the last decade, dramatic changes have taken place in the perception and understanding of the importance of the daily diet. Foods are no longer judged in terms of taste and immediate nutritional needs, but also in terms of their ability to improve the health and well-being of the consumers. The role of diet in human health has led to the recent development of the so-called functional food concept. A functional food is a dietary ingredient that has cellular or physiological effects above its normal nutritional value. It can also contain prebiotics [1].

"A prebiotic is a non-digestible food ingredient that beneficially affects the host by selectively stimulating the growth and/or activity of one or a limited number of bacteria in the colon, and thus improves host health" [2]. The definition updated by Gibson specifies that a prebiotic is "a selectively fermented ingredient that allows specific changes, both in the composition and/or activity in the gastrointestinal microflora that confers benefits upon host well-being and health" [3]. The current definition of prebiotics was suggested during the ISAPP experts' meeting in 2008, according to which "a dietary prebiotic is a selectively fermented ingredient that results in specific changes in the composition and/or activity of the gastrointestinal microbiota, thus conferring benefit(s) upon host health" [4]. For a substance to qualify as a prebiotic, it must meet certain criteria: it must be chemically characterized, exhibit health benefits that are measurable and outweigh any adverse effects, and appropriately modulate the composition or activity of the microbiota in the target host [5].

Some carbohydrates, such as fructo oligosaccharides (FOS) [6,7], inulin [8]; Van Loo, [9], transgalacto-oligosaccharides (TOS)

[10,11] and lactulose [12,13] are well-accepted prebiotics, while isomaltooligosaccharides (IMO) [14] and xylooligosaccharides [15] are candidate prebiotics. The fermentation of some oligosaccharides is not as selective as that of FOS, so their prebiotic status remains in doubt. Promising sources of prebiotics are starch products, especially resistant starch (RS) [16-18], and products of partial degradation of starch, that is, dextrins [19,20].

The objective of this study was to determine whether dextrin obtained as a result of heating starch with tartaric acid (patent claim no. 392894) is a substance with prebiotic properties [7]. Thus, it was examined whether the dextrin would be utilized as a source of carbon by probiotic and intestinal bacteria. It was also investigated whether probiotic lactobacilli and bifidobacteria cultured with intestinal bacteria in the presence of resistant dextrin would be able to dominate the intestinal isolates. In the study, the prebiotic index (PI) and the fermentation products of resistant dextrin were determined.

Materials and Methods

Materials

Potato starch and tartaric acid ($\geq$99.5%) were purchased from Sigma–Aldrich Corp. (Poznan, Poland). Hydrochloric acid and ethanol (96%) were procured from POCH (Gliwice, Poland). *Lactobacillus*,

*Corresponding author: Slizewska Katarzyna, Institute of Fermentation Technology and Microbiology, Faculty of Biotechnology and Food Sciences, Technical University of Lodz, 171/173 Wolczanska Street, 90-924 Lodz, Poland E-mail: katarzyna.slizewska@p.lodz.pl

Bifidobacterium, Clostridium, Fusobacterium, Escherichia coli and *Enterococcus* bacteria strains were isolated from feces of three healthy 70-year old men volunteers. The 24-h cultures were frozen at -20°C: (a) *Lactobacillus, Bifidobacterium* bacteria in MRS broth with 20% glycerol, (b) *Fusobacterium,* and *Clostridium* in VL broth, and (c) *Escherichia, Enterococcus* in a nutrient bouillon with 20% glycerol. Prior to experiments bacteria were activated by twofold inoculation (3%) in: (a) liquid MRS broth (*Lactobacillus, Bifidobacterium*), (b) liquid VL broth (*Fusobacterium, Clostridium*), and (c) nutrient broth (*Escherichia, Enterococcus*).

Preparation of dextrin

Enzyme-resistant tartaric acid-modified dextrin (TA–dextrin) was prepared following the method of Jochym et al. [21]. Thus potato starch was sprayed with hydrochloric acid solution (0.5% w/w) to obtain a final HCl concentration of 0.1% on a dry starch basis (dsb). The tartaric acid solution (20% w/v) was then added to obtain a final organic acid concentration of 40% dsb. Thoroughly mixed sample was dried at 110°C to obtain final moisture content below 5%. Dried sample (10 g) was placed in an anti-pressure bottle (SIMAX), capped and heated at 130°C for 2 h in an ELF 11/6 EUROTHERM CARBOLITE oven (Hope, England). Product was cooled in a desiccator and milled into powder with a particle size of <1 mm. Dextrin was then washed with 80% EtOH to remove excess of tartaric acid, and low molecular weight material formed during dextrinization, dried overnight at 50°C, and then at 110°C for 1 h, and finally milled in a cyclone lab sample mill (UDY Corp., Fort Collins, CO, USA) fitted with a 0.50 mm screen.

The dynamics of growth of mixtures of bacteria

The intestinal bacteria *Lactobacillus, Bifidobacterium, Escherichia coli, Enterococcus, Clostridium* and *Fusobacterium* were co-cultured in the presence of resistant dextrin to determine whether the beneficial bacteria *Lactobacillus* and *Bifidobacterium* can dominate their environment in the presence of a mixture of isolated intestinal bacteria. Inoculants of bacterial monocultures were prepared in such a way that after 24 h of growth the number of particular bacteria ranged from 3.50×10^7 to 4.50×10^7 CFU/mL, corresponding to the number of these bacteria in the terminal section of the ileum [22]. The monocultures of bacteria isolated from three 70-year-old persons were incubated in liquid MRS (*Lactobacillus* and *Bifidobacterium*), in liquid VL medium (*Clostridium* and *Fusobacterium*) and in liquid broth (*Escherichia coli* and *Enterococcus*). All monocultures were incubated in sterile 15 mL test tubes (Marfour) – *Lactobacillus, Escherichia coli, Enterococcus* under aerobic conditions and *Bifidobacterium, Fusobacterium* and *Clostridium* under anaerobic conditions. After incubation, the cultures were centrifuged in a MPW-350R centrifuge (Med. Instruments, Poland) at 9.000 rpm for 10 min at 22°C, the supernatant was decanted and the biomass was transferred to 100 mL of medium of Wynne et al. [23] with the addition of resistant dextrin (TA-dextrin). The cultures were incubated for 168 h under anaerobic conditions (in similar conditions as in the intestine). Following dilution in physiological salt, the cultures were plated (Koch's plate method) in duplicate immediately after inoculation (0 h) and after 24, 48, 72 and 168 h on selective media: *Lactobacillus* on Rogosa agar, *Bifidobacterium* on RCA agar with the addition of the antibiotic dicloxacillin, *Escherichia coli* on ENDO agar, *Enterococcus* on bile-aesculin agar, *Clostridium* on DRCM agar and *Fusobacterium* on Schaedler agar with an antibiotic. The plates were incubated for 48 h at 37°C; *Lactobacillus, Escherichia coli,* and *Enterococcus* under aerobic conditions and *Bifidobacterium, Fusobacterium* and *Clostridium* under anaerobic conditions in a Concept 400 anaerobic chamber (Ruskinn Biotrace, USA). The control

trial was determined by the trial without the addition of the source carbon.

Determination of prebiotic index (PI)

Prebiotic fermentation of resistant dextrins were analyzed using quantitative equation (prebiotic index – PI). The PI equation is based on the changes in key bacterial groups during fermentation. The bacterial groups incorporated into this PI equation were bifidobacteria, lactobacilli, clostridia and *Fusobacterium.* The equation assumes that an increase in the populations of bifidobacteria and/or lactobacilli is a positive effect while an increase in *Fusobacterium* and clostridia are negative [8].

The PI equation is described below:

$$PI = (Bif/Total) - (Fus/Total) + (Lac/Total) - (Clos/Total)$$

where PI is prebiotic index; Bif, bifidobacterial numbers at sample time/numbers at inoculation; Fus, Fusobacterium numbers at sample time/numbers at inoculation; Lac, lactobacilli numbers at sample time/ numbers at inoculation; Clos, clostridia numbers at sample time/ numbers at inoculation; Total, total bacteria numbers at sample time/ numbers at inoculation.

pH changes

Changes in pH were monitored with an Elmetron CP-401 pH-meter (Elmetron, Zabrze, Poland).

Determination of fermentation products by High Performance Liquid Chromatography (HPLC)

Organic acids, aldehydes and ethanol concentrations were determined by HPLC in supernatant liquid. The chromatographic analysis was performed by Finnigan Surveyor HPLC system (Thermo Scientific, Riviera Beach Fl., USA) with refractive index (RI Plus) and photodiode (PDA Plus) detectors. The column used was an Aminex HPX 87H, 300 × 7.8 mm (HPLC Organic Acid Analysis Column, Bio-Rad, Hercules CA, USA). The mobile phase was 0.005 M H_2SO_4. The separation was carried out by isocratic elution with a flow rate of 0.6 ml/min, and the column temperature was maintained at a constant 60°C.

Quantification of fermentation products was carried out using the external standard method. Lactic acid, formic acid, acetic acid, propionic acid, butyric acid, succinic acid, ethanol, and acetaldehyde of known retention times were used as external standards. For each standard, solutions were prepared, filtered through 0.22 μm syringe filters (Milipore, Belford, USA), and injected into the HPLC system to provide standard curves (concentration versus peak area), and for calculating the quantities of organic acids, aldehydes and ethanol. Linear regression curves based on peak areas were calculated for the individual standards covering a broad range of concentrations (Table 1).

Results and Discussion

It seems likely that prebiotic activity will be exhibited by dextrin obtained by simultaneous thermolysis and chemical modification of potato starch in the presence of a volatile inorganic acid (hydrochloric acid) as a catalyst of the dextrinization process and an excess amount of an organic acid (tartaric acid) as a modifying factor (patent claim no. 392894 "Preparation with prebiotic qualities"). In previous research, Kapusniak et al. [24] analyzed this dextrin in terms of the solubility and pH of its 1% aqueous solution, the content of reducing sugars,

Substance	Calibration regression equation	Retention times (min)
Lactic acid	$C=2.36 \times 10^{-7} \times p$	12.7
Acetic acid	$C=4.08 \times 10^{-7} \times p$	15.7
Formic acid	$C=2 \times 10^{-6} \times p+0.0576$	14.4
Propionic acid	$C=2 \times 10^{-6} \times p+0.0218$	18.5
Butyric acid	$C=2 \times 10^{-6} \times p+0.0182$	16.7
Succinic acid	$C=1.97 \times 10^{-7} \times p$	12.2
Acetaldehyde	$C=4.27 \times 10^{-7} \times p$	17.5
Ethanol	$C=3.85 \times 10^{-7} \times p$	22.6

C: The concentration of the analyzed substances (mg/100 ml).
P: Peak area.

Table 1: Regression equation and retention time for HPLC standards.

molecular mass distribution, weight average molecular mass using high performance size-exclusion chromatography (HPSEC), average chain length using high performance anion exchange chromatography with pulsed amperometric detection (HPAEC-PAD), and the content of the resistant fraction using the enzymatic-gravimetric method AOAC 991.43 [25], the enzymatic-gravimetric-chromatographic method AOAC 2001.03 [26], the enzymatic-spectrophotometric method [27], and the pancreatin-gravimetric method [28]. It has been shown that the use of tartaric acid in the process of starch thermolysis yields acidic dextrin characterized by high water solubility (about 68%) and a high content of reducing sugars (about 29%). It has also been found that dextrin modified with tartaric acid does not contain any traces of unreacted starch, and the percentage share of the main fraction (having a weight average molecular mass of about 1.800 g/mol) was 80%. The average length of the carbohydrate chain in dextrin obtained with tartaric acid was 8.2 as determined by means of HPAEC. A study by Kapusniak et al. [24] revealed that the content of the resistant fraction in dextrin modified with tartaric acid, determined by means of AOAC 991.43, amounted to 44.5%. However, the results obtained by the Englyst enzymatic-spectrophotometric method showed that the actual content of the resistant fraction was above 68% [27]. Kapusniak et al. [24] used the official AOAC 2001.03 method to determine the content of the resistant fraction in dextrin modified with tartaric acid. This is the latest approved method for determining the total content of dietary fiber in foods containing resistant maltodextrins. Apart from measuring the content of insoluble dietary fiber and the high molecular weight fractions

of soluble fiber, this method makes it possible to determine resistant oligosaccharides (by using high-performance liquid chromatography, HPLC). Currently, many authors define resistant starch as the starch fraction that remains undigested by amylolytic enzymes after 16 h. In the Englyst method, fractions undigested after 120 min are considered resistant. In the pancreatin-gravimetric method, similarly as in the Englyst method, samples are digested with pancreatin, but resistant fractions are determined gravimetrically only after 16 h. In the case of dextrin modified with tartaric acid, the results of determination by the pancreatin-gravimetric method (67%) were similar to those obtained in previous studies using the Englyst method (68%), but much higher than those obtained using the AOAC 2001.03 method (50%) [27,24]. The observed differences among the various methods in terms of the measured content of the resistant fraction in dextrin modified with tartaric acid was caused by the fact that, according to the latest reports, enzymatic-gravimetric methods (including AOAC 2001.03) using thermostable α-amylase can determine only part of resistant starch type 4 [29]. And based on enzymatic tests, it can be argued that dextrin obtained using an excessive amount of tartaric acid may be classified as resistant starch type 4.

In the present study, enzyme-resistant dextrin, prepared by heating of potato starch in the presence of hydrochloric (0.1% dsb) and tartaric (40% dsb) acid at 130°C for 2 h (TA-dextrin), was tested as the source of carbon for probiotic lactobacilli and bifidobacteria cultured with the intestinal bacteria isolated from the feces of three healthy 70-year-old volunteers.

In media where TA-dextrin was the source of carbon, all *Lactobacillus* and *Bifidobacterium* strains reached the stationary phase at 24 h of incubation. The number of bacteria of the genus *Lactobacillus* and *Bifidobacterium* in the stationary phase was similar and amounted to: 8.70 log cfu/mL and 8.41 cfu/mL, respectively. At 168 h of culture in a medium with dextrin modified with tartaric acid, the number of lactobacilli and bifidobacteria remained high and ranged from 7.75 to 7.91 log cfu/mL, which shows their substantial viability (Figure 1).

The control strains were cultured in media with glucose. At 24 h of incubation, in cultures with glucose the number of lactobacilli and bifidobacteria amounted to from 9.35 to 8.83 log cfu/mL. The bacteria entered the stationary phase, similarly as in media containing dextrin,

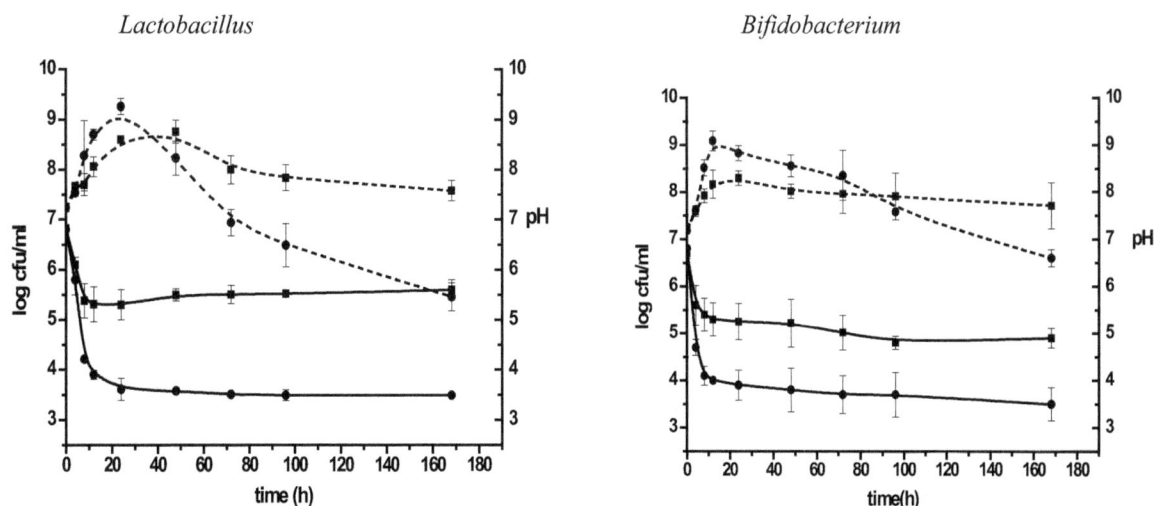

Figure 1: The growth curves (--) and changes in pH (-) for *Lactobacillus* and *Bifidobacterium* bacteria grown in the medium containing TA–dextrin (■) or glucose (control) (●). Results show means and standard deviations of n=3 replicates.

after 24 h of incubation. However, the stationary phase lasted much shorter than in media containing dextrin. At 168 h of culture, the number of viable *Lactobacillus* and *Bifidobacterium* cells cultured with glucose was much lower than that of cells cultured with resistant dextrin, and amounted to from 5.56 to 6.70 log cfu/mL (Figure 1).

In the medium containing dextrin, the acidifying activity of bifidobacteria was higher than that of lactobacilli. After incubation, a test of culture pH revealed that *Bifidobacterium* had the highest acidifying activity (pH 4.9), while *Lactobacillus* – the lowest (pH 5.6). In the control medium containing glucose, the pH of *Lactobacillus* and *Bifidobacterium* cultures decreased much more than that in the medium containing dextrin; at 168 h the pH was 3.50 (Figure 1).

In media containing TA-dextrin, the other bacteria isolated from human feces were able to grow, but the degree of dextrin utilization depended on the strain. In media with resistant dextrin, bacteria entered the stationary phase between 12 and 48 h of incubation and it lasted for 20-30 consecutive hours. In this phase, the highest cell count was found for *Fusobacterium* strains (8.50 log cfu/mL) (Figure 2). Also *Enterococcus* strains grew successfully, with the number of cells in the stationary phase reaching 8.49 log cfu/mL. Lower growth was found for *E. coli* and *Clostridium* strains (8.31 and 8.26 log cfu/mL, respectively) (Figure 2). In media containing dextrin, bacteria isolated from human

feces preserved high viability, and at 168 h the number of viable cells was by 1-2.5 log cycles larger than that of the control cells cultured with glucose.

Out of the bacteria isolated from human feces, the most acidifying ones were *Clostridium* strains, which decreased pH to about 3.50. In cultures containing dextrin, no significant differences were found between the remaining strains, as the pH ranged from 4.51 to 5.50. The strains were most active by 24 h of incubation; later on the pH values did not change significantly (Figure 2).

It was shown that all the bacteria isolated from human feces were able to grow and utilize TA-dextrin as a source of carbon, albeit to varying degrees. The highest growth was recorded for *Lactobacillus* and *Bifidobacterium*. *Bifidobacterium* strains were also characterized by the highest acidifying activity (lowering pH to 4.9), which remains consistent with the results reported by other authors [15,30,31]. The weakest growth was observed for *Clostridium* and *E. coli*. It was found that the stationary phase for *Lactobacillus* and *Bifidobacterium* strains was much longer than for other intestinal bacteria. After prolonging culture time to 72-168 h, which corresponds to retarded or pathological passage of digesta through the large intestine, the viability of intestinal bacteria in a medium with resistant dextrin was found to be lower by 1-1.5 log cycles than that of *Lactobacillus* and *Bifidobacterium*. The

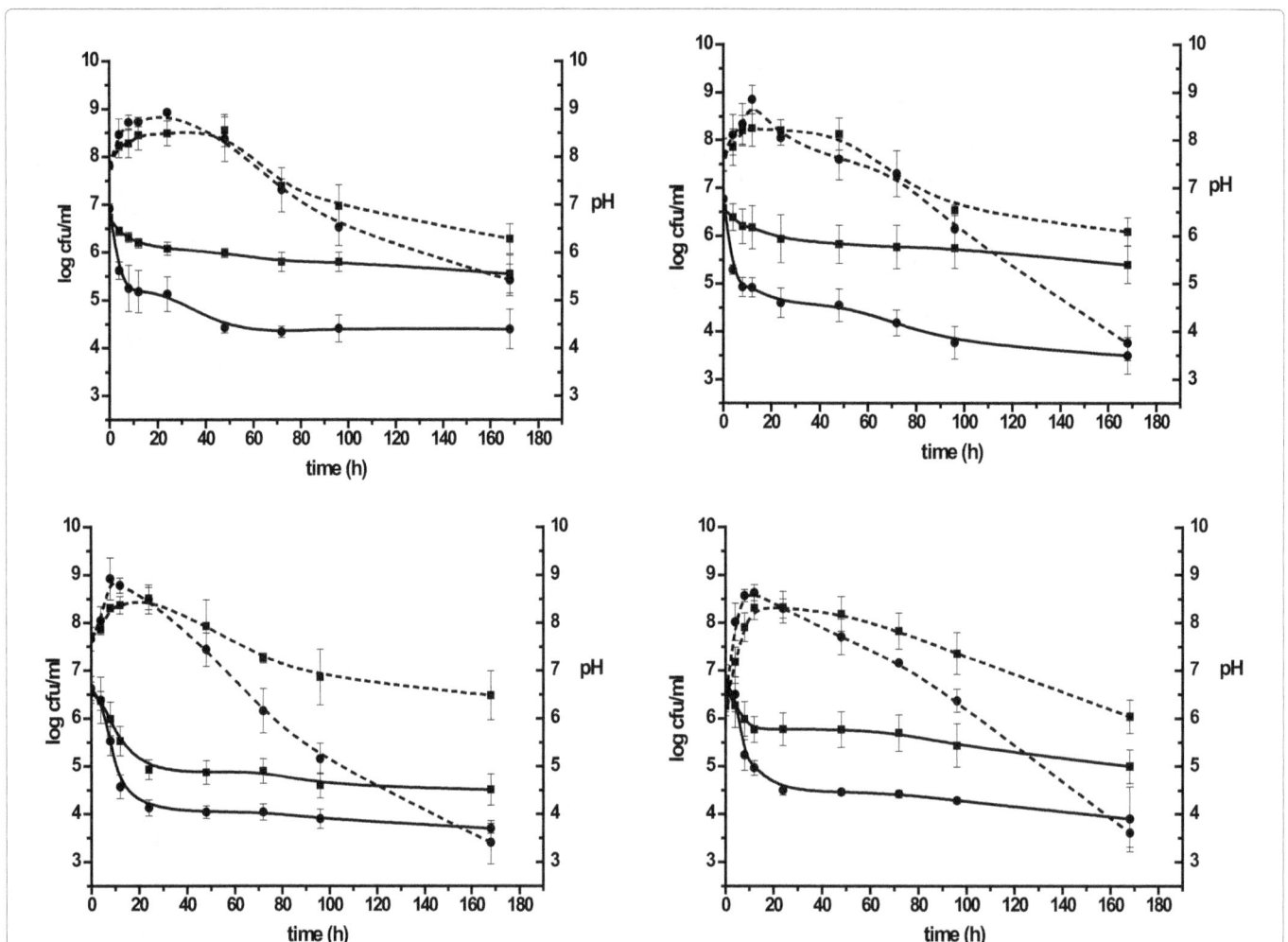

Figure 2: The growth curves (--) and changes in pH (–) for *Fusobacterium*, *Clostridium*, *Enterococcus* and *Escherichia coli* bacteria grown in the medium containing TA–dextrin (■) or glucose (control) (●). Results show means and standard deviations of n=3 replicates.

The Effect of Tartaric Acid-Modified Enzyme-Resistant Dextrin from Potato Starch on Growth...

145

Fermentation products	Lactobacillus	Bifidobacterium	Escherichia coli	Enterococcus	Clostridium	Fusobacterium
Lactic acid	112.7 ± 0.81	109.5 ± 0.42	85.0 ± 1.00	90.2 ± 1.50	55.2 ± 0.90	53.5 ± 0.21
Acetic acid	1.5 ± 0.15	18.2 ± 0.50	19.5 ± 0.15	6.5 ± 0.30	71.5 ± 0.05	14.0 ± 0.15
Formic acid	0	49.2 ± 0.25	84.5 ± 0.26	60.0 ± 0.15	0	0
Propionic acid	4.5 ± 0.01	4.6 ± 0.45	0	0	0	390.0 ± 1.85
Butyric acid	20.0 ± 0.12	11.5 ± 0.25	0	0	80.5 ± 0.51	0
Succinic acid	0	0	6.0 ± 0.10	1.7 ± 0.10	9.2 ± 0.11	28.1 ±1.00
Acetaldehyde	2.2 ± 0.15	0.3 ± 0.15	5.5 ± 0.01	5.0 ± 0.15	0.3 ± 0.02	5.1 ± 0.1
Ethanol	0.20 ± 0.01	0.15 ± 0.02	0.09 ± 0.01	0.08 ± 0.01	0.07 ± 0.01	0

SD: Standard Deviation; n=3 replicates.

Table 2: Concentration (mg/100 ml) of fermentation products after 24-h incubation of bacteria isolated from gastrointestinal tract of man in the broth containing TA–dextrin as the only one source of carbon.

number of *Lactobacillus*, *Bifidobacterium*, and other bacteria isolated from fecal samples grown in media containing 1% glucose was lower by 1-2.5 log cycles than that of corresponding bacteria grown in a medium containing dextrin. This may have been caused by the lower pH values of the controls, under which the culture environment became unfavorable to preserving high viability by the studied bacteria. This may have also been caused by the protective effects of dextrin on the bacteria.

It was shown that the PI values in media with TA-dextrin were positive; furthermore, the prebiotic index increased with the time of culture (from 0.24 at 24 h of incubation to 0.31 at 168 h), which proves that beneficial bacteria (*Bifidobacterium* and *Lactobacillus*) can dominate their environment in the presence of a mixture of intestinal bacteria cultured with the addition of resistant dextrin.

The calculated PI values for TA-dextrin were higher than those reported by Olano-Martin et al. [32] or by Kordyl [33] inulin and oligosaccharides under the same incubation conditions (anaerobiosis; 1% prebiotic addition; pH 6.8; incubation temperature of 37°C), which shows that TA-dextrin may act as a prebiotic substance.

The HPLC results indicate that the main metabolite produced by *Lactobacillus* and *Bifidobacterium* was lactic acid (Table 2). Depending on the bacterial strain, its concentration after 24 h of incubation ranged from 109.5 mg/100 mL for *Bifidobacterium* to 112.7 mg/100 mL for *Lactobacillus*.

Another fermentation product was acetic acid (1.5 mg/100 mL for *Lactobacillus* and 18.2 mg/100 mL for *Bifidobacterium*), propionic acid (4.5-4.6 mg/100 mL for *Lactobacillus* and *Bifidobacterium*), and butyric acid (20.0 mg/100 mL for *Lactobacillus* and 11.5 mg/100 mL for *Bifidobacterium*). *Bifidobacteria* generated up to 50 mg/100 mL of formic acid. Furthermore, *Lactobacillus* and *Bifidobacterium* strains fermenting TA-dextrin produced acetaldehyde (0.3 to 2.2 mg/100 mL) and ethanol (0.1-0.2 mg/100 mL).

It was found that the main metabolite produced during the fermentation of TA-dextrin by *Fusobacterium* was propionic acid, with a concentration reaching 390 mg/100 mL. *Fusobacterium* also produced lactic acid (53.5 mg/100 mL), succinic acid (28.1 mg/100 mL), acetic acid (14.0 mg/100 mL), and small amounts of acetaldehyde (5.1 mg/100 mL) (Table 1).

The fermentation of TA-dextrin by *Enterococcus* led mostly to lactic and formic acids. The concentration of those acids, determined by means of HPLC, amounted to 90.2 and 60.0 mg/100 mL, respectively [34-36]. The average concentration of acetic acid and acetaldehyde was comparable and ranged from 6.5 to 5.0 mg/100 mL. Furthermore, *Enterococcus* produced traces of ethanol (0.08 mg/100 mL) (Table 2).

According to the results, the major products of the fermentation of TA-dextrin by *Clostridium* were butyric, acetic and lactic acids. The concentration of these acids was 80.5, 71.5, and 55.2 mg/100 mL, respectively. The concentration of other metabolites, namely succinic acid and acetaldehyde, was considerably lower. Moreover, traces of ethanol were also present (Table 1).

The HPLC results showed that *E. coli* also metabolized TA-dextrin, producing a considerable amount of lactic and formic acids, at concentrations reaching 85 mg/100 mL. The concentration of another fermentation product, acetic acid, was 19.5 mg/100 mL. Succinic acid and acetaldehyde were produced at a similar level (5.5-6.0 mg/100 mL). *E. coli* produced traces of ethanol as well (Table 2).

Conclusions

The experiments showed that dextrin obtained as a result of heating potato starch in the presence of hydrochloric acid (0.1% of starch dry mass) and tartaric acid (40% of starch dry mass) at 130°C for 2 h may have prebiotic properties. The presented results are promising, but, according to the recommendations of FAO experts concerning the applications of prebiotics, it is still necessary to continue with in vivo experiments, which are now being conducted.

Acknowledgment

The study was supported by the Polish Ministry of Science and Higher Education, Grant No. N N312 3353 39.

References

1. Slizewska K, Kapusniak J, Barczynska R, Jochym K (2012) Resistant Dextrins as Prebiotic. In: Chang C-F (edn) Carbohydrates - Comprehensive studies on glycobiology and glycotechnology, In Tech, Rijeka, Croatia, pp. 261-288.

2. Gibson GR, Roberfroid MB (1995) Dietary modulation of the human colonic microbiota: introducing the concept of prebiotics. J Nutr 125: 1401-1412.

3. Gibson GR (2004) Fibre and effects on probiotics (the prebiotic concept). Clin Nutr Suppl 1: 25-31.

4. ISAPP (2008) 6th Meeting of the International Scientific Association of Probiotics and Prebiotics, London, Ontario.

5. FAO (2007) Technical Meeting on Prebiotics. Food Quality and Standards Service (AGNS), Food and Agriculture Organization of the United Nations (FAO). FAO Technical meeting Report, 15-16.

6. Roberfroid M, Gibson GR, Hoyles L, McCartney AL, Rastall R, et al. (2010) Prebiotic effects: metabolic and health benefits. Br J Nutr 104 Suppl 2: S1-63.

7. Slizewska K, Kapusniak J, Barczynska R, Libudzisz Z, Jochym K (2010) Preparation of prebiotic properties, Polish patent claim P-392895.

8. Palframan R, Gibson GR, Rastall RA (2003) Development of a quantitative tool for the comparison of the prebiotic effect of dietary oligosaccharides. Lett Appl Microbiol 37: 281-284.

9. Van Loo J, Gibson GR (2006) Inulin-type fructans as prebiotics. In: Rastall

RA (edn) Prebiotics. Development and Application, John Wiley and Sons, Chichester, pp. 57-100.

10. Davis LM, Martínez I, Walter J, Hutkins R (2010) A dose dependent impact of prebiotic galactooligosaccharides on the intestinal microbiota of healthy adults. Int J Food Microbiol 144: 285-292.

11. Sako T, Matsumoto K, Tanaka R (1999) recent progress on research and applications of non-digestible galacto oligosaccharides. Int Dairy J 9: 69-80.

12. Sekhar MS, Unnikrishnan MK, Rodrigues GS, Mukhopadhyay C (2013) Synbiotic formulation of probiotic and lactulose combination for hepatic encephalopathy treatment: a realistic hope? Med Hypotheses 81: 167-168.

13. Tuohy KM, Ziemer CJ, Klinder A, Knobel Y, Pool-Zobel BL et al. (2002) A human volunteer study to determine the prebiotic effects of lactulose powder on human colonic microbiota, Microb Ecol Health Dis 14: 165-173.

14. Kaneko T, Yokoyama A, Suzuki M (1995) Digestibility characteristics of isomaltooligosaccharides in comparison with several saccharides using the rat jejunum loop method. Biosci Biotechnol Biochem 59: 1190-1194.

15. Crittenden RG, Morris LF, Harvey ML, Tran LT, Mitchell HL, et al. (2001) Selection of a Bifidobacterium strain to complement resistant starch in a synbiotic yoghurt. J Appl Microbiol 90: 268-278.

16. Fuentes-Zaragoza E, Sanchez-Zapata E, Sendra E, Sayas E, Navarro C, et al. (2011) Resistant starch as prebiotic: A review. Starch/Starke 63: 406-415.

17. Leszczynski W (2004) Resistant Starch – classification, structure, production. Polish J Food Nutr Sci 13: 37-50.

18. Nugent AP (2005) Health properties of resistant starch. Nutr Bull 30: 27-54.

19. Betty W Li (2010) Analysis of Dietary Fiber and Non digestible Carbohydrates. In: Cho SS, Finocchiaro ET (eds) Handbook of prebiotics and probiotics ingredients, CRC Press Taylor & Francis Group, Boca Raton, pp. 1-12.

20. Mermelstein NH (2009) Analyzing for resistant starch. Food Technol 4: 80-84.

21. Jochym K, Kapusniak J, Barczynska R, Sliżewska K (2012) New starch preparations resistant to enzymatic digestion. J Sci Food Agric 92: 886-891.

22. Ouwehand A, Vesterlund S (2003) Health aspects of probiotics. IDrugs 6: 573-580.

23. Wynne AG, McCartney AL, Brostoff J, Hudspith BN, Gibson GR et al. (2004) An in vitro assessment of the effects of broad-spectrum antibiotics on the human gut microflora and concomitant isolation of a Lactobacillus plantarum with anti-Candida activities. Anaerobe 10: 165-169.

24. Kapusniak J, Barczynska R, Slizewska K, Libudzisz Z. (2008) Utilization of enzyme-resistant chemically modified dextrins from potato starch by Lactobacillus bacteria. Zeszyty Problemowe Postepow Nauk Rolniczych 530: 445-457.

25. Lee SC, Prosky L, de Vries JW (1992) Determination of total, soluble and insoluble dietary fiber in food – enzymatic-gravimetric method, MES-TRIS buffer: collaborative study. J Assoc Off Analyt Chem 75: 395-416.

26. Ohkuma K, Matsuda I, Katta Y, Tsuji K, Ohkuma K, et al. (2000) New method for determining total dietary fiber by liquid chromatography. J AOAC Int 83: 1013-1019.

27. Englyst HN, Kingman SM, Cummings JH (1992) Classification and measurement of nutritionally important starch fractions. Eur J Clin Nutr 46 Suppl 2: S33-50.

28. Shin M, Song J, Seib P (2004) In vitro digestibility of cross-linked starches – RS4. Starch/ Starke 56: 478-483.

29. Nishibata T, Tashiro K, Kanahori S, Hashizume C, Kitagawa M, et al. (2009) Comprehensive measurement of total nondigestible carbohydrates in foods by enzymatic-gravimetric method and liquid chromatography. J Agric Food Chem 57: 7659-7665.

30. Bielecka M, Biedrzycka E, Majkowska A, Juskiewicz J (2002) Effect of non-digestible oligosaccharides on gut micro ecosystem in rats. Food Res Inter 35: 139-144.

31. Dale JW, Park SF (2010) In: Molecular genetics of bacteria, Wiley-Blackwell, Chichester.

32. Olano-Martin E, Gibson GR, Rastell RA (2002) Comparison of the in vitro bifidogenic properties of pectins and pectic-oligosaccharides. J Appl Microbiol 93: 505-511.

33. Kordyl M (2010) Ability of intestinal microorganisms to metabolize prebiotic preparations, Ph D Thesis, Technical University of Lodz, Poland, Polish.

34. Crittenden RG, Playne MJ (2009) Handbook of probiotics and prebiotics, In: Lee YK, Salminen S (edn) John Wiley & Sons, New Jersey, pp 535-584.

35. Roberfroid MB (2007) Inulin-type fructans: functional food ingredients. J Nutr 137: 2493S-2502S.

36. Swennen K, Courtin CM, Delcour JA (2006) Non-digestible oligosaccharides with prebiotic properties. Crit Rev Food Sci Nutr 46: 459-471.

24

Phytochemical Analysis and Antibacterial and Cytotoxic Properties of *Barleria lupulina Lindl.* Extracts

Reshma Kumari* and Ramesh Chandra Dubey

Department of Botany and Microbiology, Gurukula Kangri University, Haridwar-249404, India

Abstract

The ethanolic and aqueous extracts of *Barleria lupulina* leaves displayed antibacterial activity against five human bacterial pathogens viz., *Escherichia coli, Pseudomonas aeruginosa, Staphylococcus aureus* and *Salmonella typhi Klebsiella pneumoniae*. The ethanolic extract was more inhibitory than the aqueous extract against all the test pathogens, which caused the maximum growth inhibition of *P. aeruginosa* at 100% concentration. In contrast, aqueous extract did not inhibit the growth of any bacterial pathogens. MIC of ethanolic extract was 2.5 mg/mL against *E. coli, S. aureus* and *P. aeruginosa*, and 10.0 mg/mL against *S. typhi* and *K. pneumonia*. GC-MS analysis displayed the presence of twelve phytochemical compounds among which benzofuranon, hexadecanoic acid, ethyl 9,12,15-octadecatrienoate, and 3,7,11,15-tetramethyl-2-hexadecanoic acid were the most prominent ones. These extracts also displayed cytopathic effects against HepG2 cell line performed by employing the 3-(4,5-dimethylthiazol-2-yl)-2,5-diphenyl-2H-tetrazolium bromide (MTT) and neutral red uptake (NRU) assay which demonstrated the varying levels of cell death of HepG2 cells by ethanolic extract. The ethanolic extract of *B. lupulina* bears a significant amount of phytochemical compounds that pose antibacterial as well as anti-cancerous properties.

Keywords: Antibacterial activity; *Barleria lupulina*; Cytopathic; GC-MS; Phytochemical analysis

Introduction

Plant-derived natural products, such as flavonoids, terpenoids, alkaloids and steroids have received considerable attention in recent years due to their diverse and effective pharmacological properties including antibacterial, antioxidant and antitumor [1]. Issasbadi et al. reported that the use of plant is a cost effective and environment friendly. Several plants of Acanthaceae family are used as medicinal plant among which *Barleria lupulina* Lindl. is an important one which comprises of several medicinal properties. It is a herb which is widely distributed throughout tropical Asia. Leaves, stems and roots of *B. lupulina* and flower of *B. prionitis* possess potential antibacterial and anti-inflammatory activities [2-4]. However, cytotoxicity of *B. strigosa* extract has been described by Nuttaporn et al. [5]. Several phytochemicals including barlerin alkaloid is derivative from *B. lupulina* which possess antimicrobial and anticancerous properties. Iridoid glucosides, bataine and alkaloids have also been reported from *B. lupulina* plant [6-8]. Traditionally, leaves of this plant are used to treat snake bites, dog bites, swelling, bleeding wounds and rheumatism. The extracts of the plant also possess anti-HSV-2 activities [9], ameliorate secondary complications of diabetes including cataract [10,11], antiarthritic [12], anti-inflammatory [13], antimicrobial [14], anti-clastogenic, anti-tumor, and anti-cancer activities besides having radiation protection [15].

The cases of cancer are increasing day-by-day due to several reasons. The emergence of multi-drug resistance among bacterial pathogens throughout the world is kindled because of over prescription of antibiotics. Therefore, use of traditional medicinal plant is required as an alternative drug to cure various types of diseases. The present investigation was designed to assess the antibacterial properties of ethanolic and aqueous extracts of *B. lupulina* against five bacterial pathogens and cytotoxic effects on HepG2 cells, and also to identify the bioactive compounds present in its leaves.

Materials and Methods

Collection of plant material

The leaves of *B. lupulina* were collected from the Botanical Garden, Department of Botany and Microbiology, Gurukula Kangri University, Haridwar (India) during the month of June 2014, and identified by the experts of the department (specimen identification No Bot. and micro/199). The plant leaves were washed with running tap water to remove the adhered dust, dirt and unwanted particles from their surfaces. The leaves were shade dried at room temperature for 15 days and separately homogenized in domestic blender to get fine powders. The powders were stored in airtight container at room temperature for further studies.

Bacterial cultures

Five standard human enteric pathogenic bacteria viz. *Staphylococcus aureus* ATCC 25923, *Escherichia coli* ATCC 27853, *Pseudomonas aeruginosa* ATCC 25922, *Klebsiella pneumoniae* MTCC 432 and *Salmonella typhi* MTCC 733 were procured from the Laboratory of the Departmental of Botany & Microbiology, Gurukula Kangri University. All the pathogens were sub cultured on nutrient agar slants and preserved at 4ºC for further study.

Preparation of crude extracts

The dried leaf powder (100 g) was subjected to hot extraction in Soxhlet continuous extraction apparatus with 300 mL of ethanol and water (1:3 ratio) for 48-72 h. The solvent was gently evaporated at room temperature to get the final volume of 100 mL. The extract was filtered and considered as 100% concentrated extract. Further, the extract was diluted with respective solvent to get 75%, 50% and 25% concentrations for antibacterial activity assay. Aqueous extract was prepared by boiling 30 g leaves powder in 150 mL distilled water till the volume was reduced

---

*Corresponding author: Reshma Kumari, Department of Botany and Microbiology, Gurukula Kangri University, Haridwar-249404, India
E-mail: reshmagupta25@gmail.com

to one fourth of its original volume. 100% concentrated ethanolic and aqueous extracts were separately evaporated under vacuum distillation unit at 60°C. Both the extracts were stored in refrigerator for further use.

Evaluation of antibacterial activity of extracts

Antibacterial assay was carried out by agar well diffusion method [16]. A loopful colony of each bacterium was impregnated into 5 mL nutrient broth and incubated at 37°C for 4-6 h with vigorous shaking. Each bacterial culture was separately swabbed uniformly on the surface of solidified Mueller Hinton agar (MHA) medium with sterile cotton swab. Agar well was made with the help of a sterile cork borer (6 mm) in the MHA agar plate. *B. lupulina* extract (100 μL from 25%, 50%, 75% and 100% concentration) was separately poured into the wells using pre-sterile micropipette tips and incubated at 37°C for 24 h. Pure solvents were separately used instead of extract as a negative control for each bacterial strain, whereas streptomycin and ciprofloxacin were used as positive control. After incubation at 37°C for 24 h the diameter of zone of inhibition of treatment and control sets were measured. Inhibition (I) of the growth of bacterial colony was calculated by using the formula: $I = T - C$, where T = diameter of total inhibition zone in treatment, C = diameter of inhibition zone in control.

Determination of minimum inhibitory concentration (MIC) and survival of bacteria

MIC of extract was carried out to find out the lowest concentration of extract that inhibits the visible growth of test bacteria by standard tube dilution method [17]. A loopful culture from the slants was separately inoculated in nutrient broth and incubated at 37°C for 24 h. Fresh medium (20 mL) was seeded with 0.25 mL of 24 h old broth culture. The extract was dissolved in dimethyl sulphoxide (DMSO) to obtain 200 mg/mL stock solution. 0.2 mL solution of test material was added to 1.8 mL of the seeded broth which was further serially diluted up to 8 tubes with respect to positive control (streptomycin and ciprofloxacin). The concentration of extract ranged from 1.25 to 20 mg/mL. Thereafter, the tubes were incubated at 37°C for 24 h and results were recorded after completion of incubation period.

Phytochemical analysis

High Pressure Thin Layer Chromatography (HPTLC) of the extract: The ethanolic extract was dissolved in ethanol (2 mg/mL) to adjust the final volume. The extract was applied with the help of Linomat syringe (100 μL) using the Linomat applicator 5 on the HPTLC plates (20.0 × 10.0 cm). 5 and 7 μL of extract samples were separately applied as a band (6 mm). Distance between two bands was 12.0 nm. Dimension of slit was 6.00 × 0.30 mm. Silica gel plate acts as stationary phase and ethyl acetate (100): acetic acid (11): formic acid (11): water (28) was used as a mobile phase. The band of plate was developed through capillary action of mobile phase in CAMAG twin trough chamber (20 × 10 cm). Thereafter, the plate was taken out from the chamber and dried in air. CAMAG HPTLC Densitometer (Scanner 201377) was used to scan plate in absorbance mode at 366 nm and the scanning data was subjected for integration through the CAMAG Visualizer (201673). The plate was heated at 120°C for 20 min. The spots of the TLC plate were detected by using the spray reagent (anisaldehyde reagent). Development of different spots revealed different R_f values of phyto-constituent, which were recorded for further study.

Gas Chromatography-Mass Spectrometry (GC-MS) analysis: The ethanolic extract was analyzed through GC-MS by using a Varian-Bruker Scion SQ mass spectrometer system equipped with DB5 capillary column (0.25 mm thickness and 30 m in length). Extract was diluted (2 mg/mL) in ethanol and 1 μL was taken as the injection

volume. An aliquot of 20:1 ratio was injected as split mode into the GC-MS. The temperature was initially hold at 40°C for 4 min, rate of 20°C/5 min to 280°C, and the running time was 45 minutes using with maintaining the temperature (280°C). Helium was used as a carrier gas with a constant flow at 1 mL/min.

Cytotoxicity assay

Preparation of test material for MTT and NRU assay: Stock solution was prepared one day in advance with 40% ethanol and MQ water for ethanolic extracts and aqueous extracts, respectively. Multiple aliquots of each sample of extracts were made.

Cell line and culture condition: Cytotoxicity assay was performed using Hep G2 cells (a perpetual cell line consisting of human liver carcinoma cells) derived from the liver tissue (National Centre for Cell Science, Pune) which had a well-differentiated hepatocellular carcinoma. Hep G2 cells were maintained and cultured in EMEM (Eagle's minimum essential media) with 10% fetal bovine serum. Desired cell growth was maintained in humidified atmosphere with 5% CO_2 saturation at 37°C throughout the experiment. Cells were plated in 96-well microtitre plate to get a cell density of 2×10^4 cells per well.

3-(4,5-dimethylthiazol-2-yl)-2,5 diphenyltetrazollium bromide (MTT) assay: Cytotoxicity of ethanolic and aqueous extracts of *B. lupulina* on Hep G2 cell line was determined by MTT assay as described by Mosmann [17] with some modification. After 24 h incubation of seeded Hep G2 cells, medium of 96 well plates was replaced with fresh medium and treated with different concentrations of extract (0-1000 μg/mL) along with negative control of untreated cells. The plate was incubated at 37°C in 5% CO_2 incubator for 24 hours. MTT (0.5 mg/mL) solution (0.5 μL) was added to each treated well. It was incubated again as earlier 2-3 hours prior to the termination of experiment. At the end, the culture supernatant containing MTT was removed and 100 μL of DMSO was added to each plate containing Hep G2 cells. The plate was gently rotated to solubilize the purple crystal formazan. Absorbance was read after 10 minutes using a plate reader at 550 nm and 660 nm (Bio-Tek Instruments Inc. Vermont, USA).

Neutral red uptake (NRU) assay: Cytotoxicity of ethanolic and aqueous extracts of *B. lupulina* on Hep G2 cell line was also determined by NRU assay. This assay was performed similar to MTT assay. Only 0.5 mg/mL of NRU and 100 μL of 1% glacial acetic acid in 40% alcohol were used in place of MTT and di-methyl sulfoxide (DMSO), respectively.

Statistical analysis

The experiments were executed in triplicates. Statistical analysis was done using GraphPad Prism5 and Microsoft Excel 2008. Data have been expressed as mean significant at $p < 0.05$ level.

Results

Evaluation of antibacterial activity of extracts

The ethanolic extract of *B. lupulina* significantly ($p < 0.05$) inhibited the growth of all the bacterial pathogens except *K. pneumonia*, whereas the aqueous extract did not inhibit the growth of any bacteria. However, ethanolic extract at 100% concentration posed more lethal effect followed 75%, 50%, and 25% concentration. *P. aeruginosa* was highly sensitive to all the concentrations of the extract exhibiting the maximum zone of inhibition. The antibiotic ciprofloxacin caused the maximum growth inhibition of all bacteria at all concentrations as compared to streptomycin (Table 1 and Figure 1).

Concentrations of extract	Zone of inhibition (mm) of different bacterial species				
	E. coli	S. typhi	S. aureus	P. aeruginosa	K. pneumoniae
100%	9.33 ± 1.52	7 ± 2.64	7.33 ± 0.57	13.33 ± 4.72	R
75%	9 ± 0.0	5.33± 2.30	6.66 ± 0.57	11 ± 3.46	R
50%	6.66 ± 0.57	3.33 ± 2.30	5.66 ± 0.57	8.66 ± 3.51	R
25%	2 ± 0.0	6 ± 0.0	4.66 ± 0.57	7.33 ± 3.05	R
Streptomycin	23 ± 1.0	12.67 ± 0.57	24.67 ± 0.57	25.67 ± 1.52	23 ± 1.0
Ciprofloxacin	31 ± 1.0	18.67 ± 0.57	31 ± 1.0	33.67 ± 0.57	13.67 ± 0.57
Mean values ± Standard deviation; R: Resistance					

Table 1: Antibacterial activity of ethanolic extract of *B. lupulina* at different concentrations.

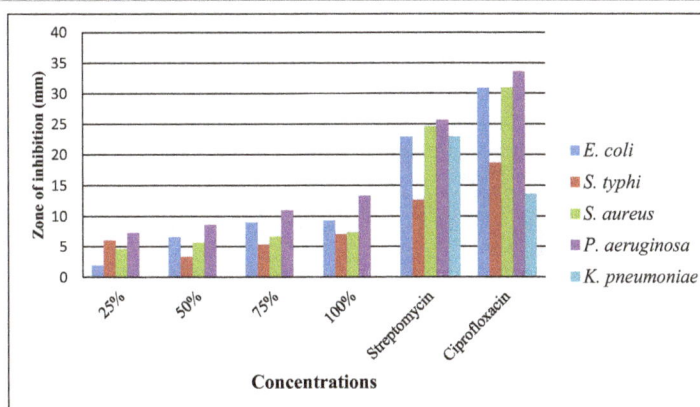

Figure 1: Mean values of diameters of zone of inhibition of bacterial pathogens caused by ethanolic extract of *B.lupulina*.

Name of extracts	MIC (mg/ml) of different bacterial species				
	E. coli	S. aureus	P. aeruginosa	S. typhi	K. pneumoniae
Ethanol	2.5	2.5	2.5	5.0	5.0
Aqueous	10.0	5.0	10.0	10.0	10.0

Table 2: Minimum inhibitory concentration (MIC) of ethanolic and aqueous leaf extracts of *B. lupulina*.

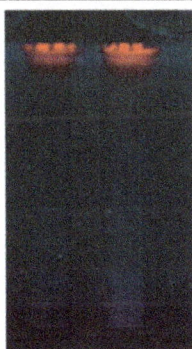

Figure 2: Different phyto-constituents separated from ethanolic extract of *B. lupulina* on different tracks at 366 nm on HPTLC plates.

Determination of minimum inhibitory concentration (MIC) of extract

The MIC of ethanolic extract against *E. coli, S. aureus* and *P. aeruginosa* was 2.5 mg/mL, whereas MIC against *S. typhi* and *K. pneumonia* was 5.0 mg/mL. In contrast, MIC of aqueous extract against all the bacterial pathogens was 10 mg/mL (Table 2).

High Pressure Thin Layer Chromatography (HPTLC) of the extract

Different phyto-constituents showing different R_f values viz., 0.17, 0.25, 0.32, 0.43, 0.60, 0.70, and 0.80 were recorded on HPTLC plates

from 5 µL sample of ethanolic extract of *B. lupulina*. Similarly chemical constituents having 0.16, 0.25, 0.32, 0.35, 0.44, 0.60, 0.73 and 0.81 R_f values were noted from 7 µL sample of extract. No extra spot was obtained under UV (254 nm) fluorescence mode. More or less similar R_f values were observed in both concentrations of extract. Spots of light pink colour appeared after spraying anisaldehyde (Figure 2) and may show presence of alkaloids and terpenoids group of compounds [18].

Gas Chromatography-Mass Spectrometry (GC-MS) analysis

Twelve compounds were identified through GC-MS analysis of ethanolic extract. Then the compounds were matched with NIST library. Some phytoconstituents viz., benzofuranon, hexadecanoic acid, ethyl 9,12,15-octadecatrienoate, and 3,7,11,15-tetramethyl-2-hexadecanoic acid etc. were identified (Table 3).

Cytotoxicity assays

MTT assay: The ethanolic extract of *B. lupulina* had the highest cytotoxicity effect on Hep G2 cells at its all concentrations as compared to that of aqueous extract. The ethanolic extract was effective at 50 µg/mL concentration, while aqueous leaf extract inhibited cell growth at 1000 µL/mL. Only 30% cell death was observed in both cases after 24 h of treatment by MTT assay (Figure 3). In MTT assay, relative percentage of live Hep G2 cells gradually declined with increase in concentration of extracts. But the significant inhibition was observed after 100 µg/mL concentration of the extract. Figure 4 showed cytopathic effect including cytoplasm vacuolation, cell shrinkage, lysis and death.

NRU assay: Cytotoxicity of both extracts were analysed through

RT*	Name of the compound	Molecular formula	Peak area	Total % of peak area
14.76	Tetradecane	$C_{14}H_{30}$	112073872	5.255
15.23	1,H-3a-7-methanoazulene	$C_{15}H_{24}$	126665448	5.939
15.79	Cis-thiopsine	$C_{15}H_{24}$	46636904	2.187
18.42	Benzofuranon	C_8H_6O	50920784	2.388
21.028	Hexadecane	$C_{16}H_{34}$	105627568	4.953
26.341	Phytol acetate	$C_{22}H_{42}O_2$	213710384	10.021
26.750	3,7,11,15, tetramethyl-2-hexadecanoic acid	$C_{21}H_{42}O_2$	36470580	1.710
27.043	3-Eicosyne	$C_{20}H_{38}$	87560240	4.106
28.732	Hexadecanoic acid	$C_{17}H_{34}O_2$	153626656	7.203
30.247	Oxiranehexadecyl (phytol)	$C_{20}H_{40}O$	891487168	41.801
30.938	Ethyl 9,12,15 Octadecatrienoate	$C_{20}H_{34}$	155076272	7.271
40.718	Squalene	$C_{30}H_{50}$	54701708	2.565
*RT: Retention time				

Table 3: Main bioactive compounds identified through GC-MS analysis of ethanolic extract of *B. lupulina* leaves.

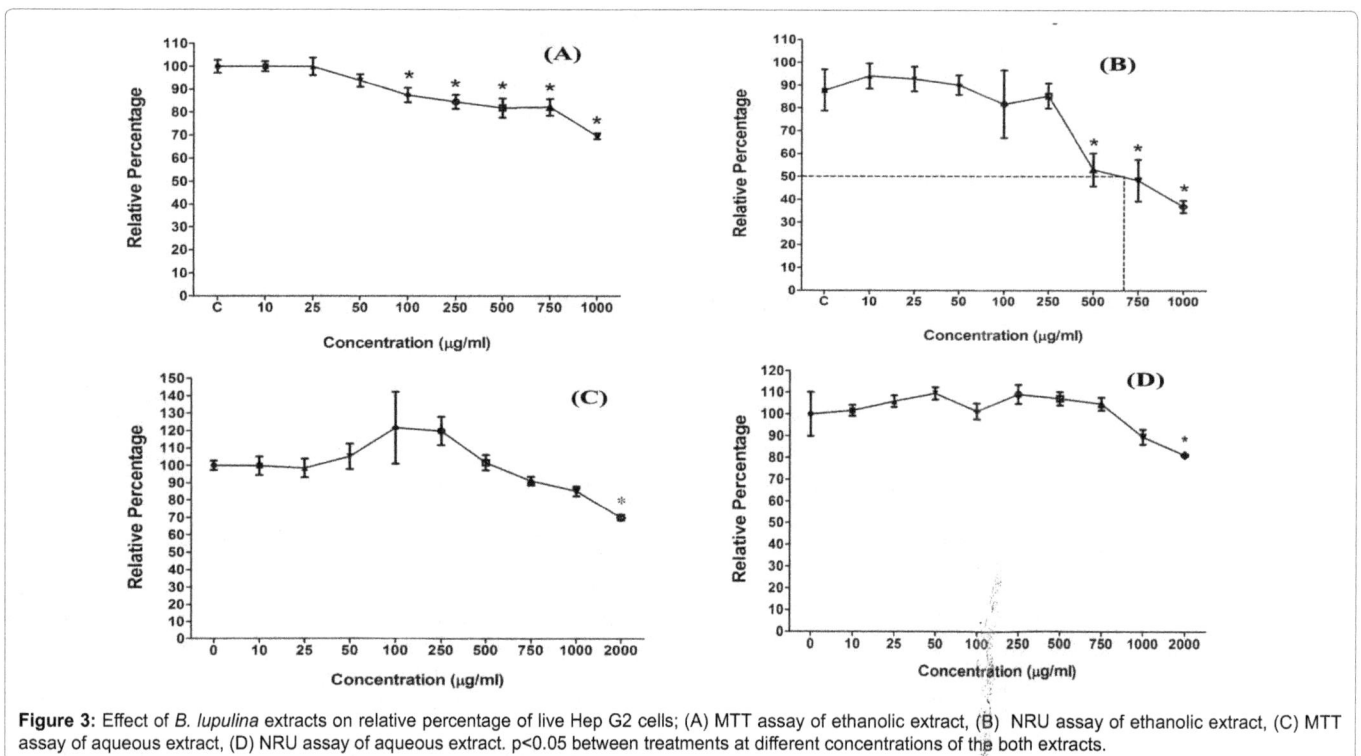

Figure 3: Effect of *B. lupulina* extracts on relative percentage of live Hep G2 cells; (A) MTT assay of ethanolic extract, (B) NRU assay of ethanolic extract, (C) MTT assay of aqueous extract, (D) NRU assay of aqueous extract. $p < 0.05$ between treatments at different concentrations of the both extracts.

NRU assay on Hep G2 cell line; more effect of ethanolic extract was recorded than aqueous extract. The highest non-cytotoxicity dose of ethanolic extract was observed at 250 μg/mL concentration. Ethanolic extract caused 52% cell death at 750 μg/mL concentration. Hence, IC_{50} value was found 650 μg/mL. Furthermore, approximately 20% cell inhibition was recorded at 1000 μg/mL of aqueous extract with different cytopathic effects (Figures 3 and 4).

Discussion

In the present investigation ethanolic extract of *B. lupulina* was found more inhibitory to growth of enteric bacterial pathogens than that of aqueous extract. However, ethanolic extract was highly effective on *P. aeruginosa* but least effective on *K. pneumoniae*. It is the first report on effect of ethanolic extract of *B. lupulina* on enteric bacteria. We found better inhibitory effect of ethanolic and aqueous extracts by agar well diffusion methods than that reported by Doss et al. [19]. This difference may be explained to be due to adoption of different assay methods. Bioactive phytochemicals would have been present in high

amount in agar wells than the filter paper discs. In contrast, ethanolic extract at 2.5 mg/mL MIC inhibited the growth of *E. coli*, *S. aureous* and *P. aeruginosa* as compared to that of aqueous extract all four pathogens were inhibited at 10 mg/mL except *S. aureus* (5.0 mg/mL). These differences may be due to more solubility of antimicrobial phyto-constituents in ethanol than water as well as extraction methods and execution of experiments.

In primary screening different phyto-constituents of ethanolic extract of *B. lupulina* were separated on different tracks on HPTLC plates that were identified as terpenoid and alkaloid groups of compounds. Sur [20] also reported six spots of ethanolic extract of *B. lupulina* under UV (254 nm) in preparative TLC. This result corresponds to the presence of steroid, terpenoid, glycoside, flavonoid, tannin and carbohydrate [20] which supports our findings. The presence of acetylbarlerin, barlerin and shanzhiside methyl ester compounds has earlier been identified from methanol extract of *B. lupulina* [21]; the R_f values of our findings are similar to that of earlier workers [20,21].

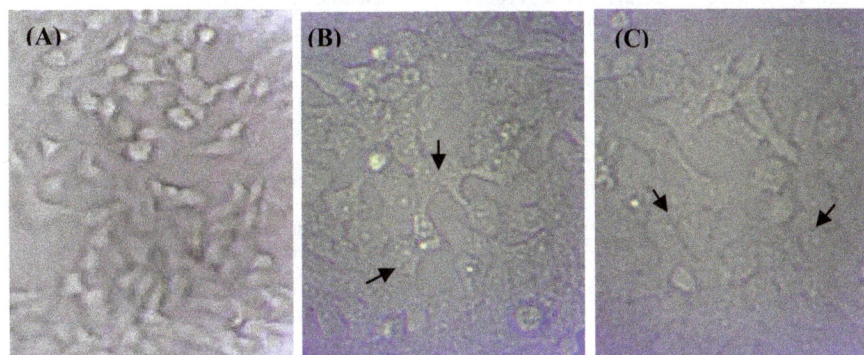

Figure 4: Cytopathic effects on Hep G2 cells treated by ethanol- and water-soluble extracts of *B. lupulina* (*see* arrows); (A) normal cells, (B) vacuolation and shrinkage of cells treated with ethanolic extract, (C) vacuolation and death of cells treated with aqueous extract.

Several chemical constituents have been identified through GC-MS analysis of the ethanolic extract (Table 3). Some compounds viz., phenol,2,4,bis (1,1-dimethylethyl)-tetradecanoic acid, 12- methyl-, methyl ester, hexadecanoic acid, methyl ester, phytol and octadecanoic acid have also been reported by Omulokoli et al. [22] and Namuli et al. [23]. Cyclobutane, 1,1-dimethyl- 2-octyl, 2-hexyl-1-octanol, 1, 2-henzenedicarboxylic acid, mono (2-ethylhexyl) ester and 1-hentetracontanol have been reported in essential oil of *B. lupulina* by Sarmad et al. [24]. But, in this communication we identified slightly different constituents. Some other phytochemical compounds viz., benzene (1-methyl decyle), benzoic acid 4-methoxy-methyl ester, propenoic acid, benzyl benzoate and 2 (4H)-benzofuranone were identified from acetone- and methanol-soluble extracts of *B. lupulina*. Presence of similar or slightly different phyto-constituents has recently been reported by Kumari and Dubey [25].

MTT and NRU dyes were used to visualize and count the viable Hep G2 cells treated with ethanolic and aqueous extracts of *B. lupulina*. The cytotoxic dose was higher but effective on Hep G2 cell line possibly due to adoption of different extraction method. IC_{50} dose could not be measured in both extract by using MTT. It is well known that methods, temperature and time of extraction, solvent type, concentration of solvent, etc. can affect the extraction of phytochemical constituents [26]. Ethanolic extract of *B. lupulina* exhibited the highest selective index (SI) (781.5) with lowest 50% inhibitory concentration (IC_{50}) dose (0.02 µg/mL) against HSV-2 cells as reported by Wirotesangthong and Rattanakiat [26]. Yoosook et al. [9] found higher IC_{50} values of *B. lupulina* than that of Wirotesangthong and Rattanakiat [27]. Therefore, Wirotesangthong and Rattanakiat accepted these differences because of use of different methods of extraction and anti-viral assay.

Multifarious cytotoxicity of *B. lupulina* extracts alone or in combination with other plant extracts has been reported by other workers [28,29]. But it is the first report on the cytotoxic effect of ethanolic extracts of *B. lupulina* on Hep G2 cell line. In our findings the Hep G2 cells treated with both ethanolic and aqueous extracts of *B. lupulina* leaves separately displayed of growth inhibition, vacuolation, cell shrinkage and cell lysis (Figure 4). Higher level cell death was caused by ethanolic extract than aqueous extracts as detected by NRU assay (Figure 4). Therefore, these differences gave different results.

Conclusion

It may be concluded that the *in vitro* data cannot be directly extrapolated to the *in vivo* conditions as several other factors are taken into account. Cytotoxicity on Hep G2 cells and antibacterial activities of ethanolic extract of *B. lupulina* leaves are being reported for the first time. Presence of several phytochemical constituents in the extracts of

B. lupulina leaves might be responsible for antibacterial effects. This study supports the ethano-botanical use of leaves of *B. lupulina* due to the presence of antibacterial and anti-cancerous phyto-constituents.

Conflicts of Interest

The authors declare no conflicts of interest.

Acknowledgements

The authors wish to thank the Head, Department of Botany and Microbiology, Gurukula Kangri University for providing laboratory facility, Aakaar Biotechnologies Private Limited (Lucknow) for Cytotoxicity test and Acharya Balkrishna, Patanjali Yog Peeth (Haridwar) for HPTLC and GC-MS analyses.

References

1. Anonymous (2006) Quality standards of Indian medical plants. Indian Council of Medical Research, New Delhi 4: 42-45.

2. Amoo SO, Finnie JF, Van Staden J (2009) *In vitro* pharmacological evaluation of three Barleria species. J Ethnopharmacol 121: 274-277.

3. Chavan CB, Shinde UV, Hogade M, Bhinge S (2010) Screening of *in-vitro* antibacterial assay of *Barleria proinitis* LINN. J Herb Med Toxicol 4: 197-200.

4. Chomnawang MT, Surassmo S, Nukoolkarn VS, Gritsanapan W (2005) Antimicrobial effects of Thai medicinal plants against acne-inducing bacteria. J Ethnopharmacol 101: 330-333.

5. Choudhury SM, Maity P, Bepari M (2015) Combined mixtures of *Calotropis gigantea* latex and *Barleria lupulina* leaf extracts ameliorate Dalton's ascitic lymphoma induced cell proliferation. Int J Eng Tech Sci Res 2: 2394-3386.

6. DeFeudis FV, Papadopoulos V, Drieu K (2003) *Ginkgo biloba* extracts and cancer: a research area in its infancy. Fundam Clin Pharmacol 17: 405-417.

7. Doss A, Parivuguna V, Vijayasanthi M, Sruthi S (2011) Antibacterial evaluation and phytochemical analysis of certain medicinal plants, Western Ghats, Coimbatore. J Res Biol 1: 24-29.

8. Issaabadi Z, Nastrollahzadeh S, Sajadi M (2016) Green synthesis of the copper nanoparticles supported on bentonite and investigation of its catalytic activity. J cleaner production.

9. Jaiswal SK, Dubey MK, Das S, Verma AR, Vijaykumar M, et al. (2010) Evaluation of flower of *Barleria prioniris* for anti-Inflammatory and anti nociceptive activity. Int J Pharma Bio Sci 1: 1-10.

10. Kaur PK, Vasisht K, Karan M (2014) HPTLC Method for Shanzhiside esters: Simultaneous quantitative analysis of barlerin, acetylbarlerin and shanzhiside methyl ester in *Barleria* species. J Chromatograph Separat Techniq 5: 246.

11. Kumari R, Dubey RC (2016) HPTLC and GC-MS profile of *B. lupulina Lindl.* extracts and their effect on enteric bacterial pathogens. J App Pharm 8: 61-68.

12. Mazumder PM, Mondal A, Sasmal D, Arulmozhi S, Rathinavelusamy P (2012) Evaluation of antiarthritic and immunomodulatory activity of *Barleria lupulina*. Asian Pac J Trop Biomed S1400-S1406.

13. Mazumder PM, Paramaguru R, Mohanty A, Sasmal D (2014) Evaluation of *in vitro* anticataract activity and aldose reductase potential of *Barleria lupulina Lindl.* Pharmacologia 5: 172-176.

14. Moghaddam MS, Kumar PA, Reddy GB, Ghole VS (2005) Effect of diabecon on sugar-induced lens opacity in organ culture: mechanism of action. J Ethnopharmacol 97: 397-403.

15. Mosmann T (1983) Rapid colorimetric assay for cellular growth and survival: application to proliferation and cytotoxicity assays. J Immunol Methods 65: 55-63.

16. Namuli A, Abdullah N, Sieo CC, Zuhainis SW, Oskoueian E (2011) Phytochemical compounds and antibacterial activity of Jatropha curcas Linn. Extracts. J Med Plants Res 5: 3982-3990.

17. Nuttaporn M, Poeaim S, Charoenying P (2015) Cytotoxicity and antimicrobial activities of leaf extracts from Barleria strigosa. J Agri Tech 11: 551-561.

18. Omulokoli E, Khan B, Chhabra SC (1997) Antiplasmodial activity of four Kenyan medicinal plants. J Ethnopharmacol 56: 133-137.

19. Perez C, Paul M, Bazerque P (1990) An antibiotic assay by the agar well diffusion method. Acta Bio Med Exp 15: 113-115.

20. Prashant T, Bimlesh K, Mandeep K, Gurpreet K, Kaur H (2011) Phytochemical screening and extraction: a review. Int Pharm Sci 198-106.

21. Sarmad M, Mahalakshmipriya A, Senthil K (2012) Chemical composition and in-vitro antimicrobial activity of Barleria lupulina essential oil. J Herbs Spice Med plants 18: 101-109.

22. Srivastava OP (1984) Techniques for the evalution of antimicrobial properties of the natural products. The use of pharmacological techniques for the evalution of antimicrobial properties of the natural products. UNESCO 72-79.

23. Suksamrarn S, Wongkrajang K, Kirtikara K, Suksamrarn A (2003) Iridoid glucosides from the flowers of Barleria lupulina. Planta Med 69: 877-879.

24. Sur PK (2015) Ameliorating effect of Barleria lupulina Lindl. extracts against (gamma)-ray (1.2 Gy) induced mitotic chromosomal aberrations of house musk shrew Suncus murinus. Int J Eng Res Man 2: 2349-2058.

25. Sur PK, Das PK (2012) Radio-protective and anti-clastogenic effects of Barleria lupulina Lindl. extract against Y (gamma)-ray (1.2 GY) induced mitotic chromosomal aberrations of laboratory mice Musmusculus and its effect on fish tumor induced after Y-irradiation. J Res Biol 2: 439-447.

26. Tuntiwachwuttikul P, Pancharoen O, Taylor WC (1998) Iridoid glucosides of Barleria lupulina. Phytochemistry 49: 163-166.

27. Wanikiat P, Panthong A, Sujayanon P, Yoosook C, Rossi AG, et al. (2008) The anti-inflammatory effects and the inhibition of neutrophil responsiveness by Barleria lupulina and Clinacanthus nutans extracts. J Ethnopharmacol 116: 234-244.

28. Wirotesangthong M, Rattanakiat S (2006) Anti-herpes simplex virus type 2 activities of some Thai medicinal plants. Thai j Pharm Sci 30: 19-27.

29. Yoosook C, Panpisutchai Y, Chaichana S, Santisuk T, Reutrakul V (1999) Evaluation of anti-HSV-2 activities of Barleria lupulina and Clinacanthus nutans. J Ethnopharmacol 67: 179-187.

25

Three-Dimensional Coating of Porous Activated Carbons with Silver Nanoparticles and its Scale-Up Design for Plant Disease Management in Greenhouses

Oleksandra Savchenko[1], Jie Chen[1,3,4], Yuzhi Hao[3], Xiaoyan Yang[3], Xiujie Li[2] and Jian Yang[2*]

[1]Department of Biomedical Engineering, University of Alberta, Edmonton, Alberta, Canada
[2]Alberta Innovates - Technology Futures, Bag-4000, Vegreville, AB, Canada
[3]Department of Electrical and Computer Engineering, University of Alberta, Edmonton, Alberta, Canada
[4]National Research Council, National Institute for Nanotechnology, Edmonton, Alberta, Canada

Abstract

Greenhouse vegetable production is significantly impacted by pathogens that cause diseases to the roots of the plants. These diseases are increasingly problematic in hydroponic vegetable production. Standard commercial practices in modern vegetable production facilities reuse the nutrient solution to reduce costs, or use dugout water for greenhouse irrigation in rural areas. This practice of recycling water may introduce or spread pathogens. Once pathogens contaminate water systems, they can spread quickly and cause dramatic losses to yield. Water filters have been used, but do not effectively kill fungi and bacteria. Therefore, better water treatment solutions are urgently needed to manage plant disease, especially for hydroponically grown vegetables. Numerous studies have demonstrated the efficacy of silver ion (Ag$^+$) and silver-based compounds for disinfection of a wide range of harmful microorganisms. In this article, we present a new filter material based on three-dimensional (3D) silver nanoparticle (AgNP)-coated substrates for water treatment. We prepared AgNP-coated active carbon materials and tested their antimicrobial efficacy against phytopathogenic bacterial and fungal spores, such as *Pseudomonas sp.*, and *Fusarium sp.* We then conducted large-scale tests in a dynamic flow setting and evaluated the effect of the filter on Pythium root rot control of hydroponically grown cucumbers. Results indicated that killing efficiencies of 3D coating were greater than 95% in the laboratory and that cucumber plants had no root infection in AgNP-AC filter treatment. The developed technique is approved to be a very efficient approach and has a great potential to be used in the greenhouse to manage plant root diseases.

Keywords: Water filtration; Plant disease management; Greenhouse vegetable production; Silver nanoparticle; Activated carbons; Greenhouse pathogens; Antibacterial filters; Hydroponic systems

Introduction

Root diseases are increasingly problematic in hydroponic vegetable production, because pathogens can quickly spread and cause losses to yield of agricultural products once they enter water systems. Highly destructive plant diseases, such as *Phytophthora* late blight, have become substantial risks to greenhouse tomatoes because of their presence in farm fields and home gardens. Pathogens are often introduced from dugout water used for greenhouse irrigation. The standard practice in modern vegetable production uses recycled nutrient solutions that further increase risk by spreading disease from infected plants throughout the entire greenhouse. Hong and Moorman [1] pointed out the challenges and opportunities of plant pathogens in irrigation water. Stewart-Wade [2] reported the detection and management of plant pathogens in recycled irrigation water in commercial plant nurseries and greenhouses. Various techniques have been applied to treat recycled nutrient solutions in greenhouses, including heat treatment, ozone treatment, UV disinfection, H_2O_2 treatment, and biofilters [1,2]. Biofilters are used in greenhouses to treat recycled nutrient solutions, but their main purpose is to convert ammonia to nitrogen gas and remove carbon dioxide and various organic contaminants, with less in-built design mechanisms for removing pathogens [3,4]. Better water treatment solutions are needed for plant disease management in the domain of hydroponically grown vegetables.

Silver nanoparticles (AgNP) range in size between 1 nm and 100 nm in diameter. Silver ions (Ag+) have been studied for many years for the disinfection of various harmful microorganisms because of their multiple modes of inhibition. Safavi et al. [5] reported that nano-silver could remove bacterial contaminants in plant tissue media. Dibrov et al. [6] showed that low concentrations of Ag$^+$ induce a massive proton leakage and result in a high degree of *Vibrio cholerae* death. Mazurak et al. [7] illustrated how silver-coated dressings can help with healing skin wounds. Maiti et al. [8] and Zahir et al. [9] demonstrated that AgNP can be synthesized by using eco-friendly reducing agents: red tomato juice and *Euphorbia prostrate* leaves, respectively. Tuan et al. [10] presented a modified sono-electrodeposition technique for making non-toxic nanosilver colloids. Karumuri et al. [11] discussed the coating of AgNP onto hierarchical structures fabricated by grafting carbon nanotubes to increase specific surface area. Richter et al. [12] explained the design of lignin nanoparticles infused with silver ions and coated with a cationic polyelectrolyte layer, which is a biodegradable and environmentally friendly alternative to silver nanoparticles.

Recently, several promising reports on using silver nanoparticles (AgNPs) against plant pathogenic fungi have been published. Jo et al. [13] and Kim et al. [14] showed various forms of silver ions and nanoparticles in killing two plant-pathogenic fungi: *Bipolaris sorokiniana* and *Magnaporthe grisea*, *Raffaelea* sp., and the other eight plant pathogenic fungi [15], respectively. Lamsal et al. [16] demonstrated in field tests, silver nanoparticles of 100 ppm reach the

*Corresponding author: Jian Yang, Alberta Innovates - Technology Futures, HW16A & 75th Street, Vegreville, AB, Canada
E-mail: jian.yang@albertainnovates.ca

highest inhibition rate before and after disease outbreaks on cucumbers and pumpkins. Nasrollahi et al. [17] examined the effect of AgNP on killing fungi (*Candida albicans*), and yeasts (*Saccharomyces cerevisiae*). To prevent the leakage of AgNP, Karumuri et al. [18] and Tuan et al. [10] proposed to coat AgNP on porous carbon structures. A review article on the antimicrobial action, synthesis, medical applications, and toxicity effects of AgNPs was published in 2012 [19]. In addition to AgNPs, Wani et al. [20] showed that zinc oxide (ZnO) and magnesium oxide (MgO) nanoparticles also exhibit antifungal effects. The mechanisms and potential application of synthesized AgNPs as alternatives to pesticides have been reviewed [21,22], but there have been few attempts to leverage AgNP technology for practical use.

In this article, we present the procedure to develop a new water filter material using an AgNP-coated substrate. Additionally, we show the results of efficacy evaluation of the treated filter substrate *in vitro* and in a growth chamber trial. The substrate was characterized using state-of-the-art equipment (i.e. Transmission Electron Microscopy, TEM). We subsequently evaluated the *in-vitro* antifungal effect of the designed filter against common bacterial and fungal pathogens. Since there are many plant pathogens in irrigation water [1], which can cause diseases in various greenhouse vegetables, it is almost impossible for us to test all pathogens. In this study, we chose bacteria *Pseudomonas* sp., *Paenibacillus* sp. and a fungus *Fusarium oxysporum* as examples to show the antimicrobial effectiveness of our designed bio-filtration material.

Materials and Methods

Sample preparation and characterization

Activated carbon (AC) has been successfully used for water purification over the past decades as an inexpensive and efficient filter material. We have combined the antibacterial properties of silver with the absorption properties and large surface area of activated carbon for water purification to kill bacteria and fungi that infect greenhouse crops. Considering the specific characteristics of AC, it was employed as the filter material of choice to be coated with silver nanoparticles. Two types of AC materials were tested - larger granules of AquaSorb® 1500 (Jacobi) from Bituminous coal, water washed (size 0.6 - 2.36 mm); and smaller diameter AC - CR1230C-AW (Carbon Resources manufacturer), from coconut shell, acid washed (size 0.6-1.7 mm).

We evaluated two types of coating methods, single coating and double coating, in order to determine which has better killing effects. Sample preparation was done according to the following protocols:

Single coating: A range of AgNP concentrations were tested to determine the optimum concentration. AgNO$_3$ solutions (3 mL) of different concentrations (3, 8, 11, 17, 20, 23 and 33 g L^{-1}) was mixed with 2 g of activated carbon, gently shaken and left overnight to penetrate into pores and undergo absorption by the carbon particles. The remains of the solution were then removed, and 10% HCl was used in an analytical reaction for Ag+ detection to prove that all silver ions were absorbed and none were left in the solution. Three millilitres of NaBH$_4$ solution (5 g L^{-1}) was added into the above AC to initiate a reduction reaction [23], and shaken evenly. Samples were then washed using deionized (DI) water.

Double coating: A single coated sample was used as the precursor. Two grams (wet weight) of the previously coated sample was mixed with 1.6 mL of AgNO$_3$ solution (10.8 g L^{-1} concentration). Samples were shaken gently and left for an hour at room temperature. Samples were then washed using DI water. A total of 1.6 mL NaBH$_4$ (6 g L^{-1})

solution was quickly added to the sample and shaken until bubbles disappeared. Samples were then washed three times with DI water and treated for removal of any remaining liquid.

Three types of sample-drying treatments were evaluated: (a) removal of the liquid using a pipette; (b) freeze-drying under vacuum; and (c) drying in the oven (60°C) for 2-4 hours. AC mixed with DI water without AgNO$_3$ was used as a control.

The coated materials were characterized using standard methods, including Scanning Electron Microscopy (SEM), Transmission Electron Microscopy (TEM), and X-ray diffraction analyses.

SEM: AgNP treated activated carbon particles were mounted onto the stubs and air-dried at room temperature in the dark, then coated with evaporative carbon in a Leica EM SCD005. Samples were observed with a Field Emission scanning electron microscope (JEOL 6301F).

TEM: Samples of AC pre-treated with AgNP were air-dried and embedded in plastic. Thin cuts (80-100 nm) were made using a glass knife to extract sample slices from this substrate. TEM images were taken using a Philips-FEI Morgagni 268 instrument, operated at 80 kV.

X-Ray: Diffraction analysis was performed on the Rigaku Ultima IV Powder X-Ray diffractometer. Samples were air-dried at room temperature and prepared in the same way as for SEM imaging.

ICP-MS: Detection of silver (Ag$^+$) ions in the filtrates was performed using ICP-Mass Spectroscopy. Samples (0.5 g) were mixed with 2 mL of DI water and shaken overnight. Subsequently, the solution was filtered through a 0.22 μm millipore filter to remove remnants of carbon and then tested for Ag$^+$.

Evaluation of the developed filter materials

Common fungal pathogens, *Botrytis* sp., *Fusarium* spp., *Pythium* spp., *Rhizoctonia* sp., and *Sclerotinia* sp. as well as bacteria (*Pseudomonas* sp.) were collected from greenhouse cucumbers, lettuce, tomatoes, and peppers. Infected plant tissue was surface disinfected in 0.5% bleach, rinsed in sterile distilled water and placed on potato dextrose agar (PDA). Fungal/bacterial isolates were purified by single spore/single colony culture techniques and stored at 4°C.

Screening test

Preliminary efficacy of AgNP-treated AC samples was tested against yeast cell culture (*Saccharomyces cerevisiae*) at the lab scale to determine the best drying condition and optimum AgNP concentration for different treatment durations as well as for selection of AC material. One colony of two-day-old *S. cerevisiae* culture was transferred from yeast-extract peptone-dextrose (YPD) agar plate to 100 mL of YPD broth and incubated in a shaker at 180 rpm at 30°C for 24 hours. One mL of diluted (OD$_{600}$ = 0.1) overnight-grown yeast culture was added to a 20 mL vial containing 0.5g of AgNPs-AC (wet weight). The mixture was shaken for 3 to 24 hours, and then 100 μL of the culture was plated on YPD agar plates, duplicated and incubated at 30°C for 24 hours and the colonies per plate were counted and recorded.

AgNP-AC against pathogenic bacteria and fungi (rotation test)

Activated carbon treated with silver nanoparticles (20 g L^{-1} or 23 g L^{-1}) were evaluated for their effect against a bacterium *Pseudomonas* sp., and a fungus *Fusarium oxysporum* followed the method described by Karumuri et al. [18]. 0.5 g of each sample was added to 5 mL spore suspension (approximately 3 × 10^6 spores mL^{-1}) in a test tube. AC treated with water was included as a control. Tubes were incubated on a shaker at 200 rpm at room temperature. A drop (10 μL) of the culture mix was

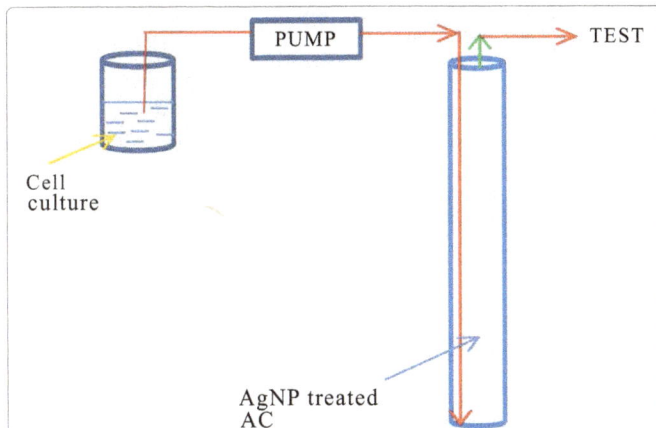

Figure 1: Schematic drawing of a filtration system used for dynamic filtration tests of the AgNP treated AC material. *Paenibacillus* spore suspension was pumped at a flow rate of 70-75 mL h^{-1} to the bottom of the 100 mL column filled with the developed material and the filtrated water samples were collected on the top of the column for plating on agar and further CFU count.

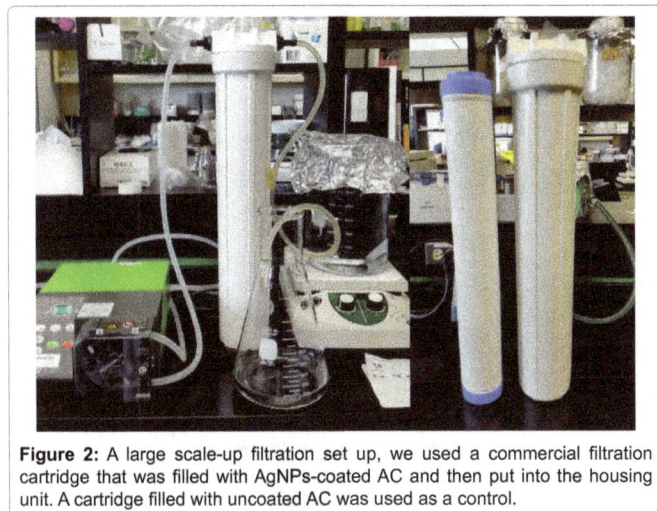

Figure 2: A large scale-up filtration set up, we used a commercial filtration cartridge that was filled with AgNPs-coated AC and then put into the housing unit. A cartridge filled with uncoated AC was used as a control.

plated to a PDA plate after growing for 1 h and 24 h, respectively, with four plates per treatment. Plates were then incubated for 3 days at 25°C in an incubator and the colonies per plate were counted and recorded to calculate the killing efficiency of the filter materials. The percent reduction of colony forming units (CFU) efficiency was calculated based on the CFU counts of bacteria on control plate:

*Efficiency (%) = (CFU of Control – CFU of Treatment)/CFU of Control * 100* (1)

AgNP-AC against bacteria (small-scale filtration test)

Smaller size carbon (CR1230C-AW) had larger cumulative surface area and showed better results on small samples in the rotation test, thus we used it for filtration tests. Sample preparation was performed using the same protocol as that at a rotation test. Dynamic-flow filtration testing was performed using bacterium *Paenibacillus polymyxa* that is morphologically different from other bacteria, with untreated AC as the control. A cylindrical column (100 mL) was filled with active carbon (CR1230C-AW) pre-treated with a single or double coating of AgNP. Scale-up tests were performed with the following system for dynamic tests (Figure 1). Bacterial (*Paenibacillus*) suspension with a concentration of about 10^5 CFU/mL was filtered through the column at a flow rate of 70-75 mL h^{-1}. The test was run non-continuously for

6 days for a total of 6 hours per day. Filtrates were collected hourly for the first 3 days, then every 3 hours on day 4, then every 6 hours on day 5 and day 6. Collected samples were plated on PDA plates and total CFU per plate was counted after incubation for two days. This test was repeated three times. The percent of antibacterial efficiency was calculated based on the original bacterial concentration:

Efficiency (%) = (1 - bacterial concentration in filtrate/original bacterial concentration)*100 (2)

AgNP-AC against bacteria using a commercial filtration cartridge

In the final scale-up filtration test, we used a commercial filtration cartridge and filled it with our developed AgNP-AC material. Based on the result of dynamic testing for 100 mL, we chose double-coated material for commercial sized scale up testing. A total of 1.7 L AC (CR1230C-AW) was double-coated with AgNPs, and then was filled in a commercial RFF-Series Refillable Cartridge (200701, 2.5 × 20 inch, AXEON Water Technologies) that was then inserted into a Slim Line Series Filter Housing (207296, AXEON Water Technologies). Bacterial (*Paenibacillus*) suspension with a concentration about 10^4 CFU mL^{-1} was filtered through the filter device. A pump (Mandel Watson 505U) was set at 2 rpm and a flow rate at about 800 mL per hour (Figure 2). Filtrate samples were collected hourly and plated on PDA plates, which were incubated at 25°C in the dark. CFU was counted 2 days after plating. A cartridge filled with uncoated AC (CR1230C-AW) was used as a control and tested using the same procedure. The percent of antibacterial efficiency was calculated based on the original bacterial concentration using the formula (2) described above.

Growth chamber trial

Preparation of filter: In the growth chamber test, we used the commercial filtration cartridge and filled it with the freshly prepared AgNP-double-coated AC material. A total of 1.7 L AC (CR1230C-AW) was double-coated with AgNPs, and then was filled in a commercial RFF-Series Refillable Cartridge (200701, 2.5 × 20 inch, AXEON Water Technologies) that was then inserted into a Slim Line Series Filter Housing (207296, AXEON Water Technologies). A filter filled with untreated AC was used as a control.

Preparation of cucumber seedlings: Cucumber seeds ('Marketmore') were seeded to 1-inch rockwool cubes and incubated at 25°C 12 h/12 h light/dark in a growth chamber for 7 days. Seedlings were transplanted to 4-inch rockwool blocks one week after seeding, watered with nutrient solution and incubated at the same condition for 10 days.

Preparation of *Pythium* inoculum: *Pythium ultimum* was cultured on potato dextrose agar for 5 days. The culture agar was cut into strips that were transferred to clean plates containing autoclaved distilled water. Plates were incubated at 25°C in the dark for 7-10 days until numerous sporangia were formed. The agar stripes were then blended to fine pieces and used as an inoculum to the nutrient solution.

Application of treatments: Cucumber seedlings (2-week-old) were transplanted to coconut coir slabs in a growth chamber programed at 24°C/20°C, 16h/8h light/darkness. Plants were irrigated with nutrient solution (NS) of following treatments (1) NS contaminated with *Pythium* spores/mycelia, filtered through the AgNPs treated AC filter, 10 plants; (2) NS contaminated with *Pythium* spores/mycelia, filtered through the AC filter without AgNPs treatment, 10 plants. Two extra controls were also included (1) NS contaminated with *Pythium* spores/ mycelia, no filter, positive control, five plants and (2) clean NS, no filter, negative control, five plants. There were five plants per slab per trough. The feeding system was set up to 500 mL water per plant per day at the first week, and then increased to 1 L per plant per day.

Data collection: Plant growth was observed weekly. Biomass and root rot disease were recorded at the end of the experiment (4 weeks after transplanting into the chamber).

Results

Pre-screening AgNP killing efficacy

Since we are developing a technology to treat greenhouse water, the common greenhouse microorganisms should all be employed in testing. In this study, we chose a range of different microorganisms (yeast, fungi and bacterium) to show the broad spectrum of applicability of AgNPs in killing any type of living species which threaten greenhouse vegetable production. For simplicity, we tested yeast (*Saccharomyces cerevisiae*) to obtain preliminary data used to optimize the synthesis.

In our studies, evaluating multiple concentrations of AgNP efficacy against yeast *S. cerevisiae* increased with silver concentration from 3 g L^{-1} to 23 g L^{-1} (data not shown here), and subsequently decreased when concentration was raised further. Evidently, there is an optimal ratio of AgNPs that can be coated on a given amount of activated carbon particles due to constraints on the limited surface area of activated carbon that AgNPs can be deposited on. Beyond this critical density, additional AgNPs can easily fall off or are easily removed. Based on our experimental results, efficiency sharply declines when the concentration of AgNPs reaches 33 g L^{-1}. Optimal efficacy was reached at concentration of 20-23 g L^{-1}, thus chose as the range over which to test against other pathogens (*Pseudomonas* and *Fusarium*).

Determining treatment duration

Time (treatment duration) was another important factor impacting antimicrobial efficacy of filter material. While 20 to 24 hours of treatment yielded excellent results, shorter times did not (data not shown here). Therefore, we tested another type of AC material with

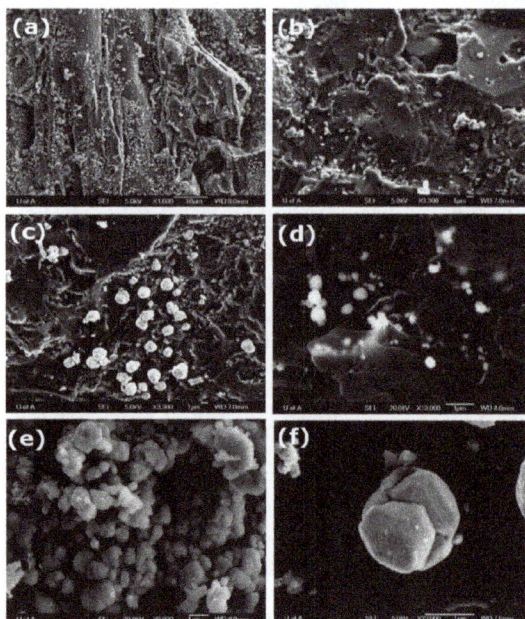

Figure 3: Scanning Electron Micrographs (SEM) of single-coated samples: (a) AgNP (20 g L^{-1})-treated activated carbon (AC); (b) untreated AC; and (c) – (f) AgNPs (23 g L^{-1}) treated AC. AgNPs are shown with red arrows.

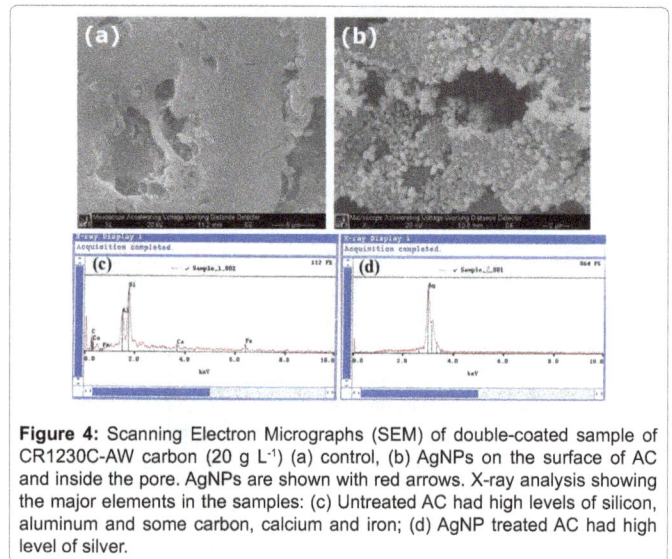

Figure 4: Scanning Electron Micrographs (SEM) of double-coated sample of CR1230C-AW carbon (20 g L^{-1}) (a) control, (b) AgNPs on the surface of AC and inside the pore. AgNPs are shown with red arrows. X-ray analysis showing the major elements in the samples: (c) Untreated AC had high levels of silicon, aluminum and some carbon, calcium and iron; (d) AgNP treated AC had high level of silver.

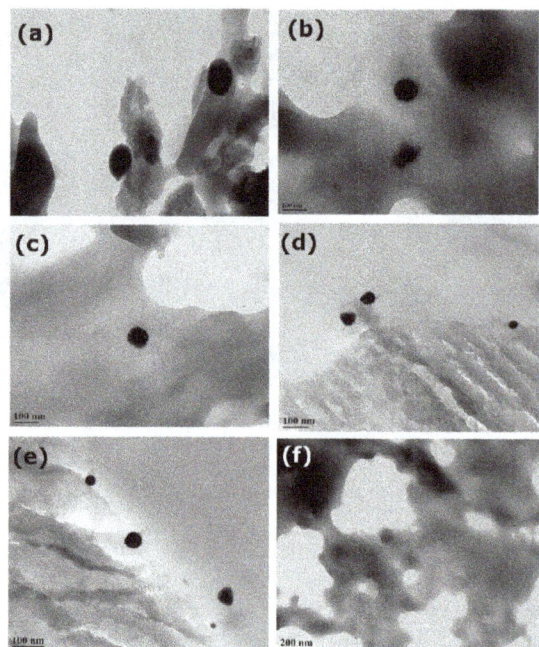

Figure 5: Transmission Electron Microscopy images: AgNP (20 g L^{-1}) treated activated carbon (AC) sample (a)-(b) bigger sized Aquasorb carbon with single coating, and (c)-(e) smaller sized CR1230C-AW with double coating, (f) control (No nanoparticles coating). Pictures show AgNPs on the surface (a), (d), (e) and embedded in the structure of AC (b) and (c).

smaller particle size and larger surface area to address this problem. Another approach to increase exposure to silver was by a second coating the material. Results showed that smaller size carbon had better cell killing efficacy. Whereas larger sized AC required 20 to 24 hours of treatment, 3 hours was enough when using smaller AC granules. The larger surface area of the carbon with smaller particle size can easily explain this phenomenon. Second coating also improved results with the larger sized AC, but smaller AC-AgNP still showed better results.

Characterizing AgNP coated AC

Scanning Electron Micrographs (SEM) showed that the distribution of AgNPs was quite uniform on the surface and inside pores of activated carbon (Figures 3a and 4b) while there were no AgNPs found on the untreated AC (Figures 3b, 4a). AgNPs of different sizes (Figure 3c) can be found inside the porous area (Figure 3d) or can aggregate (Figure 3e) to form larger particles (Figure 3f). The micrograph samples with double and single coating exhibit similar appearance (Figures 3 and 4). There was no obvious difference between samples of smaller and larger sized AC (we used the same concentration of silver to coat them in both cases). X-ray diffraction analysis confirmed the presence of AgNPs on the surface of treated active carbon (Figure 4d) and no AgNPs were found on untreated AC (Figure 4c).

Transmission Electron Microscopy (TEM) images showed that while the carbon structure was destroyed during the process of cutting in preparation for TEM, AgNPs were still present on both the surface and in the pores of the carbon (Figure 5). There was no visible difference between samples of single-coated and double-coated AC-

Analytes	Ag (ppm)
Detection Limits (DL)	0.00001
Control	<DL
SD	0.0273
SS	0.00354

Table 1: ICP-MS testing results of silver in filtrate.

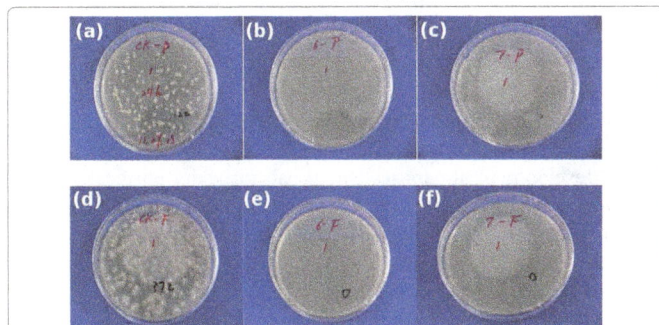

Figure 6: *Pseudomonas* sp. treated with AgNP-coated AC for 24 hours on a shaker at 200 rpm at 25°C, 3 days after plating. (a) Control; (b), (c) filtered with 20 g L^{-1} and 23 g L^{-1} treated AgNP-AC, respectively. *Fusarium oxysporum* treated using AgNP-coated AC for 24 hours on a shaker at 200 rpm at 25°C, three days after plating. (d) Control; (e), (f) filtered water with 20 g L^{-1} and 23 g L^{-1} treated AgNP-AC, respectively.

Microbe	Treatment	Mean CFU per Plate (n = 4)	
		1 hour	24 hours
Pseudomonas	20 g L^{-1}	0.0	0.0
	23 g L^{-1}	0.0	0.0
	0 g (CK)	102.3	95.0
Fusarium	20 g L^{-1}	12.5	0.5
	23 g L^{-1}	66.8	0.0
	0 g (CK)	500.0	472.8

Table 2: Bioassay of AgNPs-coated activated carbon against *Pseudomonas* and *Fusarium* treated for one or 24 hours, and also counted three days after plating and incubating at 25°C.

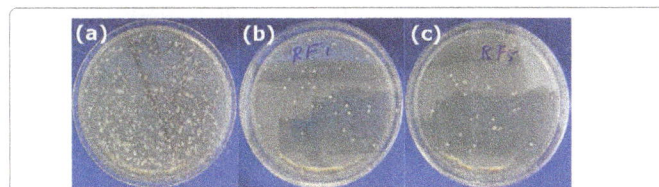

Figure 7: Agar plates two days after plating with the filtrated water collected from the dynamic scale-up test. (a) Control, original bacterial suspension; (b), (c) filtrates collected after one hour and 5 hours of filtration through AgNP-AC material, respectively.

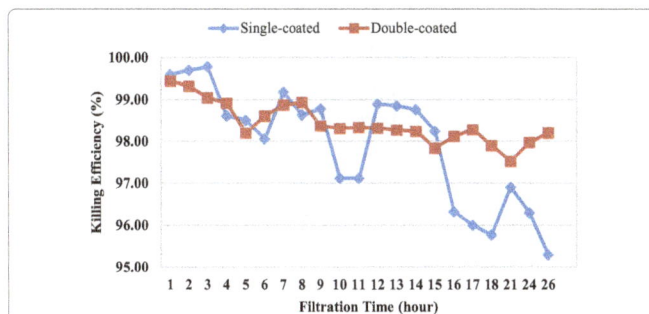

Figure 8: Efficiency test using the scale-up (100 mL) dynamic flow system to treat *Paenibacillus* spore suspension with single-coated vs. double-coated material.

AgNPs. Though some AgNPs fall off during the cutting, from Figure 5d we can see how AgNPs remain attached to the surface of AC on the edge of AC and plastic. These particles were within the range of 20-100 nm, generally round in shape, and could be distinguished on the micrographs by possessing a much higher density compared to that of carbon.

Testing filter materials against plant pathogens (Rotation test)

Our main objective was to develop a material, which is effective in killing major plant pathogens in greenhouse water. After the preliminary screening studies (for killing *S. cerevisiae* yeast cells), tests against a bacterium *Pseudomonas* sp. and a fungus *Fusarium* sp. collected from greenhouses were performed. Both (20 g L^{-1} and 23 g L^{-1}) AgNP-coated ACs completely inhibited the growth of *Pseudomonas* after one-hour treatment while the ACs significantly reduced the CFU counts after one hour treatment and almost completely killed the *Fusarium* spores after 24 hours of treatment (Table 1 and Figure 6).

Our testing results showed that the developed AgNP-AC material was very successful in killing living cells of pathogenic bacterial and fungal species in the small-scale setting of the lab. The next step was to test the developed material in dynamic flow with the expectation to implement the design in greenhouses (Table 2).

Testing in a dynamic scale-up setting (Filtration test)

Because smaller size carbon (CR1230C-AW) had larger surface area and showed better results on the small scale, we selected it for scale-up tests. Sample preparation was performed using the same protocol as for the smaller-scale tests. Dynamic testing was performed using a bacterium *Paenibacillus* as a model and untreated AC as a control. The results showed up to 90% killing efficiency (Figures 7 and 8).

Testing using commercial filtration unit

In the scale-up filtration test, we used a commercial filtration cartridge and filled it with our developed AgNP-AC material. Results showed that more than 99% of initial bacterial population were killed in the first hour and continued for 11 hours with the AgNP-coated AC substrate. The efficiency remained at 97% after 30 hours of filtration (Figure 9a). In the control (filter with AC substrate without AgNP coating), it was observed that bacteria also reduced by more than 90% in the beginning, but the efficiency rapidly dropped to about 30% after 10 hours filtration (Figure 9b) because the bacteria gradually obstructed the porous carbon. Consequently, the AC only filter has a very short life span. Results demonstrated that the addition of AgNPs to the filter substrate can significantly increase the filter life span and antimicrobial efficiency by killing bacteria which have adhered onto the filter surfaces. Again, no Ag+ was detected in the filtrated water.

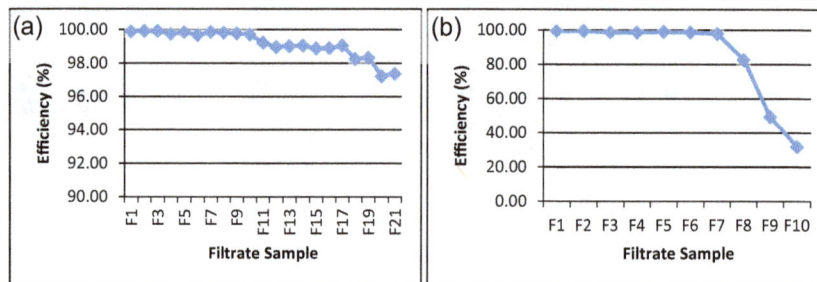

Figure 9: Summary of the scale-up test using commercial cartridge filled with 1.7 L AgNP double-coated AC material against a bacterium Paenibacillus. (a) Filtrate samples were collected hourly for the first 17 hours, and then collected every 3 hours between 18 to 21 hours; (b) Control: Filter cartridge filled with uncoated AC, filtrate samples were collected hourly for the first 10 hours. Filtrates were plated on PDA and incubated at 25°C for two days.

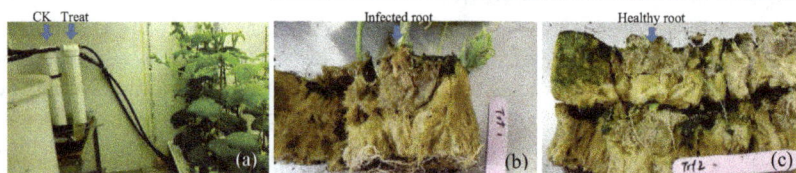

Figure 10: Six-week-old plants watered with *Pythium*-contaminated nutrient solution (NS) filtered through AgNP-AC filter (Treat) and watered with *Pythium*-contaminated NS filtered through AC without AgNP coating filter (CK) in the growth chamber trial (a); cucumber crown root showed discoloration watered with *Pythium* contaminated NS filtered with AC without AgNPs treatment (b) and healthy cucumber roots watered with *Pythium* contaminated NS filtered with AC with AgNPs treatment (c).

Treatment	Root Discoloration Rating	Mean Fresh Shoot Weight (g)
AgNP-AC Filter, Pythium Contaminated NS	0.0 a*	139.6 a
AC Filter, Pythium Contaminated NS	0.3 a	139.1 a
No Filter, Pythium Contaminated NS	1.0 a	127.8 a
No Filter, No Pythium clean NS	0.0 a	142.2 a

*Values followed by the same letter are not significantly different by LSD test (P=0.05).

Table 3: Root infection rating and fresh biomass of cucumber plants in the growth chamber trial (n=10).

Growth chamber trial

Cucumber crown root showed discoloration four weeks after the plants were watered with *Pythium* contaminated NS without filtration while slight discoloration was observed on plants filtered with AC without AgNP treatment. There was no root infection observed on plants watered with AgNP-AC treated NS and control plants (Figure 10). This growth chamber trial demonstrated that the filter treatment reduced the Pythium root disease on cucumbers although the difference on root infection was mild between treatments. Plants of positive control had higher fresh shoot biomass but the difference was not significant between treatments (Table 3).

Discussion

Nanotechnology is an emerging field of science and has been applied in various medical procedures including imaging, diagnostic, therapeutics, drug delivery and tissue engineering [24,25]. It also has great potential in controlled environmental agriculture (CEA), an advanced and intensive form of hydroponically based agriculture [26]. Because AgNPs have strong antimicrobial/anti-fungal properties, they have attracted great research attention in agricultural applications, such as the reduction of root diseases and the improvement of the growth and health of various plants [27]. It can also be used to clean ground water. Argonide Corporation based out of USA, used 2 nm diameter aluminum oxide nanofibres (NanoCeram) as a water purifier that can remove viruses, bacteria and protozoan cysts from water

(http://nanotechweb.org/articles/news/3/4/7). BASF is one of the largest chemical producers in the world and has devoted a significant proportion of its nanotechnology research fund to water purification techniques. Many countries in Europe, Asia and America have worked on the development of nanoscale materials for water purification [26]. Nanotechnology has been used in municipal and rural water treatment to remove bacteria. Nanosilver-coated polypropylene water filters have been designed and evaluated to treat *Escherichia coli* contaminated drinking water [28]. AgNP-coated materials were investigated for the removal of bacteria from groundwater [29]. Potentially, these nanomaterials can be applied to hydroponic production to control plant diseases. Furthermore, they can be applied to treat dugout water for plant irrigation in greenhouses in rural areas.

Our results showed that, although AC coated with AgNPs is effective in killing microorganisms, the sample preparation methods had a distinct impact. We tested three drying methods in the sample preparation and tested the samples against *S. cerevisiae* cell culture. Results showed that all three drying methods exhibited some killing potential for AgNP-treated AC samples to kill the yeast cells. However, samples using vacuum or oven showed greatly reduced efficacy compared to liquid pipetting. The reason is that coated nanoparticles fell off from active carbon substrates during the process of vacuum or oven drying. Furthermore, oxidation can occur using the oven-drying method [30]. Therefore, we chose the pipetting method because the other two methods are much harsher on coated nanoparticles, producing unviable samples.

High anti-bacterial efficiency of the developed material suggested that it might have broader applications including drinking water filtration. However, the impact of AgNPs on humans is still unclear. Therefore, we designed another experiment to detect whether AgNPs were present in the eluent after filtration. We used an ICP-MS detection method for checking, and silver ions were not detected in the control or the sample (data not shown here).

Based on our small-scale tests using *S. cerevisiae*, *Pseudomonas* and *Fusarium*, 99% killing efficiency was observed for large size carbon and 99% to 100% killing efficiency for small size carbon during the first

24 hours of treatments using both single-coating and double-coating method. The bacterium is more sensitive to Ag[+] than the fungus. One possible reason may be due to bacteria being single-celled while *Fusarium* has multi-celled spores and mycelia. Results for the scale-up dynamic flow test using *Paenibacillus* spore suspension showed that in both cases (either for single-coated or for double-coated filtration material), over 95% of bacterial cells and more than 80% of bacterial cells were killed after the system ran for 29 hours and 40 hours, respectively. The antimicrobial efficiencies decreased over the period of testing (Figure 9). However, the flow rate did not change significantly during the whole filtration process (data not shown here). Killing efficiency of the designed filter has been demonstrated in the growth chamber test. Cucumber plants watered with AgNP-AC treated nutrient solution had no Pythium root rot while plants watered without AgNP-AC filter treated NS showed root discoloration in four weeks testing period.

With respect to the environmental consequences of AgNP, so far no universal conclusion has been drawn about the toxicity of AgNP [31]. The only adverse effect in humans of chronic exposure to silver is argyria or argyrosis of the skin and/or the eyes [32,33]. Many studies show that only the release of ionic silver has been found to be toxic, but the release alone cannot be accounted for the toxic side effects. In the aqueous condition, the ionic silver tends to form silver chloride and silver sulphide, which are highly stable. According to the World Health Organization (WHO), the toxicity of trace amount of silver (0.2 - 0.3 μg liter-1) in drinking water is normally negligible. A recent publication [34] described that available data is inadequate for drafting a health-based guideline about the impact of AgNP in the drinking water.

Overall, our study showed that AgNP-coated AC material effectively killed pathogen cells during the filtration. Although AC with single-coated or double-coated AC had similar bacterial killing efficiency, double-coated AgNP-AC has a longer lifespan than single-coated AgNP-AC materials. In the scale-up tests, we loaded the 3D coated filtration materials into empty commercial filter cartridges and maintained the constant water flow. We observed up to 95% to 97% of bacterial cells were killed even after 29 hours of filtration. We believe that commercial filters for water treatment can be designed to fit the requirements of each greenhouse based on their water usage and contamination level.

Conclusion

Silver nanoparticles have previously shown great potential for killing many pathogenic microorganisms. In this study we have combined its disinfecting properties with an efficient, inexpensive, commonly used filter material - activated carbon. We developed a new filter material to address the problem of pathogen contamination of water in greenhouses. Our method of depositing silver nanoparticles on active carbon particles allowed for a three-dimensional coating of the activated carbon with nanoparticles covering both inside and outside surfaces of the pores, offering better antimicrobial efficiency during water filtration. The preparation of the material is easy and our tests showed that pathogen killing efficiency could reach as high as 90% to 99%. We have also tested the material in both a lab-scale setting and a scale-up dynamic setting for 100 mL and in the commercial large size cartridge. In our tests, we used four different types of microorganisms; three of them are typical pathogens in greenhouses: *Pseudomonas* sp. (G-negative bacterium), *Fusarium* oxysporum (fungus), and *Paenibacillus* (a G-positive bacterium). All tests showed that the developed AgNP-AC has great potential for plant disease management in greenhouse vegetable production. Such a design can be extended to kill other common bacterial and fungal cells/spores,

such as *Erwinia* spp., *Phytophthora* sp. and *Pythium* spp., which often occur in greenhouse vegetables and irrigation water. Our solution is an alternative strategy to overcome pesticide resistance developed by pathogens, and is an alternative strategy to treat contaminated water in greenhouses to reduce pesticide related health risks and environmental contaminations.

Acknowledgements

We would like to thank the funding support from Alberta Crop Industry Development Fund Ltd, Alberta Innovates Technology Futures, the Department of Biotechnology Government of India and IC-IMPACTS Centers of Excellence of Canada, Water for Health. We also would like to thank Mr. Ray Yang for valuable feedback about manuscript revision.

References

1. Hong CX, Moorman GW (2005) Plant pathogens in irrigation water: Challenges and opportunities. Crit Rev Plant Sci 24: 189-208.

2. Steward-Wade SM (2011) Plant pathogens in recycled irrigation water in commercial plant nurseries and greenhouses: Their detection and management. Irrig Sci 29: 267-297.

3. Tyson RV, Simonne EH, Treadwell DD, White JM, Simonne A (2008) Reconciling pH for ammonia biofiltration and cucumber yield in a recirculating aquaponic system with perlite biofilters. HortScience 43: 719-724.

4. Berghage RD (1996) Green' water treatment for the green industries: Opportunities for biofiltration of greenhouse and nursery irrigation water and runoff with constructed wetlands. HortScience 31: 690.

5. Safavi K, Mortazaeinzadah F, Estahanizadeh M, Dastjerd H (2011) The study of nano silver (NS) antimicrobial activity and evaluation of using ns in tissue culture media. International Conference on Life Science and Technology, Singapore.

6. Dibrov P, Dzioba J, Gosink KK, Häse CC (2002) Chemiosmotic mechanism of antimicrobial activity of Ag(+) in Vibrio cholerae. Antimicrob Agents Chemother 46: 2668-2670.

7. Mazurak VC, Burrell RE, Tredget EE, Clandinin MT, Field CJ (2007) The effect of treating infected skin grafts with Acticoat on immune cells. Burns 33: 52-58.

8. Maiti S, Krishnan D, Barman G, Ghosh SK, Laha JK (2014) Antimicrobial activities of silver nanoparticles synthesized from Lycopersicon esculentum extract. J Anal Sci Technol 5: 1-7.

9. Zahir A, Bagavan A, Kamaraj C (2012) Efficacy of plant-mediated synthesized silver nanoparticles against Sitophilus oryzae. J Biopest 5: 95-102.

10. Tuan TQ, Son NV, Dung HTK, Luong NH, Thuy BT, et al. (2011) Preparation and properties of silver nanoparticles loaded in activated carbon for biological and environmental applications. J Hazard Mater 192: 1321-1329.

11. Karumuri AK, Maleszewski AA, Oswal DP, Hostetler HA, Mukhopadhyay SM (2014) Fabrication and characterization of antibacterial nanoparticles supported on hierarchical hybrid substrates. J Nanopart Res 16: 1-14.

12. Richter AP, Brown JS, Bharti B, Wang A, Gangwal S, et al. (2015) An environmentally benign antimicrobial nanoparticle based on a silver-infused lignin core. Nat Nanotechnol 10: 817-823.

13. Jo YK, Kim BH, Jung G (2009) Antifungal activity of silver ions and nanoparticles on phytopathogenic fungi. Plant Dis 9: 1037-1043.

14. Kim SW, Lamsal K, Kim YJ, Kim SB, Jung M, et al. (2009) An in vitro study of the antifungal effect of silver nanoparticles on oak wilt pathogen Raffaelea sp. J Microbiol Biotechnol 19: 760-764.

15. Kim SW, Jung JH, Lamsal K, Kim YS, Min JS, et al. (2012) Antifungal effects of silver nanoparticles (AgNPs) against various plant pathogenic fungi. Mycobiology 40: 53-58.

16. Lamsal K, Kim SW, Jung JH, Kim YS, Kim KS, et al. (2011) Inhibition effects of silver nanoparticles against powdery mildews on cucumber and pumpkin. Mycobiology 39: 26-32.

17. Nasrollahi A, Pourshamsian K, Mansourkiaee P (2011) Antifungal activity of silver nanoparticles on some of fungi. Int J Nano Dimens 1: 233-239.

18. Karumuri AK, Oswal DP, Hostetler HA, Mukhopadhyay SM (2013) Silver nanoparticles attached to porous carbon substrates: Robust materials for chemical-free water disinfection. Mater Lett 109: 83-87.

19. Prabhu S, Poulose EK (2012) Silver nanoparticles: mechanism of antimicrobial action, synthesis, medical applications, and toxicity effects. Int Nano Lett 2: 32.

20. Wani AH, Shah MA (2012) A unique and profound effect of MgO and ZnO nanoparticles on some plant pathogenic fungi. J App Pharm Sci 2: 40-44.

21. Al-samarrai AM (2012) Nanoparticles as alternative to pesticides in management plant diseases - A review. Int J Sci Res Publications 2: 1-4.

22. Sahayaraj K, Rajesh S (2011) Bio-nanoparticles: Synthesis and antimicrobial applications. Science against microbial pathogens: communicating current research and technological advances, Mendez-Vilas A (ed), Formatex Research Center, 228-244.

23. Mavani K, Shah M (2013) Synthesis of silver nanoparticles by using sodium borohydride as a reducing agent. Int J Eng Res Technol 2: 1-5.

24. Srilatha B (2011) Nanotechnology in agriculture. J Nanomed Nanotechnol 2: 1-5.

25. Shanti V, Mrudula T, Deepth N, Venkateshwarlu S (2011) Novel applications of nanotechnology in life sciences. J Bioanal Biomed R1: 1-4.

26. Joseph T, Morrison M (2006) Nanotechnology in agriculture and food. Nanoforum Report by European Nanotechnology Gateway.

27. An NT, Dong NT, Hanh PTB, Nhi TTY, Vu DA, et al. (2010) Silver-N-carboxymethyl chitosan nanocomposites: Synthesis and its antibacterial activities. J Bioterr Biodef 1: 102.

28. Heidarpour F, Ghani W, Ahmadun F (2010) Nano silver-coated polypropylene water filter: II. Evaluation of antimicrobial efficiency. Dig J Nanomater Biostruct 5: 797-804.

29. Mpenyana-Monyatsi L, Mthombeni NH, Onyango MS, Momba MNB (2012) Cost-effective filter materials coated with silver nanoparticles for the removal of pathogenic bacteria in groundwater. Int J Environ Res Public Health 9: 244-271.

30. Gallardo OAD, Moiraghi R, Macchion MA, Godoy JA, Perez MA, et al. (2012) Silver oxide particles/silver nanoparticles interconversion: susceptibility of forward/backward reactions to the chemical environment at room temperature. RSC Adv. 2: 2923.

31. Yu SJ, Yin YG, Liu JF (2013) Silver nanoparticles in the environment. Environ Sci Process Impacts 15: 78-92.

32. European Commission (2014) Opinion on nanosilver: safety, health and environmental effects and role in antimicrobial resistance p: 103.

33. World Health Organization (1996) Silver in drinking-water. Guidelines for drinking-water quality, (2ndedn). WHO, Geneva.

34. World Health Organization (2011) Guidelines for drinking-water quality, (4thedn). WHO, Geneva, Chapter 8: 186.

New Way to Develop Mixture of Lactic Leavens and Cardoon Flower Powder (*Cynaraca rdunculus*) in Producing Yoghurt: Approach to Immobilization

Benahmed Djilali Adiba[1]*, Ouelhadj Akli[1], Derridj Arezki[1], Bedrani Fatiha[1] and Belkhir Fatiha[1] and Raman Yakout[2]

[1]*Faculty of Biological and Agronomical Sciences, University of Mouloud Mammeri of Tizi-Ouzou 15000, Algeria*
[2]*National School of Agronomy (ENSA), El Harrach 16000, Algeria*

Abstract

The principal objective of this study was to develop a combination between lactic leavens (*Lactobacillus thermophilus*) and the cardoon flower powder (*Cynaracardunculus*) and their application on yoghurt. The milk clotting was optimized by using the two coagulant agents (with fresh and immobilized states).

The results obtained revealed that a quantity of the cardoon flower powder has a very interesting clotting rate (2.55 min) in comparison with the use of the optimized mixture M2 (75% of cardoon flower powder and lactic leavens 25%) and the optimized quantity of the leavens (0.1 g) with (3.6 and 22.58 min) as respective rates. The immobilization of the various coagulant agents improves the milk-clotting rate. Indeed, a quantity of 6 g of the beads prepared from cardoon flower powder shows a very fast speed (1.06 min) in comparison to the same quantity of the beads prepared starting from the mixture M2 (3.71) and the immobilized leavens (73 min).

The beads prepared starting from the cardoon flower powder and the mixture M2 can completely substitute the immobilized lactic leavens according to the matrix of similarity (similarity of 70%). Moreover, on the one hand, the beads containing the cardoon flower powder improve the speed of yoghurt clotting (75 min) in comparison toimmobilized leaven yoghurt (270 min). On the other hand, the rheological properties were also improved (smooth structure and absence of syneresis phenomenon).

Keywords: Lactic leavens; Milk; Clotting; Cardoon; Immobilization; Substitution

Introduction

Dairy industry sector uses a number of processing aids, including lactic acid bacteria to develop fermented foods. Lactic acid bacteria are added to milk to start the fermentation process; they are used for the production of a wide range of dairy products such as cheese, yoghurt, butter and cream [1,2]. Given the high cost of these ferments, import presents a major impediment to the development of local production of dairy products. Therefore, Algeria's dependence on foreign suppliers drew our attention to search other producing local sources of milk clotting.

Although their numbers may be considered low, some work has already addressed this issue, like those conducted by the Transformation Team and Development of Agri-Food (TEPA) which covered pepsin extracted from proventricule chicken [3,4], given the importance of such enzymes, we decided to undertake an interesting result of research by exploring a different track this time; the coagulants of plant origin, which are enzymes extracted from the cardoon flowers (*Cynaracardunculus*).

The interest generated by the choice of this plant is due to its great distribution in Algeria, more precisely in the highlands, where soil and climate conditions are adequate for its development, especially during the rainy years. Rural population empirically uses its fresh or dried forms in manufacturing fresh cheese made from raw sheep milk called: djben [5].

Generally, enzymes, which have low stability and high costs, need to immobilize in porous solids and protective matrices. This technique allows dairy industries to increase their yields. Recognizing this, and in order to remedy this situation to better develop this plant, it is useful to consider its transformation through the acquisition of new technologies, including the immobilization of enzymes and their use in the preparation of yoghurt.

This study's main objective is to substitute fully or partially lactic ferments by promoting cardoon to bring to market a new clotting agent. Optimization of milk clotting conditions has been studied.

Materials and Methods

Plant material

Dried cardoon flowers (*Cynaracardonculus*) were bought at the local markets in Tlemcen (Algeria) in May 2014. Just after being harvested, the flowers were dried at ambient air and preserved in bags at ambient temperature.

Biological material

The freeze-dried species of lactic leavens (*Lactobacillus thermophilus*) was delivered by yoghurt manufacturing in Algiers.

The pasteurized milk (1.5% Lipid Matter, pH=6.7) was bought at

---

*Corresponding author: Benahmed Djilali Adiba, Faculty of Biological and Agronomical Sciences, University of Mouloud Mammeri of Tizi-Ouzou 15000, Algeria, E-mail: adiba.benahmed@yahoo.fr the local markets in Tizi-Ouzou (Algeria) delivered by D.B.K milk of Tizi-Ouzou. The powders of κ-carraghenan and milk (0% Lipid Matter) used in this study were delivered by the same yoghurt manufacturing.

Methods

This part of work was carried out at the Laboratory of Microbiology (University of Mouloud Mammeri of Tizi-Ouzou); Algeria during the period of May-July 2014.

The method defined by Fernandez et al. [6] and Aquilanti et al. [7] was used to prepare the plant enzymatic extract. A quantity of 3 g of dried cardoon flower was macerated in 100 ml distilled water at 45° C for six hours and then, filtered through Watman No: 4 filter paper. The obtained enzymatic extract is conserved at 4°C before undergoing clotting tests.

κ-carraghenan solution (3%)was used as an optimal concentration in term of suitable beads elaboration. It was sterilized at 70°C for three hours and conserved at 4°C before being immobilization process.

CaCl$_2$ solution (4%) was used as solidification agent for elaborating different optimized coagulant beads (cardoon flower powder, lactic leavens, and mixture).

The process of immobilization of different optimized coagulant agents was achieved by inclusion κ-carraghenan solution using searing in the sterilized area. Beads solidification was achieved in CaCl$_2$ solution at 4°C for 3 hours. Finally, beads were washed twice with distilled water and preserved at 4°C. Clotting tests were considered according to experimental conditions summarized in Table 1.

In all studies the milk was heated to 45°C, the average time of clotting and standard deviation were calculated from three replicates.

Morphology of different coagulants agent: Environmental Scanning Electron Microscope SEM (PHILIPS ESEMXL.30; Heindoven, Netherlands) was used to study the texture properties, the microstructure of differently coagulants (cardoon flower powder, lactic leavens, sour milk issued from milk clotting using cardoon beads, immobilized lactic leavens and the immobilized mixture M2). Samples were made without metallization according to the protocol recommended for 20 μm particle size.

Statistical study: In order to show the effect of the nature of coagulants agent (free or immobilized) on the time of milk clotting, we used the test of similarity, using Statistica software (version 6). This was to compare two-to-two similarity of the different time of clotting using different coagulants.

Results and Discussion

The results obtained revealed that a quantity of 0.3 g of the cardoon flower powder has very interesting clotting speed (2.55 min) (Figure 1a) in comparison with the optimized mixture of powders M2 (75% of cardoon flower powder and lactic leavens25%) (Figure 1b) and the optimized quantity of the leavens (0.1 g) with the respective speeds (3.6 min and 22.58 min) (Figure 1a).

These results are similar to those reported by Feranndez -Salguera et al. [8] working on the same cardoon enzymes. Other researchers, such as Chazarra et al. [9] worked on *Cynarascolymus*; and Yousif et al. [10] on *Solanumdudium,* demonstrated the effect of using enzymes extracted from plants.

Comparison between the various coagulant agents in terms of pH, acid clotting using lactic acid bacteria revealed a slight increase in pH

Nature of coagulants agent	Quantity (g)	ClottingConditions
Powder of cardoon flowers	(0.1-0.3)	
Lactic leavens	(0.1-0.3)	
Mixture (cardoon flowerpowder+Lactic leavens)	(0.3-0) to (0-0.1)	
Beads of cardoon flower powder	(1-6)	10 ml. of milk, T=45°C
Beads of lactic leavens	(1-6)	
Beads of optimized mixture M2	(1-6)	
Number of Recycling beads	(1-10)	
Water extract of cardoon flowers	(1-10 ml)	

M2=(25% lactic leavens+75% of cardoon flower powder).

Table 1: Various clotting tests.

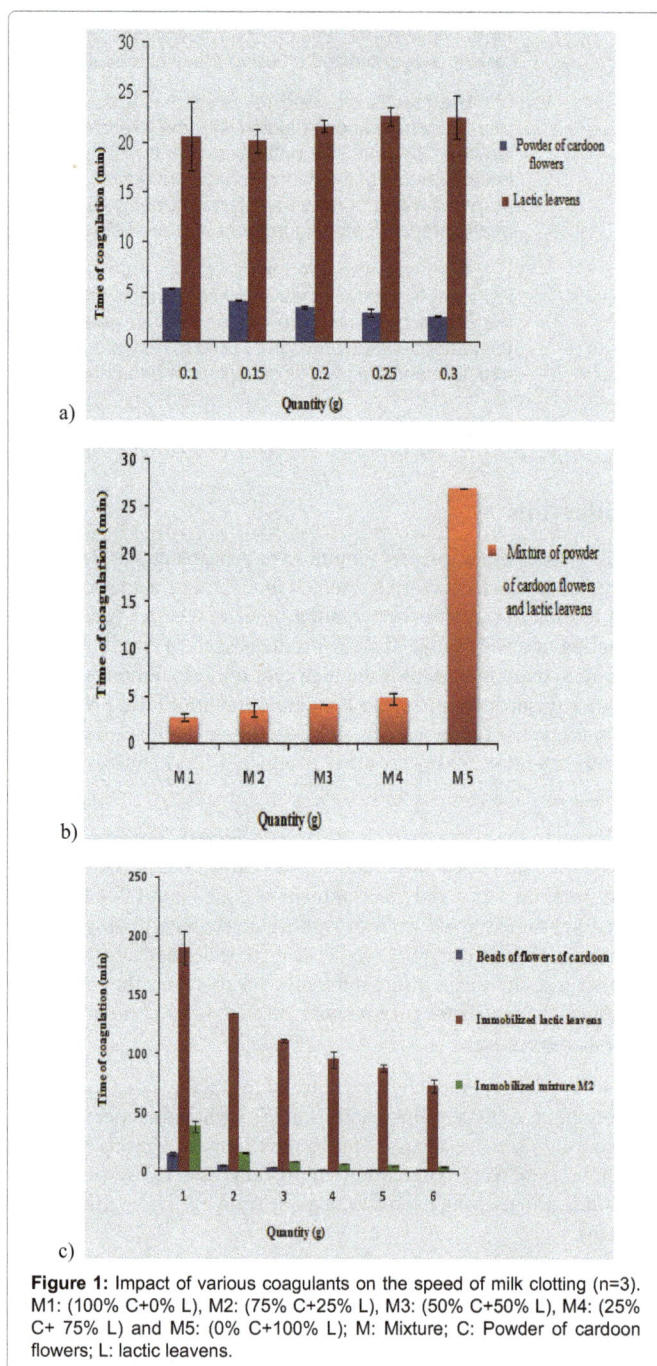

a)

b)

c)

Figure 1: Impact of various coagulants on the speed of milk clotting (n=3). M1: (100% C+0% L), M2: (75% C+25% L), M3: (50% C+50% L), M4: (25% C+ 75% L) and M5: (0% C+100% L); M: Mixture; C: Powder of cardoon flowers; L: lactic leavens.

(4.68 to 4.81). For cons, the cardoon flower powder leads to a slight decrease in pH (6.53 to 6.28). Our results are in agreement with the results of David and Forte [11], who reported that getting a good curd obtained from the enzyme clotting was observed at pH ≥5.2 and a curd from acid clotting under the action of lactic acid bacteria was observed at a pH of ≤4.6.

Also, the mixture of various coagulant agents promotes a decrease in pH. Indeed, with such a combination, there are two types of proteolytic hydrolyzed: 1) partial hydrolysis (at the surface of the casein) with the enzymes from the cardoon flowerpowder or 2) complete hydrolysis (whole casein) by enzymes ferments. It was shown that enzymes derived from plants had a broad spectrum of activity in medium acid [9] and in a basic medium [12].

Furthermore, the effect of the immobilization had the major impact on the speed of milk clotting. A quantity of 6 g of the beads prepared from the cardoon flower powder shows a very fast speed (1.06 min) in comparison with the same quantity of the beads prepared starting from the mixture M2 (3.71) and the immobilized leavens (73 min).

It is worth noting that recycling provides better enzyme activity in the case of using ball base of powder of cardoon flowers (Figure 1c). A decrease in the activity was observed from the 5th recycling. The recycling of immobilized enzymes causes an increase in the latter passing from one cycle to the other (Figure 1d).

According to the similarity test, we can replace the enzymes of *Cynaracardunculus* immobilized enzymes, immobilized lactic acid bacteria (71.4% similarity) and the balls of the mixture M2 by immobilized enzymes (69.29% similarity).

The beads prepared starting from the cardoon flower powder and the mixture M2 can substitute completely call the immobilized lactic leavens according to the matrix of similarity (similarity of 70%).

(Figure 2) shows the appearance of the curds obtained by varying the amount of the beads prepared using the powder flowers thistle. It appears that the coagulation rate is directly related to the availability and the diffusivity of the enzymes within the beads to the surface. These results are reinforced with those obtained by Arima et al. [13], which proved that residual activity of the immobilized rennet was low during reuse. Similarly, Bartolini et al. [14] showed that the enzyme activity in the immobilized state remains stable.

Recycling can generate strong fermentation activity due to the multiplication of bacteria in the balls. Thus, Stenroos et al. [15] demonstrated the possibility of reusing in several cycles beads prepared from low cell concentrations.

SEM Results

Figure 3 shows the microscopic structures of the different coagulants used in this study. SEM results show that the cardoon flower powder has a homogeneous layered sheet structure (Figure 3a) whilst lactic acid bacteria contain a heterogeneous tube structure (Figure 3b).

Curds from milk clotting using cardoon flower powder or the mixture M2 were characterized by non-porous structures as various-size dense flakes (Figure 3c and 3e).

The curd from milk enzymes had a micro-lump structure (destructured flakes associated with pores) (Figure 3d). The breakdown could be due to lack of training of the network, which provides the gel permeability but the elasticity and plasticity are virtually nil [16].

Yoghurts results

Table 2 presents the main characteristics of some yoghurt. It is worth highlighting the positive effect of that cardoon flower powder on the quality and time of yoghurt clotting.

The comparison between these yoghurts also reveals that the yoghurt (2) and (3) (Figure 4) form opaque firm gels with good cohesion (lack of syneresis phenomenon). This phenomenon has been observed in the yoghurt (1).

In order to better compare the yogurt quality, it was interesting to prepare three gels and make a thorough characterizing study of the rheological parameters (viscosity, elasticity, ...). Whilst, from pharmacological point of view, it also was interesting to characterize the two yoghurts to recommend them for people who suffer from stomach acid problem.

By comparing structures of prepared yoghurts (Figure 5), it was

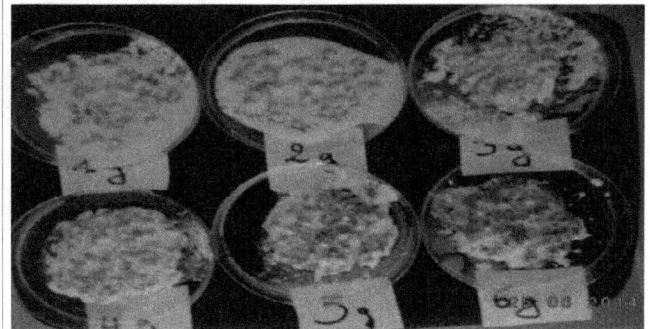

Figure 2: Curds from the milk clotting by the beads of cardoon flower powder.

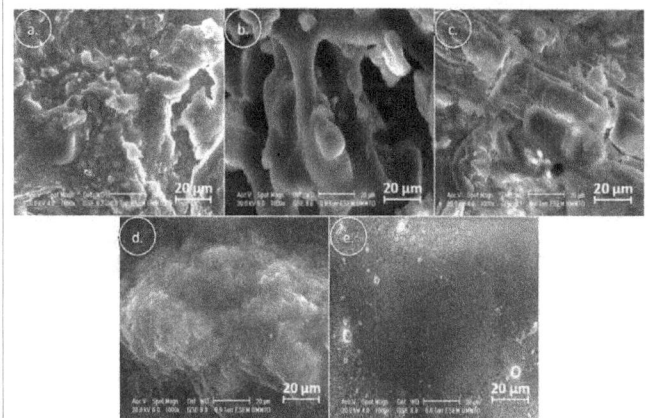

Figure 3: Microscopic structures for various sour milk and coagulants:(a): Powder of cardoon flowers; (b): lactic leaven; (c): Sour milk issued for milk clotting using cardoon flowers; (d): Sour milk issued for milk clotting using lactic leavens; (e): Sour milk issued for milk clotting using the immobilized mixture M2.

Prepared yoghurts	Time of coagulation	pH at 4°C
Yoghurt (1)	75 min	6.61
Yoghurt (2)	270 min	5.66
Yoghurt (3)	90 min	6.57
Yoghurt standards by Codex (2003)	150 to 300 min	3.6 to 4.6

Table 2: Some parameters of various prepared yoghurts.

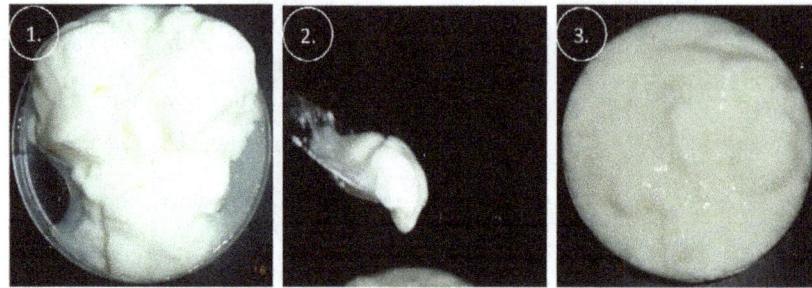

Figure 4: Aspects of yoghurt. 1) Yoghurt fermented using sour milk issued from clotting milk by beads of cardoon flowers 2) Yoghurt fermented using sour milk issued from immobilized lactic leavens Yoghurt fermented using sour milk issued from the immobilized mixture M2.

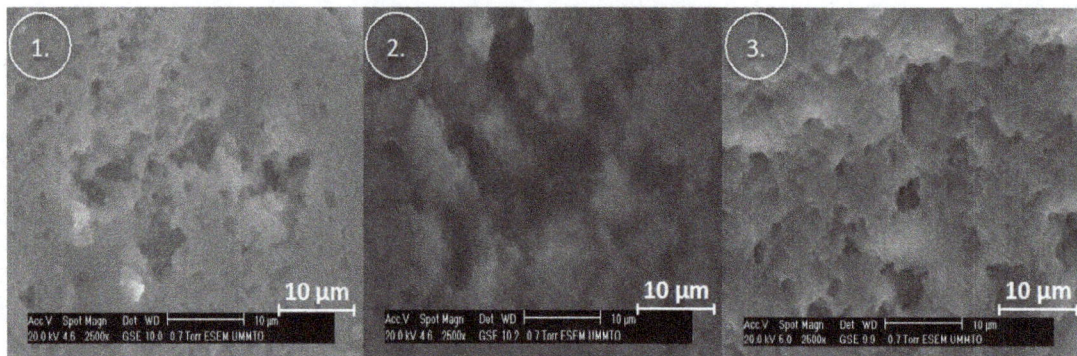

Figure 5: Microscopic structure of various yoghurts (Ø=10 μm).

confirmed that the yoghurt (1) is characterized by a smooth structure somewhat porous. By cons, yoghurt (2) has a rough structure in the form of multi-lump (destructured flakes). It also turns out that the combination of the two coagulants gives a firm gel form a dense, porous flake structure.

Conclusion

This work aimed at studying the feasibility of the immobilization of clotting by beads of cardoon flowers and comparing it to other conventional agents. Beads had an advantage on milk coagulation rate, and, moreover, it could be used ten times. This work allowed us to prepare such yoghurt, completely and thoroughly study its behavior and quality from physical and chemical aspects, nutritional values (lactic acid content, protein, etc.), and organoleptic and rheological attributes.

References

1. HylckamaVliega JET and Hugenholtzb Van J (2007) Mining natural diversity of lactic acid bacteria for flavor and health benefits. International Dairy Journal 17: 290-1297.

2. Chen GQ (2010) Plastics from bacteria 'Natural functions and applications. Microbiology Monographs 14: Münster, Germany, p. 450.

3. Boughellout H (2007) Milk coagulation by chicken pepsin. Magister memory University Mentouri, Constantine p. 69.

4. Benyahia-Krid FA, Attia H, Zidoune MN (2010) Comparative study of milk coagulation with chicken pepsin or rennet: Interactions and microstructure. J. Agriculture, Biotechnology and Ecology 3: 75-86.

5. Zikiou A (2013) The coagulation of milk by the extract of flowers cardoon (Cynaracardunculus) Magister memory in Food Science University of Constantine.

6. Fernandez-Garcia E, Imhof M, Schlichtherle-Cerny H, Bosset JO, Nunez M (2008) Terpenoids and benzenoids in La Serena cheese made at different seasons of the year with a Cynaracardunculus extract as coagulant. International Dairy Journal 18: 147-157.

7. Aquilanti L, Babini V, Santarelli S, Osimani A, Petruzzelli A, et al. (2011) Bacterial dynamics in a raw cow's milk Caciotta cheese manufactured with aqueous extract of Cynara cardunculus dried flowers. Lett Appl Microbiol 52: 651-659.

8. Fernandez F, Shridas P, Jiang S, Aebi M, CJ Waechter (2002) Expression and characterization of a human cDNA that complements the temperature-sensitive defect in dolichol kinase activity in the yeast sec59-1 mutant: the enzymatic phosphorylation of dolichol and diacylglycerol are catalyzed by separate CTP-mediated kinase activities in Saccharomyces cerevisiae. Glycobiology 12: 555-562.

9. Chazarra S, Sidrach L, Lopez-Molina D, Rodriguez-Lopez JN (2007) Characterization of the milk-clottingproperties of extracts from artichoke (Cynarascolymus L.) flowers. International Dairy Journal 17: 1393-1400.

10. Yousif, OM, Osman MF, Alhadrami GA (1996) Evaluation of date and date pits as dietary ingredients in tilapia (Oreochormisaureus) diets differing in protein sources. Biores Technol 57: 81-85.

11. David V, Forte R (1998) National guide to good practice in cheese production fermiere. Second ed. Paris: Livestock Institute. VII 21-26.

12. Raposo S, Domingos A (2008) Purification and characterization milk-clotting aspartic proteinases from Centaureacalcitrapa cell suspension cultures. Process Biochemistry 43: 139-144.

13. Arima K, Yu J, Iwasaki S (1970) Milk-clotting enzyme from Mucorpusillus. Methods in Enzymology 19: 446-460.

14. Bartolini M, Bertucci C, Cavrini V, Andrisano V (2003) Beta-amyloid aggregation induced by human acetylcholinesterase: inhibition studies. Biochemical Pharmacology 65: 407-416.

15. Stenroos SL, LinkoYY, Linko P, Harju M, Heiknen M (1982) Enzyme Eng 6: 299-301.

16. Abiazar R (2007) Complexation of milk proteins by extracts of green pods of the carob tree technological properties coagulums obtained: Doctor of Agro Paris Tech. Agro Paris Tech ABIES graduate school.

27

In Vitro and *In Vivo* Management of *Alternaria* Leaf Spot of *Brassica campestris* L.

Aqeel Ahmad and Yaseen Ashraf*

Institute of Agricultural Sciences, University of the Punjab, Lahore-54590, Pakistan

Abstract

Alternaria black leaf spot caused by *Alternaria brassicae* is one of the most destructive disease of brassicaceae crops and causes 30 to 45% overall yield loss in the world.

Plant susceptibility toward this saprophytic and necrotrophic pathogen is greatly influenced by extreme weather conditions e.g. temperature and humidity.

Six plant extracts, six Biological agents and six fungicides were evaluated both *in vitro* and *in vivo* experiment for their effectiveness to manage Alternaria leaf spot of *Brassica campestris*. In cause of *in vitro* pathogenic fungus was applied in the field at 2 g colonized mustard seeds kg-1 soil. plant extract, Biological agents and six fungicides were evaluated for their efficacy at various concentrations 5%, 10%, 15% and were sprayed in the field at 0.2% a.i. l-1. Out of all treatments, *Allium sativum, Parthenium hysterophorus, Trichoderma harzianum, Trichoderma viride,* Wisdom (50% WP) and Proctor (60% WP) were screen out in laboratory at 15% concentration. The maximum growth inhibition (in laboratory 57.83%, in field 6.07% and in greenhouse 26.32%) was recorded by *Allium sativum* followed by *Parthenium hysterophorus* (in laboratory 53.01%, in field 17.05%, in green and house 29.08%). Out all biological agents, the maximum growth inhibition (in laboratory 61.44%, in field 27.34% and in greenhouse 38.45%) by *Trichoderma harzianum* followed by *Trichoderma viride* (in laboratory 55.42%, in field 29.63%, in green and house 29.08%). Out of all fungicides, the maximum growth inhibition (in laboratory 98.79%, in field 56.08% and in greenhouse 63%) by Wisdom (50% WP) and followed by Proctor (60% WP) (in laboratory 100%, in field 51.76% and in greenhouse 55.16%). It was worth noting that the fungicides, Wisdom (50% WP) and Proctor (60% WP) have highest net value as compare to other treatments but the biological agents also show off their importance.

Keywords: Alternaria leaf spot; *In vitro* and *in vivo* management of *Alternaria* leaf spot by plant extracts; Biological agents and chemicals

Introduction

The origin of mustard (*Brassica compestris* L.) lies in south-east Asia [1]. Mustard is one of most important and oldest known oil seed crop of subcontinent with global contribution of 28.3% acreage and 19% of production [2]. Its oils contain low erusic acid and glucosinolates contents. The percentage of poly-unsaturated fatty acid and linolenic acid of the total fatty acid increase from 15-0% and from 8-12%, respectively [3].

Among the biotic stress of *Alternaria* leaf of mustard and the causal agent is *Alternaria brassicae*. It has been reported from all the continents of the world and is one among the important diseases of mustard causing up to 47% yield losses [4]. Different species of *Alternaria* on *Brassica* spp. vary in host specificity. *Alternaria brassicae* also depending on host susceptibility and environmental factors [5]. *Alternaria brassicae* infected the plant at all growth stages. Fungus infect all parts of plant as leaves, pods, branches, pods and stem but the special target point of fungus are leaves and pods. Often lesions are produced on green leaves and during sever attack in pods seeds become shrivel and early ripening or shattering [6]. Conventionally plant diseases are controlled by applying fungicides, but this practice increase input cost on the crop on one hand and on the other hand cause environmental pollution [7]. So this situation compels to focus on disease management by utilizing biological agents, plant extracts and fungicides in lowest concentration. Application of biological agents and extract is eco-friendly and a sustainable approach apart from being a promising alternative to fungicide application. In the absence of resistant cultivars, chemical fungicides provide the most reliable means of disease control. The present study was aimed at determining a cost-effective management of Alternaria leaf spot.

Material and Method

Study area and sampling

In 2012, eight *brassicae* fields were visited, randomly diseased samples were chosen. A survey was conducted at eight different locations in district Lahore for prevalence, severity and mortality of *Alternaria* leaf spot of mustard at maturity of mustard crop during cropping season. To assess disease prevalence, severity and mortality, and ten plants were selected in each quadrate in a diagonal configuration depending on the geometry of the field. The following formula was calculated percentage prevalence.

$$\text{Prevalence }(\%) = \frac{\text{Locations with disease symptoms}}{\text{Total Locations}} \times 100$$

$$\text{Severity }(\%) = \frac{\text{Sum of rating scale}}{\text{Total number of leaves observed}} \times 100$$

$$\text{Mortality }(\%) = \frac{\text{Sum of dead plants}}{\text{Total plants}} \times 100$$

Alternaria leaf spot disease was collected randomly, with at least five lesions in the leaf blade. From each of the five lesions per leaf,

***Corresponding author:** Yaseen Ashraf, Institute of Agricultural Sciences, University of the Punjab, Lahore-54590, Pakistan
E-mail: yaseenashraf00@gmail.com

fragments of tissues containing fungi structures were taken out and microscope preparations on glass slides containing a drop of blue Aman were made. The slides were, afterwards, observed in optical microscope at 400X magnification. Identification was made according to literature [8-10]. After identification, pure cultures were submitted in Pakistan First Fungal Culture Bank Institute of Agriculture Science, University of the Punjab, Lahore.

In vitro screening of treatments

Alternaria brassicae inhibitory effect was checked against different plants extract in laboratory under food poisoning technique. The food poisoning technique was adopted for invitro testing of biological agents and fungicides. For this purpose, six plants were selected Coronopus didymus (Leaves), Medicago sativa (Leaves), Zingiber officinale (Bulb), Chenopodium hirsutum (Leaves), Allium sativum (Bulb) and Parthenium hysterophorus (Leaves). Ten grams of plants relevant part were grinded with help of pistol and mortal by adding equal amount of distal water (1: 1 w/v). At last extract was filter with the help of muslin cloth. Aqueous solution (100%) was obtained. Further, the extract was diluted by adding sterilized water to get 10 percent concentration. Future plants extract were need to heat at 50°C to avoid contamination. 2 ml plant extract was poured in 20 ml MEA petriplate and gently shake both for mixing of plants extract in media. When MEA and plants extract solidified then 8 mm disc of 10 days old pathogen was placed in center of every petriplate. All pertiplates were incubated at 23°C for 10 days. Growth inhibition of pathogen, inoculated and uninoculated was calculated according to the formula given by Vincent.

% Inhibition over control: $\frac{C-T}{C} \times 100$

Where;

I=Percent inhibition

C=Growth in control

T=Growth in treatments

Biological agents (Trichoderma viride, Trichoderma harzianum, Trichoderma hamatum, Trichoderma koningii, Trichoderma reesei and Trichoderma aureoviride) were obtained from Fungal Culture Bank of Pakistan Institute of Agriculture Science, University of the Punjab, Lahore. These biological agents were screened in laboratory condition against Alternaria brassice. Culture of both pathogen and biological agents (8 mm) were collected from margin of actively growing mycelium and transfer to MEA medium on opposite site of about at 1 cm from wall of the plate. The petriplates were subsequently incubated at 25 ± 1°C. After 5 days fungal colonies were observed and recorded.

Fungicides inhibitory assessment against A. brassicae was performed in laboratory by food poisoning technique. Seven fungicides Ridomil (20% WP), Diesomil (30% WP), Topsin-M (45% WP), Thiram (35% WP), Dolomile (30% WP), wisdom (50% WP) and Proctor (60% WP) were used for confirmation of efficacy against A.brassicae under invitro conditions. All selected fungicides were tested at 0.1% concentration. Two ml of each fungicide were incorporated in sterilized petriplates and gently mixed it. After solidifying MEA media, 8 mm disc of ten days old sporulating culture of Alternaria brassice was inoculated in center of every petriplate. Controls were maintained. Inoculated petriplates were incubated at room temperature 28°C in the laboratory. The colony diameters were measured after 10 days when the control plates were full of fungal growth.

Green house experiment

Pots having 25 cm depth and 20 cm diameter were used. Each

treatment was replicated thrice. Sandy loam soil was used and each pot was filled with 10 Kg sterilized soil. Seed were sown in pots at the depth of 2 cm in January 16, 2016. Three seeds per pot were sown. Pots were irrigated twice in a week. Green house plants become 2 to 3 leaf stage then pathogen inoculated in form of suspension in February, 7 2016. Inoculum was obtained from culture of Alternaria brassicae grown on malt extract agar. Inoculum was prepared in suspension form, 10 ml sterile water added in petriplates and shake well. The numbers of spores were counted with heamocytometer and spores were adjusted to 32×10^7 sporesml^{-1}.

About 500 ml of inoculums suspension is used with sprayers that run off from top of leaf.

After two weeks leaves established disease symptoms. Older leaves were more severely infected from Alternaria leaf spot as compare to younger leaves. Initially leaves beared light brown lesion which gradually become dark brown and at last dark spot on whole leaf. In severe condition, gradually it spread to other parts of plants such as pods, stem and branches.

Disease management

The fresh plants extracts (Allium sativum, bulb and Parthenium hysterophorus, leaves) were gently washed under tap water and finally in sterile distilled water. They were separately grind in sterile water at the rate of 1 mlg^{-1} of plant material in pistal and mortal. Then it stained through double layer of muslin cloth and finally through sterilized whatman no. 1 filter paper. This formed 100% standard plant extract solution. Further its dilution performed of required concentration with sterilized water [11]. Plants extract application at 5%, 10% and 15% concentrations.

One week old culture of Trichoderma harzianum and Trichoderma viride were obtained from FCBP. The spore's concentration was adjusted to 32×10^6 spores ml^{-1} by hemocytometer. Biological agent's was applied at 5%, 10% and 15% concentrations. Fungicides with recommended dozes were used that are available in market and Spray in field at 5%, 10% and 15% concentrations.

Field experiment

The experiment was conducted by randomized block design (RBD) with three replications and the sowing was done om 10 m×15 m plots, with a spacing of 90 cm×60 m on 2015 and 2016.

Plants were inoculated with a suspension of pathogen (Alternaria brassicae) at February 7 2016. The spore suspensions of pathogens performed as in green house experiment and application according to section about 2000 ml of inoculums suspension is used with hand sprayers that run off from top of leaf.

Disease management

Plant extract preparation and application was as discussed in green house experiment. Biological agents preparation and application was as discussed in green house experiment. Fungicides preparation and application was as discussed in section greenhouse experiment.

Statistical analysis

Treatment mean and standard error were calculated from the data obtained for various parameters using package Costat version 3.03.

Results

Disease survey

After peripatetic survey prevalence, severity and mortality of

Alternaria black spot disease was recorded. The disease prevalence percentage ranged between 20 to 60 at different locations. The maximum prevalence was (60%) recorded at P.U campus and minimum (20%) recorded at Multan road (Sunder) and G.T road (Rana town). Disease severity percentage was range between 30 to 70 at different locations. The maximum severity (70%) was recorded at P.U campus and minimum (30%) was recorded at G.T road (Rana town) and Raiwind road (Bubtiya chowk). Mortality was ranged between 8 to 25 percent at different locations. The maximum mortality was (25%) recorded at P.U campus and minimum was 8% recorded at G.T road (Rana town) (Figure 1).

In vitro screening of treatment against *Alternaria brassicae*

Six plants extract (*Parthenium hysterophorus, Coronopus didymus, Medicago sativa, Chenopodium hirsutum, Zingiber officinale and Allium sativum*) were tested against *Alternaria brassicae* growth. According to result shown in Figure 2, fungal growth inhibition ranged between 53.01% to 57.83%. The maximum growth inhibition (57.83%) was recorded by *Parthenium hysterophorus,* followed by *Allium sativum* (53.01%). *Chenopodium hirsutum inhibited (39.75%) followed by Medicago sativa* (38.55%). The minimum growth inhibition (30.12%) was recorded by *Zingiber officinale followed by Coronopus didymus* (24.93%).

Six biological agents (*Trichoderma harzianum, Trichoderma reesei, Trichoderma viride, Trichoderma aureoviride, Trichoderma konngii* and *Trichoderma hamatum*) were tested for their antifungal activity against *Alternaria brassicae*. According to results shown in Figure 3, fungal growth inhibition ranged recorded between 15.23 to 61.44%. The maximum growth inhibition (61.44%) was recorded by *Trichoderma harzianum* followed by *Trichoderma viride* (55.42%) and *Trichoderma konngii* (40.96%). Meanwhile, *Trichoderma reesei* inhibited (33.01%) followed by *Trichoderma hamatum* (31.32%). The minimum growth inhibition (15.90%) was recorded by *Trichoderma aureoviride*.

Seven fungicides (Ridomil, Diesomil, Topsin-M, Thiram, Dolomile, Wisdom and Proctor) were tested against *Alternaria brassicae* growth. According to results shown in Figure 4, fungal growth inhibition ranged between 63.85 to 100 percent. The maximum growth inhibition (100%) was recorded by proctor. Wisdom inhibited (98.79%) growth at 0.1% concentration followed by Topim (81.52%). Dolomile inhibited 75.90% followed by Ridomil 72.28%. The minimum growth inhibition (69.87%) was recorded by Diesomil followed by Thirm (63.85%).

Field and green house studies for prevalence, severity and mortality of *Brasscia compestris*

From data observation, in field mean disease prevalence, severity and mortality were recorded (40%, 55% and 15% respectively) followed

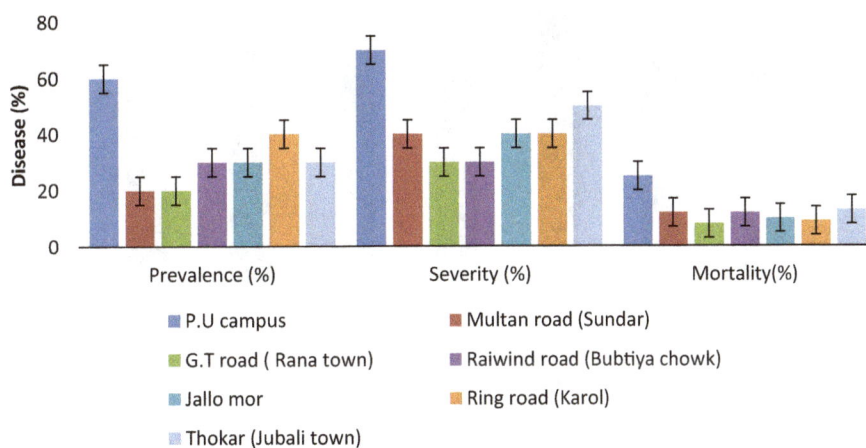

Figure 1: Prevalence (%), severity (%) and mortality (%) of *Alternaria* leaf spot of mustered at seven different survey location.

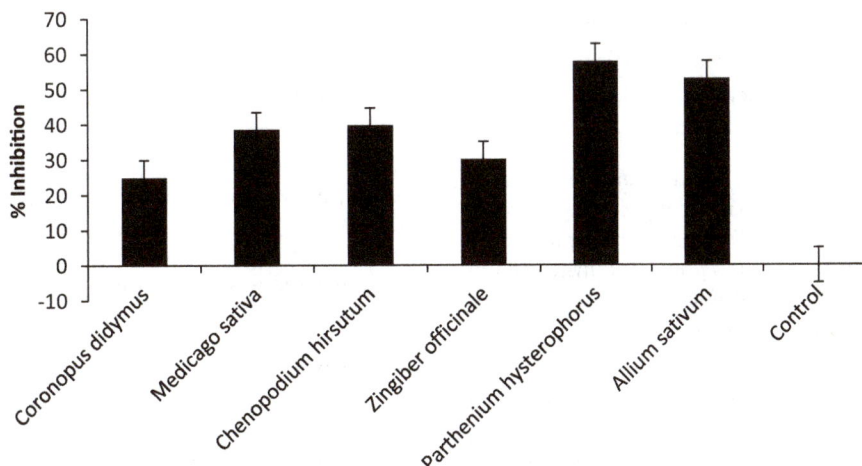

Figure 2: *In vitro* screening of plant extracts.

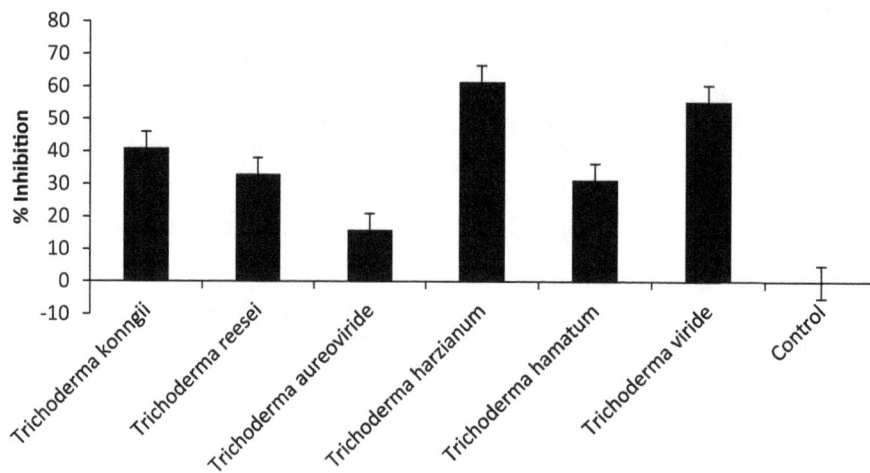

Figure 3: *In vitro* screening of Biological agents.

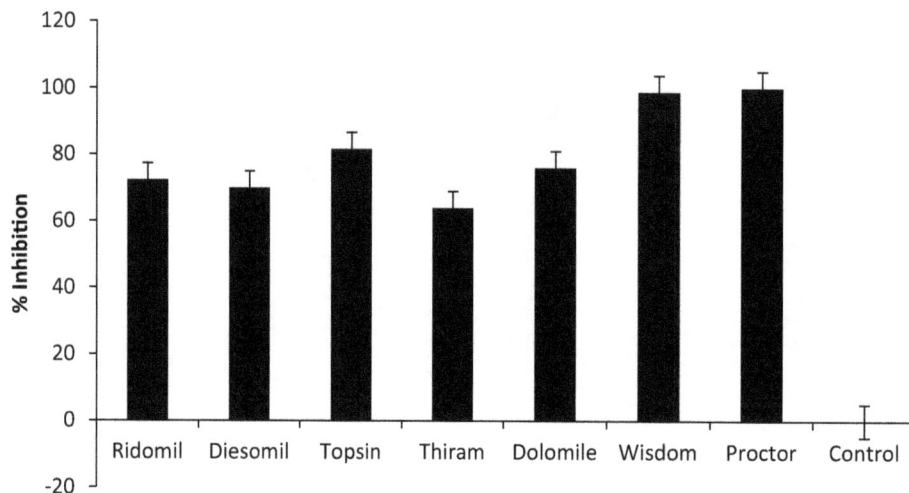

Figure 4: *In vitro* screening of Fungicide.

by control (4%, 2% and 0% respectively). In green house mean disease prevalence, severity and mortality were recorded (50%, 50% and 10% respectively) followed by control (3%, 1% and 0% respectively) (Figure 5). Field symptoms categories according to disease rating scale (Plate. 4) and plant disease index was calculated by the following formula;

$$\text{Plant Disease index } (\%) = \frac{\text{Sum of numerical rating}}{\text{Number of leaves infected}} \times 100$$

Disease management in field

Field experiment for management of *Alternaria* leaf spot of *Brasscia compestris:* After disease management according to result shown that maximum PDI was (88.02%) recorded in plot of control followed by *Allium sativum* (81.43%), *Parthenium hysterophorus* (70.7%), *Trichoderma harzianum* (60.86%), *Trichoderma viride* (58.57%), Proctor (36.44%) and minimum PDI was recorded by Wisdom (32.12%).

The maximum leaf spot disease (56.08%) was reduces in plot treated with 5%, 10% and 15% concentration of Wisdom, followed by Proctor (51.76%), *Trichoderma viride* (29.63%), *Trichoderma harzianum* (27.34%), and *Parthenium hysterophorus* (17.5%) *Allium*

sativum (6.77%); whereas no any disease reduction was recorded at untreated plot (Table 1).

Green house disease management: After disease management according to result shown that maximum PDI was (86.66%) recorded in plot of control followed by *Allium sativum* (60.34%), *Parthenium hysterophorus* (56.86%), *Trichoderma harzianum* (48.5%), *Trichoderma viride* (48.14%), Proctor (31.5%) and minimum PDI was recorded by Wisdom (23.66%).

The maximum leaf spot disease (63%) was reduces in plot treated with Wisdom, followed by Proctor (55.16%), *Trichoderma viride* (38.16%), *Trichoderma harzianum* (38.45%), and *Parthenium hysterophorus* (38.45%) *Allium sativum* (26.32%); whereas no any disease reduced was recorded in plants not sprayed with fungicides (Table 2).

Discussion

Disease prevalence ranged between 20 to 60% followed by severity 30 to 70% and mortality 25 to 8% at different locations. Maximum disease prevalence, severity and mortality (60, 70 and 25% respectively) were recorded at P.U campus followed by Thokar (30, 50 and 12%

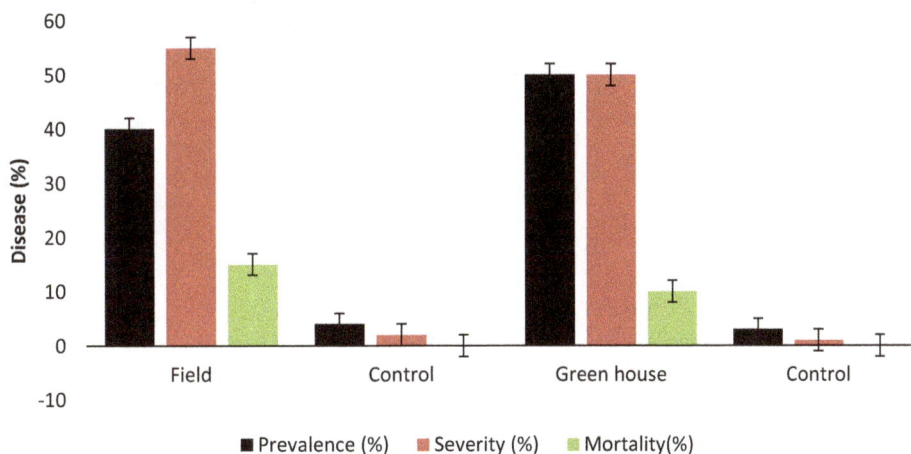

Figure 5: *Alternaria* leaf spot disease in field and green house.

Nature of Treatments	Alternaria Leaf spot												%PDI	% Decrease
	Replicate 1				Replicate 2				Replicate 3					
	5%	10%	15%	Mean	5%	10%	15%	Mean	5%	10%	15%	Mean		
Allium sativum	80	72	64	70 ± 2.42	81	73	65	73 ± 3.97	79	71	62	70.66 ± 3.10	81.43 ± 5.02	6.77 ± 0.02
Parthenium hysterophorus	74	64	49	60.4 ± 2.04	75	65	50	63.4 ± 3.31	73	62	49	61.33 ± 2.21	70.7 ± 4.41	17.5 ± 0.09
Trichoderma harzianum	63	54	44	51.7 ± 1.54	64	55	45	54.7 ± 2.30	62	52	44	52.67 ± 1.90	60.86 ± 3.92	27.34 ± 0.91
Trichoderma viride	64	49	43	50 ± 1.52	65	50	44	53 ± 2.21	61	48	42	50.34 ± 1.12	58.57 ± 2.41	29.63 ± 2.08
Wisdom	41	34	09	26 ± 0.01	42	35	10	29 ± 0.04	40	33	10	27.67 ± 0.03	32.12 ± 0.02	56.08 ± 3.41
Proctor	39	44	19	28.6 ± 0.08	40	45	20	31.6 ± 0.00	41	45	23	36.33 ± 0.41	36.44 ± 0.05	51.76 ± 2.52
Untreated Control	91	84	90	87.4 ± 3.95	92	85	90	88.4 ± 5.04	90	86	88	88 ± 5.41	88.2 ± 5.92	0 ± 0.00

± Standard error

Table 1: Field studies for *Alternaria* leaf spot of *Brassica compestris* by using plant extract, biological agents and fungicides.

Nature of Treatments	Alternaria Leaf spot												%PDI	% Decrease
	Replicate 1				Replicate 2				Replicate 3					
	5%	10%	15%	Mean	5%	10%	15%	Mean	5%	10%	15%	Mean		
Allium sativum	69	61	53	61 ± 1.98	71	63	55	63 ± 3.49	66	57	50	57.67 ± 3.14	60.34 ± 3.90	26.32 ± 0.02
Parthenium hysterophorus	63	57	45	55 ± 1.92	65	55	45	55 ± 2.59	70	60	46	58.67 ± 3.31	56.86 ± 3.38	29.8 ± 0.08
Trichoderma harzianum	52	47	41	46.66 ± 2.02	54	45	40	46.34 ± 3.25	60	50	40	50 ± 2.52	48.21 ± 3.0	38.45 ± 1.12
Trichoderma viride	53	46	40	47.34 ± 2.08	55	46	41	47.34 ± 3.25	61	47	41	49.66 ± 2.21	48.5 ± 2.9	38.16 ± 1.42
Wisdom	34	26	11	23.67 ± 0.02	32	26	10	22.67 ± 1.35	37	27	10	24.66 ± 1.91	23.66 ± 1.95	63 ± 3.23
Proctor	32	37	19	29.33 ± 0.03	30	35	17	27.33 ± 2.05	40	45	22	35.67 ± 2.97	31.5 ± 2.06	55.16 ± 2.51
Untreated Control	89	81	88	86 ± 4.95	90	80	88	86 ± 9.82	87	85	90	87.33 ± 5.07	86.66 ± 6.56	0 ± 0.00

±Standard error

Table 2: Green house studies for *Alternaria* leaf spot of *Brassica compestris* by using plant extract, biological agents and fungicides.

respectively), Multan road (20, 40 and 8% respectively), Ring road (40, 40 and 12% respectively), Jallo mor (30, 40 and 10% respectively), Raiwind raod (30, 30 and 9% respectively) and minimum disease was (20, 30 and 13% respectively) recorded at G.T road (Rana town). Similar results were reported by Maltoni et al. [12].

The fungus, *Alternaria brassicae* was isolated and identified by its morphological characteristics as described by Lelivet. It was confirmed as cause of *Alternaria* leaf spot disease of *Brassica compestris*. The findings were conformed from Fungal Culture Bank of Pakistan. *Allium sativum* and *Parthenium hysterophorus* was screen out by food poisoning technique. According to data they inhibited maximum (53.01% and 57.83% respectively) fungal growth followed by

Trichoderma harzianum and *Trichoderma veridie* (61.44% and 55.42% respectively) and wisdom and Proctor (98.79% and 100% respectively).

Chemical control is ultimate and easy solution of disease. But biological management is more acceptable to environment and human being. Lot of literature is available regarded to justify biological management of *Alternaria* leaf spot of *B. compestris*. Spray of soil isolates of *Trichoderma viride* at 45 and 75 days after sowing could manage *Alternaria* blight of Indian mustard (*Brassica juncea*) as effectively as mancozeb and other fungicides, which have been incorrigible later in multiplication trials (AICRP-RM, 2007). Botanicals viz., bulb extract of *Allium sativum* has been reported to effectively manage *Alternaria* blight of Indian mustard [13].

The maximum leaf spot reduction was (63%) recorded of Wisdom followed by Proctor (55.16%), *Trichoderma viride*, (38.16%), *Trichoderma harzianum* (38.45%) and minimum disease reduction was recorded from *Parthenium hysterophorus* (29.8%) followed by *Allium sativum* (26.16%).

Chemical management is most favorable and widely used method against disease. One sided it completely control disease but other hand it has many unseen sides' effects which directly or indirectly transfer to human beings. Meanwhile, it also has hazardous effect in our environment. Although fungicide remains more effective in reducing diseases in plants, increasing public concern about environmental health is proving to be major hindrance in use of chemical pesticides including fungicides.

References

1. Vavilov NI, Dorofeev VF (1992) Origin and Geography of Cultivated Plants. Transl.by Doris Love. Cambridge University Press p: 498.

2. Bandopandopadyay L, Basue D, Sikdar SR (2013) Identification of genes involved in wild Crucifer *Rorippa indica* resistance response on mustard aphid *Lipaphis erysimi* challenge. PLoS One 8: e73632.

3. Tayo TO, Morgan DG (1979) Factors influencing flower and pod development in oil-seed rape (*Brassica napus*). J Agric Sci Cambridge 92: 363-373.

4. Kolte SJ (1985) Diseases of Annual Edible Oilseed Crops, Vol. II, Rapeseed-Mustard and Sesame Diseases. CRC Press Inc. Boca Raton, Florida p: 135.

5. Paul VH, Rawlinson CJ (1992) Diseases and Pests of Rape. Gelsenkirchen-Buer, Germany: Verlag Th. Mann.

6. Conn KL, Tewari JP, Awasthi RP (1990) A Disease Assessment Key for Alternaria Black Spot in Mustard. Can Plant Dis Sur 70: 19-22.

7. Chattopadhyay AK, Bagchi BN (1994) Relationship of disease severity and yield due to leaf blight of mustard and spray schedule ofmancozeb for higher benefit. J Mycopathol Res 32: 83-87.

8. Gilman JC (1945) A manual of Soil Fungi. 2nd ed. The Lowa State University Press, Ames, Lowa, USA.

9. Domsch KH, Gams W, Andreson TH (1980) Compendium of soil fungi. Volume 1. Academic press. A subsidiary of Harcourt braceJovanovich, publishers. London, New York, Toronto, Sydney, SanFrancisco.

10. Barent HL, Hunter BB (2003) Illustrated Genera of Imperfect Fungi. APS Press, The American Phyto pathological society St. Paul, Minnesota.

11. Shekhawat PS, Prasada R (1971) Antifungal properties of some plant extract and inhibition of spores germination. Indain phytopathol 24: 800-802.

12. Maltoni ML, Magnani S, Baruzzi G (2000) Screening forAlternaria black spot resistance in strawberry plants 62: 95-97.

13. Patni CS, Kolte SJ (2006) Effect of some botanicals in management of Alternaria blight ofrapeseed-mustard. Ann Pl Prot Sci 14: 151-156.

28

Virulence Diversity in *Rhizoctonia Solani* causing Sheath Blight in Rice Pathogenicitya

Ramji Singh[1], Shiw Murti[2], Mehilal[3], Ajay Tomer[4] and Durga Prasad[5]

[1]*Department of Plant Pathology, SardarVallabhbhi Patel University of Agriculture &Technology, Meerut-250110, Uttara Pradesh, India*
[2]*Agriculture officer Oriental Bank of Commerce, India*
[3]*Scientist CPRI, Meerut, India*
[4]*Lovely Professional University Jalandhar Punjab, India,*
[5]*MB Agriculture College Saharsa, Bihar Agriculture University, Bihar, India*

Abstract

Knowledge of variations in *Rhizoctonia Solani* causing rice sheath blight disease in different geographic regionsis still scarce and may be a useful tool for examining the nature and spread of population, disease epidemiology and host-pathogen interaction within rice patho-system. Molecular markers provide a basis for identifying patterns, dispersal and colonization in spatial and temporal distribution of pathogenic population and in development of species concepts by providing information about the limits of genetically isolated group in relation to patterns of morphological variation and behavior of any pathogen. Twenty-five isolates of *Rhizoctonia Solani* causing sheath blight in rice were collected from Kerala, New Delhi, Punjab, Uttarakhand and Uttar Pradesh of India and subjected for determination of virulence diversity. There was great diversity in the population of *R. Solani* which deferred greatly according to color and texture of colony, number and size of sclerotia, time taken for sclerotia formation and also the place and manner of sclerotia formation in the colony. At molecular level, also there was great diversity in the *R. Solani* population. A total of 80 PCR bands were detected among 25 isolates of *R. Solani*. The number of alleles per locus, varied from 1to7. Highest [1] PCR products were obtained with primers-OPW-13 and OPA-04 whereas lowest PCR products [2] were obtained with primer UBC-310 and OPB-08. There was only one monomorphic band, which was present in related primer UBC-373.The similarity coefficients among the *R. Solani* population ranged between 0.53 to 0.94. One isolate of *R. Solani* from Uttar Pradesh (RS-16) and another isolate from Punjab (RS-1) were most distantly related. The *R. Solani* isolates (RS-11&RS-12) and (RS-20& RS-21) all belonging to Uttar Pradesh were genetically most closely to each other.

Keywords: *Rhizoctonia Solani*; Isolates; Virulence diversity; Sheath blight; PCR bands; Similarity coefficients; Alleles; Locus

Introduction

Sheath blight, caused by *Rhizoctonia Solani* Kuhn, earlier a minor disease, is now ranked second only tothe blast disease. The pathogen is regarded as an unspecialized with indefinite pathogenic races [3]. Attempts to control rice sheath blight have been dependent upon cultural practices, resistant cultivars, fungicides, and to some extent, biological methods. No resistant cultivar is available for practical use and the present intensive rice cultivation practices offer a favorable condition for sheath blight development. In addition, there is considerable pressure from environmental scientists to decrease emphasis on chemical control. However, the use of fungicides is unavoidable, but certainly, their use can be minimized as a long term solution to the crop health problem. Breeding for disease resistance may be most practical and feasible, but it could not be a final solution because of virulence diversity in the pathogenic population. Under thesecircumstances, management of sheath blight will require sharply focused approach including host resistance. Spray of fungicides and micronutrients at boot leaf and heading stage of rice plant has been resulted in controlling quite a few rice diseases and lead to significant increase in yield [4]. Management of this disease through host resistance will require a comprehensive knowledge of prevalent races of the target pathogen, which can be achieved by exploring the virulence diversity in the population of targeted pathogen.

The genetic difference underlying *R. Solani* population has provided an useful means of examining the nature and spread of population within rice Patho-system. Knowledge of variationsin the population of *Rhizoctonia* spp. causing rice sheath blight disease in different geographic regions is still scarce. Many problems associated with studying different level of diversity in *Rhizoctonia* are best addressed through the use of molecular markers, at the species level. Molecular markers aid in development of species concepts by providing information about the limits of genetically isolated group in relation to patterns of morphological variation and behavior. At the population level, molecular markers provide a basis for identifying patterns, dispersal and colonization in spatial and temporal distribution. The most convincing validation of Anastomosis group (AG) and inter specific group (ISG) concepts has come from molecular studies [5]. In view of these facts, investigation on virulence diversity of *Rhizoctonia Solani* causing sheath blight in rice were therefore undertaken.

Materials and Methods

Sources and Maintenance of culture

Twenty-five isolates of *R. Solani* were isolated and maintained. Of these, five isolates were obtained from Indian type culture Collection

*Corresponding author: Ramji Singh, Department of Plant Pathology, Sardar Vallabhbhi Patel University of Agriculture &Technology, Meerut- 250110 (UP) India
E-mail: singh.ramji@gmail.com

(ITCC), IARI, New Delhi and remaining twenty isolates were collected from different parts of India (Table 1). Pathogen cultures were isolated from diseased leaf sheaths of infected rice plants on PDA media in petri plates and incubated at 28 ± 1°C for 2-3 days. After three days of incubation, fine radiating mycelium growth were initiated form the edges of infected leaf bits. A small bit of actively growing mycelium was then transferred into the potato dextrose agar slants to obtain pure culture, which were further purified by germinating asingle sclerotium and maintained on PDA slants.Purity of *R. Solani* was ascertained by proving Koch's postulates after inoculating them on the highly susceptible cultivar, Pusa Basmati-1. Production of sheath blight symptoms by the respective isolates was taken as an indication of conformity of purity for the test fungus.

Determination of cultural and morphological variability

Cultural and Morphological variability were determined by taking observation on colony colour, growth pattern, and growth rate.The color of colony was determined with the help of munsel's Soil Color Chart (Munsell Color Company, Inc.1954). Growth pattern was recorded by visual observation according to growth of aerial hyphae. Growth rate of each isolate was measured, using plastic scale at the interval of 24 h, 48 h, 72 h and 96 h.

Determination of sclerotial characteristics

Sclerotial characteristics viz, color, texture, number, size, time taken for initiation of sclerotia formation, pattern of production (central, peripheral and scattered) and location (Aerial-Sclerotia formed, on the aerial mycelium; Surface-Sclerotia formed on the surface mycelium; Embedded –Sclerotia formed on the substrate) of sclerotia were recorded [6]. The diameter of the sclerotia was measured with the help of randomly harvested 20 sclerotia with the help of Digital Vernier Calipers (Mitutoyo Corp made in Japan). No sclerotia forming isolates were categourized as poor, 1-10 fair, 11-20 moderate, 21-40 good, 41-60 very good and >60 as excellent.

Molecular variability

Preparation of genomic DNA: DNA extraction from mycelial mat of *R. Solani* grown on Potato Dextrose Broth (PDB) was done by CitimideTetradecylTrimethyl Ammonium Bromide (CTAB) methods [7]. DNA samples were further purified by taking a stock solution of RNase-A which was prepared @ 10 mg/ml in 10 mMTris-HCl (pH 8.0) and 15 mM sodium chioride. The solution was boiled for 10 minutes to destroy DNase and allowed for cooling slowly at room temperature. From the stock, 2 µl, of RNase was added to the crude DNA samples and incubated at 37°C for one hour. DNA concentration of sample and purity was determined by taking ultraviolet absorbance at 260 nm and 280 nm, in a spectrophotometer followed by agrose gel electrophoresis

Quantification of DNA was done by using Spectrophotometer (Nano Drop,NA- 1000) . One µl sterilized double distilled water was placed in Spectrophotometer and wiped with tissue paper and one µl of sample DNA was placed in spectrophotometer. Reading was taken at wavelengths of 260 nm.Concentration of DNA varied (399.1-5085.2 ng) in different isolates of *R. Solani*. Working solutions having concentration of 25 ng / µl were prepared for optimization of PCR reaction by Gel electrophoresis method by loading 7 µl of mixture (5 µl DNA+2 µl loading dye)of each isolates along with 1Kb marker in 1.0 per centagrose gel.

Standardization of RAPD protocolandOptimization of Polymerase Chain Reaction (PCR): Composition of PCR reaction was optimized by varying concentration of different component of reaction mixture, which included template DNA (25 ng, 30 ng, 50 ng, 75ng, 100ng), Taq DNA polymerase (0.5units, 1.0units,1.5units) and MgCl$_2$ (0.4mM, 0.5mM,0.6mM). Different PCR profiles were tested for obtaining best amplification of nucleic acid of *R.Solani* isolates. The PCR reaction was performed in Bioneer thermal cycler programme. RAPD condition for *R. Solani* were standardized and final PCR composition was: 10X Taq buffer containing 15 mM MgCl$_2$ (2.5µl) dNTP mix (dATP, dGTP, dCTP, dTTP) 1.0 µl, primer (0.4µM)1.0 µl, Taq DNA polymerase (3U/ 1µl) 0.5µl.Total reaction volume was made up to 25 µl using sterile double distilled water. Different PCR profiles were tested for best amplification of DNA. The standardized temperature profiles of 94°C for 5 minutes followed by 38 cycles of 94°C of 1minute, 36°C for 1 minute, 72°C for 2 minutes, with an elongation of 72°C for 10 minutes which gave best results hence, was used for DNA amplification [8].

Primer survey and selection: Preliminary primer screening was carried out with 25 primer (GCC Company limited) out of them, only 12 primer gave best amplification and each primer was testedwith two isolates for assessing amplification and was employed for molecular variation analysis (Table 2). The primer that gave reproducible and scorable amplification were further used in the analysis of genetic variability of the isolates.

Agarose gel electrophoresis: 2 µl of loading dye was added to the 25 µl of amplification products obtained after the PCR reaction, and loaded into individual wells of 1.2 per cent horizontal agarose gel in TAE buffer, pre-stained with Ethidium Bromide (1µg/ml). Electrophoresis was carried out at 60 volts for 2 h in TAE buffer. One Kb ladder (MBI, Fermantas) was used as a marker. The gel was observed in a PerkinElmer Geliance 200 Imaging system.

Scoring of RAPD-PCR amplification

Each amplification product was considered as RAPD marker and was scored across all samples. Data was entered in a matrix in which

Isolate code	Place of collection	State	Crop (Host)
RS-1	Ludhiana	Punjab	Rice
RS-2	Ludhiana	Punjab	Rice
RS-3	ITCC ,IARI	New Delhi	Rice
RS-4	ITCC,IARI	New Delhi	Rice
RS-5	ITCC,IARI	New Delhi	Rice
RS-6	ITCC,IARI	New Delhi	Urd bean
RS-7	ITCC,IARI	New Delhi	Maize
RS-8	Pattambi	Kerala	Rice
RS-9	Moncopu	Kerala	Rice
RS-10	Pant nagar	Uttarakhand	Rice
RS-11	Faizabad	Uttar Pradesh	Rice
RS-12	Faizabad	Uttar Pradesh	Moong bean
RS-13	Meerut	Uttar Pradesh	Rice (PB-1)
RS-14	Meerut	Uttar Pradesh	Rice (Masoori)
RS-15	Meerut	Uttar Pradesh	Rice
RS-16	Meerut	Uttar Pradesh	Rice (HKR-1)
RS-17	Meerut	Uttar Pradesh	Rice(PB-9)
RS-18	Meerut	Uttar Pradesh	Rice(Sungandha-5)
RS-19	Meerut	Uttar Pradesh	Rice(Sungandha-7)
RS-20	Meerut	Uttar Pradesh	Rice(PRH-10)
RS-21	Meerut	Uttar Pradesh	Rice(Pusa-677)
RS-22	Faizabad	Uttar Pradesh	Rice(ND-10)
RS-23	Ghaziabad	Uttar Pradesh	Rice
RS-24	Meerut	Uttar Pradesh	Potato
RS-25	Meerut	Uttar Pradesh	Rice(PR-53)

Table 1: List of *Rhizoctonia solani* isolates with their place of collection.

all observed bands or characters state was listed. The RAPD pattern of each isolate was evaluated assigning character state to all bands that could be observed in gel with a particular primer. The characters state "1" was given if this band could reproducibly detect in all RAPD analysis with an isolate/primer combination. The character state "0" was assigned if band was lacking or it was not possible to determine its presence with certainty.

Molecular diversity analysis using Jaccard similarity coefficient: Similarity index matrix generated, based on proportion of common amplified band between two isolates [9] using F=(2 MxY) / (Mx + My) where, F is similarity index, Mx is the number of band in isolates x, My is the number of band in isolate y, and Mxy is the number of bands common to both x and y. From the number of common bands in the fingerprint pattern on the two isolates, the formula of Nei&Li [9] was used to calculate a similarity coefficient for pairs of isolates. Pair, which has a higher coefficient, is more closely related than those with a lower coefficient. In the construction of the matrix, it is assumed that corresponding bands arise by amplification of the same genetic locus. Similarity matrixes are very useful for showing relationship that is not apparent from the gel photograph. Cluster analysis of the data was made, based on the similarity indices between pathogen isolates, using NTSYSpc version 2.0 software [10] and a dendogram was constructed.

Results

Cultural variability of *Rhizoctonia Solani*

Colony colour: Out of twenty-five isolates, five isolates of (RS-1, RS-2, RS-8, RS-21 and RS-25) *Rhizoctonia Solani* were of whitish brown color, seven isolates (Rs-3, RS-6, RS-13, RS-14, RS-15, RS-17 and RS-19) were of light brown color, six isolates (RS-5, RS-7, RS-10, RS-11, RS-16 and RS-23) were of yellowish brown color, four isolates (RS-4, RS-12, RS-20 and RS-24) were of dark brown color, two isolates (RS-18, RS-22) were of pale brown color, and one isolate (RS-9) was of milky browncolor (Table 3).

Growth pattern: Of the twenty-five isolates of *Rhizoctonia Solani*, six isolates (RS-2, RS-3, RS-7, RS-8, RS-11, RS-13) were found to be growing abundantly and accordingly categorized into thegroup-1. Thirteen isolates(RS-1, RS-5, RS-6, RS-9, RS-12, RS-14, RS-15, RS-18, RS-19, RS-20, RS-21, RS-22, RS-24) were found to be of moderate growth pattern and categorized into group-2, whereasremaining six isolates (RS-4, RS-10, RS-16, RS-17, RS-23, RS-25) were found to be of slightly low growth pattern hence categorized into group-3 (Table 3).

Growth rate: Among twenty five isolates, fast growing (>65 mm) 3 isolates (RS-3, RS-7and RS-17) were categorized in group-1, medium growing (60-65 mm) 3 isolates (RS- 8,RS-11, and RS-13) were categorized into group-2 and remaining nineteen isolates (40-59 mm) *i.e.*RS-1,RS-2,RS-5,RS-6,RS-9,RS-10,RS11,RS12,RS-14,RS-15,RS-16,RS-18,RS-19,RS-0,RS21,RS-22,RS-23,RS-24,RS-25 were categorized into group-3 (Table 3).

Colour of sclerotia: All twenty five isolates of *Rhizoctonia Solani* showed great variation in the colour of sclerotium (Table 5). Based on the pigmentation of the sclerotium, isolates were assigned into four groups*i.e.* dark brown which included isolates RS-4, RS-7, RS-9, RS-11, RS-17, RS-18, RS-23, and RS-25, dark yellowish brown which included isolates RS-2, RS-5, RS-13, RS-16, RS-19,and RS-21, olive brown which included isolates RS-8, RS-10, RS-12, RS-14, RS-15, RS-20 and RS-22 and light brown which included isolates RS-1, RS-3 and RS-6. The isolate RS-24 did not bear any sclerotia.

Texture of sclerotia: Based on texture of sclerotia the isolates were classified into two groups *i.e.* smooth and rough (Table 5). Of all twenty five, ten isolates *viz-* RS-1, RS-2, RS-3, RS-8, RS-10, RS-12, RS-14, RS16,- RS-22 and RS-23 were having smooth category of sclerotium and remaining 14 isolates viz-RS-4, RS-5, RS-6, RS-7, RS-9, RS-11, RS-13, RS-15, RS-17, RS-18, RS-19, RS-20, RS-21 and RS-25 were having rough categoryof sclerotium.

Number of sclerotia: Among the 25 isolates,no sclerotium was formed in isolates RS-24 and was categorized into group-1 (poor). None of the isolates was categorized into group-2 (fair) and group 3 moderate. Group-4 (good) included six isolates (RS-2, RS-8, RS-12, RS-16, RS-18 and RS-19), Group -5 (very good) included seven isolates (RS-4, RS-7, RS-9, RS-13, RS-17, RS-23 and RS-25) and group-6 (excellent) included eleven isolates (RS-1,RS 3,RS5,RS-6,RS-10, RS-11, RS-14, RS-15,RS-20, RS-21 and RS-22,).

Diameter of sclerotia: Based on diameter of sclerotia, the isolates were categorized into two groups. Group -1 had diameter range form 1.21-1.75 mm and group -2 had diameter range from 1.75-2.94 mm. Diameter of the sclerotia was observed up to maximum range in isolates RS-17(2.94 mm) and minimum in RS-1(1.24 mm). Seven isolates (RS-1,RS-2, RS-4, RS-10, RS-11,RS16 and RS-19) were categorized into group-1 with diameter range between 1.21-1.75 mm. Seventeen isolates (RS-3, RS-5, RS-6, RS-7, RS-8, RS-9, RS-12, RS-13, RS-14, RS-15, RS-17, RS-18, RS-20, RS-21, RS-22, RS-23 and RS-25) were categorized into group-2 (Table 4.4) having diameter range between 1.75-2.94 mm.

Location of sclerotia formation: The location of sclerotia was observed on the basis where actually the sclerotia were formed in the fungal colony (Table 4.5). Based on the location of sclerotia formation, these isolates were categorized into three groups. First group included those isolates where sclerotium formed within the aerial mycelium

S. No.	Name of primers	Sequences
1.	OPA-01	CAGGCCCTTC
2.	OPA-02	TGCCGAGCTG
3.	OPA-03	AGTCAGCCAC
4.	OPA-04	AATCGGGCTG
5.	OPA-10	GTGATCGCAG
6.	OPA-19	CAAACGTCGG
7.	OPA-20	GTTGCGATCC
8.	OPB-01	CCGTCGGTAG
9.	OPB-08	GTCCACACGG
10.	OPB-18	CCACAGCAGT
11.	OPN-20	CGTGCTCCGT
12.	OPU-02	CTGAGGTCTC
13.	OPU-06	ACCTTTGCGG
14.	OPU-14	TGGGTCCCTC
15.	OPV-12	ACCCCCCACT
16.	OPV-20	CAGCATGGTC
17.	OPW--13	CACAGCGAACA
18.	UBC-220	GTCGATGTCG
19.	UBC-248	GAGTAAGCGG
20.	UBC-249	GCATCTACCG
21.	UBC-310	GAGCCAGAAG
22.	UBC-338	CTGTGGCGGT
23.	UBC-373	CTGAGGAGTC
24	UBC-374	GGTCAACCCT
25	OPZ-20	ACTTTGGCGG

Table 2: List of Primers used for survey of amplification.

Isolates	Colony colour	Growth pattern			Colony growth diameter (mm) at different intervals				
		Abundant	Moderate	Slight	24h	48h	72h	96h	Mean
RS-1	White brown	-	+	-	5.60	47.00	90.00	90.00	58.15
RS-2	White brown	+	-	-	8.60	35.00	67.33	90.00	50.23
RS-3	Light brown	+	-	-	30.00	90.00	90.00	90.00	75.00
RS-4	Dark brown	-	-	+	0.00	46.66	79.66	90.00	54.08
RS-5	Yellowish brown	-	+	-	0.00	33.33	85.66	90.00	52.25
RS-6	Pale brown	-	+	-	0.00	35.00	83.33	90.00	52.08
RS-7	Yellowish brown	+	-	-	28.30	66.33	90.00	90.00	68.66
RS-8	White brown	+	-	-	16.00	49.00	90.00	90.00	61.25
RS-9	Milkish brown	-	+	-	0.00	48.00	73.42	89.00	52.61
RS-10	Yellowish brown	-	-	+	0.00	55.33	89.00	90.00	58.58
RS-11	Yellowish brown	+	-	-	15.00	58.3	90.00	90.00	63.33
RS-12	Dark brown	-	+	-	0.00	37.66	90.00	90.00	54.42
RS-13	Light brown	+	-	-	16.00	59.00	90.00	90.00.	63.75
RS-14	Light Brown	-	+	-	8.00	41.00	81.00	90.00	55.00
RS-15	Cream brown	-	+	-	6.00	39.66	84.46	90.00	55.03
RS-16	Yellowish brown	-	-	+	0.00	19.00	51.33	90.00	40.08
RS-17	Pale brown	-	-	+	25.00	60.00	90.00	90.00	66.25
RS-18	Pale brown	-	+	-	0.00	16.33	56.33	90.00	40.67
RS-19	Light brown	-	+	-	0.00	24.31	68.34	90.00	45.66
RS-20	Dark brown	-	+	-	0.00	27.00	80.33	90.00	49.33
RS-21	White brown	-	+	-	0.00	43.35	90.00	90.00	55.84
RS-22	Pale brown	-	+	-	0.00	43.00	75.00	90.00	52.00
RS-23	Yellowish brown	-	-	+-	0.00	60.00	86.66	90.00	59.17
RS-24	Dark brown	-	+	-	18.00	22.10	44.33	90.00	43.61
RS-25	White brown	-	-	+	0.00	6.33	61.00	78.60	36.48

Table 3: Cultural characteristics of different isolates of *R. solani.*

Isolates	Time taken for initiation of sclerotial formation (Days)	Average sclerotia diameter (mm)	No. of sclerotia per Petridish
RS-1	4	1.24	Excellent
RS-2	5	1.21	Good
RS-3	3	2.38	Excellent
RS-4	4	1.65	Very good
RS-5	5	1.75	Excellent
RS-6	5	1.97	Excellent
RS-7	5	2.31	Very good
RS-8	5	1.75	Good
RS-9	5	2.21	Very good
RS-10	6	1.59	Excellent
RS-11	3	1.73	Excellent
RS-12	5	1.79	Good
RS-13	4	1.81	Very good
RS-14	5	2.01	Excellent
RS-15	4	2.12	Excellent
RS-16	5	1.46	Good
RS-17	4	2.94	Very good
RS-18	5	1.86	Good
RS-19	4	1.54	Good
RS-20	5	1.90	Excellent
RS-21	6	1.80	Excellent
RS-22	5	1.87	Excellent
RS-23	5	2.18	Very good
RS-24	-	--	No
RS-25	6	1.90	Very good
CD(P=0.05)		0.036	

Table 4: Sclerotial characteristics of different isolates of *R. solani.*

(RS-1, RS-2, RS-5, RS-7, RS-9, RS-11, RS-13, RS-20 and RS-22). Second group included those isolates where sclerotia formed at the surface of the mycelium (RS-4, RS-6, RS-8, RS-12, RS-15, RS-16, RS-17, RS-18 RS-19, RS-21, RS-23 and RS-25). The isolates, RS-3, RS-8, RS-10, RS-14 and RS-18 had sclerotia embedded in the fungal mycelium itself.

Manner of sclerotia formation within the petridish: Based on number of sclerotium formed in the Petri dishes isolates of *Rhizoctonia Solani* were classified into three groups (Table 5). Sclerotium formed in the central ring was placed into group-1, which consisted of eight isolates (RS-3, RS-6, RS-9, RS-13, RS-15, RS-17, RS-21, and RS-22). Nine isolates (RS-2, RS-5, RS-10, RS-11, RS-12, RS-18, RS-23 and RS-25) were those, where sclerotium were formed in the peripheral manner and classified into group-2. Those isolates where sclerotium were formed scattered in the petridish *i.e.,* neither peripheral nor central were placed into a separate groups-3 which included seven isolates (RS-1, RS-4, RS-7, RS-8, RS-14, RS-19 and RS-20).

Time taken for sclerotia formation

All isolates showed great variation in the time taken for initiation of sclerotia formation also whichranged from 3 to 6days (Table 4).Two isolates (RS-3,and RS-11) took 3 days for initiation of sclerotia formation, six isolates (RS-1, RS-4, RS-13, RS-15, RS-17 and RS-19) took 4days for initiation of sclerotia formation, thirteen isolates (RS-2, RS-5,RS-6, RS-7, RS-8,RS-9, RS-12, RS-14, RS-16, RS-18, RS-20, RS-22 and RS-23) took 5days for initiation of sclerotia formation and three isolates (RS-10, RS-21 and RS-25) took 6days for initiation of sclerotia formation.

Molecular variability of *R. Solani* isolates

DNA amplification using RAPD: Among 25 RAPD primers, only

Isolates	Manner of sclerotia Formation.			Location of sclerotia			Texture of sclerotia		Colour of sclerotia
	Central	Peripheral	Scattered	Arial Surface	Embedded		Smooth	Rough	
RS-1	-	-	+	+	+	-	+	-	Light brown
RS-2	-	+	-	+	-	-	+	-	Dark yellowih brown
RS-3	+	-	+	-	-	+	+	-	Light brown
RS-4	-	-	+	-	+	-	-	+	Dark brown
RS-5	-	+	-	+	-	-	-	+	Dark yellowish brown
RS-6	+	-	+	-	+	-	-	+	Light brown
RS-7	-	-	+	+	-	-	-	+	Dark brown
RS-8	-	-	+	-	+	+	+	-	Olive brown
RS-9	+	-	-	+	-	-	-	+	Dark brown
RS-10	-	+	-	-	-	+	+	-	Olive brown
RS-11	-	+	-	+	-	-	-	+	Dark brown
RS-12	-	+	-	-	+	-	+	-	Olive brown
RS-13	+	+	-	+	-	-	+	-	Dark yellowish brown
RS-14	-	-	+	-	-	+	+	+	Olive brown
RS-15	+	+	-	-	+	-	-	+	Olive brown
RS-16	-	+	-	-	+	-	+	-	Dark yellowish brown
RS-17	+	-	+	-	+	-	-	+	Dark brown
RS-18	-	+	-	-	+	+	-	+	Dark brown
RS-19	-	-	+	-	+	-	-	+	Dark yellowish brown
RS-20	-	-	+	+	-	-	-	+	Olive brown
RS-21	+	+	-	-	+	-	-	+	Dark yellowish brown
RS-22	+	+	-	+	-	-	+	-	Olive brown
RS- 23	-	+	-	-	+	-	+	-	Dark brown
RS-24	-	-	-	-	-	-	-	-	No formation of sclerotia
RS-25	-	+	-	-	+	-	-	+	Dark brown

+=Present; - =Abesent.

Table 5: Formation, location, texture and colour of sclerotia of different isolates of R. solani.

S.No.	Primers	Total no. of bands	No. of polymorphic bands	No. of monomer-phic bands	Per cent Polymorphism
1.	OPA 19	6	6	0	100
2.	OPU 6	3	3	0	100
3.	OPV 12	5	5	0	100
4.	OPV 20	5	5	0	100
5.	OPW 13	7	7	0	100
6.	OPA 02	6	6	0	100
7.	OPA 04	7	7	0	100
8.	OPN 20	4	4	0	100
9.	OPA 01	5	5	0	100
10.	OPA 10	4	4	0	100
11.	OPA 09	4	4	0	100
12.	OPA 20	6	6	0	100
13.	OPB 01	3	3	0	100
14.	UBC 310	2	2	0	100
15.	UBC 203	3	3	0	100
16.	UPC248	3	3	0	100
17.	OPB08	2	2	0	100
18.	UBC373	5	4	1	80
19	OPU-14	0	0	0	0
20	UBC-310	0	0	0	0
21	OPU-02	0	0	0	0
22	OPW-13	0	0	0	0
23	UBC-338	0	0	0	0
24	UBC-374	0	0	0	0
25	UBC-248	0	0	0	0
Total		80	79	1	

Table 6: Primer sequence, number of polymorphic and monomorphic bands, present polymorphism.

18 resulted in the PCR products and rest of the 7 couldn't resulted in any PCR product. Out of 18 primers, resulting in PCR products, 17 primers gave polymorphic bands whereas one primer *i.e.* UBC-373 resulted in one monomorphic bands also.

Number of alleles in RAPD: During this experimentation, a total of 80 alleles were detected among 25 isolates of *R. Solani*. The number of alleles per locus varied from 1 to 7. (Table 4.6). The highest number of alleles were observed due to primer OPW 13 and OPA -04 (seven alleles) followed by OPA-02, OPA-19 and OPA- 20 (six alleles), OPV-12 (five alleles), OPV- 20 (five alleles), OPA-01 (five alleles), OPN-20 (four alleles), OPA-10 (four alleles), OPA-09 (four alleles), OPU-6 (three alleles), OPB-01 (three alleles), UBC-203 (three alleles), UPC-248(three alleles), UBC-310 (two alleles) and OPB-08 (two alleles). This provides the summarized data regarding the number of unique alleles and their distribution in various isolates. There was only 01 monomorphic band, which was present in related primer UBC-373

Polymorphism in RAPD primers: The total number of bands obtained were 80, of which 79 bands were polymorphic and one was monomorphic, of which 17 were 100% polymorphic, with an average of 20.25 bands per primer, the number of polymorphic allele ranged from 1 to 7 and 100% polymorphism were obtained in the case of 17 primers and remaining 1 primer *i.e.*UBC373 did not give 100% polymorphism rather this could resulted in 80% polymorphism.

Similarity Vs dissimilarity analysis for RAPD primers: The cluster denogram revealed that all the 25 isolates of *R. Solani* can be grouped into two major clusters at a cut off similarity coefficient level 0.53 (Figure 1). Among these, cluster Ist was having only one isolates *i.e.* RS-16, whereas the cluster IInd was largest with 24 isolates. The IInd cluster having 24 isolates was further sub-divided into two groups at a similarity coefficient level of 0.57. Among these two groups the group I included 7 isolates *viz.* RS-1, RS-3, RS-14, RS-25, RS-5, RS-9 and RS-18 at a similarity coefficient level of 0.65. Group IInd was consisted of remaining 17 isolates at a similarity coefficient 0.57. The group Ist was again divided into two sub-group at a similarity coefficient 0.65 of which the isolated RS-1 was alone in one sub-group and remaining 6 were in other sub-group. These remaining 6 isolates were again sub-divided into two groups at a similarity coefficient 0.71 wherethe isolate RS-18 was alone in one sub-group and remaining 5 isolates were in another sub-group. These 5 isolates were further sub-divided into two sub-groups at a similarity coefficient of 0.76. Among these 5 isolates, the RS-9 & RS-5 were in one sub-group & remaining 3 isolates were in other sub-group. These remaining 3 isolates were again divided into two groups at a similarity coefficient 0.80 into two groups where RS-3 was alone in one group & RS-14 & RS-25 were in another sub-group with 86% level of similarity (Table 7).

The group two of IInd cluster, which consisted of 17 isolates, was further sub-divided into two groups at a similarity coefficient of 0.59. Out of these, there was only one isolate *i.e.* RS-24 into one group and remaining 16 isolates of IInd group where reassigned to two groups at a similarity coefficient of around 0.65. Between these two groups the isolates RS-20, RS-21, RS-22 and RS-23 were in one group with a similarity coefficient 0.85. Whereas remaining 12 isolates were in another group with a similarity coefficient level of around 0.67. These 12 isolates were reassigned to two groups with a similarity coefficient 0.67. Out of these two groups, in one group, there was 3 isolate *i.e.* RS-4, RS-15 and RS-19 with a similarity coefficient of 0.71 and in another group, remaining 9 isolates were placed with a similar similarity coefficient of 0.71. These 9 isolates were again reassigned to two groups having 3 isolates *i.e.* RS-6, RS-7, and RS-8 in one group with similarity coefficient of 0.75 and remaining 6 isolates in another group with similarity coefficient of 0.76. The highest level of similarity coefficient *i.e.* 0.94 was observed between the isolates RS-11 and RS-12 both belonging to Uttar Pradesh followed by 0.92 between RS-20 and RS-21 these also belongs to Uttar Pradesh. The isolates RS-16, which belong to, Uttar Pradesh was found to highly distantly relate to isolate RS-1, which belongs to Punjab.

Discussion

From the experimental results, it is quite evident that all the 25 isolates of *R. Solani* were highly different from each other with respect to all the characters examined viz. color, growth, and texture of mycelium, number, size and color of sclerotium and also with respect to their genetic materials. Cultural and morphological characteristics in any living organism are governed and controlled by genetic and environmental factors. These characteristics also determine the virulence and aggressiveness of the pathogen. Variability is a common phenomenon occurring in plants and in the pathogen as well.Earlier also *Rhizoctonia Solani* has been thoroughly studied for variation in its different aspects and has shown great variability in its cultural, morphological ,physiological and pathogenic charectaristics. Singh *et al.* [11] studied variability among 46 isolates from hil areas (Uttrakhand) and plain areas of U.P. in India. They analyzed intra-field variability in *R. Solani* through RAPD fingerprinting and found high variability among them. Neeraja*et al.* [12] observed 18 isolates of *R. Solani* from different geographic location of India for RAPD analysis. The similarity values of RAPD profile ranged from 0.41 to 0.85 with an average of 0.66 among all the isolates. Lal and Kandhari [1] determined the pathogenic variability of twenty five isolates of *Rhizoctonia Solania* causing sheath blight of rice on a highly susceptible cultivar,Pusa Basmati-1 in the

Phytotron Facility at IARI, New Delhi. The isolates were charectarized into two virulent groups. Thirteen isolates (RS-1, RS-2,RS-3, RS-4, RS-7, RS-9, RS-10, RS-12, RS-13, RS-14, RS-19, RS-22, RS-23, RS-25) showed 7.0 disease score, and were grouped as moderately virulent, whereas twelve isolates, belonging to the virulent group (RS-3, RS-5, RS-6, RS-8, RS-11, RS-15, RS-16, RS-17, RS-18, RS-20, RS- 21, RS-24) had disease score of 9.0. Maximum relative lesion height (75.96%) and disease severity (59.16%) was observed in isolate RS-18 (Haryana) whereas isolate RS-25 (New Delhi) caused minimum relative lesion height (55.81%) and disease severity (13.35%). All the isolates showed maximum relative height at tillering stage but subsequently decreased at panicle initiation stage. Although in present studies the pathogenic aspects were not investigated but the cultural and morphological variations noticed are of great academic significance provided they are properly correlated with virulence, which may be the objectives of some other studies.

Molecular virulence diversity of *Rhizoctonia Solani*:

Based on cluster dendogram prepared according to PCR products obtained in different isolates and due to different primers, all the 25 isolates of *R. Solani* have been grouped into two major clusters at a similarity coefficient of 0.53. Between these two clusters, the cluster Ist was having only one isolates *i.e.* RS-16, whereas the cluster IInd was largest with 24 isolates. The IInd cluster having 24 isolates was further sub-divided into two groups with a similarity coefficient of 0.57. Between these two groups, the group I included 7 isolates *viz.* RS-1, RS-3, RS-14, RS-25, RS-5, RS-9 and RS-18 at a similarity coefficient of 0.65. Group IInd was consisted of remaining 17 isolates at a similarity coefficient 0.57. The group Ist was again divided into two sub-group at a similarity coefficient 0.65 of which the isolated RS-1 was alone in one sub-group and remaining 6 were in other sub-group. These remaining 6 isolates were again sub-divided into two groups at a similarity coefficient 0.71. Where the isolate RS-18 was alone in one sub-group and remaining 5 isolates were in another sub-group. These 5 isolates were further sub-divided into two sub-groups at a similarity coefficient of 0.76. Among these 5 isolates the RS-9 & RS-5 were in one sub-group & remaining 3 isolates were in other sub-group. These remaining 3 isolates were again divided into two groups at a similarity coefficient 0.80 into two groups, where RS-3 was alone in one group & RS-14 & RS-25 were in another sub-group with 86% level of similarity.

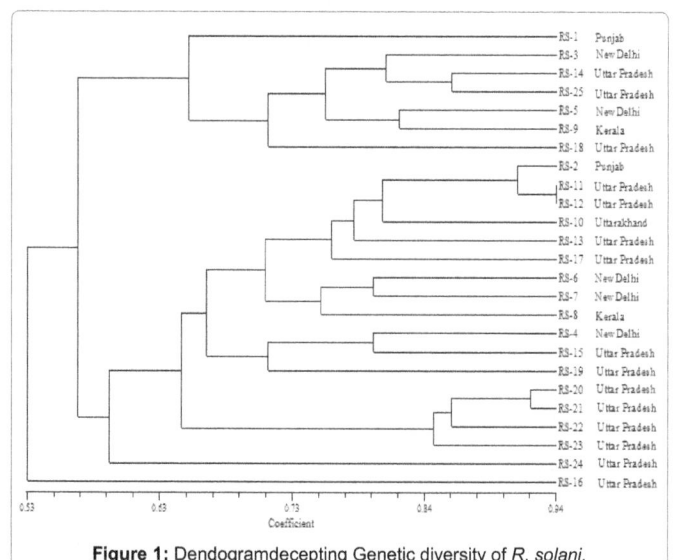

Figure 1: Dendogramdecepting Genetic diversity of *R. solani*.

Isolates	Rs-1	Rs-2	Rs-3	Rs-4	Rs-5	Rs-6	Rs-7	Rs-8	Rs-9	Rs-10	Rs-11	Rs-12	Rs-13	Rs-14	Rs-15	Rs-16	Rs-17	Rs-18	Rs-19	Rs-20	Rs-21	Rs-22	Rs-23	Rs-24	Rs-25
Rs-1	1.000																								
Rs-2	0.571	1.000																							
Rs-3	0.612	0.633	1.000																						
Rs-4	0.673	0.653	0.735	1.000																					
Rs-5	0.714	0.490	0.735	0.592	1.000																				
Rs-6	0.490	0.755	0.714	0.694	0.490	1.000																			
Rs-7	0.408	0.714	0.592	0.653	0.449	0.796	1.000																		
Rs-8	0.531	0.633	0.633	0.694	0.408	0.796	0.714	1.000																	
Rs-9	0.735	0.510	0.755	0.694	0.816	0.510	0.388	0.469	1.000																
Rs-10	0.531	0.796	0.592	0.735	0.612	0.673	0.714	0.551	0.510	1.000															
Rs-11	0.571	0.918	0.633	0.694	0.490	0.755	0.796	0.673	0.429	0.796	1.000														
Rs-12	0.551	0.898	0.653	0.714	0.510	0.776	0.816	0.694	0.449	0.816	0.939	1.000													
Rs-13	0.469	0.776	0.571	0.633	0.510	0.776	0.776	0.612	0.408	0.776	0.816	0.755	1.000												
Rs-14	0.633	0.571	0.816	0.673	0.796	0.571	0.490	0.490	0.776	0.612	0.531	0.551	0.551	1.000											
Rs-15	0.633	0.653	0.694	0.796	0.673	0.612	0.612	0.612	0.694	0.735	0.653	0.673	0.633	0.755	1.000										
Rs-16	0.510	0.571	0.408	0.551	0.429	0.449	0.531	0.490	0.408	0.571	0.612	0.551	0.551	0.429	0.510	1.000									
Rs-17	0.612	0.755	0.755	0.735	0.653	0.714	0.714	0.673	0.592	0.755	0.796	0.776	0.735	0.694	0.694	0.531	1.000								
Rs-18	0.653	0.673	0.796	0.653	0.694	0.673	0.551	0.592	0.714	0.551	0.633	0.653	0.531	0.694	0.612	0.490	0.755	1.000							
Rs-19	0.510	0.694	0.612	0.714	0.551	0.571	0.653	0.490	0.612	0.735	0.694	0.714	0.633	0.592	0.714	0.551	0.735	0.612	1.000						
Rs-20	0.490	0.714	0.592	0.653	0.531	0.633	0.551	0.510	0.510	0.714	0.714	0.694	0.653	0.531	0.612	0.612	0.673	0.591	0.734	1.000					
Rs-21	0.449	0.673	0.551	0.653	0.531	0.633	0.592	0.551	0.510	0.714	0.673	0.653	0.694	0.531	0.653	0.653	0.633	0.551	0.693	0.948	1.000				
Rs-22	0.429	0.694	0.490	0.551	0.469	0.653	0.653	0.571	0.367	0.694	0.694	0.714	0.673	0.429	0.551	0.633	0.571	0.489	0.632	0.857	0.857	1.000			
Rs-23	0.408	0.633	0.673	0.653	0.531	0.714	0.673	0.633	0.551	0.594	0.633	0.653	0.612	0.571	0.612	0.571	0.673	0.673	0.653	0.836	0.877	0.816	1.000		
Rs-24	0.490	0.633	0.592	0.571	0.490	0.633	0.592	0.592	0.469	0.551	0.592	0.571	0.571	0.531	0.531	0.571	0.673	0.673	0.571	0.591	0.591	0.571	0.632	1.000	
Rs-25	0.571	0.592	0.796	0.653	0.776	0.551	0.469	0.510	0.714	0.592	0.551	0.531	0.571	0.857	0.735	0.490	0.673	0.673	0.612	0.673	0.632	0.571	0.673	0.551	1.000

Table 7: Jaccard's similarity coefficient of 25 isolates of *Rhizctonia solani* based on DNA bands characters.

The group two of IInd cluster, which consisted of 17 isolates, was further sub-divided into two groups at a similarity coefficient of 0.59. Out of these, there was only one isolate *i.e.* RS-24 into one group and remaining 16 isolates of IInd group where reassigned to two groups at a similarity coefficient of around 0.65. Between these two groups, the isolates RS-20, RS-21, RS-22 and RS-23 were in one group with a similarity coefficient 0.85. Whereas remaining 12 isolates were in another group with a similarity coefficient level of around 0.67. These 12 isolates where reassigned to two groups with a similarity coefficient 0.67. Out of these two groups, in one group, there was 3 isolate *i.e.* RS-4, RS-15 and RS-19 with a similarity coefficient of 0.71 and in another group remaining 9 isolates were placed with a similar similarity coefficient of 0.71. These 9 isolates were again reassigned to two groups having 3 isolates *i.e.* RS-6, RS-7, and RS-8 in one group with similarity coefficient of 0.75 and remaining 6 in another group with similarity coefficient of 0.76. The highest level of similarity coefficient *i.e.* 0.94 was observed between the isolates RS-11 and RS-12 both belonging to Uttar Pradesh followed by 0.92 between RS-20 and RS-21 these also belong to Uttar Pradesh. The isolates RS-16, which belong to Uttar Pradesh, were found to highly distantly relate to isolate RS-1, which belongs to Punjab.

More recently, the use of molecular markers has given a boost to analysis of accurate variation among various isolates of the pathogen. Random Amplified Polymorphic DNA (RAPD) technique has been used consistently to determine the genetic variation and subsequently correlating it with the variation in the pattern of the pathogen. Random polymorphic PCR approaches are being increasingly used to generate molecular marker, which are useful for taxonomic characterization of fungal populations. The main advantage of these approaches is that, previous knowledge of DNA sequence is not required, so that any random primer can be tested to amplify any fungal DNA. RAPD primers are chosen empirically and tested experimentally to find RAPD banding patterns, which are polymorphic between the isolate, studied. The RAPD method successfully used to differentiate and to identify fungi at the transpacific level [11,13-15] and interspecific level [16]. RAPD-PCR analyzes loci with few each primer. Differences between isolates from different areas are therefore more easily detected. The RAPD technique has been used to detect genetic variation amongst strains / isolates with in a species [2,17].

It can be concluded that there is great genetic diversity in the population of *R. Solani* causing sheath blight in rice or infecting other hosts. Genetic diversity in the pathogen has been clearly visible in its morphological features like, color, texture,growth pattern and growth rate of the mycelium along with color, size and number of sclerotium. There is all possibility that genetic diversity existing in *R.Solani* population may be one of the main reason that why sheath blight disease has become a problem worldwide and very difficult to manage through host resistance [18].

References

1. Lal M, Janki K (2009) Cultural and Morphology variability in *Rhizoctonia solani* isolates causing Sheath Blight of Rice. J MycoPlthol 39: 7-81

2. Boyette ML, Carris LM (1997) Molecular relationship among varieties of *Tilletiafusca (T. bromi)* complex and related species. Mycol Res 99: 1119-1127.

3. Kotasthane AS, Agrawal T, Shalini, Saluja M (2004) A specialized approach for managing on unspecialized fungus causing sheath blight disease in rice . International Symposium an Rainfed Rice Ecosystems: Perspective and Potential, IndraGhandi Agri Univ, Raipur, Chhattisgarh, India. pp: 40.

4. Eechhout E, Rush MC, Black well M (1991) Effect of rate and timing of fungicide applications on incidence and severity of sheath blight and grain yield of rice. Plant Dis 75: 1254-1261.

5. Vilgalys R, Cubeta MA (1994) Molecular systematic and population biology of *Rhizoctonia*. Annu Rev Phytopathol 32: 135-155.

6. Burpee LL, Sander PL, Sherwood RT (1980) Anastomosis group amongislates of *Ceratobasdium cornigerum* (Bourd) Rogers and related fungi. Mycologia 72: 689-701.

7. Murray MG, Thompson WF (1980) Rapid isolation of high molecular weight plant DNA. Nucleic Acid Res 8: 4321-4325.

8. Pascual CB, Toda T, Raymondo AD, Hyakumachi M (2002) Characterization by convectional techniques and PCR of *Rhyzoctonia solani* isolates causing banded leaf blight in Maize. Plant Pathol 49: 108-118.

9. Nei M, Li WH (1979) Mathematical model for studying genetic variation in terms of restriction endonucleases. Proc Natl Acad Sci USA 76: 5269-5273.

10. Rohif FJ (2000) Numerical Taxonomy and Multivariate Analysis System. Version 2.11a.Exeter software Setauket, New York.

11. Singh V, Singh US, Singh KP, Singh M, Kumar A (2002) Genetic diversity of *Rhizoctonia solani* Isolates from Rice:Differentiation by Morphological characteristics, Pathogenicity Anastomosis Behaviour and RAPD Fingerprinting. J Myco Pl Pathol 32: 332-334.

12. Neeraja CN, Vijayabhanu N, Shenoy VV, Reddy CS, Sharma NP (2002) RAPD analysis of Indian isolates of rice sheath blight fungus *Rhizoctonia solani*. J Plant Biochemistry & Biotechnology 11: 43-48.

13. Guthrie PAI, Magill CW, Frederkisen RA, Odavory GN (1992) Random amplified polymorphic DNA mrkers: asystem for identifying and differentiating isolates of *Clletotrichum graminicola*. Phytopathology 82: 832-835.

14. Assigetse KB, Fernandez D, Dubois MP, Geiger JP (1994) Differentiation and of *Fusarium oxysporum* f. sp *vasinfectum races* on cotton by random amplified polymorphic DNA (RAPD) analysis. Phytopathology 84: 622-626.

15. Nicholson P, Rezanoor HN (1994) The use of random amplification polymorphic DNA to identify pathotype and detected the variation in *Pseudocerocosporella herpotrichoides*. Mycol Res 98: 13-21.

16. Lehmann PF, Lin D, Lasker BA (1992) Genotypic identification and characterization of species and strains within the genus Candida by using random amplified polymorphic DNA. J Clin Microbiol 30: 3249-3254.

17. Cooke DEL, Kennedy DM, Guey DC, Russell J, Unkley SE, et al. (1996) Relatedness of group in species of *Phytophthora* as assessed by Random amplified polymorphic DNA (RAPD) and sequences of ribosomal DNA. Mycol Res 100: 297-303.

18. Sharma D, Thrimurty VS (2005) Protection and therapeutant ability of validamycin against sheath blight of rice. Environment and Ecology 23: 933-994.

29

Nutrigenic Efficiency Change and Cocoon Crop Loss due to Assimilation of Leaf Spot Diseased Mulberry Leaf in Silkworm, *Bombyx Mori L.*

Sajad-Ul-Haq[1*,] **S.S Hasan**[1], **Anil Dhar**[2] **and Vishal Mittal**[2]

[1]*School of Sciences, Indira Gandhi National Open University, New Delhi-1100068*

[2]*Central Sericultural Research and Training Institute, Central Silk Board, Pampore,*

192 121, Jammu and Kashmir, India

[*]**Corresponding author:** Sajad-Ul-Haq, School of Sciences, Indira Gandhi National Open University, New Delhi-1100068
E-mail: sajadulhaqzargar@yahoo.in

Abstract

Healthy growth and development of silkworm largely depending on the quality of mulberry leaves. The activity of sericulture is declining due the reduction of mulberry production area in sericulture practicing countries, which lead to adverse effects on silkworm rearing and cocoon production. Screening for nutritional trait change by feeding leaf spot diseased of mulberry leaf to silkworm, *Bombyxmori L.* (Lepidoptera: Bombycidae) is an essential prerequisite for better understanding of reduced food consumption, nutritional efficiency loss and low efficiency conversion. The aim of this study was to identify efficiency and cocoon crop loss due to the consumption of leaf spot diseased mulberry leaves to bivoltine silkworm breeds using the hybrid races as SH6 and NB_4D_2. The 1stday of 5th stage silkworm larvae of bivoltine strains were subjected to standard gravimetric analysis until spinning for two tothree consecutive generations covering two different seasons on 11 nutrigenic traits. Highly significant ($p \leq 0.01$) differences were found among all nutrigenic traits of bivoltine silkworm strains in the treated worms compared to the control worms, where healthy leaves were given. Higher nutritional efficiency conversions were found in the bivoltine silkworm strains on efficiency of conversion of ingesta to cocoon and shell than leaf spot diseased leaf fed worms of the same races. Comparatively smaller consumption index, respiration, metabolic rate with superior relative growth rate, and quantum of food ingesta and digesta requisite per gram of cocoon and shell were found; the highest amount was in healthy leaf fed worms than in the diseased leaf fed worms. The significant weight loss in both the races ranged from the 3.38% to 34.28% in the diseased leaf fed larva as compared to the healthy leaf fed larva Furthermore, based on the overall loss in nutrigenic traits utilized as index or 'biomarkers', the two bivoltine silkworm strains (SH6 and NB_4D_2) were identified as having the high potential for nutrition efficiency conversion, when healthy leaves were provided to the silkworms. The data from the present study advances our knowledge to study the loss of nutritional efficiency conversion due to the leaf spot disease fed mulberry leaves and their effective commercial consequences in the sericulture industry progress and management.

Keywords: Efficiency conversion; Nutrigenic trait; bivoltine breed; leaf spot disease; *Bombyxmori L.*

Introduction

To achieve the goal of production of good quality silkworm cocoon crop, certain factors play important role. The most important factor is the mulberry leaf, contributing about 38.2 % followed by climate (37.0%), rearing techniques (9.3%), silkworm race (4.2%), silkworm egg (3.1%) and other factors (8.2 %) in producing good quality cocoons (1). Hence, quality of mulberry leaf is one of the basic prerequisite of sericulture and plays a pivotal role for successful silkworm cocoon crop. Healthy mulberry leaves influences the growth, development and quality of cocoons formed and thus decide the superiority of silk to a greater extent. Hence production of good quality leaves in terms of nutrition is very important in ensuring quality besides quantity. The mulberry silkworm, *Bombyxmori L.* (Lepidoptera: Bombycidae) is a monophagous insect that feeds exclusively on the mulberry (Morus spp.) foliage for its nutrition and produces the natural proteinaceous silk. Intensive and cautious domestication over centuries has apparently privileged this commercial insect the opportunity to increase in nutrition efficiency. Nutritional intake has direct impact on the overall genetic traits such as larval and cocoon weight, amount of silk production, pupation, and reproductive traits. The sericulture activity is declining due the reduction of mulberry production area in sericulture practicing countries on silkworm rearing and silk production. This consensus is more pronounced in countries more advanced in sericulture compared to developing countries in Asia and Pacific regions. Thus, among many factors attributed to reduction in silk production, the major one is the loss of nutrition efficiency conversion in Bivoltine silkworm strains in temperate areas because of feeding the worms with the diseased leaves and mostly the leaf spot diseased leaves. This disease is predominant during the rearing period of the silkworms in the temperate region Therefore, one of the key considerations in developing Bivoltine hybrids for temperate regions could be the need for nutrition feeding the bivoltine strains with healthy mulberry leaves. The recent advances in silkworm breeding and those with nutrition efficiency loss have opened up new avenues to evolve different management strategies to control the leaf spot disease, the deadly disease of mulberry. Sericulture in India is practiced to a limited extent in the temperate environment of Jammu and Kashmir. The existing situation provides scope for creating biovoltine hybrids (crossbreeds) as a commercial venture as hybrids above 90% of total silk production

(2). Crossbreed nutritional efficiency conversion is low when compared to the existing bivoltine (3). Earlier studies have demonstrated fundamental interaction of nutrition/physiology on gene expression (4). Similarly, nutrition or diet/physiology play important roles in insect gene expression (5), some earlier studies addressed the importance of nutritional aspects, but nutrigenics is often neglected in the selection of silkworm breeds with respect to type of nutrition consumption and efficiency conversion for evaluating the loss due to the foliar diseases on nutritionally efficient silkworm hybrids. However, a clear understanding of the genetic basis and variability in the gene expression of productive and qualitative traits during the analysis of loss of nutrigenetic traits are an important step for the production of good quality leaf and to increase the efficiency conversion.

The purpose of this study is to obtain new data about screening the nutritional efficiency loss in bivoltine silkworms due to leaf spot disease, only to augment current knowledge on gene interaction between nutrition efficiency conversion and loss of quantitative traits due to feeding leaf spot diseased leaves to the bivoltine silkworms under varied conditions.

Materials and Methods

Bivoltinesilkworm hybrid races

The two Bivoltine silkworm breeds used for the study were SH6 and NB_4D_2. These strains with varied phenotypic quantitative traits, maintained at Silkworm Breeding and Genetics section of Central Sericulture Research and Training Institute, Gallander, Pampore, India, were utilized for the study.

Silkworm rearing

The disease-free layings (DFLs) from each strain were reared and cocoons were harvested and maintained until eclosion of moths. Healthy female moths emerging on the peak day of eclosion were allowed to mate for 3-4 hours and held until oviposition. The eggs were incubated at $25 \pm 1°C$ temperature and 70- 80% relative humidity (RH) after surface treatment with 2% formalin solution. 20 to 30 eggs were chosen from each brood and pasted onto to egg sheets. Three such egg sheets for each breed were prepared, wrapped in white tissue paper and boxed with black paper to synchronize the embryonic development. On the day of hatching, the eggs were exposed to light in order to obtain uniform hatching, and finely chopped fresh mulberry leaves were fed to the young eclosions. The whole process from silkworm egg incubation to completion of rearing activities was carried out under hygienic conditions in a silkworm-rearing house that had been thoroughly disinfected with bleach followed by formalin solution. Silkworm rearing was conducted for each breed in plastic trays by feeding them with the healthy and Leaf spot diseased Goshoerami, TR-10, Ichinose and Chinese white variety of mulberry leaves from the well maintained irrigated mulberry plots in rearing room of Entamology and Pathology section of the Institute. A standard rearing procedure was adopted as recommended by Krishnaswami et al. (6). The young larvae (1st-3rd instars) were reared at 26-28 °C with 80-90% RH, and late age larvae (4th and 5th instars) were maintained at 24-26°C with 70-80% RH until the resumption of 4th molt. Each batch was divided into six, one of which was maintained as reserved stock under standard rearing conditions and the other five were subjected to standard gravimetric analysis. Reserve batch of 200 worms of each breed (NB_4D_2 and SH6) were fed with the healthy leaves of the selected mulberry varieties in the separate trays in the same rearing room.One batch fed with the healthy leaves and the other three were treated with the 4th grade leaf spot diseased respective leaves.

Estimation of nutritional traits

The nutrigenetic traits estimation study was carried out in July 2012 to August 2012 covering the summer rearing period of the temperate area, in a completely randomized block design. Silkworm rearing was conducted following the standard method under the recommended temperature and relative humidity until the 4th molt. On the 1st day of fifth instar, 50 healthy silkworm larvae per hybrid in three replications of 150 larvae each were selected for estimation on nutritional traits analysis. Accurately weighed fresh mulberry leaves were fed 3 times a day to the experimental batches and the control. Simultaneously, an additional batch of larvae for each breed was maintained to determine the dry weight on subsequent daily increments in larval weight were recorded separately as suggested by Maynard and Loosli (7). Silkworm rearing continued using appropriate plastic trays. The healthy larvae were counted daily in each replicate, and any missed larvae were replaced from the reserve batch. Left over leaves and excreta were collected on each subsequent day, separated manually and dried in a hot air oven daily at about 100-115°C until they reached constant weight using an air-tight electronic balance. When the larvae finished feeding they were shifted to the mountage for spinning at normal ambient temperature of 25 ± 2 °C and RH $65\pm5\%$. Cocoons were harvested 4-5 days later after completion of cocoon spinning. Harvested cocoons were accessed for quantitative traits using the equations detailed below. The dry weight of left over leaves, excreta, larvae, cocoon, and shell in each of the breed was recorded. The nutrigenetic traits interaction was obtained by utilizing standard gravimetric analysis methods for three consecutive seasons. During the silkworm nutritional study, data were collected on the biomass of larvae and cocoons for the 11 nutrigenetic traits on ingesta, digesta, excreta, reference ratio (RR), relative growth rate (RGR), efficiency conversion of ingesta (ECI) and digesta (ECD) for larva, cocoon, and shell in both the treatments as described by standard gravimetric methods (8-11), the equations with brief description of the nutrigenetic traits evaluated given below.

Ingesta (g)

Total intake of the dry weight (g) of mulberry leaves by silkworm larvae during the 5th stage up to spinning or ripening stage: (Dry weight of leaf fed – Dry weight of left over leaf).

Digesta (g)

Total assimilated dry food from the intake or ingesta of dry weight of mulberry leaves by silkworm larva during the 5th stage until spinning or ripening: (Dry weight of leaf ingested – dry weight of litter).

Excreta (g)

Refers to the non-utilized mulberry leaves in the form of litter from the ingested mulberry leaves of a silkworm: (Ingesta – Digesta).

Reference ratio

An indirect expression of absorption and assimilation of food. Expresses the ingesta required per unit excreta produced: (RR= Dry weight of food ingested / Dry weight of excreta).

Relative growth rate

Refers to larval gain biomass and indicates the efficiency of conversion of nutrition into larval biomass: (RGR = Weight gain of the larva during feeding period / 5th stage mean fresh larval weight (g) x 5th stage larval duration in days)

Efficiency conversion of ingesta to larva (%)

Associated with the efficiency conversion of ingested nutrition into biomass or body matter at different stages and expressed in percentage. ECI to larva was the efficiency of conversion of ingested food into larva: (ECI larvae = Maximum dry weight of larva / Dry weight of ingesta x 100).

Efficiency conversion of digesta to larva (%)

The expression of efficiency conversion of digesta into larval biomass: (ECD larvae = Maximum dry weight of larva / Dry weight of digesta x 100).

Efficiency conversion of ingesta to cocoon (%)

This is the most economically important trait used by the sericulture industry. It was the expression of efficiency conversion of ingesta into cocoon, also referred to as the leaf-cocoon conversion rate. This nutrigenetic trait was kept as the ultimate index for assessing the superiority of breed for nutritional efficiency in this investigation: (ECI cocoon = Dry weight of cocoon / Dry weight of ingesta x 100).

Efficiency conversion of digesta to cocoon (%)

It was the expression for efficiency conversion of digesta into cocoon: (ECD cocoon = Dry weight of cocoon / Dry weight of digesta x 100).

Efficiency conversion of ingesta to shell (%)

This was the expression efficiency conversion of ingesta into shell. It is also referred to as the leaf-shell conversion rate and is the ultimate index to evaluate superiority of breed in nutritional efficiency: (ECI shell = Dry weight of shell / Dry weight of ingesta x 100).

Efficiency conversion of digesta to shell (%)

The expression of efficiency conversion of digesta into shell: (ECD shell = Dry weight of shell / Dry weight of digesta x 100). The data obtained was analyzed using technique of ANOVA as given by Ronald E Walpole to test the effectiveness of leaf spot diseased leaves on the rearing and nutrigenic traits of the treated silkworms.

Results

Negative correlations were found against all the 11 traits and no positive correlations were found among these traits (Figures 1).

Performance on nutrigenic traits

Considerable variations were found for 11 nutrigenetic traits and the average larval weight gain (from fourth to fifth instar) among the bivoltine breeds on nutritional parameters and efficiency loss due to feeding the worms with the leaf spot diseased leaves. Data were obtained for ingesta, digesta, excreta, RR, RGR, ECI, and ECD to larval biomass, ECI and ECD to cocoon and shell, I/g and D/g to cocoon and shell for two diseased leaf fed bivoltine breeds under standard nutritional estimation including control breed. There was evidence of clear declines in consumption efficiency in food conversion to biomass for major nutrigenic traits in experimental bivoltine breeds over the control (healthy leaf fed silkworms) worms. (Tables 1-2).

Ingesta, digesta, excreta and reference ratio

In the experiment, the highest decrease in ingesta 42.6% was found in NB_4D_2 race after feeding the larva with leaf spot infected Chinese white leaves followed by SH6 race when fed with the leaf spot diseased TR-10 leaf. The least decrease was found in SH6 worms fed with diseased Chinese white leaves. The highest decrease of 43%, 38.5% and 37.9% in digesta was found in SH6 worms fed with the infected leaves of Chinese white, KNG and TR-10 varieties respectively. Decrease in digesta for the NB_4D_2 was least compared to SH6. The highest decrease in excreta of more than 50% was found in NB_4D_2 and less than 30% decrease was found in SH6 when treated with the healthy leaf compared to control. Reference ratio value was the highest in SH6 breed worms fed with C. White and TR-10 Leaf spot infected mulberry leaf followed by the NB_4D_2 breed worms.

Nutrigenic traits	GSH			KNG			C. W			TR-10		
	I*	H	%D/(χ2)	I*	H	%D/(χ2)	I*	H	%D/(χ2)	I*	H	%D/(χ2)
Ingesta/L (g)	2.36 {8.62}	3.44 {11.0}	31.5 (0.71)	2.16 {8.45}	3.50{10.7}	37.4 (0.38)	2.21 {8.55}	3.85{11.3}	42.6 (0.42)	2.25 {8.82}	3.70 {10.6}	39.2 (0.51)
Digesta/L (g)	0.84 {4.89}	0.9{5.56}	8.33 (0.98)	0.98 {5.25}	1.07 {5.49}	8.06 (0.99)	0.73 {5.65}	0.94{5.92}	22.8 (0.97)	0.75 {4.97}	0.87 {5.36}	13.7 (0.99)
Excreta/L (g)	1.55 {7.13}	2.74{9.51}	43.7 (0.53)	1.27{6.4}	2.80{9.60}	54.7 (0.34)	1.24 {6.38}	2.75{9.55}	55.1(0.34)	1.54 {7.11}	2.73 {9.50}	43.7 (0.55)
RR (g)	2.30{8.84}	2.9{9.78}	18.3 (0.93)	2.32{8.70}	3.30{10.3}	29.2 (0.66)	1.74 {8.45}	2.52 {9.96}	30.8 (0.78)	2.16 {7.58}	3.0{9.12}	27.8 (0.82)

RGR	0.43 {4.11}	0.47{4.65}	9.86(1.0)	0.52 {3.86}	0.66{5.52}	21.7 (0.99)	0.46 {3.73}	0.90{3.93}	50.9 (0.81)	0.74 {4.93}	0.97 {5.64}	23.4 (0.97)
ECIL (%)	53.6 {42.4}	56.3{43.9}	4.86(0.9)	59.5 {40.1}	61.0 {41.9}	2.30 (0.82)	45.7 {38.9}	55.4 {43.0}	17.5 (0.08)	49.5 {41.6}	53.4 {45.8}	7.20(0.77)
ECDL (%)	18.6 {21.4}	20.2{22.4}	7.82 (0.72)	17.7 {21.6}	23.0 {23.1}	23.8 (0.15)	19.6 {20.7}	25.4 {23.1}	22.8 (0.05)	22.9 {19.6}	25.5 {21.4}	10.0(0)
ECIC (%)	45.5 {47.0}	48.3 {48.6}	5.64(0.8)	41.6 {50.5}	45.0 {51.3}	7.15 (0.82)	39.5 {44.7}	46.6 {46.9}	15.2 (0.23)	44.3 {42.5}	51.5 {40.0}	14.0 (0.19)
ECDC (%)	13.3 {25.5}	14.6 {6.7}	8.33 (0.94)	13.6 {24.8}	15.0 {28.8}	12.4 (0.62)	12.5 {26.2}	15.5 {30.1}	19.1 (0.52)	11.4 {28.6}	13.4 {30.3}	15.3 (0.73)
ECIS (%)	16.4 {23.8}	17.7 {24.8}	7.66 (0.74)	20.3 {26.7}	21.0{27.6}	5.61 (1.0)	16.5 {23.9}	18.2 {25.2}	9.35(0.9)	20.4 {26.8}	21.5 {27.6}	5.37 (0.90
ECDS (%)	7.66 {14.6}	8.46 {15.9}	9.53 (0.69)	7.54 {15.9}	8.30 {16.7}	9.62 (0.8)	6.44 {16.0}	7.56 {16.8}	14.8 (0.76)	8.54 {16.9}	9.46 {17.9}	9.76 (0.85)
SEm±/F-test	0.43	0.68	**	0.52	0.58	**	0.49	0.70	**	0.35	0.44	**
CD at 5%	1.26	2.0		1.52	1.71		1.46	2.05		1.04	1.29	
CV (%)	3.94	5.83		4.66	4.80		4.66	5.91		3.22	3.64	
SD±	17.94	18.72		18.67	19.22		16.19	18.01		16.40	19.23	

Table 1: Effect of leaf spot diseased mulberry leaf on various nutrigenic parameters of NB_4D_2 race of silkworm *Bombyxmori L.* in the fifth instar. H= healthy,I= Infected, %D= Percent Decrease, Values in smallBrackets () are the χ2-test values, values in long Brackets{} are arc sinevaluesD/L, digesta per larva; I/L, ingestaperLarva; ECD, efficiency conversion of digesta; ECI, efficiency of conversion of ingesta; RR, reference ratio; RGR, relative growth rate;

Nutrigenic traits	GSH			KNG			C W			TR-10		
	I*	H	%D/(χ2)	I*	H	%D/(χ2)	I*	H	%D/(χ2)	I*	H	%D/(χ2)
Ingesta/L(g)	2.29 {8.70}	3.52 {10.8}	34.8 (0.64)	2.46 {9.01}	3.46 {10.7}	29.1 (0.76)	2.24 {11.9}	4.31 {8.59}	48.1 (0.07)	3.44{11.1}	3.74{10.6}	8.83 (0.96)
Digesta/L(g)	0.76 {4.99}	0.86 {5.32}	11.7(1.0)	0.73 {4.88}	1.28 {6.41}	43.0(0.77)	0.53 {5.75}	1.01 {4.17}	47.5 (0.05)	0.63{5.32}	0.86 {4.53}	37.5 (0.95)
Excreta/L(g)	1.59 {7.25}	2.68 {9.41}	40.5 (0.62)	1.76 {7.62}	2.55 {9.18}	31.0(0.80)	1.67 {10.9}	3.58 {7.42}	53.4 (0.05)	2.79{9.58}	2.78 {10.9}	0.43(1.0)
RR (g)	2.31 {8.73}	2.84 {9.69}	18.7 (0.94)	2.43 {8.96}	3.20{10.2}	23.9 (0.85)	1.98 {11.3}	3.90{8.09}	49.1 (0.08)	2.32{0.72}	3.46 {11.3}	49.3(0.51)
RGR	13.9 {3.95}	15.1 {4.30}	8.01(1.0)	14.6 {3.89}	16.3 {4.26}	10.7(1.0)	12.5 {4.37}	15.4 {3.84}	18.9 (0.98)	13.0{4.30}	14.3 {4.37}	9.85(1.0)
ECIL (%)	0.48 {21.8}	0.56 {22.8}	15.4 (0.94)	0.46 {22.4}	0.56 {23.8}	16.8(0.86)	0.45 {23.3}	0.58 {20.7}	22.9 (0.45)	0.45 {22.2}	0.57 {23.1}	25.9 (0.86)
ECDL (%)	16.7 {40.6}	18.5 {43.0}	9.91 (0.68)	18.4 {40.5}	20.5 {41.8}	10.4 (0.86)	20.4 {43.8}	23.4 {40.1}	12.8 (0.26)	18.5 {44.6}	21.8 {43.8}	17.8 (0.04)
ECIC (%)	42.5 {24.0}	46.5 {25.4}	8.76 (0.87)	42.3 {25.3}	44.5 {26.9}	5.03 (0.70)	41.6 {28.9}	47.9 {26.8}	13.2 (0.63)	40.4 {27.8}	49.3 {28.9}	22.2 (0.39)
ECDC (%)	52.3 {46.2}	54.6 {47.6}	4.27 (0.88)	53.2 {46.8}	58.3 {49.7}	8.73 (0.61)	48.5 {46.9}	53.3 {44.1}	9.05 (0.59)	50.3 {48.0}	55.3 {46.9}	9.90(0.44)
ECIS (%)	16.6 {16.2}	17.3 {16.8}	3.77 (0.94)	20.2 {15.6}	22.6 {16.5}	10.7 (0.93)	17.8 {15.5}	20.1 {17.0}	11.4 (0.71)	15.2 {18.0}	21.4 {15.5}	41.1 (0.24)

ECDS (%)	7.87 {24.0}	8.39 {24.5}	6.24 (0.98)	7.28 {26.6}	8.13 {28.3}	10.5 (0.48)	7.24 {24.9}	8.57 {26.6}	15.5 (0.78)	6.99 {27.5}	9.56 {24.9}	36.7 (0.01)
SEm±/f-test	0.27	0.28	**	0.34	0.33	**	0.29	0.20	**	0.30	0.33	**
CD at 5%	0.79	0.80		1.00	0.96		0.84	0.57		0.88	0.95	
CV (%)	2.91	2.81		3.61	3.24		2.83	2.12		2.94	3.52	
SD±	17.03	17.95		17.29	18.52		17.56	16.76		18.59	16.23	

Table 2: Effect of leaf spot diseased mulberry leaf on various nutrigenic parameters of SH6 race of silkworm Bombyxmori L. in the fifth instar. %D= Percent Decrease, H= healthy, I= Infected, Values in short Brackets () are the $\chi2$-test values, values in long Brackets{} are arc sine values D/L, digesta per larva; I/L, ingesta per Larva; ECD, efficiency conversion of digesta; ECI, efficiency of conversion of ingesta; RR, reference ratio; RGR, relative growth rate.

Silkwom Race	Parameters	Variety											
		GSH			KNG			C W			TR-10		
		H	I*	% Loss	H	I*	% Loss	H	I*	% Loss	H	I*	% Loss
NBD	Average wt. in4th Instar	5.05	4.22	3.38 (0.83)	5.94	5.34	10.2 (0.97)	6.93	4.55	34.28 (0.35)	7.72	6.31	18.3(0.73)
	Average larval wt. in 5th instar	11.6	10.4	9.28 (0.91)	14.06	12	14.7 (0.75)	12.6	10.74	14.65 (0.78)	13.3	11.1	17.1(0.61)
SH6	Average wt. in4th Instar	5.59	4.78	3.97 (0.90)	5.36	4.73	11.8 (0.95)	6.83	4.86	28.88(0.5)	5.36	4.52	15.6(0.3)
	Average larval wt. in 5th instar	12.9	11.2	13.5 (0.81)	13.02	11.2	14.1 (0.77)	113	93.52	17.48 (0.03)	134	108	6.17(0)
CV (%)/f-test		1.60	4.34	**	2.86	3.61	**	2.95	8.90	**	1.72	8.09	**
Cd at 5%		1.19	2.96		2.37	2.65		2.65	6.57		0.57	6.70	

Table 3: Larval weight of NB_4D_2 and SH6 silkworm breed in 4th and 5th instars. Values in Brackets are the $\chi2$-test value.

Relative growth rate

The highest relative growth rate of more than 50% was found in NB_4D_2 race when fed with the diseased Chinese white leaf, followed by NB_4D_2 when fed with 4th grade leaf spot diseased Goshoerami leaf.

Efficiency of conversion of ingesta and digesta to larval biomass

The efficiency of mulberry leaf ingested and digested in conversion to silkworm larval biomass or body varied prominently among the bivoltine breeds (Table 1 and 2) when fed with leaf spot infected mulberry leaf. The highest decrease of 17.5% efficiency conversion of ingesta was found in NB_4D_2 when fed with leaf spot diseased C. White leaf and 23.8% decrease in efficiency conversion of digesta for larva was recorded in NB_4D_2 fed with Leaf spot diseased KNG leaf.

Efficiency of conversion of ingesta and digesta to cocoon and shell

The significant decrease of 19.1% in efficiency conversion to cocoon was shown in NB_4D_2 race when fed with C. white Leaf spot diseased leaf followed by more than 15% decrease in TR-10 fed diseased leaf in NB_4D_2 and SH6. The 14.6% significant decrease in efficiency conversion of digesta to cocoon was seen in SH6 when C. white leaf

was fed and 3.77% least significant in NB_4D_2, when fed with diseased Goshoerami leaf. The significant decrease in efficiency conversion of ingesta to shell ranged from 14.8% in NB_4D_2 to 3.36% in SH6 when both fed with the leaf spot diseased Goshoerami leaf. In the fourth instar the significant larval weight loss of 41.3% due to leaf spot disease was found in NB_4D_2 when fed with diseased mulberry and 56.7% weight loss in fifth instar. The least decrease was found in fifth instar when the worms become the voracious feeders followed by 36.4% and 33.1% in SH6 race when fed with leaf spot infected Chinese white and Goshoerami leaves The least decrease of 3.70% in larval weight loss was found in NB_4D_2 when fed with Goshoerami Leaf Spot diseased leaves.

Based on all morphological, nutrigenic traits and lower consumption of mulberry leaves and maximum efficiency of conversion of nutrients, with highly significant ($p \leq 0.001$) differences in bivoltine breeds for 11 nutrigenetic traits, two bivoltine silkworm breeds, NB_4D_2 and SH6 were identified as potential nutritionally efficient breeding resources for comparative studies in effect of diseased and healthy leaves on larval, cocoon and shell parameters. Silkworm breeding can be defined as the science of improving the genetic entity of silkworms in relation to their economic utility. Silk producing countries in Asia and Pacific regions experience serious problems in the field of silkworm rearing on healthy leaf. This investigation intends to serve as a guideline to organize or revive healthy silkworm feeding programs, as well as a quick reference to sericulture loss. It also offers a brief background on the change in silkworm nutrition and physiology. It also outlines the necessary facilities and tools required to establish modern silkworm breeding programs for the sustenance of sericulture in the temperate regions. The study of the interactions between nutrition and quantitative traits, the major physiological and biochemical traits of silkworm showed a greater decline in consumption with decrease of food efficiency conversion into biomass in experimental bivoltine breeds compared to the healthy leaf fed silkworms of the same breed. A similar result was reported for polyvoltine and commercial hybrid silkworms by Maribashetty et al. (12) and Meneguim et al. (13) respectively. Such dietary factors and related metabolic interactions on specific gene expression were also reported by Walker and Blackburn. Nutrition affects nearly all biological processes including the rates of biochemical and physiological reactions (14,15), and eventually can affect the larval quality or quantity of cocoon crops in the silkworm. Several reports (15-18) demonstrated that silkworms were more responsive to nutrition supplement during the 4th and 5th stages, which are recommended for the recognition and selection of nutritional efficient changes due to leaf spot disease. Hence, the nutrition utilization study was confined to the 5th instar larvae, since 80-85% of total leaf consumption was observed in this stage of silkworm development (16,18) and the silk gland stimulation starts at this stage. For instance, bivoltine breeds reared in temperate countries are known to be less nutritionally efficient, which is also true with cross breeds that have evolved for a tropical climate (3,19-20). It is essential to analyze nutrigenetic traits reduction to understand the racial difference among bivoltinegermplasm breeds for commercial purposes. Recently, the effects of nutrigenetic traits for bivoltine germplasm breeds also have been shown by Ramesha et al. (3). The success of the sericulture industry depends upon several factors, including production of quality mulberry leaves. This factor is of vital importance, since it accounts for 60% of total expenditures (21). It is well understood that the majority of the economically important genetic traits of silkworm are qualitative in nature, and phenotypic expression is greatly influenced by the environmental and leaf factors such as temperature, relative humidity, light, diseased leaf and nutrition (3,6,22-23). Therefore, it is essential to gauge the degree of phenotypic difference of the economical traits to understand the genetic steadiness under the controlled nutrition conditions. The problem of balancing and fixing the desirable traits for a given environment is a challenging task for the silkworm breeder. Hence, understanding the range of a reaction of the selected breeds to variable nutritional conditions especially to feeding with diseased leaf, is important for the breeder to utilize them appropriately in hybrid programs. In order to achieve greater success in this regard, it is important to understand the level of nutrition efficiency in bivoltine silkworm breeds and to analyze the least significant loss to cocoon crop because of leaf spot disease. The main objective of this study was to identify the conversion index and efficiency of conversion of ingesta, when larva were fed with the leaf spot diseased mulberry leaves, to biomass through standard gravimetric method for three successive generations on different races and varieties of mulberry, is supported by earlier observations (24-27). Our emphasis was on the phenotypic manifestation of 11 nutrigenetic traits. The results revealed highly significant (p ≤ 0.001) variability among the two bivoltine breeds with respect to 11 nutrigenic traits over, when the larvae were fed ith the leaf spot diseased mulberry leaf. Although earlier studies showed that somebivoltine breeds are moderately nutritionally efficient (12,28-29), We concluded that bivoltine breeds with minimum consumption index and maximum

efficiency of conversion of ingesta/cocoon identified strains NB_4D_2 and SH6 as potential bivoltine breeding resource material for the development of nutritionally efficient breeds/hybrids in Asia especially Indian subcontinent, pacific regions and other temperate regions(30-36).

Acknowledgement

Authors are highly thankful to Dr. K. A Sahaf, Director CSR&TI, Govt. of India, Pampore, Kashmir, Indiafor providing all the necessary facilities and encouragement for the present work. We duly acknowledge the contributions of Prof. Vijayshri, Director, School of Sciences, IGNOU, Main Campus, New Delhi and Dr. IrfanIllahi, Scientist-B CSR&TI, Pampore and Prof. NeeraKapoor, (Coordinator School of Sciences, IGNOU), Prof. Amrita Nigam and Prof. JaswantSokhi eminent professors of the School of Sciences, IGNOU, Main Campus, MaidanGarhi, New Delhi for their generous and enthusiastic help and constructive suggestions.

References

1. Hamano K, Miyazawa K, Mukiyama F (1986)Racial difference in the feeding habit of the silkworm, Bombyxmori. Journal of Sericultural Science of Japan 55: 68-72.

2. Ramesha C, Seshagiri SV, Rao CGP (2009) Evaluation and identification of superior polyvoltine crossbreeds of mulberry silkworm, Bombyxmori L. Journal of Entomology 6: 179-188.

3. Ramesha C, Anuradha CM, Lakshmi H,SugnanaKumari S, Seshagiri SV et al. (2010)Nutrigenetic traits analysis for identification of nutritionally efficient silkworm germplasm breeds. Biotechnology 9: 131-140.

4. Phillips CN, Tierney AC, Roche, HM (2008) Gene–nutrient interactions in the metabolic syndrome. Journal of Nutrigenetics and Nutrigenomics 1: 136-151.

5. Rharrabe K, Sayah F, La Font R(2010) Dietary effects of four phytoecdysteroids on growth and development of the Indian meal moth, Plodiainterpunctella. Journal of Insect Science 10:13

6. Krishnaswami S, Narasimhanna MN, Surayanarayana SK, Kumararaj S (1973) Manual on sericulture 2: Silkworm rearing. UN Food and Agriculture Organization: 54-88

7. Maynard AL, Loosli KJ (1962) Animal Nutrition, 5th Edition. McGraw Hill.

8. Waldbauer GP (1968) The consumption and utilization rate of food by insects. Advanced Insect Physiology 5: 229-288.

9. Scriber JM, Feeny P(1979) Growth of herbivorous caterpillars in relation to feeding specialization and to the growth form of their food plant. Ecology 60: 829-850.

10. Kogan M, Parra JRP (1981) Techniques and applications of measurements of consumption and utilization of feed by phytophagous insects. In: Bhaskaran G, Friedman S, Rodrigues JG, Editors. Current Topics in Insect Endocrinology and Nutrition. Plenum Press: 337-352.

11. Slansky F, Scriber JM (1985) Food consumption and utilization. In: Kerkut AA, Gilbert LI, Editors. Comprehensive Insect Physiology, Biochemistry and Pharmacology. Pergamon Press: 87-163.

12. Maribashetty VG, Aftab Ahmed CA, Chandrakala MV, Rajanna GS (1999) Consumption and conversion efficiency of food and water in new multivoltine breeds of silkworm, Bombyxmori L. Indian Journal of Sericulture 38: 140-144.

13. Meneguim AM, Lustri C, Oliveira DD, Yada IFU, Pasini A(2010)Bromatological characterization of mulberry cultivars, Morus spp., and determination of nutritional indexes of Bombyxmori L. (Lepidoptera: Bombycidae). Neotropical Entomology 39: 506-512.

14. Parra JRP, Kogan M (1981) Comparative analysis of methods for measurements of food intake and utilization using the soybean looper,

Pseudoplusiaincludens and artificial media. Entomological Experimental Application 30: 45-57.

15. Paul DC, SubbaRao G, Deb DC(1992) Impact of dietary moisture on nutritional indices and growth of Bombyxmori and concomitant larval duration. Journal of Insect Physiology 38: 229-235.

16. Ueda S(1965) Changes in some quantitative factors and their mutual relationships concerning the growth and development in the fifth larval instar of the silkworm, Bombyxmori L. Bulletin of Sericultural Experiment Station in Japan 19: 331-341.

17. Mano Y, Asaoka K, Ihara O, Nakagawa H, HirabayashiTet al. (1991) Breeding and evaluation of adaptability of silkworm, Bombyxmori to new low cost artificial diet, LPY lacking mulberry leaf powder. Bulletin of National Industrial Sericultural Entomological Science 3: 31-56.

18. Rahmathulla VK, HaqueRufaie SZ, Himanthraj MT, Vindhya GS, Rajan RK (2005) Food ingestion, assimilation and conversion efficiency of mulberry silkworm, Bombyxmori L. International Journal of Industrial Entomology 11: 1-12.

19. Rahmathulla VK, Vindya GS, Sreenivasa G, Geethadevi RG (2003) Evaluation of the consumption and nutritional efficiency in three new bivoltine hybrids (CSR series) silkworm Bombyxmori L. Journal of Experimental Zoology 6: 157-161.

20. Ramesha C, Raju PJ (2009b) Analysis of nutrigenetic traits for identification of nutritionally efficient germplasm breeds of bivoltine silkworm, Bombyxmori L. National Conference on Recent Trends in Animal Physiology: 27

21. Datta RK, Nanavaty M (2005) Global silk industry: A complete source book. Universal Publishers.

22. Thaigarajan, Bhargava VSK, Ramesh Babu M, Nagaraju B(1993) Differences in seasonal performance of 26 strains of silkworm Bombyxmori (Bombycidae). Journal of the Lepidopterists' Society 47: 321-337.

23. Zhang YH, Xu AY, Wei YD, Li MW, Hou CX, et al. (2002) Studies on feeding habits of silkworm germplasm resources for artificial diet without mulberry. ActaSericologicaSinica 28: 333-336.

24. Hassanein MH, El Shaaraway MF, El Garthy AT (1972) Food assimilation and out put of the silk in the different races of the silkworm, Bombyxmori L. Bulletin of the Entomological Society of Egypt 56: 333-337.

25. Anantharaman KV, Magadum SB, Datta RK (1994) Feed efficiency of silkworm, Bombyxmori L. hybrid (NB$_4$D$_2$ x KA). Insect Science Application 15: 111-116.

26. Trivedy K, Nair KS (1999) Feed conversion efficiency of improved multi x biovoltine hybrids of silkworm, Bombyxmori L. Indian Journal of Sericulture 38: 30-34.

27. Kumaresan P, Sinha RK, Sahni NK, Sekar S(2000) Genetic variability and selection indices for economic quantitative traits of multivoltine mulberry silkworm, Bombyxmori L. genotypes. Sericologia 40: 595-605.

28. Remadevi OK, Magadum SB, Shivashankar N, Benchamin KV (1992) Evaluation of the food utilization efficiency in some polyvoltine breeds of silkworm, Bombyxmori L. Sericologia 32: 61-65.

29. Datta LC, Saikia MK, Datlo SK (1996) Nutritional efficiency of two multivoltine breeds of Bombyxmori L. native to Assam. Indian Journal Sericulture 35: 32-34

30. Chandrashekharaiah, Ramesh Babu M (2003) Silkworm breeding in India during the last five decades and what next?. Proceedings of the MulberrySilkworm Breeders Summit: 6-13

31. Ding N, Zhang XM, Jiang MQ, Xu WH, Wang ZE, Xu MK (1992)Genetical studies on the dietary efficiency of the silkworm, Bombyxmori L. CanyeKexue 18: 71-76.

32. Gokulamma K, Reddy YS (2005) Role of nutrition and environment on the consumption, growth and utilization indices of selected silkworm races of Bombyxmori L. Indian Journal of Sericulture 44: 165-170.

33. Hassanein MH, El Shaaraway MF, El Garthy AT (1972) Food assimilation and output of the silk in the different races of the silkworm, Bombyxmori L. Bulletin of the Entomological Society of Egypt 56: 333-337.

34. Junliang X, Xiaofeng W(1992) Research on improvement of efficiency of transferring leaf ingested into silk of the silkworm, Bombyxmori L. Abstract.International Congress on Entomology: 623.

35. PrabhakarMK, Reddy DNR, Narayanaswamy KC(2000) Consumption and utilization of mulberry leaves by the silkworm, Bombyxmori L. Bulletin of Indian Academy Sericulture 4: 52-60.

36. Rajesh D, Haemanand A(2005)Nutrigenomics – A future-omics. Advanced Biotech 4: 26-31.

30

Isolation, Identification and Antifungal Activities of *Streptomyces aureoverticillatus* HN6

Lanying Wang[1], Mengyu Xing[1], Rong Di[2]* and Yanping Luo[1]*

[1]*Key Laboratory of Protection and Development Utilization of Tropical Crop Germplasm Resources, Ministry of Education, College of Environment and Plant Protection, Hainan University, Haikou, Hainan, 570228, P. R. China*
[2]*Department of Plant Biology and Pathology, School of Environmental and Biological Sciences, Rutgers, the State University of New Jersey, 59 Dudley Road, New Brunswick, New Jersey, 08901, USA*

Abstract

Banana *Fusarium* wilt caused by *Fusarium oxysporum* f. sp. *cubense* race 4 (FOC4) is destroying numerous banana plantations in southern China. In order to select an effective biocontrol agent for this devastating disease, eighty nine actinomycete isolates were collected from soil samples in the Botanical Garden of Chinese Academy of Tropical Agricultural Sciences in the tropical Hainan Province, China. These isolates were evaluated for their antagonistic activity against FOC4. Our results showed that eight isolates exhibited strong anti-FOC4 activity. One of the isolates, HN6, resulted in an inhibition zone of 35 mm in diameter in the antagonistic test. The mycelia of HN6 were extracted with methanol, and the extract was tested against eight indicator pathogens by the mycelium growth rate method. The HN6 extract demonstrated broad-spectrum antifungal activity, with an EC_{50} less than 0.08 mg/ml. Based on the morphological, biochemical, physiological and cultural characteristics and the 16S rRNA gene sequence, the HN6 isolate was identified as *Streptomyces aureoverticillatus*. HN6 isolate can be potentially developed into a biocontrol agent for banana *Fusarium* wilt and other plant diseases.

Keywords: Banana *Fusarium* wilt; Actinomycete; Antifungal Activity; *Streptomyces aureoverticillatus*; Isolate HN6

Introduction

Banana *Fusarium* wilt, also called banana Panama wilt, is a destructive disease that affects bananas in tropical and subtropical areas worldwide. *Fusarium oxysporum* f. sp. *cubense* race 4 (FOC4) is one of the major pathogens causing banana *Fusarium* wilt [1]. Banana is one of the most important agricultural crops in the world, many disease management strategies have been applied to control this disease including resistant cultivars, fungicide application, crop rotation and soil management [2]. However, FOC4 is still expanding at a speed of 30-50 km each year, destroying numerous banana plantations in south China [3].

Extensive application of chemical pesticides in agriculture has led to numerous side effects for the environment and human health. In search for environmentally friendly and safe substitute, biological pesticides are more appropriate compared to chemical pesticides. Actinomycetes play an important role in the discovery of antibiotics, antitumor agents and biopesticides [4-7]. Actinomycetes produce a wide range of bioactive secondary metabolites that are known to have anticancer, antibacterial, antifungal, antialgal, antimalarial and anti-inflammatory activities [8]. Approximately two-thirds of the naturally occurring antibiotics have been isolated from actinomycetes [9]. Soil from the tropical Hainan Island (Hainan Province) in China presents a unique ground for the discovery of microbial inhabitants with extensive biodiversities that have been found to produce many biologically active natural products [7,10]. In order to identify biological control agents against FOC4, we isolated and screened eighty nine actinomycetes from soil samples in the Botanical Garden on Hainan Island, China. These isolates were tested for their antagonistic activities against FOC4 and for their broad-spectrum antifungal activities against totally eight indicator plant pathogens. Out of these isolates, HN6 was shown to exert the highest anti-FOC4 activity. HN6 was identified to belong to the *Streptomyces* genus based on its morphological, biochemical, physiological and cultural characteristics and its 16S rRNA gene sequence.

Materials and Methods

Sample collection and processing

Soil samples were collected from the Botanical Garden of Chinese Academy of Tropical Agricultural Sciences in Hainan Province (Hainan Island) in China. They were collected from 10-15 cm depth in soil, placed into sterile plastic bags and transported aseptically to the laboratory. The collected soil samples were mixed thoroughly and passed through a 2 mm sieve to remove gravel and debris. They were air dried for one week at room temperature.

Isolation of actinomycetes

To prepare soil suspension, 5 g of soil sample was transferred into a sterile bottle. After adding 45 ml of sterile distilled water to the sample, the bottle was shaken for 30 min. Five sets of ten-fold serial dilutions were prepared from the original supernatant, and 0.1 ml

---

*Corresponding authors: Yanping Luo, Key Laboratory of Protection and Development Utilization of Tropical Crop Germplasm Resources, Ministry of Education, College of Environment and Plant Protection, Hainan University, Haikou, Hainan, 570228, P. R. China
E-mail: yanpluo@126.com

Rong Di, Department of Plant Biology and Pathology, School of Environmental and Biological Sciences, Rutgers, the State University of New Jersey, 59 Dudley Road, New Brunswick, New Jersey, 08901, USA
E-mail: di@aesop.rutgers.edu

of each diluted sample was used to spread on Gause's no.1 synthetic medium aseptically [11]. Each sample was spread onto three plates and incubated at 28°C for 5-10 days. The plates were observed periodically for the growth of actinomycetes. The pure colonies were selected, isolated and maintained in Gause's no.1 synthetic medium at 4°C for subsequent studies.

Screening for FOC4 antagonistic actinomycetes

The isolated actinomycetes were inoculated into Gause's no.1 medium and incubated at room temperature for 5 days. Actinomycete cakes (Φ=5 mm) were obtained and inoculated onto one side of the PDA (potato dextrose agar) plate. The ATCC76255 strain of *Fusarium oxysporum* f. sp. *cubense* race 4 (FOC4) was inoculated onto the other side of the PDA plate, with a distance of 4 cm. They were cultured at 25°C for 5 days. The antagonistic belt (inhibition zone) was recorded by measuring the distance (mm) between the edge of the fungal mycelium and the actinomycete cakes. All isolates were tested in three independent replicates. Of all the isolates, eight of the best antagonistic actinomycete isolates were selected, identified macroscopically and microscopically by Gram's staining and used for further studies.

Characterization of isolate HN6

The selected actinomycete isolate HN6 was further identified by morphological, biochemical and physiological characterization and by 16S rRNA gene sequencing. The morphological characteristics for colony, aerial and substrate mycelia and spores were studied after culturing the individual isolates on Gause's no.1 medium at 28°C for at least 7 days [12]. HN6 was investigated for various physiological and biochemical properties including the utilization of cellulose, inositol, mannitol and seven different sugars as carbon source, starch hydrolysis, glutin liquefaction, litmus milk reaction and H_2S production [13,14]. The cell wall type and whole-cell sugar analysis for chemical grouping of HN6 was performed as previously described [15].

The 16S rRNA gene was amplified by PCR with Taq DNA polymerase and the conserved primers F (5'-AGAGTTTGATCCTGGCTCAG-3') and R (5'-ACGGCTACCTTGTTACGACTT-3') [16]. The conditions for thermal cycling were as follows: denaturation of the fungal DNA at 95°C for 5 min followed by 35 cycles at 94°C for 1 min, primer annealing at 50°C for 1 min and DNA elongation at 72°C for 2 min. At the end of the PCR reaction, the reaction mixture was held at 72°C for 10 min [17]. PCR amplification was detected by agarose gel electrophoresis in Shanghai Yingjun Biotechnology Co., Ltd. The PCR product obtained was sequenced by a Sanger-based, automated sequencer (Applied Biosystems). The sequence was compared for similarity with the reference species of bacteria in the genomic databank, using the NCBI BLAST available at http://www.ncbinlm-nih.gov/. The phylogenetic tree was constructed by neighbor-joining method in Mega4.0 software [18].

Antifungal activities of HN6 methanol extract

The fermentation medium for the antagonistic actinomycete HN6 isolate included soluble starch 9.0 g, soybean meal 3.0 g, K_2HPO_4 0.5 g, $MgSO_4 \cdot 7H_2O$ 0.5 g, $FeSO_4 \cdot 7H_2O$ 0.01 g, NaCl 0.5 g and distilled water in 1000 ml, pH 6.0. After the fermentation growth, the actinomycete biomass was harvested by centrifugation at 5.2 g at 20°C for 10 minutes. The mycelia were washed three times with sterile distilled water under aseptic conditions. They were then resuspended in a small amount of methanol and ground with mortar and pestle and broken under 20 KHZ ultrasonic waves for 5 min. Methanol was added to the ground cells in the ratio of 1:1 (w/v) and the mixture was shaken vigorously overnight.

The extract was then filtered through a blotting paper. The filtrate was evaporated using a rotary evaporator at 50°C. The concentrated extract was then transferred into glass screw-capped tubes and stored at 4°C for further use.

The HN6 extract was diluted with methanol to 6.60, 1.32, 0.66, 0.66, 0.132, 0.066, 0.044 mg/ml preparations. One milliliter of each extract was added to 19 ml PDA medium at 45°C, and quickly portioned into three Petri dishes (Φ=5 mm). Each concentration was replicated three times. One milliliter of methanol adding into 19 ml PDA medium was used as control.

To conduct mycelium growth rate test eight indicator plant pathogens *Fusarium oxysporum* f. sp. *cubense* race 4 (ATCC 76255), *Botryodiplodia theobromae* (ATCC 10936), *Colletotrichum gloeosporioides* (ATCC 20358), *Colletotrichum gloeosporioides* Penz (ATCC 16330), *Colletotrichum musae* (ATCC 96167), *Corynespora cassiicola* (ATCC 36294), *Periconia circinata* (ATCC 32727) and *Rhizoctonia solani* (ATCC 58938) were grown on PDA Petri dishes. The fungal mycelia were taken from the periphery of stock cultures [19,20]. Plugs of mycelia were removed with a 5 mm cork borer from the advancing margin of the fungal colonies, placed in the center of each PDA Petri dish containing the HN6 methanol extract at different concentrations, with the mycelia facing the medium. The cultures were incubated at 25 ± 2°C for 3 days. Fungal toxicity was expressed by the inhibitory percentage of the mycelia growth compared to the control. The colony diameter was measured in millimeters, excluding the plug. An average was taken from three measurements made on each Petri dish.

Results

Isolation and screening for antagonistic actinomycetes

Eighty nine actinomycetes were isolated from soil samples from the Botanical Garden of tropical Hainan Province, China by the gradient dilution method, and identified by their morphology. To study their antagonistic activities, these isolates were tested by the antagonistic belt (inhibition zone) method against *Fusarium oxysporum* f. sp. *cubense* (FOC4), the causal agent of banana *Fusarium* wilt. Out of 89 actinomycetes, 8 isolates including HN3, HN6, HN14, HN24, HN29, HN59, HN62, HN75 showed strong inhibition of FOC4 growth, with the diameters of inhibition zones of more than 18 mm (Table 1). HN6 exhibited the strongest antagonistic activity, with an inhibition zone of 35 mm.

Morphological, physiological and biochemical characteristics of HN6 isolate

Due to the strong antagonistic activity against FOC4, HN6 isolate was selected for further studies on its morphological, physiological and biochemical characteristics.

Isolate	Inhibition zone (mm)*
HN3	20 ± 0.8
HN6	35 ± 0.6
HN14	18 ± 1.2
HN24	19 ± 1.0
HN29	20 ± 1.1
HN59	19 ± 0.9
HN62	18 ± 1.4
HN75	18 ± 1.2

Table 1: Inhibiting zones of 8 actinomycete isolates against FOC4 five days after incubation. *Values are means of three replications.

When HN6 was cultured on different media, its aerial mycelium and substrate mycelium displayed different colors (Table 2). The morphology of HN6 spore and mycelium was observed on Gause's no.1 medium at 28°C. The colonies were small. The aerial mycelia were white in color, and the substrate mycelia were beige. They did not produce diffusible pigment on this medium. Further observation by microscope and scanning electron microscope revealed that the aerial mycelia were slender and much more branched than the substrate mycelia (Figure 1a and 1c). The aerial mycelia branched at 90 degree from the main

Media	Aerial mycelium	Substrate mycelium	Diffusible pigment
Gause's no. 1	white	beige	-
Kligler's no. 1	grey	grey	-
Czapek	grey	incarnadine	-
Glucose asparagine	white	bright orange	-
Glucose yeast extract	shark blue	light grey	caramel
PDA	light crane ash	beige	-

Table 2: Colors of isolate HN6 on different media. "-" indicates no diffusible pigment produced.

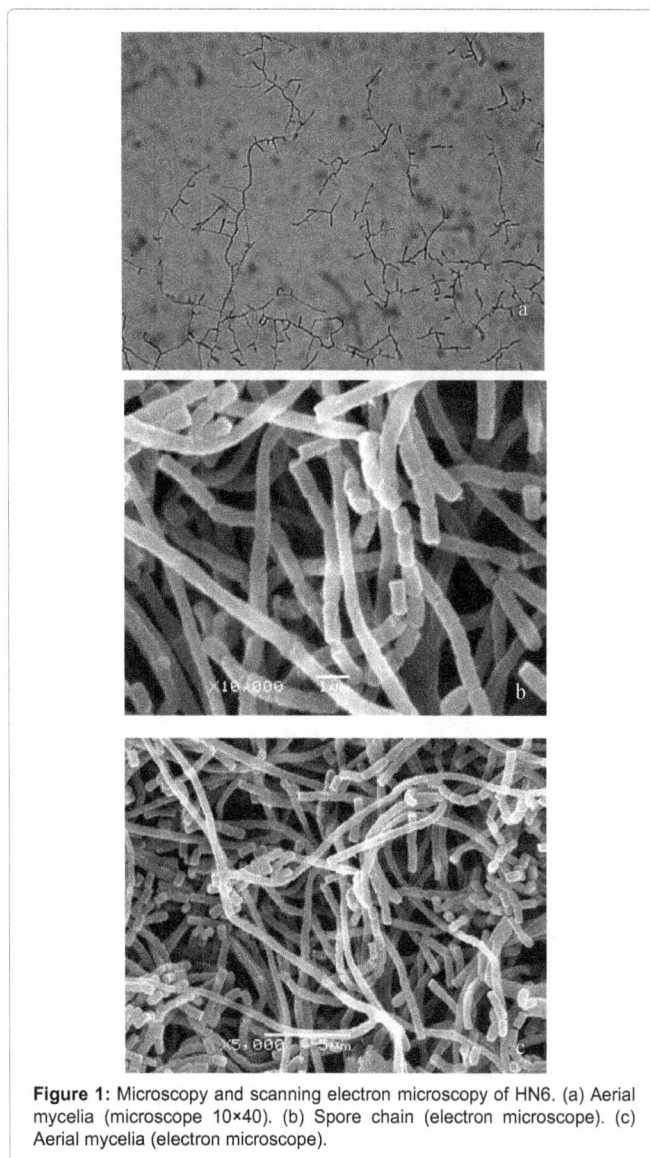

Figure 1: Microscopy and scanning electron microscopy of HN6. (a) Aerial mycelia (microscope 10×40). (b) Spore chain (electron microscope). (c) Aerial mycelia (electron microscope).

Tests	Results	Tests	Results
Starch hydrolysis	++	L-Arabinose utilization	++
Litmus milk reaction	-	Fructose utilization	++
H2S production	-	Glucose utilization	++
Glutin liquefaction	++	Galactose utilization	++
Cellulose utilization	-	Rhamnose utilization	++
Inositol utilization	++	Sucrose utilization	++
Mannitol utilization	++	Xylose utilization	+

Table 3: Physiological and biochemical characteristics of HN6. "++" strongly positive; "+" positive; "-" negative.

mycelium (Figure 1a). The spore chains broke into cylindrical conidial spores after maturing (Figure 1b).

The physiological and biochemical characteristics of HN6 were tabulated in Table 3. The HN6 isolate was able to hydrolyze starch and liquefy glutin. It could fully utilize fructose, inositol, L-arabinos, mannitol, rhamnose, sucrose, glucose and galactose as the carbon source. The HN6 isolate could partially utilize xylose, and not able to utilize cellulose. It could not react with litmus milk, nor decompose the sulfur-containing amino acid to H_2S. The morphological, physiological and biochemical characteristics of HN6 indicated that it shared the characteristic features of the genus *Streptomyces*. The cellulose TLC analysis showed that the cell wall of HN6 belonged to type I, and that its whole cell sugar belonged to type C.

Cloning and sequencing of HN6 16S rRNA gene and its phylogenetic analysis

To further classify HN6 isolate, its 16S rRNA gene was PCR-amplified and the 1422 bp-long PCR fragment was sequenced. The sequence was deposited in GenBank (NCBI) with the accession number of FJ911617. The HN6 16S rRNA gene sequence was analyzed by the BLAST server of NCBI. It was confirmed that HN6 belongs to the *Streptomyces* species. The HN6 isolate was 99% similar to *Streptomyces aureoverticillatus* (AY999774). The phylogenetic tree was constructed with bootstrap values (Figure 2). A neighbor-joining tree based on the 16S rRNA gene sequences showed that HN6 occupied a phylogenetic position alongside *Streptomyces aureoverticillatus* (AY999774). Combining the morphological, physiological and biochemical characteristics, HN6 was determined to belong to the species of *Streptomyces aureoverticillatus*.

Antifungal activity of HN6 methanol extract

To further elucidate the antifungal activities of HN6, its methanol extract was tested against eight indicator pathogens by the mycelium growth rate method. Table 4 shows the regression equation values (with all correlation coefficient values >0.95) of HN6 extract against the eight indicator pathogens, indicating an ideal positive correlation. The median lethal concentration (EC_{50}) was less than 0.08 mg/ml to every indicator pathogen, indicating a broad spectrum antifungal activity of HN6 methanol extract. *Botryodiplodia theobromae* (ATCC10936) was the most sensitive to the HN6 methanol extract with an EC_{50} of 0.016 mg/ml. *Colletotrichum gloeosporioides* (ATCC20358), *Rhizoctonia solani* (ATCC58938), *Colletotrichum gloeosporioides* (ATCC16330), *Colletotrichum musae* (ATCC96167), *Corynespora cassiicola* (ATCC36294), *Fusarium oxysporum* f. sp. *cubense* race 4 (ATCC76255) and *Periconia circinata* (ATCC32727) were also sensitive to the HN6 methanol extract with the EC_{50} of 0.020, 0.024, 0.025, 0.027, 0.053, 0.073 and 0.078 mg/ml respectively.

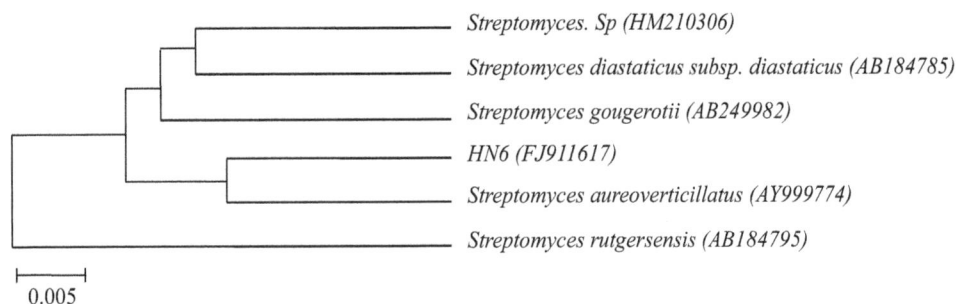

Figure 2: The phylogenetic relation of HN6 isolate (FJ911617) to *Streptomyces aureoverticillatus* (AY999774) and other closely related *Streptomyces* spp. The phylogenetic tree was constructed based on sequences of 16S rRNA genes using the neighbor-joining analysis of 1000 sampled data sets. Species with values of >50% are shown.

Pathogen (ATCC #)	Regression equation (y)	Correlation coefficient (r)	EC50 (mg/ml)	95% confidence interval (mg/ml)
B. theobromae (10936)	11.26+3.48x	0.960	0.016	0.0081-0.3804
C. cassiicola (36294)	7.45+1.92x	0.991	0.053	0.0289-0.4901
C. gloesporioides (16330)	12.69+4.78x	0.961	0.025	0.0195-0.4080
C. gloesporioides (20358)	12.39+4.37x	0.950	0.020	0.0181-0.3902
C. musae (96167)	11.48+4.14x	0.956	0.027	0.0184-0.4312
F. oxysporum (76255)	14.71+8.93x	0.970	0.073	0.0169-0.2135
P. circinata (32727)	14.70+8.76x	0.982	0.078	0.0146-0.5312
R. solani (58938)	12.62+4.70x	0.950	0.024	0.0064-0.3725

Table 4: Antagonistic activity of HN6 methanol extract against eight indicator plant pathogens.

Discussion

The Botanical Garden of Chinese Academy of Tropical Agricultural Sciences in Hainan Province, China, is an integrated ecological system with rich biodiversities. We isolated totally 89 actinomycetes from the soil in this garden. Many isolates had inhibition zones of 0-18 mm against the banana *Fusarium* wilt causal agent FOC4. The HN6 isolate was shown to be the most inhibitive to FOC4.

The morphological, physiological and biochemical characteristics of HN6 classified it into *Streptomyces* spp. The 16S rRNA gene sequence classified HN6 as *Streptomyces aureoverticillatus*. Publications have shown that *Streptomyces* spp. produce valuable bioactive metabolites with broad spectrum activities such as antibacterial, antifungal, antibiotic, antiparasitic, antitumor, antiviral, insecticidal and, herbicidal [21,22]. *Streptomyces aureoverticillatus* isolated from a marine sediment was discovered to produce novel anticancer and anti-infection agents which are novel macrocyclic lactam secondary metabolites [23]. As far as we know, there had not been one isolate of *Streptomyces aureoverticillatus* that had been shown agriculturally applicable activities. There is also no publication demonstrating the antimicrobial properties of these types of macrocyclic lactam metabolites. Recently, a new actinomycete strain NA4 was isolated from a deep-sea sediment from the South China Sea [24]. NA4 was identified as *Streptomyces acvourensis* by morphological, physiological and phylogenetic analysis based on its 16S rRNA gene sequences [24]. NA4 was shown to be antifungal against several soil-borne plant pathogens due to its production of bafilomycins B1 and C1 [24]. Future studies are needed to characterize the secondary metabolites of HN6 isolate and its antagonistic mechanisms in order to fully extend its potential as a biocontrol agent for FOC4 and other plant pathogens.

Conclusions

A *Streptomyces aureoverticillatus* isolate HN6 has been isolated from soil samples on Hainan Island, China. HN6 exhibits broad spectrum antifungal activity and is strongly inhibitive to *Fusarium oxysporum* f. sp. *cubense* race 4 (FOC4), the causal agent of banana *Fusarium* wilt in bio-assays.

Acknowledgement

This study was funded by Hainan Applied Technology R&D, the Demonstration Projects (ZDXM2014115), and Hainan Province Natural Science Foundation (314054).

References

1. Sun Y, Yi X, Peng M, Zeng H, Wang D, et al. (2014) Proteomics of Fusarium oxysporum race 1 and race 4 reveals enzymes involved in carbohydrate metabolism and ion transport that might play important roles in banana Fusarium wilt. PLoS One 9:e113818. doi: 10.1371/journal.pone.0113818.

2. Wen T, Huang X, Zhang J, Zhu T, Meng L, et al. (2015) Effects of water regime, crop residues, and application rates on control of Fusarium oxysporum f. sp. cubense. J Environ Sci (China) 31: 30-7. doi: 10.1016/j.jes.2014.11.007.

3. Yang L, Sun L, Ruan X, Qiu D, Chen D, et al. (2015) Development of a single-tube duplex real-time fluorescence method for the rapid quantitative detection of Fusarium oxysporum f. sp. Cubense race 1 (FOC1) and race 4 (FOC4) using TaqMan probes. Crop Protection 68: 27-35.

4. Strobel G, Daisy B, Castillo U, Harper J (2004) Natural products from endophytic microorganisms. J Nat Prod 67: 257-268.

5. Fiedler HP, Bruntner C, Riedlinger J, Bull AT, Knutsen G, et al. (2008) Proximicin A, B and C, novel aminofuran antibiotic and anticancer compounds isolated from marine strains of the actinomycete Verrucosispora. J Antibiot (Tokyo) 61: 158-63.

6. Atta HM, Dabour SM, Desoukey SG (2009) Sparsomycin antibiotic production by Streptomyces sp. AZ-NIOFD1: taxonomy, fermentation, purification and biological activities. Am-Euras J Agric Environ Sci 5: 368-377.

7. Xue L, Xue Q, Chen Q, Lin C, Shen G, et al. (2013) Isolation and evaluation of rhizosphere actinomycetes with potential application for biocontrol of Verticillium wilt of cotton. Crop Protection 43: 231-240.

8. Ravikumar S, Inbaneson SJ, Uthiraselvam M, Priya SR, Ramu A, et al. (2011)

Diversity of endophytic actinomycetes from Karangkadu mangrove ecosystem and its antibacterial potential against bacterial pathogens. J Pharm Res 4: 294-296.

9. Dhanasekaran D, Thajuddin N, Panneerselvam A (2009) Distribution and ecobiology of antagonistic streptomycetes from agriculture and coastal soil in Tamilnadu, India. J Cult Collect 6: 10-20.

10. Chun J, Kim SB, Oh YK, Seong CN, Lee DH, et al. (1999) Amycolatopsis thermoflava sp. nov., a novel soil actinomycete from Hainan Island, China. Int J Syst Bacteriol 49 Pt 4: 1369-1373.

11. Shirling EB, Gottlieb D (1966) Methods for characterization of Streptomyces species. Int J Syst Bacteriol 16: 313-340.

12. Kelly KL (1965) A universal color language. Color Eng 3: 2-7.

13. Valan Arasu M, Duraipandiyan V, Agastian P, Ignacimuthu S (2009) In vitro antimicrobial activity of Streptomyces spp. ERI-3 isolated from Western Ghats rock soil (India). J Med Mycol 19: 22-28.

14. Balachandran C, Duraipandiyan V, Balakrishna K, Ignacimuthu S (2012) Petroleum and polycyclic aromatic hydrocarbons (PAHs) degradation and naphthalene metabolism in Streptomyces sp. (ERI-CPDA-1) isolated from oil contaminated soil. Bioresour Technol 112: 83-90.

15. Hasegawa T, Takizawa M, Tanida S (1983) A rapid analysis for chemical grouping of aerobic actinomcetes. J Gen Appl Microbiol 29: 319-322.

16. Gupta VK, Shivasharanappa N, Kumar V, Kumar A (2014) Diagnostic evaluation of serological assays and different gene based PCR for detection of Brucella melitensis in goat. Small Ruminant Res 117: 94-102.

17. Weisburg WG, Barns SM, Pelletier DA, Lane DJ (1991) 16S ribosomal DNA amplification for phylogenetic study. J Bacteriol 173: 697-703.

18. Tamura K, Dudley J, Nei M, Kumar S (2007) MEGA4: Molecular Evolutionary Genetics Analysis (MEGA) software version 4.0. Mol Biol Evol 24: 1596-1609.

19. Amadioha AC (2001) Fungicidal activity of some plant extracts against Rhizoctonia solanii in cowpea. Arch Phytopath Pflanz 33: 509-517.

20. Zhao S, Ren FE, Liu J, Qin J, Pan H (2012) Screening, identification and optimization of fermentation conditions of an antagonistic actinomycetes strain to Setosphaeria turcica (in Chinese). Acta Microbiol Sinica 50: 1228-1236.

21. Baltz RH (2008) Renaissance in antibacterial discovery from actinomycetes. Curr Opin Pharmacol 8: 557-563.

22. Liu H, Qin S, Wang Y, Li W, Zhang J (2008) Insecticidal action of Quinomycin A from Streptomyces sp KN-0647, isolated from a forest soil. World J Microbiol Biotechnol 24: 2243-2248.

23. Mitchell SS, Nicholson B, Teisan S, Lam KS, Potts BC (2004) Aureoverticillactam, a novel 22-atom macrocyclic lactam from the marine actinomycete Streptomyces aureoverticillatus. J Nat Prod 67: 1400-1402.

24. Pan HQ, Yu SY, Song CF, Wang N, Hua HM, et al. (2015) Identification and characterization of the antifungal substances of a novel Streptomyces cavourensis NA4. J Microbiol Biotechnol 25: 353-357.

31

Study of the Potential Application of Lactic Acid Bacteria in the Control of Infection caused by *Agrobacterium tumefaciens*

Limanska N[1], Korotaeva N[1], Biscola V[2,3], Ivanytsia T[1], Merlich A[1,2], Franco BDGM[3], Chobert JM[2], Ivanytsia V[1] and Haertlé T[2]*

[1]*Microbiology, Virology and Biotechnology Chair, Biological Department, Odessa National I.I. Mechnikov University (ONU), 65082 Dvoryanska str., 2, Odessa, Ukraine*

[2]*UR 1268 INRA Biopolymères Interactions Assemblages, BP 71627, rue de la Géraudière, 44316 Nantes Cedex 3, France*

[3]*Department of Food and Experimental Nutrition, Faculty of Pharmaceutical Sciences, University of São Paulo, 580, Professor Lineu Prestes, 13 B, Sao Paulo, SP, 05508-000, Brazil*

Abstract

Inhibition of crown gall on test plants in case of co-inoculation with lactic acid bacteria (LAB) has been investigated. From nine LAB strains tested, eight reduced amount of galled carrot explants by 36.4-87.7% and decreased the intensity of disease manifestation. The antagonistic activity against *Agrobacterium tumefaciens*, *in vitro*, was due to the low pH of organic acids produced by LAB. However, in the same pH, different LAB cultures displayed various levels of inhibition *in vivo*. *Lactobacillus plantarum* ONU 12 with the best results in tumor inhibition on carrots, showed high antagonistic activity on surfaces of kalanchoe and grapevines. Depending on the method of inoculation, the culture of *L. plantarum* ONU 12 could protect from 72.7% to 100% of wounded kalanchoe tissues. Evaluation of number of surviving cuttings and amount of buds that grew indicated that co-inoculation with agrobacteria and LAB removed completely the negative influence of phytopathogen on grapevines and reduced the number of infected cuttings by approximately 80%. One-hour treatment with *L. plantarum* ONU 12 helped to decrease the number of infected plants by approximately 68%. The studied strain *L. plantarum* ONU 12 can be proposed for further evaluation of possibility of practical use in plant protection.

Keywords: Antagonism; Crown gall; *Lactobacillus; Enterococcus*

Introduction

Lactic acid bacteria (LAB) are widely used in many fields of human activity and especially in food preservation and in medicine. Some authors have proposed to apply lactobacilli for improvement of crop plant yields and for protection of agriculturally important species against certain phytopathogens. Visser et al. [1] reported antagonistic action of *Lactobacillus plantarum* L292 against *Pseudomonas syringae* observed *in vivo* on haricot beans resulting in significant reduction of disease symptoms. Inhibition of *Xanthomonas campestris* growth with *L. plantarum* strains was described by Trias et al. [2] and Dalirsaber Jalali et al. [3]. Other LAB such as *Enterococcus mundtii* suppressed the growth of *Erwinia carotovora*. Bacterial mixtures containing mostly lactobacilli were efficient against *Ralstonia solanacearum* [4].

Recent publications described also the antimycotic effects of LAB. These effects are often species- and strain-specific. For instance, certain strains of *Weissella cibaria, L. plantarum, Leuconostoc mesenteroides* and *Lactococcus lactis* reduced rot caused by *Penicillium expansum* in wounds of stored apples while other tested strains did not exhibit such activities [2]. El-Mabrok et al. [5] reported the inhibition of phytopathogenic fungi *Colletotrichum gloeosporioides* by strains of *L. plantarum*. Lactobacilli were used for decreasing disease symptoms or production of metabolites responsible for food intoxication caused by *Fusarium* [6], *Aspergillus flavus* [7], *Aspergillus ochraceus* [8], *Aspergillus niger* and *Penicillium expansum* [9].

The main metabolites from LAB found to be active against phytopathogens were organic acids and hydrogen peroxide; the microbial competition was also described [2,7,10].

Interesting results of aforementioned authors stimulated to study the possibilities of use of LAB against known phytopathogen *Agrobacterium tumefaciens* (*Rhizobium radiobacter* according to the nomenclature proposed by Young et al. [11], which can be devastating

for the nurseries of stone fruits and ornamental plants. The symptoms of the caused disease (crown gall) are tumor formations on stems and roots of infected plants resulting in deficiency of nutrients and water supply (reviewed in Burr and Otten, [12]). Tissue proliferations induced by pathogenic agrobacteria also include cane gall caused by *A. rubi* (*R. rubi*) and hairy-root as the result of infection with *A. rhizogenes* (*R. rhizogenes*) [11,13]. Strains of *A. tumefaciens* biotype 3 were reclassified as a distinct species *A. vitis* (*R. vitis*) on the basis of their host (*Vitis spp.*) and their genetic peculiarities [11,14]. Besides by *A. vitis*, crown gall on grapevine in some cases can be caused by *A. tumefaciens* biotype 1 strains [15,16].

The pathogenesis of crown gall is unique and includes the transfer of the part of Ti-plasmid from *A. tumefaciens* into the chromosome of the plant [17]. As the result, plant cells start to produce an increased amount of hormones leading to uncontrolled tissue proliferation [18] and to synthesize the unusual compounds such as are opines – derivatives of sugars and special amino acids used by bacteria as nutritional sources [19,20].

Representatives of several bacterial genera have been used for efficient biocontrol of crown gall. Some avirulent agrobacteria produce highly specific bacteriocin agrocin as *A. rhizogenes* K84 does – the

*****Corresponding author:** Thomas Haertlé, UR 1268 INRA Biopolymères Interactions Assemblages, BP 71627, rue de la Géraudière, 44316 Nantes Cedex 3, France, E-mail: haertle@nantes.inra.fr most widely used antagonistic strain against *A. tumefaciens* [21]. The genes responsible for agrocin synthesis and self-immunity are localized on a plasmid, which can be transferred into pathogenic strains and make them resistant [22]. To overcome biocontrol failure, a stable Tra⁻ deletion mutant of K84 – the strain K1026 was constructed [23]. Biopreparations based on K84 have been traditionally used for many years [20,24]. Being highly effective in many dicotyledonous plants, *A. rhizogenes* K84 does not affect crown gall agents in grapevines [25]. One of the most active antagonists decreasing level of tumour formation in grape is *A. vitis* F2/5 [26-28]. Its inhibitory effect cannot be explained by agrocin production and competition for attachment sites but it is directly related to interactions with grapevine cells resulting in cambium necroses resembling hypersensitivity response [29-31]. *A. vitis* F2/5 does not affect the growth of pathogen population in plant tissue but inhibits the tissue transformation [32].

Several other efficient antagonists strains belonging to species *A. vitis* are known. *A. vitis* E26 effectively inhibited crown gall on peach and cherry caused by *A. tumefaciens* and on grapevine caused by *A. vitis* [33]. The antagonistic effect was explained by bacteriocin production [34,35].

A. radiobacter HLB-2 supressed grape crown gall by bacteriocin production and competing for attachment sites and nutrients [36,37]. *A. vitis* VAR03-1 has shown a significant reduction effect of gall formation in tomatoes, roses and grapevines [38,39]. Recently described novel antagonistic strain *A. vitis* ARK-1 reduced the tumor incidence in grapevine plants and stably survived on roots [40].

The other important antagonists against crown gall agents belong to *Pseudomonas* genus. Crown gall formation on grapevine and raspberry was efficiently inhibited with *P. aureofaciens* and *P. fluorescens* [41,42]. The similar clear antagonistic activity was revealed in *P. corrugate* isolated from grapevine xylem sap. Besides this bacterium, the inhibitory effect against *A. vitis* was demonstrated by strains of *Enterobacter agglomerans* and *Rahnella aquatilis* isolated from the same source [43]. It was revealed also that a specific antimicrobial compound with wide spectrum of action was responsible for the suppressive effect of *Rahnella aquatilis* [44]. Treatments with *Pseudomonas putida*, *Burkholderia phytofirmans* and *Azospirillum brazilense* strains producing 1-aminocyclopropane-1-carboxylate deaminase, which degrades the precursor of ethylene in plants, inhibited the tumor formation in tomatoes [45].

Bacilli also have been investigated as possible biocontrol agents against crown gall. *Bacillus subtilis* and *Bacillus* spp. reduced gall size in treated plants and densities of internal populations of *R. vitis* [42]. A biopreparation based on *B. subtilis* could significantly reduce disease incidence. Additional use of resistance inducers for plants was strongly recommended to decrease crown gall severity under the field conditions [46].

Trials of using LAB against *A. tumefaciens* described in literature have not been found, although this trend of biological control in plant protection is very attractive because of some important characteristics of these agents, such as their capability of inhibiting other microorganisms by competition and production of antimicrobial compounds such as organic acids and bacteriocins. Besides some LAB, such as *Lactobacillus plantarum*, are considered by the Food and Agriculture Organization (FAO) as GRAS (Generally Recognized As Safe) for the application in biopreservation systems.

Hence, the aim of this work was to test *in vivo* the antagonistic activity of some LAB strains isolated from various sources against *A. tumefaciens*.

Materials and Methods

Bacterial strains

The tested LAB strains originated from the collection of cultures of Microbiology, Virology and Biotechnology Chair of ONU, Odessa, Ukraine and from the collection of BIA-FIP laboratory, INRA, Nantes, France (Dr. S. Migaw and Dr. M. Barbosa). Pathogenic strain *A. tumefaciens* C58 was kindly provided by the collection of microorganisms of D.K. Zabolotny Institute of Microbiology and Virology (Dr. F.I. Tovkach), Kiev, Ukraine. All LAB strains were stored in MRS broth [47] and *A. tumefaciens* C58 in Luria Bertani (LB) broth [48] at -80°C with 20% glycerol. Strains *L. plantarum* ONU 87, ONU 206 and ONU 991 were isolated from dairy products and *L. plantarum* ONU 12, ONU 311, ONU 312, ONU 313 from grape berries collected in Ukraine; *Enterococcus faecium* C8 was isolated from Azerbaijan cheese, and *Enterococcus durans* 3y from Tunisian fish.

Inoculation of carrot explants

Carrots (*Daucus carota* L.) were purchased on local markets of France and Ukraine, washed with commercial "Javel" (sodium hypochlorite) solution, rinsed in tap water, immersed in ethanol, flamed, peeled and sliced in discs [49]. The disks were placed in sterile Petri dishes. *A. tumefaciens* was cultivated overnight in LB broth at 28°C, and the final concentration of cells reached up 2-8 x 10^9 CFU/mL. LAB cell suspensions were obtained by inoculation of each strain in MRS broth followed by incubation overnight at 37°C (final concentration 1-5 x 10^{10} CFU/mL). The inoculum was obtained by mixing the culture of *A. tumefaciens* with the LAB cultures in equal volumes (1/1). To inoculate the carrots, 100 μL of the mixture were applied on the basal surfaces of the disks. Besides mixtures "*A. tumefaciens* C58/LAB", agrobacterial culture mixed with sterile saline solution (0.85% NaCl, w/v) at a ratio 1:1 was applied on explants as a positive control, and LAB cultures – as negative controls. After 21 days, explants were observed for the presence of tumors and fermentation. Amount of galled explants was calculated and manifestation of crown gall symptoms was evaluated by the modified method of Ryder et al. [48] as follows: "++++" 100% cambial ring covered with tumors; "+++" 75% of cambial ring with tumors; "++" 50% of cambial ring with tumors; "+" less than 25% of cambial ring with tumors.

Production of antimicrobial metabolites

Cell Free Supernatant (CFS) instead of bacterial cultures was applied on carrot explants together with agrobacteria. LAB strains with the best results in the previous experiments (*L. plantarum* ONU 311, *L. plantarum* ONU 312, *L. plantarum* ONU 12, *E. faecium* C8 and *E. durans* 3y) were used. To obtain the CFS, LAB were grown in MRS broth at 37°C for 24 h, and cells were harvested by centrifugation (8000 g, 4°C, 10 min).

The production of antimicrobial compounds was also evaluated *in vitro* by agar-well diffusion assay according to Batdorj et al. [49] using CFS of the same LAB strains applied in the *in vivo* tests. CFS was obtained after incubation of LAB at 30°C and 37°C for 24 h in MRS broth. For the agar-well diffusion test, Brain Heart Infusion (BHI) soft agar (0.8%, w/v) or LB soft agar (0.8% agar, w/v) were inoculated with 10^6 CFU/mL of the indicator strain, *L. ivanovii* ATCC 19119 (as a classic test strain for the study of LAB bacteriocinogenic activity) or *A. tumefaciens* C58, respectively. The concentrations of bacterial cultures were assesed spectrophotometrically and serial dilutions were carried out to obtain the needed concentrations of cells. 50 μL of CFS with initial acidic pH and pH adjusted to neutral with 1 N NaOH were poured into wells made on the surface of the plates containing each indicator strain. After 24 h of incubation at 37°C for *L. ivanovii* ATCC 19119 or at 28°C for *A. tumefaciens*, the presence of inhibition zones was observed [50].

Co-incubation of agrobacteria in a mixture with LAB suspensions with different initial pH (4.1-4.5 and 5.0-5.5) during one hour was carried out at 28°C. After the incubation period, bacterial mixtures were diluted ten-fold, plated on LB medium, incubated overnight and colonies of agrobacteria were counted.

Inoculation of *Kalanchoe daigremontiana* Mill

Five methods of inoculation were applied. (1) 50 μl of *L. plantarum* ONU 12 overnight culture were injected together with 50 μl of agrobacterial overnight culture into upper tissues of leaves by sterile syringe. *Agrobacterium tumefaciens* culture with equivalent volumes of sterile distillated water (SDW) were injected as positive controls. The culture of lactobacilli was applied as a negative control. (2) 100 μl of a mixture "*A. tumefaciens* C58/*L. plantarum* ONU 12" were spotted on one-cm wounds made on leaves. (3) Wounds were treated with agrobacteria and after 30 min with *L. plantarum* ONU 12 culture. (4) Scars on leaves were treated with LAB and after 30 min inoculated with the phytopathogen. (5) Roots and the aerial parts (crowns) of plants were wounded and dipped for one hour in agrobacterial culture (positive control), SDW, culture of *L. plantarum* ONU 12 (negative controls) and in the mixture "C58/ONU 12". Plants were cultivated under greenhouse conditions. Leaves were observed for crown gall symptoms on the 60th day after inoculation. Treated roots and crowns were observed after six months. Tumor tissues were excised and weighted.

Inoculation of grapevine cuttings

Cuttings of *Vitis vinifera* L. cv. Pinot noir, Vostorg, Moldova (in equal quantities each) cultivated in the south of Ukraine were gathered during March of 2011-2013. Cuttings with freshly cut basal parts soaked for one hour in agrobacterial culture were used as positive controls. Cuttings were also treated for one hour with *L. plantarum* ONU 12 culture and with the mixture "C58/ONU 12". One variant of the treatments was soaking for one hour in *L. plantarum* ONU 12 culture and after inoculation with *A. tumefaciens* C58 culture for 15 min.

As negative controls, cuttings soaked for one hour in tap water were brought to assay. Other negative controls were MRS (pH 4.1) and a mixture "MRS/LB" (1:1) with pH 5.5-6.0 indicating the effect of nutritional media with pH of subsequent bacterial cultures on grape cutting development.

Cuttings were planted in commercial pot soil with abundance of peat and cultivated under greenhouse conditions for 30 days. After,

amount of surviving cuttings and number of buds that grew were evaluated as percentages from the total quantity of tested cuttings and buds. Green shoots on survived cuttings were measured, and mean lengths were calculated in each variant. Cuttings were tested for the presence of pathogens by a bacterial culture method followed by polymerase chain reaction (PCR) assay with the primers to *ipt* oncogene of pathogenic agrobacteria according to Haas et al. [51].

Statistical analyses

Carrots were inoculated in three independent experiments with 20-22 explants in each variant for bacterial cultures and CFS. For kalanchoe inoculation, a total of 90 plants of each variant was used in three independent experiments. 30-50 grape cuttings were treated in each variant in three independent experiments carried out during springs of 2011-2013. Agar-well diffusion assay was carried out in five repeats for each variant. CFU/mL in bacterial suspensions were evaluated by counting colonies grown in five repeats. The obtained results were presented in percentages and standard errors (SE) for qualitative attributes (number of infected plants and explants with necroses, amount of buds that grew) and in mean values with 95% confidence intervals (CI) for quantitative attributes (lengths of green shoots, amount of CFU/mL). Significant differences between the control and test samples were estimated in *t*-test (p < 0.05) and marked in the tables with data. Software "Microsoft Excel" was used for calculations and graphics.

Results

Tests on carrot explants

The best results for tumor growth inhibition on carrot discs co-inoculated with *A. tumefaciens* C58 and LAB cultures were obtained with 5 strains from 9 tested. *L. plantarum* ONU 12, *E. faecium* C8, *L. plantarum* ONU 312, *E. durans* 3y and *L. plantarum* ONU 311 suppressed completely tumor development on the majority of carrot discs, suggesting a high antagonistic activity of the mentioned lactobacilli and enterococci (Table 1).

The reduction in galled samples was observed in case of all tested LAB strains except *L. plantarum* ONU 991 (Figure 1).

The obtained results suggest that the antagonistic activity against *A. tumefaciens* within *L. plantarum* species is strain-specific.

Co-inoculation with LAB cultures shifted the level of crown gall manifestation to smaller area of galled tissue ("+" level compared with the positive control where "++" – "++++" levels were prevalent) (Table 2).

If after the treatments with active antagonistic strain *L. plantarum* ONU 12 some discs still appeared infected, less than 25% of cambial ring on disc were covered with tumors in such explants, while in case of non-active strain *L. plantarum* ONU 991 the range of tissue proliferation resembled that in the positive control.

The fermentation of carrot by some LAB strains occurred but it

Inoculum	Percentage of discs with tumors	Inoculum	Percentage of discs with tumors
A. tumefaciens C58 (positive control)	61.0 ± 6.0	*A. tumefaciens* C58 + *L. plantarum* ONU 311	16.4 ± 4.7d
A. tumefaciens C58 + *L. plantarum* 991	54.5 ± 6.1	*A. tumefaciens* C58 + *E. durans* 3y	13.6 ± 4.2d
A. tumefaciens C58 + *L. plantarum* ONU 313	38.8 ± 5.9d*	*A. tumefaciens* C58 + *L. plantarum* ONU 312	10.7 ± 3.8d
A. tumefaciens C58 + *L. plantarum* ONU 206	30.0 ± 5.9d	*A. tumefaciens* C58 + *E. faecium* C8	9.1 ± 3.5d
A. tumefaciens C58 + *L. plantarum* ONU 87	27.2 ± 5.4d	*A. tumefaciens* C58 + *L. plantarum* ONU 12	7.5 ± 3.2d

Note: *d - significant differences between values of the control and the test sample (p<0.05, t-test)

Table 1: Tumor formation on carrot explants after co-inoculation with A. tumefaciens C58 and LAB (%).

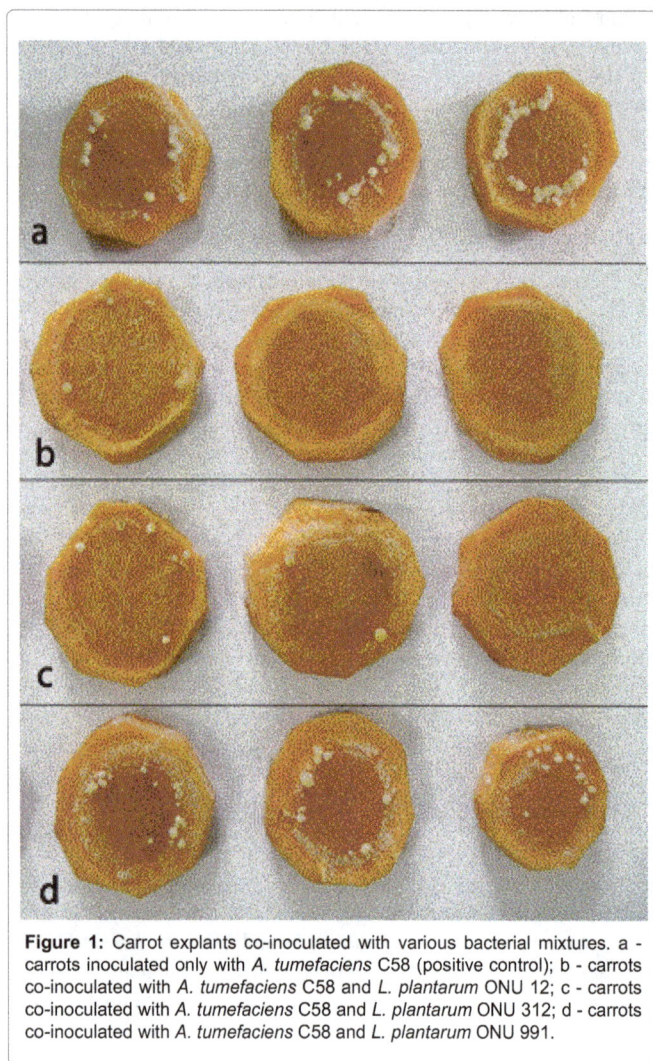

Figure 1: Carrot explants co-inoculated with various bacterial mixtures. a - carrots inoculated only with *A. tumefaciens* C58 (positive control); b - carrots co-inoculated with *A. tumefaciens* C58 and *L. plantarum* ONU 12; c - carrots co-inoculated with *A. tumefaciens* C58 and *L. plantarum* ONU 312; d - carrots co-inoculated with *A. tumefaciens* C58 and *L. plantarum* ONU 991.

Inoculum	+	++	+++	++++
A. tumefaciens C58 (positive control)	30.0	40.0	20.0	10.0
A. tumefaciens C58 + *L. plantarum* 991	38.3	27.7	11.1	22.2
A. tumefaciens C58 + *L. plantarum* ONU 313	84.6	7.7	0	7.7
A. tumefaciens C58 + *L. plantarum* ONU 206	77.8	16.7	0	5.5
A. tumefaciens C58 + *L. plantarum* ONU 87	83.3	5.5	11.1	0
A. tumefaciens C58 + *L. plantarum* ONU 311	60.0	20.0	20.0	0
A. tumefaciens C58 + *E. durans* 3y	55.5	22.2	11.1	11.1
A. tumefaciens C58 + *L. plantarum* ONU 312	85.7	14.3	0	0
A. tumefaciens C58 + *E. faecium* C8	66.6	33.3	0	0
A. tumefaciens C58 + *L. plantarum* ONU 12	100.0	0	0	0

Table 2: Manifestation of crown gall symptoms on carrot discs in presence of LAB (%).

did not attain more than 7.5 ± 3.2% of discs (*E. faecium* C8) and 6.1 ± 2.9% (*L. plantarum* ONU 991), and was not detected in case of *L. plantarum* ONU 12, *L. plantarum* ONU 312, *L. plantarum* ONU 87 and *L. plantarum* ONU 206. Thus, number of fermented discs reached only 4.5 ± 0.5% when *L. plantarum* ONU 313 was applied, 3.0 ± 0.4 in case of *Enterococcus durans* 3y, 1.6 ± 0.2% in case of *L. plantarum* ONU 311. Fermentative activity was less intense in case of the treatments of carrot explants with mixtures "*Agrobacterium*/LAB".

Production of antimicrobial metabolites

When CFS instead of cultures was applied, the results were similar to inoculation with alive LAB suspensions. Thus, treatments of explants with *A. tumefaciens* C58 in a mixture with CFS of *L. plantarum* ONU 12 caused crown gall symptoms only in 10.6 ± 3.6% of explants, in case of *E. faecium* C8 CFS – in 8.3 ± 4.8% of carrot discs, when treated with *L. plantarum* ONU 312 CFS – in 11.6 ± 4.0%, in case of *E. durans* 3y CFS – in 15.0 ± 4.6% of explants and in 18.3 ± 4.9% of discs treated with CFS of *L. plantarum* ONU 311.

In case of the *in vitro* tests of production of antimicrobial compounds, it was observed that none of the strains was able to inhibit *A. tumefaciens* C58 when the CFS pH 6.5 was used. Otherwise, when the acidic CFS was applied into the agar wells, clear zones of inhibition were observed, suggesting that the activity against the phytopathogen was due to the production of organic acids. The pH of the acidic CFS obtained after cultivation of the selected LAB at 30°C or 37°C varied between 4.1 and 4.7 for lactobacilli and 4.7 and 5.0 for enterococci. Incubation of agrobacteria in a mixture with LAB suspension with initial pH 5.0-5.5 caused 1-2 fold decrease in viable pathogen cell quantity. If initial pH of LAB suspensions was 4.1-4.5, one-hour of co-incubation was sufficient for 4-folds decrease in amount of viable cells (Table 3).

The experiments carried out with *L. ivanovii* ATCC 19119 as indicator strain showed that the strains *E. faecium* C8 and *E. durans* 3y inhibited its growth, when incubated at 30°C or 37°C, suggesting the production of antimicrobial metabolites other than organic acids, such as bacteriocins. However, these metabolites were inactive against *A. tumefaciens*.

Tests on *Kalanchoe daigremontiana* Mill

Strain *L. plantarum* ONU 12 with the best result in crown gall inhibition on carrot explants was used for further investigations.

When the mixture of bacterial cultures "C58/ONU 12" was injected in kalanchoe leaf tissues, no tumors were formed in any of repeats (Figure 2).

Treatment of the scars with this mixture resulted in tumor formation just in one case that was evaluated as 1.1% from the total amount of infected plants.

If scars were first infected with *A. tumefaciens* C58 and after 30 min inoculated with LAB culture, the percentage of galled plants decreased in 3.5 times (20.0 ± 4.2% comparing with 73.3 ± 4.6% in positive control) and mean weight of tumors was 4.4 times less (0.0309 ± 0.0181 g comparing with 0.1325 ± 0.0582 g in the positive controls).

Crown gall was not detected on the scars treated with LAB cultures 30 min before inoculation with the pathogen.

Variant	Amount of viable cells (CFU/mL)		
	Before incubation	After incubation	
		pH 5.0-5.5	pH 4.1-4.5
C58 + ONU 311	(6.4 ± 0.8) × 10⁹	(4.2 ± 0.7) × 10⁷	(1.5 ± 0.4) × 10⁵
C58 + ONU 312		(2.2 ± 0.5) × 10⁷	(4.8 ± 1.2) × 10⁵
C58 + ONU 12		(1.6 ± 0.3) × 10⁷	(5.2 ± 1.1) × 10⁵
C58 + ONU 87		(2.4 ± 0.8) × 10⁷	(2.1 ± 0.2) × 10⁵

Table 3: Effect of LAB on the survival of A. tumefaciens C58 cells after one hour of co-incubation.

Figure 2: Kalanchoe leaves infected by injections (left) and by scars (right): a, d - inoculated only with *A. tumefaciens* C58; b, e - co-inoculated with *A. tumefaciens* C58 and *L. plantarum* ONU 12; c, f - inoculated with *L. plantarum* ONU 12.

Figure 3: Kalanchoe plants infected via wounded crowns and roots: a - inoculated only with *A. tumefaciens* C58; b - co-inoculated with *A. tumefaciens* C58 and *L. plantarum* ONU 12; c – soaked in water; d - inoculated with *L. plantarum* ONU 12.

After the treatments with a mixture "*A. tumefaciens* C58/*L. plantarum* ONU 12" and in some cases with LAB suspension, necroses were observed but they were restricted to wounded sites.

Treatment with LAB simultaneously with the inoculation with agrobacteria via root system allowed to protect plants in 100% of cases (Figure 3).

The obtained results indicated the high efficacy of using *L. plantarum* ONU 12 to inhibit crown gall on kalanchoe.

Tests on grapevine cuttings

Inoculation with *A. tumefaciens* C58 resulted in the induction of necroses on basal ends of 100% of the treated cuttings in positive controls. 24% of inoculated cuttings died (Figure 4) and the surviving grapevines showed 9.4% smaller amount of buds that grew (Figure 5).

Inoculation with *L. plantarum* ONU 12 had a positive effect on the amount of buds that grew, but the effect was small (11.2%). But the highest stimulating effect on buds that grew was revealed in case of the treatments with a mixture "*A. tumefaciens* C58/*L. plantarum* ONU 12". Soaking the cuttings in the mixture "*A. tumefaciens* C58/*L. plantarum* ONU 12" during 1 h resulted in 45.0% increase in amount of buds that grew. One-hour treatments with LAB culture followed by subsequent inoculation with *A. tumefaciens* C58 during 15 min lead to 35.0% increase. The mixture of nutritional media MRS and LB in the same ratio as in the bacterial cultures did not demonstrate any stimulation of the treated cuttings.

When inoculation with *A. tumefaciens* C58 was carried out simultaneously with the treatment with *L. plantarum* ONU 12 culture, the negative effect of phytopathogen was not observed. All the evaluated characteristics of plants did not differ from that in the negative control (water). Thus, the number of surviving cuttings reached 85.3 ± 3.5% in negative control, 64.8 ± 5.6% in positive controls infected with pathogenic agrobacteria – and 89.7 ± 5.0% in cuttings co-inoculated with *A. tumefaciens* C58 and *L. plantarum* ONU 12. Mean lengths

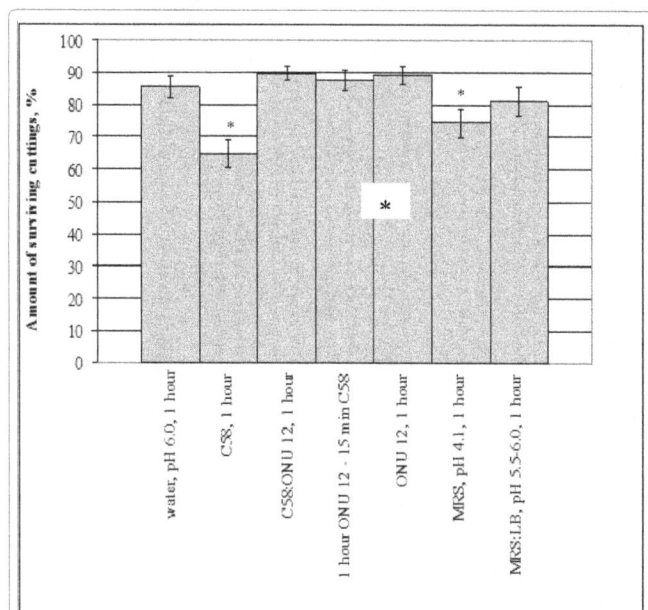

Figure 5: Effect of the treatments of grape cuttings with *A. tumefaciens* C58 and *L. plantarum* ONU 12 on amount of buds that grew (expressed as a mean ± SE): * - significant differences between means after soaking in SDW and soaking in lactobacilli alone and in combinations of bacteria (p<0.05, *t* test).

of green shoots formed in grape cuttings treated with *A. tumefaciens* C58 and *L. plantarum* ONU 12 in different combinations didn't differ significantly from the control (data not shown).

PCR with DNA of bacteria isolated from the treated cuttings showed the presence of pathogenic agrobacteria in tissues of 87.2% of samples inoculated in positive controls (Table 4). After a treatment with a mixture "*A. tumefaciens* C58/*L. plantarum* ONU 12" the amount of infected cuttings decreased to 17.0%. One-hour preliminary treatment with *L. plantarum* ONU 12 culture followed by a 15 min inoculation with *A. tumefaciens* C58 helped to decrease the amount of infected plants to 27.5% [52].

Discussion

Investigations of LAB use in plant protection against bacterial and fungal infections, for plant growth stimulation and for soil treatment have been largely described [1-3,5-9,53]. However, the possibility of crown gall biocontrol with LAB has not been evaluated yet. For the preliminary tests, we used the carrot disc model, what allowed to obtain rapid results and to carry out the simultaneous estimation of possibility of LAB strains to cause fermentation of plant tissues. It is mostly because of their fermentative activity and consecutive acidification of fermented media that LAB are used for food preservation [54]. For plant protection it was necessary to find strains with low fermentative activity not causing significant damage of wounded tissues.

Whereas *in vitro* tests have showed that the main inhibitory effect was based on low pH, different strains of LAB with the same pH of overnight cultures still varied in levels of their antagonistic activities *in vivo* on carrot explants showing the presence of other factors involved in suppression of phytopathogens. Still on this stage of the experiment it is unclear what is the mechanism of such inhibition. The suppression of phytopathogens can occur on the next stage as well when bacteria of both species are applied on carrot disk surface and LAB compete for the nutrients, attachment sites and excrete metabolites such as organic acids (lactic acid) and oxygen peroxide, which affect negatively

Figure 4: Survival of grape cuttings after the treatments with *A. tumefaciens* C58, controls and *L. plantarum* ONU 12 (expressed as a mean ± SE): * - significant differences between means after soaking in SDW and soaking in agrobacteria and MRS medium with pH 4.1 (p<0.05, *t* test).

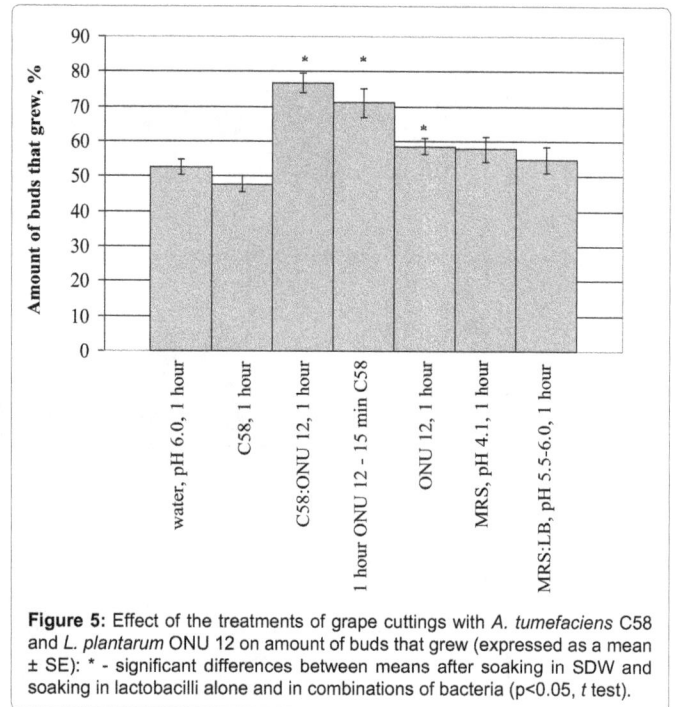

Variant	Amount of tested cuttings	
	Total	Infected (%)
A. tumefaciens **C58**	94	87.2
A. tumefaciens C58:L. plantarum ONU 12	88	17.0
1 hour L. plantarum ONU 12 then 15 min A. tumefaciens C58	80	27.5

Table 4: Presence of pathogenic agrobacteria in grapevine tissues after the experimental inoculation.

the survival of agrobacteria on plant surfaces. Different levels of antagonistic activity among strains of L. plantarum species described remain in agreement with the results of Visser et al. [1], Trias et al. [2], El-Mabrok et al. [5], who described the strain-specific effect of LAB against phytopathogens.

Strain L. plantarum ONU 12 with the highest antagonistic activity on carrot explants was also effective in suppressing crown gall on Kalanchoe daigremontiana and agrobacterial infection on grapevine cuttings. Depending on the method of inoculation, its suspension protected 100% of wounded tissues (via injections of the mixture "A. tumefaciens C58/L. plantarum ONU 12" or via soaking damaged roots and crowns in such a mixture), or decreased the number of galled plants by 98.5% (treatments of leaf scars with the mixture "A. tumefaciens C58/L. plantarum ONU 12") and by 72.7% in case of inoculation with agrobacteria followed by the treatment with LAB after 30 min interval.

Inoculation of grapevine with A. tumefaciens C58 culture reduced the number of viable cuttings, amount of buds that grew and mean length of green shoots from the cuttings that survived. A. tumefaciens in some cases can be a crown gall agent on grapevine, the same as A. vitis does [15,16]. Certain pathogenic strains originally isolated from grapevine as A. tumefaciens FACH was aggressive on kalanchoe but did not cause tumor formation on grapevine [51]. As for A. tumefaciens C58, its high tumorigenic activity on grapevines was reported by Holden et al. [54]. In our experiments, necroses instead of tumors were formed on all inoculated in the positive control grape cuttings. It was revealed by Bazzi et al. [27] that high concentrations of pathogenic A. vitis (approximately 10^8 CFU/mL) caused vast necroses of woody parenchyma instead of gall formation [55]. Plants with necrotic tissues exhibited poor growth and high mortality [28]. In the case of A. vitis, the necrogenic response of grape tissues can be explained by activity of polygalacturonase degrading cell walls [56] and other necrosis factors inducing inoculum-dependent damages on shoot explants and grape leaves [12]. A. tumefaciens C58 genome does not carry pehA gene responsible for polygalacturonase synthesis [51] and the mechanism of necrogenesis in case of A. tumefaciens is different from that induced by A. vitis [12]. Thus, strain A. tumefaciens A281 being super virulent on many plant species induced a necrogenic response (84%) rather than gall formation on several grapevine cultivars and such necroses seemed to be cultivar specific and related to genes of Ti-plasmid. Exogenous auxin increased the level of necrogenesis and this implied that plant hormones influenced the process of necrogenesis limiting the tumor formation. Deng et al. [57] proposed the hypothesis that A. tumefaciens-induced necrotic response is a consequence of increased levels of auxin or precursors of auxin affecting grapevine cells. The greatest degree of necrogenesis was observed in case of inoculation of basal ends [58]. Treatments of basal ends of cuttings in our experiments resulted also in vast necroses and subsequent high mortality of grapevines.

Co-inoculation with agrobacteria and LAB removed completely the negative effect of phytopathogen on grapevines as it was shown by comparison of numbers of survived cuttings, amounts of buds and mean lengths of green shoots. If grapevine cuttings were inoculated

simultaneously with A. tumefaciens C58 and L. plantarum ONU 12, amount of infected cuttings decreased by 81% whereas preliminary inoculation with L. plantarum ONU 12 reduced this number by 68.4% showing the effect of pre-treatments with LAB suspensions.

As for the number of buds that grew, higher stimulating effect of lactobacilli in a mixture with a pathogen was rather unexpected. Locke et al. [58] showed similar effect on buds distal to tumors caused by the attenuated culture of agrobacteria due to cytokinins produced by tumors (reviewed in Binns, [59]). In our study, tumors were not formed but the stimulation activity was observed.

Stimulating activity of the overnight culture of L. plantarum ONU 12 on number of buds that grew was not associated with the components of MRS medium but only with the substances synthesized by LAB. Indeed, soaking in MRS, pH 4.1 as a negative control simulating the medium for lactobacilli with the lowest possible pH of the bacterial culture after appropriate time of incubation influenced negatively the grapes. The number of survived cuttings and mean length of green shoots decreased. Grapevine is not an acid tolerant plant, although vineyards with acid subsoils with pH lower than 5.0 are known (reviewed in Kirchhoff et al., [60]). Significant decrease in reduction of root and shoot biomass was observed in soils with pH below 4.5 [61,62]. Hence, it may be proposed that the positive effect of metabolites of L. plantarum ONU 12 compensated in executed experiments the unfavorable influence of low pH of overnight LAB culture.

Our previous investigations showed the stimulating effect of L. plantarum ONU 12 on development of tomato seedlings. In that case we used LAB cells washed from MRS with metabolites synthesized overnight and resuspended in SDW with final pH 5.0-5.2 [63]. Thus, the treatment with washed cells in case of tomato seeds and with overnight culture in MRS in case of grapevine cuttings had stimulating effect on test plants in both experiments.

Positive effects of treatments of the plants with lactobacilli under field [6] and laboratory conditions [63] has been described. Improving of root and shoot lenghts [53,63], including plants on plots infested with Fusarium oxysporum [6] indicated the possibility of treatments with lactobacilli not only as antimicrobial agents but also as plant stimulating microorganisms.

Taking into account the marked antagonistic activity in vivo, L. plantarum ONU 12 can be proposed as a microorganism with the high potential for protection of plants against A. tumefaciens. However, experiments with more of other plant varieties and species in green house and under field conditions would be still needed for deeper evaluation of all potential of practical use of studied LAB.

Acknowledgments

The authors express their gratitude for support of this study by the Bilateral French-Ukrainian collaboration program "Dnipro" (2011-2012) granted by the Ministry of Foreign and European Affairs of France and the State Agency of Science, Innovations, and Information of Ukraine, and for support in frame of the programme "Science in Universities" NU3-2011 granted by the Ministry of Education, Science, Sport and Youth of Ukraine. Authors express their gratitude to Dr. Matheus de Sousa Barbosa, Dr. Sarah Migaw and Dr. Ganna Yamborko for kindly providing the strains E. faecium C8, E. durans 3y, L. plantarum ONU 87, ONU 206 and ONU 991.

References

1. Visser R, Holzapfel WH, Bezuidenhout JJ, Kotze JM (1986) Antagonism of lactic acid bacteria against phytopathogenic bacteria. Applied and Environmental Microbiology 52: 552-555.

2. Trias R, Bañeras L, Montesinos E, Badosa E (2008) Lactic acid bacteria from fresh fruit and vegetables as biocontrol agents of phytopathogenic bacteria and fungi. International Microbiology 11: 231-236.

3. Dalirsaber Jalali M, Khosro I, Ghasemi MF, Tabrizi SSKh (2012) Antagonism of Lactobacillus species against Xanthomonas campestris isolated from different plants. Journal of Applied Environmental and Biological Sciences 9: 480-484.

4. Lwin M, Ranamukhaarachchi SL (2006) Development of biological control of Ralstonia solanacearum through antagonistic microbial populations. International Journal of Agriculture and Biology 8: 657- 660.

5. El-Mabrok ASW, Hassan Z, Mokhtar AM, Aween MM (2012) Efficacy of Lactobacillus plantarum C5 cells and their supernatant against Colletotrichum gloeosporioides on germination rate of chilli seeds. Research Journal of Biological Sciences 7: 159-164.

6. Hoda AH, Yomna AM, Shadia MA-A (2011) In vivo efficacy of lactic acid bacteria in biological control against Fusarium oxysporum for protection of tomato plant. Life Science Journal 8: 462-468.

7. Xu J, Ran L, Yang B, Li Z (2002) Inhibition of Lactobacillus species on the germination of Aspergillus flavus spore. Wei Sheng Yan Jiu 31: 47-49.

8. El-Taher EM, El-Ghany ATM, Alawlaqi MM, Ashour MS (2012) Biosecurity for reducing ochratoxin A productivity and their impact on germination and ultrastructures of germinated wheat grains. Journal of Microbiology and Biotechnology and Food Sciences 2: 135-151.

9. Corsetti A, Gobbetti M, Rossi J, Damiani P (1998) Antimould activity of sourdough lactic acid bacteria: identification of a mixture of organic acids produced by Lactobacillus sanfrancisco CB1. Applied Microbiology Biotechnology 50: 253-256.

10. Wang H, Yan Y, Wang J, Zhang H, Qi W (2012) Production and characterization of antifungal compounds produced by Lactobacillus plantarum IMAU10014. PloS ONE. doi: 10.1371/journal.pone.0029452.

11. Young JM, Kuykendall LD, Martinez-Romero E, Kerr A, Sawada H (2001) A revision of Rhizobium Frank 1889, with an emended description of the genus, and the inclusion of all species of Agrobacterium Conn 1942 and Allorhizobium undicola de Lajude et al. 1998 as new combinations: Rhizobium radiobacter, R. rhizogenes, R. rubi, R. undicola and R. vitis. International Journal of Systematic and Evolutionary Microbiology 51: 89-103.

12. Burr TJ, Otten L (1999) Crown gall of grape: biology and disease management. Annual Review Phytopathology 37: 53-80.

13. Kersters K, De Ley J (1984) Genus III. Agrobacterium Conn: Bergey's Manual of Systematic Bacteriology. Vol. 1, Williams and Wilkins Co., Baltimore-London.

14. Ophel K, Kerr A (1990) Agrobacterium vitis sp. nov. for strains of Agrobacterium biovar 3 from grapevines. International Journal of Systematic and Evolutionary Microbiology 40: 236-241.

15. Süle S, Burr TJ (1998) The effect of resistance of rootstocks to crown gall (Agrobacterium spp.) on the susceptibility of scions in grapevine cultivars. Plant Pathology 47: 84-88.

16. Szegedi E, Bottka S, Mikulás J, Otten L, Süle S (2005) Characterization of Agrobacterium tumefaciens strains isolated from grapevine. Vitis 44: 49-54.

17. Chilton MD, Drummond MH, Merlo DJ, Sciaky D, Montoya AL, et al. (1977) Stable incorporation of plasmid DNA into higher plant cells: the molecular basis of crown gall tumorigenesis. Cell 11: 263-271.

18. Inzé D, Follin A, van Lijsebettens M, Simoens C, Genetello C, et al. (1984) Genetic analysis of the individual T-DNA genes of Agrobacterium tumefaciens: further evidence that two genes are involved in indole-3-acetic acid synthesis. Molecular and General Genetics 194: 265-274.

19. Guyon P, Chilton MD, Petit A, Tempe J (1980) Agropine in «null-type» crown gall tumors: evidence for the generality of the opine concept. Proceedings of the National Academy of Sciences of the United States of America 77: 2693-2697.

20. Filo A, Sabbatini P, Sundin GW, Zabadal TJ, Safir GR, et al. (2013) Grapevine crown gall suppression using biological control and genetic engineering: a review of recent research. American Journal of Enology and Viticulture 64: 1-14.

21. Kerr A (1972) Biological control of crown gall: seed inoculation. Journal of Applied Bacteriology 35: 493-497.

22. Vicedo B, Penyalver R, Asins MJ, Lopez MM (1993) Biological control of Agrobacterium tumefaciens, colonization, and pAgK84 transfer with Agrobacterium radiobacter K84 and the Tra⁻ mutant strain K1026. Applied and Environmental Microbiology 59: 309-315.

23. Jones DA, Ryder MH, Clare BG, Farrand SK, Kerr A (1988) Construction of a Tra⁻ deletion mutant of pAgK84 to safeguard the biological control of crown gall. Molecular and General Genetics 212: 207-214.

24. Junaid JM, Dar NA, Bhat TA, Bhat AH, Bhat MA (2013) Commercial biocontrol agents and their mechanism of action in the management of plant pathogens. International Journal of Modern Plant & Animal Sciences 1: 39-57.

25. Kerr A, Roberts WP (1976) Agrobacterium: correlations between and transfer of pathogenicity, octopine and nopaline metabolism and bacteriocin 84 sensitivity. Physiological Plant Pathology 9: 205-211.

26. Staphorst JL, van Zyl FGH, Strijdom BW, Groenewold ZE (1985) Agrocin-producing pathogenic and nonpathogenic biotype-3 strains of Agrobacterium tumefaciens active against biotype-3 pathogens. Current Microbiology 12: 45-52.

27. Burr TJ, Reid CL (1994) Biological control of grape crown gall with non-tumorigenic Agrobacterium vitis strain F2/5. American Journal of Enology and Viticulture 45: 213-219.

28. Bazzi CM, Alexandrova E, Stefani F, Anaclerio F, Burr TJ (1999) Biological control of Agrobacterium vitis using non-tumorigenic agrobacteria. Vitis 38: 31-35.

29. Burr TJ, Reid CL, Tagliati E, Bazzi C, Süle S (1997) Biological control of grape crown gall by strain F2/5 is not associated with agrocin production or competition for attachment sites on grape cells. Phytopathology 87: 706-711.

30. Herlache TC, Zhang HS, Ried CL, Carle SA, Zheng D, et al. (2001) Mutations that affect Agrobacterium vitis-induced necrosis also alter its ability to cause a hypersensitivity response on tobacco. Phytopathology 91: 966-972.

31. Creasap JE, Reid CL, Giffinet MC, Aloni R, Ullrich C, et al. (2005) Effect of wound position, auxin, and Agrobacterium vitis strain F2/5 on wound healing and crown gall in grapevine. Phytopathology 95: 362-367.

32. Kaewnum S, Zheng D, Reid CL, Johnson KL, Gee JC et al. (2013) A host-specific biological control of grape crown gall by Agrobacterium vitis strain F2/5: its regulation and population dynamics. Phytopathology 103: 427-435.

33. Liang ZH, Wang HM, Wang JH (2001) Preliminary study on effectiveness and the stability of E26 on controlling crown gall disease. Journal of China Agricultural University 6: 91-95.

34. Wang HM, Wang HX, Ng TB, Li JY (2003) Purification and characterization of an antibacterial compound produced by Agrobacterium vitis strain E26 with activity against A. tumefaciens. Plant Pathology 52: 134-139.

35. Yun LJ, Wang HM, Wang JH (2004) A bacteriocin with a broad spectrum activity produced by grapevine crown gall biocontrol strain E26. Scientia Agricultura Sinica 37: 1860-1865.

36. Chen XY, Xiang WN (1986) A strain of Agrobacterium radiobacter inhibits growth and gall formation by biotype III strains of A. tumefaciens from grapevine. Acta Microbiologica Sinica 26: 193-199.

37. Pu X-A, Goodman RN (1993) Tumour formation by Agrobacterium tumefaciens is suppressed by Agrobacterium radiobacter HLB-2 on grapevine plants. American Journal of Enology and Viticulture 44: 249-254.

38. Kawaguchi A, Inoue K, Nasu H (2007) Biological control of grapevine crown gall by nonpathogenic Agrobacterium vitis VAR03-1. Journal of General Plant Pathology 73: 133-138.

39. Kawaguchi A, Inoue K, Ichinose Y (2008) Biological control of crown gall of grapevine, rose, and tomato by nonpathogenic Agrobacterium vitis strain VAR03-1. Phytopathology 98: 1218-1225.

40. Kawaguchi A, Inoue K (2012) New antagonistic strains of non-pathogenic Agrobacterium vitis to control grapevine crown gall. Journal of Phytopathology 160: 509-518.

41. Khmel IA, Sorokina TA, Lemanova NB, Lipasova VA, Metlitski OZ (1998) Biological control of crown gall in grapevine and raspberry by two Pseudomonas spp. with a wide spectrum of antagonistic activity. Biocontrol Science and Technology 8: 45-57.

42. Eastwell KC, Sholberg PL, Sayler RJ (2006) Characterizing potential bacterial biocontrol agents for suppression of Rhizobium vitis, causal agents of crown gall disease of grapevines. Crop protection 25: 1191-1200.

43. Bell CR, Dickie GA, Chan JWYF (1995) Variable response of bacteria isolated

from grapevine xylem to control grape crown gall disease in planta. American Journal of Enology and Viticulture 46: 499-508.

44. Chen F, Li JY, Guo YB, Wang JH, Wang HM (2009) Biological control of grapevine crown gall: purification and partial characterization of an antibacterial substance produced by Rahnella aquatilis strain HX2. European Journal Plant Pathology 124: 427-437.

45. Toklikishvili N, Dandurishvili N, Vainstein A, Tediashvili M, Giorgobiani N, et al. (2010) Inhibitory effect of ACC deaminase-producing bacteria on crown gall formation in tomato plants infected by Agrobacterium tumefaciens or A. vitis. Plant Pathology 59: 1023-1030.

46. Biondi E, Bini F, Anaclerio F, Bazzi C (2009) Effect of bioagents and resistance inducers on grapevine crown gall. Phytopathologia Mediterranea 48: 379-384.

47. de Man JC, Rogosa M, Sharpe ME (1960) A medium for the cultivation of lactobacilli. Journal of Applied Bacteriology 23: 130-135.

48. Bertani G (1951) Studies on lysogenesis. I. The mode of phage liberation by lysogenic Escherichia coli. Journal of Bacteriology 62: 293-300.

49. Ryder MH, Tate ME, Kerr A (1985) Virulence properties of strains of Agrobacterium on the apical and basal surfaces of carrot root disks. Plant Physiology 77: 215-221.

50. Batdorj B, Trinetta V, Dalgalarrondo M, Prévost H, Dousset X, et al. (2007) Isolation, taxonomic identification and hydrogen peroxide production by Lactobacillus delbrueckii subsp lactis T31, isolated from Mongolian yoghurt: inhibitory activity on food-borne pathogens. Journal of Applied Microbiology 103: 837-848.

51. Eastwell KC, Willis LG, Cavileer TD (1995) A rapid and sensitive method to detect Agrobacterium vitis in grapevine cuttings using the polymerase chain reaction. Plant Disease 79: 822-827.

52. Haas JH, Moore LW, Ream W, Manulis S (1995) Universal PCR primers for detection of phytopathogenic Agrobacterium strains. Applied and Environmental Microbiology 61: 2879-2884.

53. Higa T, Kinjo S (1989) Effect of lactic acid fermentation bacteria on plant growth and soil humus formation. Proceedings of 1st International Conference on Kyusei Nature Farming.

54. Daeschel MA, Fleming HP (1984) Selection of lactic acid bacteria for use in vegetable fermentations. Proceedings of the symposium "Selection parameters of microorganisms for use in the fermentation of plant foods and beverages".

55. Holden M, Krasatanova S, Xue B, Pang S, Sekiya M, et al. (2003) Genetic engineering of grape for resistance to crown gall. Acta Horticulturae 603: 481-484.

56. Burr TJ, Bishop AL, Katz BH, Blanchart LM, Bazzi C (1987) A root specific decay of grapevine caused by Agrobacterium tumefaciens and A. radiobacter biovar 3. Phytopathology 77: 1474-1427.

57. Pu XA, Goodman RN (1992) Induction of necrogenesis by Agrobacterium tumefaciens on grape explants. Physiological and Molecular Plant Pathology 41: 241-254.

58. Deng W, Pu XA, Goodman RN, Gordon MP, Nester EW (1995) T-DNA genes responsible for inducing a necrotic response on grapevines. Molecular Plant-Microbe Interactions 8: 538-548.

59. Locke SB, Riker AJ, Duggar BM (1938) Growth substance and the development of crown gall. Journal of Agricultural Research 57: 21-39.

60. Binns AN (2008) A brief history of research on Agrobacterium tumefaciens: 1900-1980s: Agrobacterium: from biology to biotechnology. Springer, New York.

61. Kirchhof G, Blackwell J, Smart RE (1991) Growth of vineyard roots into segmentally ameliorated acidic subsoils. Plant and Soil 134: 121-126.

62. Bates TR, Dunst RM, Taft T, Vercant M (2002) The vegetative response of "Concord" grapevines to soil pH. Horticultural Science 37: 890-893.

63. Limanska N, Ivanytsia T, Basiul O, Krylova K, Biscola V, et al. (2013) Effect of Lactobacillus plantarum on germination and growth of tomato seedlings. Acta Physiologiae Plantarum 35: 1587-1595.

Isolation, Identification, *In Vitro* Antibiotic Resistance and Plant Extract Sensitivity of Fire Blight causing *Erwinia amylovora*

Mohammed Amirul Islam[1]*, Md Jahangir Alam[1], Samsed Ahmed Urmee[1], Muhammed Hamidur Rahaman[2], Mamudul Hasan Razu[2] and Reaz Mohammad Mazumdar[3]

[1]*Department of Genetic Engineering and Biotechnology, Shahjalal University of Science & Technology, Sylhet, Bangladesh*
[2]*Department of Genetic Engineering and Biotechnology, University of Rajshahi, Rajshahi-6205, Bangladesh*
[3]*BCSIR Laboratories Chittagong, Bangladesh Council of Scientific and Industrial Research, Chittagong-4202, Bangladesh*

Abstract

Background: *Erwinia amylovora* is the causal organism of fire blight. The fire blight is widely spread in bacterial disease of plants from both epidemiological and economic points of view. Furthermore, the situation is worsening by the advent of increased antibiotic resistance in these bacteria. The study was aimed to determine the *in vitro* antibiotic and herbal sensitivity of *E. amylovora* isolated from plants available in Sylhet, Bangladesh.

Methods: In this study, bacterial isolates taken from five fire blight infected plants like apple, pear, lemon, orange and olive plants were identified based on morphological, cultural and biochemical characteristics. All the isolates were tested for antibiotic sensitivity against five commonly used antibiotics and herbal sensitivity against five plants extract.

Results: Morphological, physiological and biochemical study of pure culture of suspected organism revealed *E. amylovora* bacteria which was found 100% resistant to Cefotaxime and 81.89% to Bacitracin. Chloramphenicol was found most effective as all the isolates were sensitive to it. Besides that, most of the isolates were susceptible to plant extracts and found maximum sensitive to *Allium sativum* and *Syzygium cumini* whereas resistant to *V. amurensis*.

Conclusion: It can be concluded that the investigation of herbal treatment can be implicated for fire blight disease in contrast of antibiotic test in future.

Keywords: Fire blight; *Erwinia amylovora*; Antibiotic sensitivity and plant extract

Introduction

Most damaging disease in the fruit growing world is fire blight caused by *Erwinia amylovora* (Burrill) [1], a gram negative, facultative anaerobic, rod shaped bacterium belongs to Enterobacteriaceae family (EPPO, 2006) which generally infects plants from the Rosaceae family [1]. It was considered to be native to North America and later detected in New Zealand in 1920 is now present in 43 countries [2]. Although the life cycle of the bacterium is still not well understood, it is known that it can survive as endophyte or epiphyte for variable periods of time depending of environmental factors [3]. The development of fire blight symptoms follows the seasonal growth development of the host plant. It begins in the spring with the production of the primary inoculum and the blossoms infection, continuing on summer with the shoots and fruits infection. Economic importance of this disease is caused losses of 68 million dollars in North-West America, 10 million dollars in one region of New Zealand, and 500,000 trees were destroyed in Lebanon and in Italy [4]. Since the discovery of fire blight in Morocco in May 2006 [5] the disease spread to most of the pome fruit producing regions, inducing severe damage. There is no single control measure for fire blight that will totally eradicate the disease, provide an absolute cure, or fully protect an orchard. However, fire blight damage can be kept to a minimum by using large number of chemicals like copper compounds, antibiotics, carbamates and miscellaneous compounds.. But the main disadvantage of chemicals like copper compounds is their phytotoxicity on host plants, especially pears [6] whereas antibiotics have lead to the selection of resistant bacterial populations and therefore their use is strictly limited or even forbidden in a number of countries [7]. For this reason many researcher trying to establish alternative controlling pathway of the pathogen since 1989 by using plant extract instead of chemical [8,9]. Moreover, using of plant extracts is eco-friendly and may reduce cost of cultivation. Considering all these viewpoints, our objective of the research work was to identify the bacteria on the basis of morphological, physiological, biochemical test and make a comparative study *in vitro* between antibiotic and plant extract sensitivity of the organism.

Materials and Methods

Collection and processing of samples

Total number of 21 diseased plant samples was collected from different nurseries of Sylhet city according to standard pathological procedure. Then, 1 ml of fruits rinsed water and fruit juices sample was taken to a test-tube containing 9 ml of sterile water and thoroughly mixed to get a 10^{-1} dilution of the water sample. Again, 1 ml of 10^{-1} dilution was transferred again to another 9 ml of sterile water in another test-tube and thoroughly mixed to get a 10^{-2} dilution. In such way serial dilution of water samples were made up to 10^{-4}.

*Corresponding author: Mohammed Amirul Islam, Department of Genetic Engineering and Biotechnology, Shahjalal University of Science & Technology, Sylhet, Bangladesh, E-mail:amirul.geb@gmail.com

Scientific name	Local name	Used parts	Used volume(μl)
Allium cepa	Onion	Bulb	50
Allium sativum	Garlic	Bulb	50
Syzygium cumini	Kalajam	Leaf	50
Vitis amurensis	Grape	Leaf	50
Litchi chinensis	Litchi	Leaf	50

Table 1: Herb samples used for *in vitro* herbal sensitivity experiment.

Samples Code	G	Growth at 39°C	Growth at 4% NaCl	Fl under UV	TSI Gas	TSI H_2S	I	C	N	OF	U	Gel	Ma	Su
F1	Rod -	-	+	-	+	-	-	+	-	F	-	+	+	+
F2	Rod -	-	+	-	+	-	-	+	-	F	-	+	+	+
F3	Rod -	-	+	-	+	-	-	+	-	F	-	+	+	+
F4	Rod -	-	+	-	+	-	-	+	-	F	-	+	+	+
F5	Rod -	-	+	-	+	-	-	+	-	F	-	+	+	+
F6	Rod -	-	+	-	+	-	-	+	-	F	-	+	+	+
F7	Rod -	-	+	-	+	-	-	+	-	F	-	+	+	+
F8	Rod -	-	+	-	+	-	-	+	-	F	-	+	+	+
F9	Rod -	-	+	-	+	-	-	+	-	F	-	+	+	+
F10	Rod -	-	+	-	+	-	-	+	-	F	-	+	+	+
F11	Rod -	-	+	-	+	-	-	+	-	F	-	+	+	+

G = Gram Test, Fl= Fluorescent, TSI = Triple Sugar Iron, I= Indole Test, C = Citrate Test, OF = Oxidation Fermentation Test, N= Nitrate Test, U= Urease Test, Gel= Gelatine Liquefaction, Ma= Mannitol Fermentation, Su= Sucrose Fermentation.

Table 2: Physiological & Biochemical tests for identification of bacterial isolates causing Fire Blight.

Isolation, purification and preservation of the isolates

Isolation of *E. amylovora* was done on Nutrient Agar (NA) or Leavan media which was prepared by dissolving 1g yeast extract, 2.5 g peptone, 2.5g NaCl, 25 g sucrose into 500ml of distilled water, pH adjusted to 7.0-7.2 and sterilized by autoclaving at 121°C, 15 psi for 15 minutes. Then, transferred the suspected single colony from NA plate by sterile loop and inoculated on the King's medium agar B (KB) which was prepared by peptone 20g, glycerol 10 mL, K_2HPO_4 1.5g, $MgSO_4$ · 7 H_2O 1.5 g, agar 15 g, distilled water 1000 mL, pH adjusted to 7.0-7.2 and sterilized by autoclaving at 121° for 20 minutes [10]. The plates were then incubated at 27°C for 2-3 days and observed daily for bacterial growth. Suspected colonies of *E. amylovora* (white, circular, mucoid, and curved) were selected and further purified again on KB agar at 27°C. This operation was repeated three to four times to be sure that pure cultures were obtained for identification tests [11] and preserved it for next investigation.

Identification of the isolates

For identification of *E. amylovora*, colony morphology was studied when it appeared in KB and NA media. Moreover, Gram's staining was performed as described by [12] and growth of pure cultured isolates was measured at 39°C and 4% NaCl containing NA media. Besides that, fluorescent test was done and under UV light at 366 nm after 48 h of incubated plate.

Biochemical tests

Biochemical tests like oxidative-fermentative test, nitrate reduction test, citrate utilization test, urease test, sucrose fermentation test, TSI (Triple Sugar Iodine) test, mannitol fermentation test, gelatine hydrolysis were done.

Antibiotic susceptibility test

In antibiotic susceptibility test 250 μl of microbial inoculums of *E. amylovora* strain from cultured nutrient broth was spread on the surface of Mueller-Hinton agar (CM337-OXOID) by using a sterile L-shaped glass rod. Then *in vitro* five different commercially available antimicrobial discs Streptomycin (10 μg), Gentamycin (10 μg), Chloramphenicol (30 μg), Cefotaxime (5 μg), Bacitracin (10 μg) were applied on the inoculated plates according to the Kirby-Bauer [13]. During susceptibility test same bacterial density was maintained by using Spectrophotometer at OD_{600}. After incubation, the plates were examined and the diameters of the zone of complete inhibition were measured in mm.

Plant extracts sensitivity studies

The antimicrobial activity of five plants extracts were used in the herbal sensitivity experiments. Name and parts of the plants are given in Table 1. After cleaning the selected parts of plant by sterilized distilled water, extracts were collected in a falcon tube and centrifuged at 4000 rpm for few minutes. Besides that wells of 5 mm diameter were punched into the agar plates with the help of sterilized cork borer and 50 μl of the plant extracts were added by using a micropipette to the wells made in the agar plate along with well diffusion 0.1 ml of diluted inoculum of the freshly cultured experimental strains. Inhibitory response of the herbal extracts was recorded according to the normal growth response of the bacteria after incubation at 28°C for 24 hour and zone of inhibition was measured by mm.

Results and Discussion

Fire blight is alarming hazardous threat to citrus fruits and for economy. So, proper understanding of the pathogenic specialization of this pathogen is necessary. Bacteria were isolated from plant samples and identified using cultural, physiological and biochemical test.

Isolation and identification of bacteria

Morphological studies of *E. amylovora* colonies were done after incubating pure culture at 28°C for overnight on NA and KB media (Table 2 and Figure 1A). Physical appearance of isolates showed whitish, circular, domed, smooth, mucoid colonies on NA media (Figure 1B)

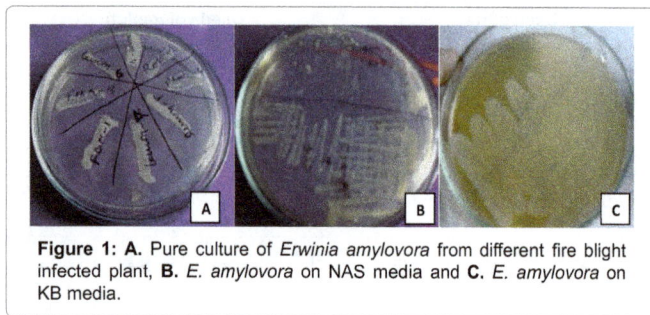

Figure 1: A. Pure culture of *Erwinia amylovora* from different fire blight infected plant, B. *E. amylovora* on NAS media and C. *E. amylovora* on KB media.

Name of Antibiotics	Disc Conc.	No. of isolate	Sensitivity pattern of Erwinia amylovora(11)		
			%R	%I	%S
Streptomycin (S)	10 µg	20	18.1	27.3	54.6
Bacitracin (B)	10 µg	20	81.89	18.1	-
Chloramphenicol (C)	30 µg	20	-	18.1	81.89
Cefotaxime (CTX)	30 µg	20	100	-	-
Gentamycin (GEN)	10 µg	20	9.99	54.6	36.4

R= resistant, I= intermediate and S= susceptible

Table 3: Antibiogram of isolated *Erwinia amylovora* causing Fire Blight.

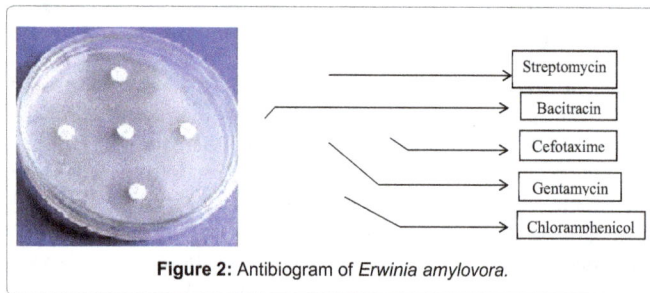

Figure 2: Antibiogram of *Erwinia amylovora*.

Plant samples	F1	F2	F3	F4	F5	F6	F7	F8	F9	F10	F11
	Erwinia amylovora										
A.sativum	20	15	14	20	40	14	13	5	16	12	11
A. cepa	-	-	-	-	-	2	-	-	-	-	-
S. cumini	10	2	3	-	3	-	-	-	5	5	-
V. amurensis	-		-	-	-	-	-	-	-	1	-
L. chinensis	-	-	-	-	-	-	-	-	-	-	-

Table 4: Isolates with inhibition zone (mm) for different plant samples.

whereas creamy white, circular, intending to spread colonies was found on KB media (Figure 1C) which resemble with [14] and [15] respectively. Among 21 isolates, a total of 20 were indicated as gram negative rod shaped bacteria whereas no growth was observed at 39°C and 12 had growth and 9 isolates didn't grow on 4% salt concentration. Moreover, *E. amylovora* exhibited non-fluorescent under UV light at 366 nm after 48 h which allowed the distinction from fluorescent *Pseudomonads*.

Biochemical tests

For biochemical characterization, a series of tests were performed with the suspected gram negative bacteria and results are given in Table 3. After analyzing the results for all bacterial isolates, it was confirmed that 11 isolates were *E. amylovora*. Identification was done by morphological, physiological and biochemical tests according to the EPPO standards diagnostic protocol (EPPO/CABI). In this table, TSI positive means both the slant and butt turned into yellow due to

bacterial fermentation and produce gas and some produce H_2S gas. Mannitol and Sucrose fermentation positive means both acid and gas were produced and OF result F means they are fermentative bacteria.

Antibiotic susceptibility test

Antibiotics were first used to control fire blight in the 1950's and quickly became an important tool for disease management. The sustainability of antibiotics for disease prevention has been threatened by emergence of antibiotic-resistant populations of *E. amylovora*, which has reduced the efficacy of some of the antibiotics in certain locations. It has been evident that bactericidal bind irreversibly to the bacterial ribosome and blocking the synthesis of proteins to make it inactive whereas mutation in chromosomal gene which encodes the production of ribosomal protein makes resistant [16,17]. In our study eleven isolates were screened for drug resistance profile which indicated the sensitivity pattern against commonly used antibiotics in fire blight (Figure 2). All the isolates offered high degree of resistance against the commonly used antibiotics. By comparing the zone created by the isolates with the standard zone of inhibition we found all isolates were 100% resistant to a single drug except Chloramphenicol (C) was the most sensitive. Among the antibiotics, resistant *E. amylovora* isolates were 9.99% , 81.89%,18.1%, 100% 0% against the antibiotic GEN,B, S, CTX and C (Table 3). So, study among the five antibiotics test low resistance showed in Streptomycin and Gentamycin which is in accordance with victoria et al. [18] and Spitkoin and Alvarado [19]. Besides that, tested pathogen revealed highly resistant against Bacitracin and Cefotaxime antibiotics and same results were also found by Kumar, Singh and Robert et al. [20,21] respectively. Moreover, highly susceptible was showed in Chloramphenicol antibiotics against *E. amylovora* in this research which is not resemblance with Weixin et al. [22] who concluded that hydrolysis of Chloramphenicol by *E. coli* conferred resistance.

Plant extracts sensitivity studies

Plants remain one of the main sources of natural products for new therapies particularly in poor countries, because most of them are cost less, affect a wide range of antibiotic resistant microorganisms, and another reason is there is an erroneous impression that herbal medicines have fewer adverse effects [23]. In present research antibacterial activity of aqueous extracts of all the five plants are presented in Table 4 in which highly significant antibacterial activity was observed in *A. sativum* and *S. cumini*, respectively against the tested pathogen. Our findings agree with other observations [24] who concluded that antimicrobial activity of allicin from garlic (*Allium sativum*) exhibit strong activity against *E. carotovora*. Even similar results was also found stated that methanol extract of *S. cumini* to be more effective on both gram positive and gram negative bacteria, and especially against gram positive bacteria such as *Staphylococcus aureus* and *Enterococcus faecalis*.

Conclusion

It can be concluded that, present study showed all the isolates of *E. amylovora* were resistant to at least one or more of commonly used antibiotics. So, emphasis must be placed on the development of effective bactericides and their proper use with knowledge of the appropriate dosage. On the other hand, herbal sensitivity test has paved the way the viable introduction of plants for the treatment of disease causing microorganism in cheaper cost and eco-friendly way. Therefore, it will be more beneficial to put emphasis on biological control of *E. amylovora* through plant extracts instead of antibiotics. Moreover, further researches are necessary to find more plant species

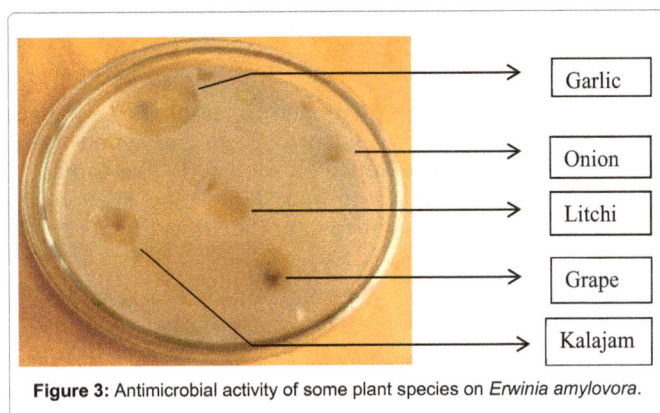

Figure 3: Antimicrobial activity of some plant species on *Erwinia amylovora*.

and purify the antimicrobial substances present in the crude plant extracts effective against *E. amylovora* (Figure 3).

References

1. Ordax M, Marco-Noales E, López MM, Biosca EG (2006) Survival strategy of Erwinia amylovora against copper: induction of the viable-but-nonculturable state. Appl Environ Microbiol 72: 3482-3488.

2. Van der Zwet T (2002) Present worldwide distribution of fire blight. Acta Horticulturae (ISHS). 590: 33-34.

3. Thomson SV (2000) Integrated orchard and nursery management for the control of fire blight. In: Fire blight the disease and its causative agent, Erwinia amylovora [ed. by Vanneste JL] Wallingford, UK: CABI 9-36.

4. Vanneste JL (2000) What is fi re blight? Who is Erwinia amylovora How to control it? In: Fire blight: the disease and its causative agent Erwinia amylovora. (Vanneste J.L., edn.). CABI publishing, Wallingford, UK 1-6

5. Fatmi M, Bougsiba M, Saoud H (2008) First report of fire blight caused by Erwinia amylovora on pear, apple and quince in Morocco. Plant Disease 92: 314.

6. MÄƒruÅ£escu L, Saviuc C, Oprea E, Savu B, Bucur M, et al. (2009) In vitro susceptibility of Erwinia amylovora (Burrill) Winslow et. al. to Citrus maxima essential oil. Roum Arch Microbiol Immunol 68: 223-227.

7. Cesbron S, Thomson SV, Paulin JP (2000) Acibenzolar-S-methyl induces the accumulation of defense-related enzymes in apple and protects from fire blight. European Journal of Plant Pathology 106: 529-536.

8. Mosch J, Mende A, Zeller W, Rieck M, Ullrich W (1993). Plant extracts with a resistance induction effect against fire blight (Erwinia amylovora). Acta Horticulturae 338: 389-395.

9. Mosch J, Zeller W, Rieck M, Ullrich W (1996) Further studies on plant extracts with a resistance induction effect against Erwinia amylovora. Acta Horticulturae 411: 361-366.

10. KING EO, WARD MK, RANEY DE (1954) Two simple media for the demonstration of pyocyanin and fluorescin. J Lab Clin Med 44: 301-307.

11. Jones AL, Geider K (2001) Erwinia amylovora group. In: Laboratory guide for identification of plant pathogenic bacteria. 3rd ed. (Schad N. W., Jones J.B. and Chun W., ed.). APS press, St. Paul, MN, USA, 40–54.

12. Merchant IA, Packer RA (1967) Veterinary Bacteriology and Virology. 7th edn., The Iowa University Press, Ames, Iowa, USA. pp. 286-306.

13. Bauer AW, Kirby WM, Sherris JC, Turck M (1966) Antibiotic susceptibility testing by a standardized single disk method. Am J Clin Pathol 45: 493-496.

14. Billing E, Baker LAE, Crosse JE, Garret CME (1961) Characteristics of English isolates of Erwinia amylovora (Burrill) Winslow et al. J. Appl. Bacteriol 24: 195-211.

15. Paulin JP, Samson R (1973) Fire blight in France. II. Characters strains Erwinia amylovora (Burrill) Winslow et al. 1920 isolated household Franco-Belge. Annals of Phytopathology 5: 389-397

16. Chiou CS, Jones AL (1995) Expression and identification of the strA-strB gene pair from streptomycin-resistant Erwinia amylovora. Gene 152: 47-51.

17. Jones AL, Schnabel EL (2000) The development of streptomycin-resistant strains of Erwinia amylovora. In Fire blight: the disease and its causative agent, Erwinia amylovora (J. Vanneste, ed.). CAB International, Wallingford, UK. pp.235-251.

18. Victoria Donat, Elena G Biosca, Arantza Rico, Javier Penalver, Marisa Borruel, Dionisio Berra, Txaran Basterretxea, Jesus Murillo, Maria M Lopez (2004) Erwinia amylovora strains from outbreaks of fire blight in Spain: phenotypic characteristics(Abstr.). Annals of Applied Biology, 146:105-114. DOI: 10.1111/j.1744-7348.2005.04079.x

19. Spitko RA, Alvarado M (1999) Use of Agry-Gent (Gentamicin sulfate) as a control material for fire blight of apple in the U.S. Acta Hort. (ISHS) 489: 629-630.

20. Sujeet Kumar, Bhoj Raj Singh (2013) An overview of mechanisms and emergence of antimicrobials drug resistance. Advances in Animal and Veterinary Sciences 1: 7-14.

21. Bonomo RA, Donskey CJ, Blumer JL, Hujer AM, Hoyenm CK, et al. (2003) Cefotaxime-resistant bacteria colonizing older people admitted to an acute care hospital. J Am Geriatr Soc 51: 519-522.

22. Tao W, Lee MH, Wu J, Kim NH, Kim JC, et al. (2012) Inactivation of chloramphenicol and florfenicol by a novel chloramphenicol hydrolase. Appl Environ Microbiol 78: 6295-6301.

23. Ozoula IR, Idogun SE, Tafamel GE (2010) Acute and sub-acute toxicological assessment of aqueous leaf extract of Bryophyllum pinnatum (Lam) in Sprague-Dawley rats. Am J. Pharmacol Toxical 5: 145-151.

24. Curtis H, Noll U, Stormann J, Slusarenko A (2004) Broad-spectrum activity of the volatile phyto anti cipinallicin in extracts of garlic (L.) against plant pathogenic bacteria, fungi and Oomycetes. Physiol. Mol. Plant Pathol 65: 79-89.

PERMISSIONS

LIST OF CONTRIBUTORS

Ahmad Ali Shahid and Muhammad Ali
Institute of Agricultural Sciences, University of the Punjab, Lahore, Pakistan

Muhammad Shahbaz
Department of Pest warning and Quality control of pesticides, Govt. of Punjab, Lahore, Pakistan

Coca Morante M
Plant Science and Plant Production Department, University of San Simón, Cochabamba, Bolivia

Wenera Ghopur, Pezilet Behti, Burabiyem Obulkhasim and Ghopur Mijit
College of Life Science and Technology, Xinjiang University, China

Mahmutjan Dawut
College of Life Science and Technology, Xinjiang University, China
Graduate School of Science, Hokkaido University, Japan

Abdukadir Abliz
College of Life Science and Technology, Xinjiang University, China
Media Institute of Urumqi Vocational University, China

Ajay Tomer, Ramji Singh and Manoj Maurya
Department of Plant Pathology, Sardar Vallabhbhi Patel University of Agriculture & Technology, Meerut (UP), India

Negash Hailu, Chemeda Fininsa and Tamado Tana
School of Plant Sciences, Haramaya University, Dire Dawa, Ethiopia

Girma Mamo
Melkasa Agricultural Research Institute, Adama, Ethiopia

Badiaa Essghaier, Cyrine Dhieb, Hanene Rebib, Saida Ayari, Awatef Rezgui Abdellatif Boudabous and Najla Sadfi-Zouaoui
Microorganisms and Active Biomolecules Laboratory, Faculty of Sciences of Tunis, University Campus, Tunisia

Rajyasri Ghosh
Mycology and Plant Pathology Laboratory, Department of Botany, Scottish Church College, Kolkata, India

Alemu Nega, Fikre Lemessa and Gezahegn Berecha
Department of Horticulture and Plant Sciences Jimma University Jimma, Ethiopia

Issa Wonni and Léonard Ouédraogo
Institut de l'Environnement et de Recherches Agricoles (INERA), 01 BP 910 Bobo Dioulasso, Burkina Faso
Laboratoire Mixte Internationnal, observatoire des agents phytopathogènes en Afrique de l'Ouest, Biosécurité et Biodiversité (LMI Patho-Bios), 01 BP 910 Bobo Dioulasso, Burkina Faso

Gustave Djedatin and Valérie Verdier
Institut de Recherche pour le Développement, UMR IPME, IRD-CIRAD-UM2, 911 Avenue Agropolis BP 64501, 34394 Montpellier Cedex 5, France

Sumit Kumar, Dubey RC and Maheshwari DK
Department of Botany and Microbiology, Gurukula Kangri University, Haridwar – 249404, India

Ul-Haq S and Hasan SS
School of Science, Indira Gandhi National Open University, New Delhi, India

Dhar A, Mital V and Sahaf KA
Central Sericultural Research and Training Institute, Central Silk Board, Govt. of India, Pampore, (J&K) 192 121, India

Shima Bagherabadi, Doustmorad Zafari and Mohammad Javad Soleimani
Department of Plant Protection, College of Agriculture, University of Bu Ali Sina, Hamedan, Iran

Kifle Belachew, Demelash Teferi and Legese Hagos
Ethiopian Institute of Agricultural Research, Jimma Agricultural Research Centre, Plant Pathology Research Section, Jimma, Ethiopia

Petr Karlovsky
Molecular Phytopathology and Mycotoxin Research, University of Goettingen, Grisebachstr.6 37077, Germany

Sylvestre Gerbert Dossa C
Molecular Phytopathology and Mycotoxin Research, University of Goettingen, Grisebachstr.6 37077, Germany
International Rice Research Institute, Division of Plant Breeding, Genetics and Biochemistry, Metro Manila, Philippines

Kerstin Wydra
Erfurt University of Applied Sciences, Altonaer Str. 25, 99085 Erfurt, Germany

Mehari Desta
Woldia University, Woldia, Ethiopia

Mohammed Yesuf
Melkassa Agricultural Research Center, Ethiopia Institute of Agricultural Research (EIAR), Nazareth, Ethiopia

S Vanitha
Department of Sericulture, Centre for Plant Protection Studies, Tamil Nadu Agricultural University, Coimbatore-641 003, Tamil Nadu, India

R Ramjegathesh
Department of Plant Pathology, Centre for Plant Protection Studies, Tamil Nadu Agricultural University, Coimbatore-641 003, Tamil Nadu, India

Mostafa Fatma AM, Nour El Deen AH and Ibrahim
Agricultural Zoology Department, Faculty of Agriculture, Mansoura University, Mansoura, Egypt

Khalil AE and Dina S
Nematology Division, Plant Pathology Res. Institution, Giza, Egypt

Patience Ukamaka Ishieze, Ugwuoke Kelvin I and Aba Simon C
Department of Crop science, University of Nigeria, Nsukka

Abd El-Ghany TM, Masrahi YS, Mohamed A, Al Abboud, Alawlaqi MM and Nadeem I Elhussieny
Biology Department, Faculty of Science, Jazan University, 114, KSA

Leta A
Arsi University College of Agriculture and Environmental Sciences, Ethiopia

Lamessa F
Jimma University College of Agriculture and Veterinary Medicine, Ethiopia

Ayana G
Ethiopian Institute of Agricultural Research, Malkassa Agricultural Research Center, Ethiopia

Heshmat Aldesuquy, Zakaria Baka and Nahla Alazab
Botany Department, Faculty of Science, Mansoura University, Egypt

Chijamo Kithan and Daiho L
Department of Plant Pathology, School of Agricultural Sciences and Rural Development, Nagaland University, Medziphema Campus, Nagaland, India

Slizewska Katarzyna
Institute of Fermentation Technology and Microbiology, Faculty of Biotechnology and Food Sciences, Technical University of Lodz, 171/173 Wolczanska Street, 90-924 Lodz, Poland

Barczynska Renata, Kapusniak Janusz and Kapusniak Kamila
Institute of Chemistry, Environmental Protection and Biotechnology, Jan Dlugosz University, 13/15 Armii Krajowej Avenue, 42-200 Czestochowa, Poland

Reshma Kumari and Ramesh Chandra Dubey
Department of Botany and Microbiology, Gurukula Kangri University, Haridwar-249404, India

Oleksandra Savchenko
Department of Biomedical Engineering, University of Alberta, Edmonton, Alberta, Canada

Xiujie Li and Jian Yang
Alberta Innovates - Technology Futures, Bag-4000, Vegreville, AB, Canada

Yuzhi Hao and Xiaoyan Yang
Department of Electrical and Computer Engineering, University of Alberta, Edmonton, Alberta, Canada

Jie Chen
Department of Biomedical Engineering, University of Alberta, Edmonton, Alberta, Canada
Department of Electrical and Computer Engineering, University of Alberta, Edmonton, Alberta, Canada
National Research Council, National Institute for Nanotechnology, Edmonton, Alberta, Canada

Benahmed Djilali Adiba, Ouelhadj Akli, Derridj Arezki, Bedrani Fatiha and Belkhir Fatiha
Faculty of Biological and Agronomical Sciences, University of Mouloud Mammeri of Tizi-Ouzou 15000, Algeria

Raman Yakout
National School of Agronomy (ENSA), El Harrach 16000, Algeria

Aqeel Ahmad and Yaseen Ashraf
Institute of Agricultural Sciences, University of the Punjab, Lahore-54590, Pakistan

Shiw Murti
Agriculture officer Oriental Bank of Commerce, India

Mehilal
Scientist CPRI, Meerut, India

Ajay Tomer
Lovely Professional University Jalandhar Punjab, India

Durga Prasad
MB Agriculture College Saharsa, Bihar Agriculture University, Bihar, India

Sajad-Ul-Haq and S. S Hasan
School of Sciences, Indira Gandhi National Open University, New Delhi-1100068

Anil Dhar and Vishal Mittal
Central Sericultural Research and Training Institute, Central Silk Board, Pampore, 192 121, Jammu and Kashmir, India

Lanying Wang, Yanping Luo and Mengyu Xing
Key Laboratory of Protection and Development Utilization of Tropical Crop Germplasm Resources, Ministry of Education, College of Environment and Plant Protection, Hainan University, Haikou, Hainan, 570228, P. R. China

Rong Di
Department of Plant Biology and Pathology, School of Environmental and Biological Sciences, Rutgers, the State University of New Jersey, 59 Dudley Road, New Brunswick, New Jersey, 08901, USA

Limanska N, Korotaeva N, Ivanytsia V and Ivanytsia T
Microbiology, Virology and Biotechnology Chair, Biological Department, Odessa National I.I. Mechnikov University (ONU), 65082 Dvoryanska str., 2, Odessa, Ukraine

Merlich A
Microbiology, Virology and Biotechnology Chair, Biological Department, Odessa National I.I. Mechnikov University (ONU), 65082 Dvoryanska str., 2, Odessa, Ukraine
UR 1268 INRA Biopolymères Interactions Assemblages, BP 71627, rue de la Géraudière, 44316 Nantes Cedex 3, France

Chobert JM and Haertlé T
UR 1268 INRA Biopolymères Interactions Assemblages, BP 71627, rue de la Géraudière, 44316 Nantes Cedex 3, France

Franco BDGM
Department of Food and Experimental Nutrition, Faculty of Pharmaceutical Sciences, University of São Paulo, 580, Professor Lineu Prestes, 13 B, Sao Paulo, SP, 05508- 000, Brazil

Biscola V
UR 1268 INRA Biopolymères Interactions Assemblages, BP 71627, rue de la Géraudière, 44316 Nantes Cedex 3, France
Department of Food and Experimental Nutrition, Faculty of Pharmaceutical Sciences, University of São Paulo, 580, Professor Lineu Prestes, 13 B, Sao Paulo, SP, 05508- 000, Brazil

Mohammed Amirul Islam, Md Jahangir Alam and Samsed Ahmed Urmee
Department of Genetic Engineering and Biotechnology, Shahjalal University of Science & Technology, Sylhet, Bangladesh

Muhammed Hamidur Rahaman and Mamudul Hasan Razu
Department of Genetic Engineering and Biotechnology, University of Rajshahi, Rajshahi-6205, Bangladesh

Reaz Mohammad Mazumdar
BCSIR Laboratories Chittagong, Bangladesh Council of Scientific and Industrial Research, Chittagong-4202, Bangladesh

Index

A

Acalypha Leaf Extract, 38, 41-42

Active Monomers, 10

Africa, 8, 29-30, 44, 46, 49-55, 75, 81-82, 86-87, 122, 126

Alternaria Alternate, 67, 69, 74

Alternaria Solani, 4-6, 88, 92-93

Antagonistic Endobacteria, 10

Antibacterial Activity, 147-149, 152, 202

Antibacterial Filters, 153

Antibiotics, 15, 36, 94-97, 140, 146-147, 186, 200, 202

Antifungal, 10, 16, 31-34, 36-37, 61, 63-64, 66, 68, 95-96, 129-130, 132, 135, 138-139, 154, 159, 167, 170, 186-190, 198

Antimicrobial Activity, 31-32, 152, 159, 190, 201-203

Audpc, 21, 23, 26-29, 88-90

B

Barleria Lupulina, 147, 151-152

Bio-agents, 99, 101-102, 135-137, 139

Biochemical Activities, 98

Biocontrol, 10, 16, 20, 30-31, 35-37, 56, 61, 93-97, 103-104, 119-120, 133, 138-139, 186, 189, 191-192, 196, 198

Biosurfactant, 56-63

Black Rot of Coffee, 75, 80

C

Cardoon, 161-164

Ceratobasidium Noxium, 75-76

Cercospora Zeae-maydis, 44, 49

Cfus, 17-18

Chalcone Synthase, 10-12, 14

Chemical Control, 1, 129, 137, 169, 171

Chitinase, 31-38, 58-59, 61, 95, 98-99, 103

Chocolate Spot Disease, 127, 129-134

Climate Change Resilience, 21-27, 29

Clotting, 161-164

Cluster Analysis, 69, 72-73, 173

Coleus, 94-95, 97

Corticium Koleroga, 75, 77, 79-80

Cucumber, 3, 43, 105-109, 111-112, 153, 155, 158-159

Culture Filtrate, 81-86

Curvularia Lunata Var. Aeria, 135-136

Cytopathic, 147, 149-151

D

Defence Related Enzyme, 38

Deoiled Cakes, 17-20

Destructive Disease, 4, 88, 165, 186

Disease Incidence, 1-3, 47, 49, 64, 66, 76-77, 88-90, 92, 96, 105-112, 127, 129-130, 192

Disease Intensity, 4, 38-40, 92

Disease Severity, 21-30, 39-40, 47, 66, 76, 88-90, 92, 105-109, 111, 128-130, 167, 170, 176

Diversity Analysis, 69, 173

E

Eco-organic, 64, 66, 68

Enterococcus Sp. Bs13, 56, 58-61

Enzyme Activity, 12-13, 38-40, 43, 98-99, 103, 127, 163

Etlingera Linguiformis, 135

F

Faba Bean, 127-134

Fungicide, 1-3, 31, 66, 88-93, 95-97, 127-132, 135, 137, 165-166, 168, 170, 177, 186

Fungicides, 1-3, 29, 31, 64, 68, 88-94, 97, 105, 129, 133-139, 165-171

G

Gc-ms, 56-57, 59, 147-151

Ginger, 38-43, 65

Greenhouse Pathogens, 153

Grey Leaf Spot, 44, 46-47, 49-50

H

Hcn, 56, 58-59, 61, 63, 94-96

Hydroponic Systems, 153

I

Immobilization, 161-164

In Vivo Disease Control Test, 10

Incidence, 1-4, 6, 21, 28, 44, 46-47, 49, 54, 64, 66, 75-78, 88-90, 92-94, 96-97, 105-112, 127, 129-130, 139, 177, 192

Induced Resistance, 30, 38, 43, 96, 98-99, 102, 104, 127, 129, 133-134

Intracellular Enzymes, 31

Issr Markers, 69-70, 72-74

J

Jatropha, 17-20, 152

L

Lactic Leavens, 161-164

Leaf Spot Disease, 4-5, 8, 49-50, 64-67, 69, 165, 168-169, 179, 183-184

Longevity, 17-19

M

Macrophomina Phaseolina, 56, 59, 63, 94, 97

Meloidogyne Incognita, 98, 101-104, 133

Milk, 123, 161-164, 187-188

Moderately Halophilic Bacteria, 31-32, 34, 36-37

Morus Spp, 64

N

Native Potatoes, 4

Neem, 17-20, 43, 68

O

Organic Farming, 64, 66-68

Oxidative Enzymes, 127, 130-131

P

Pellicularia Koleroga, 75, 80

Phaseolus Vulgaris, 21, 29, 68, 122, 126

Phytochemical Analysis, 147-148, 151

Phytoextracts, 64, 66

Phytotoxicity, 81-86, 200

Plant Extracts, 64-68, 93, 133, 135-139, 151, 165, 167, 190, 200-203

Potato, 1, 3-9, 11, 32, 43, 58, 63, 68-70, 72-73, 76, 88, 93, 119, 127, 130, 133-134, 141-143, 145-146, 154-155, 172, 187

Prevalence, 44, 47, 50, 75, 77, 79, 122, 124-125, 165-169

Pseudomonas Flourescens, 17, 19-20

Pseudomonas Fluorescens, 17-20, 38, 94-97, 104, 113-114, 117, 136, 138

Pythium Aphanidermatum, 38-39, 43

R

Resistance, 3, 5, 7, 20, 22, 27-30, 38-39, 43, 49, 51-53, 55, 58, 63, 79, 81, 85-87, 92-94, 96, 98-99, 101-104, 126-127, 129-134, 147, 149, 159-160, 170-171, 177, 192, 198-200, 202-203

Rhizoctonia Solani, 1, 3, 68, 94, 97, 120, 129-130, 133, 171-173, 176-178, 187-188

Rice, 1-3, 10, 36, 44, 51-55, 61, 81-87, 113, 119-120, 133, 139, 171-172, 176-178

Root Rot, 1, 21, 68, 77, 94-97, 104, 153, 156, 159

S

Salicylic Acid, 38-39, 41-43, 127-134

Sar, 38-39, 98, 131

Seed Detection, 122

Severity, 3-4, 7, 17, 21-30, 39-40, 44, 46-47, 49, 51, 54, 64, 66, 75-78, 88-90, 92, 105-109, 111-112, 126-130, 139, 165-170, 176-177, 192

Sheat Blight of Rice, 1

Shikimic Acid, 127-133

Siderophore, 58-59, 61, 94-97

Soil Type, 98-99, 101, 103

Structure Identification, 10

Substitution, 161

Sugar-beet, 98-99, 101-104

Survival, 17, 19-21, 75, 87, 148, 152, 194, 196-197, 203

T

Thread Blight, 75-80

V

V. Mungo, 56, 58, 60-62

Virulence, 52, 55, 81-82, 84-87, 171, 176, 199

X

Xanthomonas Axonopodis Pv. Phaseoli, 21, 122, 126

Xanthomonas Oryzae Pv. Oryzicola, 51, 55

Z

Zea Mays, 29, 44, 113-114, 117, 119-121